Implantable Medical Electronics

Vinod Kumar Khanna

Implantable Medical Electronics

Prosthetics, Drug Delivery, and Health Monitoring

Second Edition

Springer

Vinod Kumar Khanna
Chandigarh, India

ISBN 978-3-031-99867-6 ISBN 978-3-031-99868-3 (eBook)
https://doi.org/10.1007/978-3-031-99868-3

This Springer imprint is published by the registered company Springer Nature Switzerland AG
The registered company address is: Gewerbestrasse 11, 6330 Cham, Switzerland

This book is dedicated to:
My late father Shri Amarnath Khanna for giving me education and wisdom
My mother Shrimati Pushpa Khanna for her tender care and affection
My grandsons Avyaan and Hansh, and daughter Aloka for bringing joy and happiness in the family
and
My wife Amita for her unstinting support and cooperation

Preface to the Second Edition

Implantable medical electronic devices are crucial in monitoring physiological functions and supporting life-sustaining treatments. Pacemakers and defibrillators address heart conditions while insulin pumps manage diabetes. Spinal cord stimulators help manage chronic pain. Deep brain stimulators improve the quality of life of patients with various neurological and psychiatric disorders. Implanted sensors within the body provide real-time data on vital parameters, enabling proactive healthcare interventions. Implantable hearing devices restore hearing, and retinal implants or bionic eyes restore partial vision in individuals with severe retinal diseases, particularly those who have lost photoreceptor cells. The impact of implantable electronics is widespread and transformative, extending beyond cardiac care to monitor and regulate various bodily functions and deliver medication at identified locations inside the body.

Given the tremendous influence of implantable electronics on human well-being, the book's first edition received an overwhelming response, necessitating the bringing out of a new edition. This revised edition builds upon the success of the initial version. I have included additional study findings in this revised edition that reinforce the original work. The main aim of bringing out the revised edition is the inclusion of two essential and revolutionary topics as Chaps. 23 and 24, with the chapter on "drug delivery systems" becoming Chap. 25:

Chapter 23. Smart Contact Lenses and Eye Implants

Advances in noninvasive and minimally invasive implantable devices, such as contact lenses and surgically inserted intraocular lenses, enhanced with in-built sensors for pressure, glucose, and disease biomarkers, and supported by wireless communication capability, have made them immensely popular for medical diagnostics.

Chapter 24. Wireless Capsule Endoscopy

Introduced in clinical practice in 2000, capsule endoscopy has become a prime modality for gastrointestinal examination. It relieves patients of the discomfort of conventional endoscopy and has become a proven tool for noninvasive direct small bowel visualization.

Apart from the above, additional sections in Chap. 25 on drug delivery implants cover robotic capsules for drug delivery to the gastrointestinal tract. Also, alternative texts (alt texts) have been incorporated for all figures and illustrations to help improve the accessibility and usability of screen readers or other assistive devices.

These revisions and updates provide a more comprehensive understanding of implantable medical electronics and answer some pertinent questions that escaped attention in the original publication. It is earnestly hoped that this redesigned edition will provide a more robust reading experience for readers new to the subject and those who have followed my prior work.

Vinod Kumar Khanna
Independent Researcher
Chandigarh, India

Retired Chief Scientist
CSIR-Electronics Engineering
Research Institute
Pilani, India

Preface to the First Edition

About This Book

Implantable medical electronics brings together the essentials of electronics and medical devices implanted in the human body for overcoming various disease conditions such as for pacing the activities of the heart, the brain, and the network of nerves interconnecting the brain or the spinal cord with vital organs of the body; for performing measurements of vital biological parameters to keep surveillance over the health status of a person; and for delivering precise doses of drugs at targeted sites at predetermined dose rates and times. Notable among these diseases are the abnormal rhythms of the heart, the movement disorders, chronic back pain, epileptic seizures, respiratory deficiencies, urine and fecal incontinence problems, deafness due to sensorineural hearing loss, blindness caused by photoreceptor degeneration in the retina, and so forth. In many of these diseases, conventional pharmacological therapies have failed either totally or have been unsuccessful in eliciting a satisfactory response. Thus apart from taking drugs with associated side effects and correction by surgical interventions with attendant risks, there is a definite possibility of getting relief from serious diseases by means of implanted electronic pulse generators as pacing devices, along with implanted/wearable sensors and wireless body sensor networks, telecommunication, and advanced electronic gadgetry as health watchdogs.

This book is written at an introductory level to meet the requirements of students of both electronic and biomedical engineering and electronic instrumentation as well as professionals and researchers engaged in this fast-advancing multidisciplinary field. Providing definitions of important terms wherever necessary, it will be of great interest to the general reader interested in this emerging area of immense value to humankind at large.

Why This Book Was Written?

Implantable medical electronics is a fast-growing area, which has not only brought succor to a vast chunk of population but has also helped in pulling out many from the claws of death. Every day new technologies and new devices are reported, bringing immense relief to the suffering people. However, the information on implantable electronics lies scattered at different places. A cohesive compendium of knowledge on this field, synthesizing the information dispersed among different databases, is therefore called for. So, the foremost reason that this book was needed was that the latest information on implantable medical electronics is presently found only in research journals and on web pages. Therefore, a broad-spectrum compilation making this information available as an all-in-one resource between the covers of a single book was urgently required.

Another major reason for the need of this book was that the information given in electronics journals and magazines is provided at a level which is not comprehensible to medical students. The same is true about the information in medical journals and magazines, which is not easily grasped by students of electronics engineering. Therefore, a book steering the midway course and providing the information at a level understandable to both categories of students was the need of the hour.

Perhaps the paramount motivation was that the book was mandated by the need of an interdisciplinary book striking a balance between electronics and biomedical engineering. Keeping this in view, definitions of both the electronics and medical terms are provided, as and when it is felt that the subject will be elusive to any of the above categories of audience.

For Whom This Book Was Written?

The book is written to cater to the needs of the graduate students of electronics engineering, electronic instrumentation, and biomedical engineering. Apart from these students, another class of targeted audience is that of academic researchers in this field. They will benefit from the up-to-date bibliographies that will refer them to the excellent original literature and thus help them in pursuing their exploratory ventures. Additionally, professional engineers, practicing doctors and paramedical staff, and interested lay readers too are likely to acquire useful knowledge about this interesting field.

Layout of the Book

The book is divided into two parts: basic concepts and principles (Chaps. 2–11) and applications (Chaps. 12–23). In this book, an all-embracing perspective of the necessary electronics background is provided in the preliminary chapters (2–11).

Chapter 12 deals with implantable neural amplifiers and Chap. 13 with implantable sensors. Chapters 14–22, describe pacing techniques used for the heart, the brain, the spinal cord, and the network of nerves running through the body and interlinking the brain/spinal cord with its main organs. In the final chapter, Chap. 23, glimpses of controlled drug delivery systems are presented. Chapter-wise organizational arrangement of the book will be elaborated in the introductory chapter.

Pilani, Rajasthan, India Vinod Kumar Khanna

Acknowledgments

This book would never have come to reality without the blessings from a lot of people. To name all these people without forgetting anybody is a difficult task.

To be able to write a book is to have traversed a long and twisting road, full of bumps and depressions. To write the acknowledgments section is to try to tell which stepping stones en route were the most important ones.

I will begin with thanking the "reader" of this book. Perhaps the astute reader is the one that always gave me inspiration for writing the book on this burning topic, which signifies a paradigm shift from the traditional approaches of drug therapy and corrective surgery. This paradigm shift is leading a transformative wave, redefining medical benchmarkings and outcomes.

I thank my editors for providing me an opportunity to work on this project and providing me the requisite guidance and help, as and when necessary.

I am thankful to Director, CSIR-CEERI, Pilani for constant encouragement in my work.

I thank all colleagues at CSIR-CEERI for their moral support, companionship in despair, and motivation to move on.

I wish to thank my family, my daughter, and my wife without whose support I would never have been able to complete this task and, more importantly, not even begun it.

Thank you all!

Contents

About the Author

Vinod Kumar Khanna is an independent researcher from Chandigarh, India. He is a retired chief scientist from the Council of Scientific and Industrial Research (CSIR)—Central Electronics Engineering Research Institute (CEERI), Pilani, India, and a retired professor from the Academy of Scientific and Innovative Research (AcSIR), Ghaziabad, India. He is a former emeritus scientist at CSIR and a professor emeritus at AcSIR, India. His broad research areas include designing, fabricating, and characterizing power semiconductor devices and micro- and nanosensors.

Academic Qualifications

He received an M.Sc. degree in Physics with a specialization in Electronics from the University of Lucknow in 1975 and a Ph.D. degree in Physics from Kurukshetra University in 1988 for his thesis, "Development, Characterization and Modeling of the Porous Alumina Humidity Sensor."

Research/Teaching Experience and Accomplishments

His research experience spans 40 years, from 1977 to 2017. Starting his career as a Research Assistant in the Department of Physics, University of Lucknow, from 1977 to 1980, he joined CSIR-CEERI, Pilani (Rajasthan), in April 1980. There, he worked on several CSIR-funded and sponsored research and development projects. His major research fields included power semiconductor devices, microelectronics/MEMS, nanotechnology-based sensors and dosimeters.

In the power semiconductor devices area, he worked on the high-voltage and high-current rectifier (600 A, 4300 V) for railway traction, high-voltage TV deflection transistor (5 A, 1600 V), power Darlington transistor for AC motor drives (100 A, 500 V), fast-switching thyristor (1300 A, 1700 V), power DMOSFET, and IGBT. He contributed to developing sealed tube Ga/Al diffusion for deep junctions, surface electric field control techniques using edge beveling and contouring of large-area devices, and the design of floating field limiting rings. He extensively characterized minority-carrier lifetime in power semiconductor devices as a function of process steps. He also contributed to developing the P-I-N diode neutron dosimeter and PMOSFET-based gamma-ray dosimeter.

In the area of sensor technology, he worked on the nanoporous aluminum oxide humidity sensor, ion-sensitive field-effect transistor-based microsensors for biomedical, food, and environmental applications, microheater-embedded gas sensor for automotive electronics, MEMS acoustic sensor for satellite launch vehicles, and capacitive MEMS ultrasonic transducer for medical applications.

As an AcSIR faculty member, he was the course coordinator of MEMS/IC Technology for the advanced semiconductor electronics program (2011–2013) and taught "MEMS Technology" to students pursuing M.Tech. degree. As an adjunct faculty member at BESU, Kolkata, he taught "MEMS Technology and Design" to M.Tech. (Mechatronics) students. He was invited by IIT, Jodhpur, to deliver lectures on "Semiconductor Fundamentals and Technology" to B.Tech. students during February 2011. He guided B.Tech./M.Tech. theses of students from BITS, Pilani; VIT, Vellore; and Kurukshetra University. He also guided a Ph.D. thesis on "MEMS acoustic sensor," MNIT, Jaipur.

Semiconductor Facility Creation and Maintenance

He was responsible for setting up and managing diffusion/oxidation facilities, edge beveling and contouring, reactive sputtering, and carrier lifetime measurement facilities. As the Head of the MEMS and Microsensors Group, he managed the maintenance of a six-inch MEMS fabrication facility for R&D projects and the augmentation of processing equipment under this facility at CSIR-CEERI.

Scientific Positions Held

During his tenure of service at CSIR-CEERI from April 1980 until superannuation in November 2014, he was promoted to various positions, including one merit promotion. He retired as Chief Scientist and Professor (AcSIR) and as Head of the MEMS and Microsensors Group. Subsequently, he worked for three years as Emeritus Scientist, CSIR, and Emeritus Professor, AcSIR, from November 2014 to November 2017. After completing the emeritus scientist scheme, he now resides in Chandigarh. He is a passionate author and enjoys reading and writing.

Membership of Professional Societies

He is Fellow and Life Member of the Institution of Electronics and Telecommunication Engineers (IETE), India, and Life Member of the Indian Physics Association (IPA), Semiconductor Society, India (SSI), and Indo-French Technical Association (IFTA).

Foreign Travel

He has traveled widely. He had participated in and presented research papers at the IEEE Industry Application Society (IEEE-IAS) Annual Meeting in Denver, Colorado, USA, in September–October 1986. His short-term research assignments include deputations to Technische Universität Darmstadt, Germany, in 1999; at Kurt-Schwabe-Institut fur Mess-und Sensortechnike e.V., Meinsberg, Germany, in 2008; and at Fondazione Bruno Kessler, Trento, Italy, in 2011, under collaborative programs. He was a member of the Indian Delegation to the Institute of Chemical Physics, Novosibirsk, Russia, in 2009.

Scholarships and Awards

He was awarded a National Scholarship by the Ministry of Education and Social Welfare, Government of India, on the basis of a Higher Secondary result, 1970; CEERI

Foundation Day Merit Team Award for projects on fast-switching thyristor (1986); for power Darlington transistor for transportation (1988), for P-I-N diode neutron dosimeter (1992); and for high-voltage TV deflection transistor (1994); Dr. N. G. Patel Prize for best poster presentation in the 12th National Seminar on Physics and Technology of Sensors, 2007, BARC, Mumbai; CSIR-DAAD Fellowship in 2008 under Indo-German Bilateral Exchange Programme of Senior scientists, 2008. He is featured in the Stanford–Elsevier prestigious list of the world's top 2% scientists (2022, Elsevier Data Repository, V4, doi:10.17632/btchxktzyw.4). He is named as a highly ranked Scholar-Lifetime: #2 in Nanoelectronics by ScholarGPS (https://scholargps.com/scholars/25423546982929/vinod-kumar-khanna).

Research Publications and Books

He has published 194 research papers in leading peer-reviewed national and international journals and conference proceedings. He has authored 23 books and contributed six chapters to edited books. He has five granted patents to his credit, including two US patents.

Acronyms, Abbreviations, and Symbols

Aβ	Amyloid-β
AC	Alternating current
AC-LSK	Auxiliary carrier load shift keying
A/D	Analog-to-digital (converter)
ADC	Analog-to-digital converter
AED	Antiepileptic drug
AFE	Analog front end
Ag/AgCl	Silver/ Silver chloride
AH	Aqueous humor
Ah	Ampere hour
AI	Artificial intelligence
Al_2O_3	Aluminum oxide
ALIC	Anterior limb of the internal capsule
ALU	Arithmetic and logic unit
AM	Amplitude modulation
A_M	Midband gain
AMD	Age-related macular degeneration
APP	Amyloid precursor protein
ARM	Advanced RISC machines
ARMD	Age-related macular degeneration
ASIC	Application-specific integrated circuit
ASK	Amplitude shift keying
ATP	Adenosine triphosphate, antitachycardia pacing
ATPase	Adenosine triphosphatase
AuHNW	Gold hollow nanowire
AV	Atrioventricular (node)
AZA	Auto-zeroing amplifier
BACE1	Beta-site amyloid precursor protein cleaving enzyme 1
BAP	Bio-artificial pancreas
BASK	Binary amplitude shift keying
BFSK	Binary frequency shift keying

BGA	Ball-grid array
BJT	Bipolar junction transistor
BPEG	British Pacing and Electrophysiology Group
bps	Bits per second
BPSK	Binary phase shift keying
C	Capacitance
CAN	Controller area network
CAP	Carbonated apatite
CAT	Catalase
CCCS	Current-controlled current source
CCD	Charge-coupled device
CCS	Current-controlled stimulation
CDIP	Ceramic dual in-line package
CE	Capsule endoscopy
CF_x	Carbon monofluoride
CG	Control gate
ChCS	Charge-controlled stimulation
ChO_x	Cholesterol oxidase
CHS	Chopper stabilized
CI	Cochlear implant
CISC	Clean intermittent self-catheterization
CL	Contact lens
cm	Centimeter
CMOS	Complementary metal–oxide–semiconductor
CMRR	Common mode rejection ratio
CNS	Central nervous system
CPU	Central processing unit
Cr/Au	Chromium/gold
CRC	Cyclic redundancy check
CRPS	Complex regional pain syndrome
CSF	Cerebrospinal fluid
CT	Computerized tomography
C_T	Timing capacitor
CTAT	Complementary to absolute temperature
Cu	Copper
D	Duty cycle
DAC	Digital-to-analog converter
dB	Decibel = 10 $\log_{10}$ (power P_2/power P_1)
DBS	Deep brain stimulation
DC	Direct current
DCS	Dorsal column stimulation
DI	Deionized
DIBL	Drain-induced barrier lowering
DIP	Dual-in-line package
DMA	Direct memory access

DNA	Deoxyribonucleic acid
DOD	Depth of discharge
D/P	Diaphragmatic/phrenic (nerve)
DRAM	Dynamic random access memory
DREZ	Dorsal root entry zone
dsDNA	Double-stranded deoxyribonucleic acid
DSP	Digital signal processing
DVS	Dynamic voltage scaling
E_C	Critical electric field
ECC	Error-correcting code
ECT	Electroconvulsive therapy
EEG	Electroencephalography
EEPROM	Electrically erasable programmable read-only memory
EGD	Esophagogastroduodenoscopy
EIA	Electronic Industries Association
EIRP	Effective isotropic radiated power
EMF	Electromotive force
EMG	Electromyography
EMI	Electromagnetic interference
EPS	Extrapyramidal symptoms
ESD	Electrostatic discharge
ET	Essential tremor
f	Frequency
FASTROM	Factory advanced service technique read-only memory
FBSS	Failed back surgery syndrome
FCBGA	Flip chip ball-grid array
FCC	Federal Communications Commission
FDA	Food and Drug Administration, USA
FDTD	Finite-difference time domain
FET	Field-effect transistor
FG	Floating gate
FGMOSFET	Floating gate metal–oxide–semiconductor field-effect transistor
FIOL	Fluorescent aptamer-conjugated intraocular lens
FM	Frequency modulation
FPU	Floating-point unit
FSK	Frequency shift keying
FXTAS	Fragile X-associated tremor ataxia syndrome
G	Gate, Giga
g	Gram
GABA	Gamma-amino butyric acid
GCT	Gate control theory
GERD	Gastroesophageal reflux disease
GHz	Gigahertz
GI	Gastrointestinal tract
GIDL	Gate-induced drain leakage

g_m	Transconductance
GOD	Glucose oxidase
GPi	Globus pallidus internus
GUI	Graphic user interface
HAP	Hydroxyapatite
HMI	Human–machine interface
H_2O_2	Hydrogen peroxide
Hz	Hertz
I	Current
IC	Integrated circuit
ICD	Implantable cardioverter defibrillator
ICL	Implantable contact lens
IDDS	Implantable drug delivery system
I_{DQQ}	Quiescent power supply current
I_{DS}	Drain-source current
IEEE	Institute of Electrical and Electronics Engineers
I_{in}	Input current
IM	Intermodulation
IMD	Implanted medical device, intermodulation distortion
I/O	Input/output
IOH	Inverse opal hydrogel
IOL	Intraocular lens
IOP	Intraocular pressure
I_{out}	Output current
IPG	Implanted pulse generator
IV	Intravenous
I/V	Current-to-voltage (converter)
J	Joule
K	Kelvin
Kbps	Kilobits per second
kDa	Kilo Dalton
kHz	Kilohertz
L	Inductance, liter
LBT	Listen before talk
LC	Inductance–capacitance
LCD	Liquid crystal display
LCP	Liquid crystal polymer
LCR	Inductance–capacitance–resistance
LDD	Lightly doped drain
LED	Light-emitting diode
Li	Lithium
$Li_xAg_2V_4O_{11}$	Lithiated silver vanadium oxide
$LiBF_4$	Lithium tetrafluoroborate
$LiCoO_2$	Lithium cobalt oxide
$LiFePO_4$	Lithium iron phosphate

LIGA	Lithographie, galvanoformung, abformung (German words for lithography, electroplating, and molding)
LPF	Low-pass filter
LPS	Li_3PS_4 (lithium thiophosphate)
LSB	Least significant bit
LSK	Load shift keying
LUMI	Luminal unfolding microneedle injector
m	Meter
mA	Milliampere
Mbps	Megabits per second
MCCG	Magnetically controlled capsule gastroscopy
MCE	Magnetic capsule endoscopy
MDD	Major depressive disorder
MEDS	Medical data services
MEMS	Microelectromechanical systems
MGCE	Magnetically-guided capsule endoscopy
mg dL^{-1}	Milligram per deciliter
MHz	Mega hertz
MICS	Medical implant communication service
MIT	Massachusetts Institute of Technology
MITS	Medical implant telemetry system
mM	Millimolar
MMP-9	Matrix metalloproteinase-9
MnO_2	Manganese dioxide
MOD	Modulo
MOSFET	Metal–oxide–semiconductor field-effect transistor
MPE	Maximum permissible exposure
MRI	Magnetic resonance imaging
ms	Millisecond
MSB	Most significant bit
mV	Millivolt
mW	Milliwatt
nA	Nanoampere
NAD^+	Nicotinamide adenine dinucleotide, oxidized form
NADH	Nicotinamide adenine dinucleotide, reduced form
NASPE	North American Society of Pacing and Electrophysiology
NBG	North British Generic (code)
NBTI	Negative-bias temperature instability
nm	Nanometer
nmol L^{-1}	Nanomole per liter
NMOS	N-Channel MOSFET
NOS	Nitric oxide synthase
NVM	Nonvolatile memory
OCD	Obsessive–compulsive disorder
OCT	Optical coherence tomography

OFC	Orbitofrontal cortex
OOK	On–off keying
OP-AMP	Operational amplifier
OTA	Operational transconductance amplifier
OTP	One-time programmed memory
pA	Pico ampere
PAA	Peroxyacetic acid
PAM	Pulse amplitude modulation
PBN	Parabrachial nucleus
PC	Personal computer
pCO_2	Partial pressure of carbon dioxide (in blood)
PD	Parkinson's disease
PDIP	Plastic dual in-line package
PDM	Pulse duration modulation
PDMS	Polydimethylsiloxane
PDN	Pull-down network
PEEK	Polyetheretherketone
PEG	Polyethylene glycol
PEGDAAm	Diacrylamide-group-modified poly(ethyleneglycol)
pH	Potential of hydrogen $= -\log_{10}[H^+]$
pHEMA	Poly(2-hydroxyethyl methacrylate)
PHM	Pulse harmonic modulation
PI	Polyimide
PLCC	Plastic leaded chip carrier
PMMA	Poly(methyl methacrylate)
PMOS	P-Channel MOSFET
pO_2	Partial pressure of oxygen (in blood)
PPGA	Plastic pin-grid array
PPN	Pedunculopontine nucleus
PPSK	Passive phase shift keying
PROM	Programmable ROM
PSK	Phase shift keying
PSRR	Power-supply rejection ratio
PTAT	Proportional to absolute temperature
PTFE	Polytetrafluoroethylene
PUN	Pull-up network
PVDF	Polyvinylidenedifluoride
PVP	Polyvinylpyridine
PWM	Pulse width modulation
Q	Charge, quality factor
QRS complex	A grouping of three deflections in ECG recording
Quad	Four
R, r	Resistance
RAM	Random access memory
RC	Resistance–capacitance

RCs	Respiratory control centers
R-D	Reaction–diffusion (model)
RDAC	Resistive digital-to-analog converter or digipot
RF	Radio frequency
RISC	Reduced instruction set computer
R_L	Load resistor
RMS	Root mean square
RoboCap	Robotic capsule
ROI	Return on investment
ROM	Read-only memory
RP	Retinitis pigmentosa
RSD	Reflex sympathetic dystrophy
RSNOs	S-Nitrosothiols
R_T	Timing resistor
RV	Right ventricle
RWTT	Reflected wave transit time
SA	Sinoatrial (node)
SAB	Surface-activated bonding
SAR	Successive-approximation-register, specific absorption rate
SAW	Surface acoustic wave
SCCwm	Subcallosal cingulate white matter
SCD	Sudden cardiac death
SCI	Serial communications interface
SCS	Spinal cord stimulation
SDSR	Self-driven synchronous rectifier
SHDN	Shut down
Si	Silicon
SILC	Stress-induced leakage current
SiO_2	Silicon dioxide
SMPS	Switch-mode power supply
SNHL	Sensorineural hearing loss
SNM	Sacral neuromodulation
SNS	Sacral nerve stimulation
SOC	State of charge
SoC	System on chip
SPI	Serial peripheral interface
SRAM	Static random access memory
Steel: 316	Steel containing 16 % chromium, 10 % nickel, and 2 % molybdenum
Steel: 316L	Low-carbon version of 316 stainless steel resistant to highly corrosive environments
STN	Subthalamic nucleus
SVO	Silver vanadium oxide $Ag_2V_4O_{11}$
SWCNT	Single-walled carbon nanotube

T	Temperature, time, time period
TBE	Tick-borne encephalitis
TCE	Therapeutic capsule endoscopy
TCL	Traditional contact lens
T1D	Type 1 diabetes
TDDS	Transdermal drug delivery system
TiN	Titanium nitride
TNF	Tumor necrosis factor
TRD	Treatment-resistant depression
TRM	Tissue response modifier
UART	Universal asynchronous receiver/transmitter
U/F	Urgency–frequency
UI	Urge incontinence
UR	Urinary retention
USB	Universal Serial Bus
USRP	Universal Software Radio Peripheral
V	Volt
V_{BE}	Base-emitter voltage
V_{cc}	Positive supply voltage (BJT)
VCCS	Voltage-controlled current source
VCS	Voltage-controlled stimulation
VCVS	Voltage-controlled voltage source
VDA	Voltage-differencing amplifier
V_{dd}	Drain voltage, positive supply voltage (FET)
V_{DG}	Drain-gate voltage
VF	Ventricular fibrillation
V_{f}	Feedback voltage
V_{G}	Gate voltage
V_{GS}	Gate-source voltage
VIC	Voltage-to-current converter
Vim	Ventralis intermedius, ventral intermediate nucleus
V_{IN}	Input voltage
VLSI	Very large-scale integration
VNS	Vagus nerve stimulation
V_{out}	Output voltage
V_{p}	Peak value of the input voltage
V_{ref}	Reference voltage
V_{SS}	Source voltage, negative supply voltage/ground (FET)
VT	Ventricular tachycardia
W	Watt
WBAN	Wireless body area network
WCE	Wireless capsule endoscopy
Z	Impedance
ZrO_2	Zirconium oxide
α	Temperature coefficient

η	Size ratio
μA	Microampere
μA mM^{-1} cm^{-2}	Microampere $millimolar^{-1}$ $centimeter^{-2}$
μg/mL	Microgram per milliliter
μM	Micromolar
μm	Micrometer
μS	Micro Siemen
μs	Microsecond
μV	Microvolt
μW	Microwatt
τ	Time constant
Ω	Ohm

Chapter 1
Introduction, Scope, and Overview

Abstract This chapter provides a bird's-eye view of the subject and contents of the book, to serve as a springboard to reading the book. It concisely presents through ab initio concepts, what this book is about, what is the material covered in each chapter, and how the successive chapters are interwoven and interrelated.

Keywords Medical electronics · Clocking and timing circuits · Pulse generators · Neural amplifiers · Sensors · Cybersecurity · Cardiac pacemakers · Brain stimulation · Spinal cord stimulation · Vagus nerve stimulation · Sacral nerve stimulation · Phrenic nerve stimulation · Cochlear implants · Retinal implants · Smart eye lenses · Capsule endoscopy · Drug delivery

1.1 Electronics

Electronics is the branch of science and technology dealing with the study of electronic phenomena or phenomena related to the behavior of electrons, namely those concerned with controlling the transport of electrons in vacuum, gaseous media, and semiconducting solids with the help of devices which are able to oppose their motion, amplify and switch the electronic currents, store electrons, and gainfully manipulate their various effects in different applications.

1.2 Medical Electronics

Medical electronics is a subdivision of electronics that delves into the theory, design methodology, and assembly of instrumentation and paraphernalia geared toward health care. These instruments and equipment are intended for diagnosis and therapeutic treatment of various diseases, e.g., for inducing anesthesia, controlling cardiac and respiratory functions, and performing surgical operations, as well as for carrying out research for improving such apparatuses for the betterment of human life.

V. K. Khanna, *Implantable Medical Electronics*,
https://doi.org/10.1007/978-3-031-99868-3_1

1.3 Implantable Medical Electronics

Implantable medical electronics is a self-contained segment of medical electronics that addresses the provision of comfort and well-being to patients suffering from diseases, which cannot be healed by prescription of medicines or corrected by surgical interventions alone. These are the diseases which require the implantation of an electronic device or system into the body of the patient. The implanted device may be one of the following types, although not necessarily restricted to the types mentioned below:

1. This implanted device could be a low-noise amplifier for measuring the potentials of neurons. Percutaneous connections for signal recording restrict the patient's mobility, serve as an easy path for transmission of infection from the outside environment via the skin to the brain, and are liable to corruption by external noise/interference, warranting the need of an implanted wireless system.
2. A second possibility is that this device could be a pulse generator delivering rhythmic pulses of desired polarities and magnitudes to the heart muscle, targeted regions of brain, or particular nerves catering to specific organs of the body, e.g., for removing fecal or urinary incontinence or for facilitating respiratory function regulation. Such implanted pulse generators could also provide relief to people suffering from hearing or vision loss.
3. Alternatively, the implanted device could be a sensor for surveillance of the vital parameters of a patient. This kind of device is the diligent watchman of patient's health status.
4. Another interesting area in which implantable electronics has revolutionized disease mitigation efforts is that of focused drug delivery at sites where the drug is actually required or needed most to lessen its side effects on other regions of the body through which the drug has to unavoidably travel before reaching its destination.

Thus, implantable electronics has four main offshoots. Figure 1.1 illustrates the four facets of implantable medical electronics. First offshoot is insertion of an implantable neural amplifier in the body for accurate recording of neural signals for neuroengineering studies. Second is the use of implantable pulse generators for pacing the activities of diseased organs. Third is the use of implantable sensors for observing the influence of therapy besides monitoring biological parameters of the patient. Fourth is the use of drug delivery systems to supervise the supply of accurate doses of medicine to affected parts. All the above three areas of implantable electronics constitute the subject matter of this book.

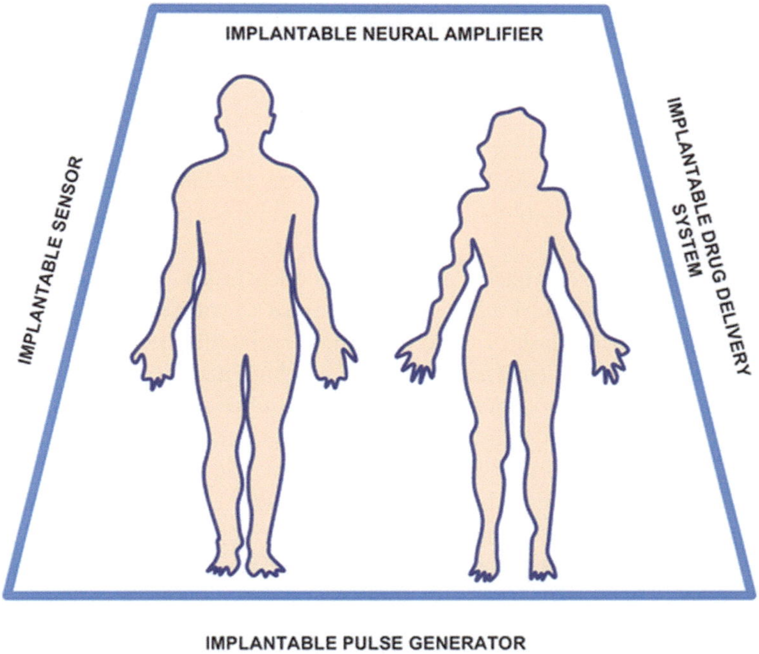

Fig. 1.1 The four facets of implantable electronics

1.4 Organization of the Book

Although every chapter starts with an abstract, the following summary of chapters will help in providing readers with the linking glue that binds each chapter with its predecessor and in appreciating their mutual correlation.

Chapter 2 differentiates medical devices from medicinal products. Classification of medical devices into classes I, II, and III is described, and their salient features are brought out. The ideas of invasiveness and noninvasiveness of/by medical devices are presented. Following the introduction of implantable medical devices, active and passive implantable devices are defined. An outline of historical development is provided to enable the reader to know the breakthroughs in different technological disciplines, which led to rapid strides in implantable electronics. Then the electrical system of the human body and the concepts of bioelectricity, membrane, and action potential are briefly explained.

Chapter 3 acquaints the reader with the generic structure that all implantable devices tend to adopt. Most of the devices consist of an external and an internal part which converse via a wireless link. Such a scheme is essential because after the implant has been surgically inserted in the body, it must be possible to supply power as well as perform corrective programming actions from outside without disturbing the patient. The external part serves as a controller of the implanted internal part, and the doctor can always adjust the settings of the internal part to suit patient needs.

Chapter 4 deals with the prime design issues discombobulating the IC design engineers who toil hard to evolve new designs of ICs for implantable devices. The crux of the problem is that the IC designer has to reduce power consumption to as low level as achievable and look for design tricks and fabrication processes that guarantee reliable operation with extremely low failure chances because a patient's life is at stake.

Chapter 5 describes the basics of stimulation of neurons. The stimulating electrodes can be monopolar or bipolar. The waveforms of stimulating signals may be either monophasic or biphasic. The operation of a few building block circuits that are widely used in stimulators such as DACs, ADCs, voltage and current sources, current mirrors, voltage-to-current converters, voltage multipliers, boost converters, etc., is snappishly explained. The three types of stimulation modes, viz., current, voltage, and charge-controlled schemes, are described. Active and passive charge balancing techniques are distinguished. The significance of charge balancing is emphasized.

Chapter 6 treats the essentials of clocking and timing circuits. It is placed as a prologue to pulse generators. The clock circuits coordinate the operation of different circuit segments in an implantable IC, serving as a metronome for making rhythm. The timing circuits decide, judge, or control the time of occurrence of an event by introducing time delays or providing an oscillatory or flip-flop function. RC and crystal-based oscillators, multivibrators, timer ICs, and related applications are described.

Chapter 7 is devoted to pulse generator circuits, which are battery-powered circuits to produce electrical discharges with rhythmic waveforms. These circuits are used to supply pulses in which the electromotive force can be made to change in a prespecified pattern with respect to time. The circuits used in implantable devices are generally microcontroller based. The shape of the pulses and their amplitude, duration, and frequency are variable at the discretion of the doctor, and these parameters can be programmed using the external part of the implantable device so as to administer a well-planned therapy to the patient.

Chapter 8 concentrates on biomaterials and biocompatibility issues with implantable devices. Tissue responses to biomaterials are discussed, notably blood–material interactions, provisional matrix formation, fibrous encapsulation, etc. Various metallic biomaterials (titanium and its alloys, gold alloys, stainless steel), bioceramics (aluminum, titanium, and zirconium oxides), and biocompatible polymers (PEEK, PVDF, PDMS, etc.) are listed. As the selection of proper biomaterial is the key factor underlying the success of an implantable device, a judicious choice of material is the starting step in implant fabrication.

Chapter 9 sheds light on the batteries for implantable devices. These being the power supplies for implantable devices must be safe and reliable, and must have long lifespan. They must be stable sources indicating their discharge state. Mainly, lithium batteries are used with lithium as the anode, but differing in cathode material (I_2, MnO_2, CF_x, $AgV_2O_{5.5}$). Electrolyte, cell design, and other mechanical features may also vary. Secondary lithium ion batteries can be recharged in the implanted state.

Chapter 10 reviews the advancements in wireless power supply and data communications with implanted devices. In wireless powering of implants, an external transmitter generates an electro- and/or magnetic field. The implanted medical device garners the energy of this field. A stable DC supply voltage is obtained by such energy harvesting. Depending on the frequency of operation, the separation of the transmitter and the implant, and the antennas used, the energy is transferred purely via magnetic or electric near fields or through electromagnetic fields. Wireless telemetry helps in relaying signals like electrocardiograms, blood pressure, glucose level, etc., to external equipment for analysis and guiding future therapy. For a cardiac pacemaker, inductive coupling between a small coil enclosed within the implanted pacemaker and a bigger coil lying on the patient's chest enables the transference of data to and from the pacemaker. Load-shift keying is a widely used modulation technique for transmitting data from the implant to the external device on an inductive link, whereas communication in the reverse direction is done via ASK, FSK, or PSK. Several advantages are derived if the communication with the implant is done at a higher carrier frequency. The resultant increase in bandwidth renders possible achieving a higher bit rate. Further, a higher frequency leads to electromagnetic waves of larger energy content, thereby outspreading the usable range of the system. But higher frequencies are detrimental to human health, and therefore, frequency cannot be subjectively increased. The newly standardized 400 MHz MICS band is available for this purpose.

Chapter 11 sets out to explore the developments in protecting the implanted devices from threats of breach of their security and privacy by roguish and wicked attackers. These worries arise because implanted devices are designed to communicate wirelessly with a nearby external device programmer who can remotely get access to read data and change settings without the requirement for fresh surgery. This arrangement is done because repeated surgeries involve great inconvenience and discomfort to the patient. But the convenience provided by wireless access also exposes the implants to the risks of hacking and inputting parameters that may pose life risk to the patient. This calls for addressing cybersecurity issues for the entire life cycle of the implant, from the initial design phase through deployment and end of life.

Chapter 12 is dedicated to a special type of amplifiers called neural amplifiers, which are used for recording the signals received from neurons. Chronic recording of neural signals is necessary for elucidating neurophysiology for building neuroprosthetic systems. Required right away are low power, implantable microsystems with wireless transmission capabilities and able to record signals at the same time from a sizeable number of neurons over an extended range of three-dimensional space- and time-based scales. A low-noise amplifier able to reject DC offsets is a critical element of neural recording systems. Both clock-based amplifiers (switched-biasing, chopper-stabilized, and auto-zeroing types) and continuous-time amplifiers are touched upon. A versatile building block circuit, viz., the operational transconductance amplifier, is described.

Chapter 13 provides the reader an exposition to the exhilarating field of implantable sensors for monitoring physiological variables. Notable examples of these variables are blood pressure, heart rate, and levels of oxygen and carbon dioxide in the blood. Apart from these parameters, blood glucose levels and in vivo nitric oxide concentration can also be controlled. In a closed-loop system, a sensor detects a change in the variable and signals to an actuator to prompt a response. This paradigm offers several opportunities for detecting early changes. The aim is to effect upstream interventions to make the patient comfortable and avert hospitalization. Implantable sensors have several advantages, e.g., continuous monitoring, objectively measured metrics without prejudiced assessment, and the ability to give a patient-specific profile that can be analyzed relatively easily. These sensors are always confronted with problems of foreign body response, oxygen deficiency, and stability of immobilized enzyme.

Chapter 14 provides a journey into the world of cardiac pacemakers, which are tiny, ultralightweight devices supplying pacing pulses to the heart muscle, either in the right atrium (single-chamber pacemaker) or in the right atrium and right ventricle (dual-chamber pacemaker) or in the right atrium and both the ventricles (biventricular pacemaker), to aid the blood pumping mechanism of an otherwise slow heart. Pacemaker devices of today represent real technological marvels, which can supply the pulses in demand mode when a pulse is missing, as also in rate-responsive mode depending on the physical exercise of the patient.

Chapter 15 extends the discussion of pacemakers to life-saving ICD devices, which impart a huge electrical jolt to the randomly quivering heart muscle to help it to recover from its irregular rhythms. Timely delivery of these jolts has pulled out many patients from an almost certain sudden cardiac death. As the high-voltage shocks have to be given immediately on sensing the abnormal activities of the heart, the benefit of an implantable defibrillator is unrivaled. Devices capable of both pacing and defibrillation functions are also used.

Chapter 16 pays attention to the use of electrical pulses to stimulate identified regions (usually the thalamus, subthalamic nucleus, and globus pallidus) located deep inside the human brain to alleviate the debilitating motor symptoms of movement disorders. These are observed in Parkinson's disease, dystonia, and essential tremor in the form of stiffness, rigidity, and problems in walking. The pulses are applied through a small device called the brain pacemaker, which looks similar to the cardiac pacemaker. The kingdom of deep brain stimulation has aggrandized to include neurological and psychiatric diseases like Alzheimer's disease, major depression, and obsessive–compulsive neurosis.

Chapter 17 brings home to the reader's mind that electrical pulses not only are useful for pacing tasks but can also block chronic pain signals. Spinal cord stimulation or neurostimulation works by applying electrical signals to spinal nerves from a programmable pulse generator placed subcutaneously via an incision in the upper buttock region or in the abdomen. The pleasant sensation produced by these pulses impedes the ability of pain perception, making the patients less dependent on painkillers.

Chapter 18 tries to present vagus nerve stimulation, a technique used for preventing epileptic seizures by sending pulses to vagus nerves, the two nerves that start from the brain stem near the bottom of the neck and propagate downwards. The pulses are sent from a pacemaker-like device, which is surgically implanted in the chest region. Doctors performing vagus nerve stimulation noticed that besides controlling seizures, the stimulations also influenced brain wave patterns. Buoyant changes in the mood of the patient were observed. From such investigations, it was inferred that this stimulation could also help in coping with diseases like depression. Research has also indicated the possibility of management of obesity and rheumatoid arthritis by this technique.

Chapter 19 concentrates on phrenic nerve or diaphragm pacing using a pacemaker-like implanted device. There are two phrenic nerves, which start from the neck and move downwards between the lung and the heart to the diaphragm. This technique provides respiratory support to patients whose diaphragm, lungs, and phrenic nerves are functional by regularizing their breathing pattern. Patients suffering from spinal cord injury, quadriplegia or tetraplegia, and sleep apnea episodes are benefitted.

Chapter 20 describes sacral nerve stimulation, a technique used to stimulate the sacral nerves accessed via the S3 foramen by mild electrical pulses supplied by an implanted pulse generator. The sacral nerve controls the urinary bladder. Therefore, its stimulation helps to overcome problems of overactive bladder, such as urge incontinence and urgency–frequency. Fecal incontinence and bowel problems can also be mitigated by sacral neuromodulation. This neuromodulation is performed in two stages: a testing phase in which the patient receives stimulation from an external stimulator and maintains a diary showing the results achieved; and a permanent implantation phase. The permanent implantation is done only when both the patient and the doctor are satisfied.

Chapter 21 covers the important subject of cochlear implants. These implants have proved their significance for people with moderate to profound loss of hearing known as sensorineural loss. This loss is characterized by damage to the tiny hair cells in the portion of the inner ear named cochlea. Cochlear implants should not be confused with the hearing aids, which are merely audio amplifiers making the sound louder in intensity. The cochlear implants sidestep the nonfunctional portion of the inner ear and directly stimulate the auditory nerve to provide the sensation of sound.

Chapter 22 provides a synopsis of retinal implants that are prosthetic aids for people who have been blinded by degenerative diseases such as retinitis pigmentosa or macular degeneration in which light-sensing photoreceptor cells in the retina are damaged. Both the subretinal and epiretinal implant approaches are explained. An array of microelectrodes is placed either beneath the retina or above its surface. By electrically stimulating the remnant cell circuitries in the retina, these implants are able to restore vision to blind patients.

Chapter 23 delves into electronic or smart lenses possessing capabilities beyond traditional lens-vision correction responsibilities. These lenses have biosensors for detecting analytes in the tear fluid, antennas for wireless communication, and drug delivery systems to release medication into the eye for therapeutic purposes. Electronic lenses have been developed for the detection of metabolites like glucose, lactate, and cholesterol, disease biomarkers for central nervous system disorders and Alzheimer's disease, and intraocular pressure monitoring for glaucoma management.

Chapter 24 explores the present status and research trends of wireless capsule endoscopy, a non-invasive medical procedure in which a camera-equipped capsule swallowed by the patient transmits pictures of the digestive tract to an external recorder wirelessly for scrutiny of doctors to diagnose diseases such as bleeding, polyps, ulcers, tumors, and other abnormalities. As the capsule movements are determined by the natural peristaltic movements of the gut, an enhanced version uses magnetic manipulation systems, employing external magnetic fields to control capsules.

Chapter 25, the concluding chapter of the book, apprises the reader of the present scenario on implantable drug delivery systems. These specially engineered systems supply controlled quantities of pharmaceutical compounds to the targeted sites as per the therapeutic plan, i.e., at prefixed places of release at predetermined rates and predefined instants of time. The realm of drug delivery has received impetus from developments in micro- and nanotechnologies. Robotic capsules are swallowable devices designed to enhance medication delivery or take images within the body.

1.5 Discussion and Conclusions

On the whole, the reader is taken on a voyage across the multidisciplinary field of implantable electronics. The book has a straightforward structure. For ease of comprehension, the beginning chapter familiarizes the reader with the basic terminology and organization of the book. This is done in preparation for Chaps. 2–11 which expose the reader, step by step, to the necessary rudimentary electronics, material science, battery technology, wireless powering of the implants from outside, and data exchange with them, as also security issues. The later Chaps. 12–22 focus mainly on applications involving implantable neural amplifiers for recording neuron potentials, implantable sensors for health monitoring, and implantable pacemaker-like devices for stimulating different nerves to overcome various diseases. Eye implants are discussed in Chap. 23 followed by wireless capsule endoscopy in Chap. 24. Implantable drug delivery methods are described in Chap. 25.

Review Exercises

1.1 Define the following terms: (a) electronics, (b) medical electronics, and (c) implantable medical electronics.

1.2 Describe the four principal types of medical applications where the necessity of an implantable electronic device or system is felt.

1.3 Why is it difficult to measure neural signal by connecting a wire across the skin of the patient? How does the implanted device help in this measurement?

1.4 What generic structure is followed in the construction of implantable devices? Why is such a structure necessary?
1.5 What is the main problem baffling the IC circuit designers for implantable devices? Explain.
1.6 Distinguish a clock circuit from a timer circuit.
1.7 What are the medical applications in which a pulse generator is used? What output parameters of a pulse generator can be varied?
1.8 What are the three principal classes of biomaterials used in implantable devices? Give examples of each class.
1.9 What are common batteries used in implantable devices? Which type of battery can be recharged from outside?
1.10 What modulation techniques are used for data exchange with the implanted device: (1) from the external device to the implant and (2) from the implant to the external device.
1.11 What are the advantages of using a high carrier frequency in data communication with the implant? What are the disadvantages?
1.12 What are the types of security infringements that an implanted device is likely to be influenced by? What precautions are necessary for avoiding these security breaches?
1.13 Why are special types of amplifiers required for neural signal measurements? What are their specialities?
1.14 What are the advantages of implanting sensors in the human body? What are the problems faced with these devices?
1.15 What does a cardiac pacemaker do? Name the main types of pacemakers used.
1.16 Can defibrillation of ventricular arrhythmias not be done from outside? If yes, how does the implanted defibrillator help in life-threatening situations?
1.17 What are the typical disorders, which are amenable to treatment by deep brain stimulations? Name three areas of the brain at which the stimulation pulses are applied?
1.18 Name the disorders, which are treated by the following stimulation techniques: (a) spinal cord stimulation, (b) vagus nerve stimulation, (c) phrenic nerve stimulation, and (d) sacral nerve stimulation.
1.19 How does a cochlear implant differ from a hearing aid? For what type of ear damages is the cochlear implant suitable?
1.20 What are the degenerative diseases in which a retinal implant aids the blind patients? What are the two chief classes of retinal implants?
1.21 What are robotic capsules used for?
1.22 What is capsule endoscopy?
1.23 Why is it necessary to place a drug delivery system inside the body? Can the same purpose not be achieved by conventional drug delivery methods?

Part I
Basic Concepts and Principles

Chapter 2
Diagnostic and Therapeutic Roles of Implantable Devices in the Human Electrical Machine: A Quick Primer

Abstract Innovations in electronic engineering have flagged the march toward realization of implantable biomedical microsystems. These microsystems are capable of interfacing with interior body parts. By such interfacing, they can monitor, manipulate, and control the functions of body parts in the anticipated manner. Distinguished precedents of such systems are the cardiac pacemakers, deep brain stimulators, those used for controlling respiratory and bladder functions, cochlear and retinal prosthesis, and many others prescribed for sicknesses that are unmanageable by medication. Headway in implantable electronics received a boost only after the invention of the bipolar transistor in 1948 and its market availability in the early 1950s. The miniaturization and low power obligation of this device rendered possible workable telemetry systems for measurement of biological parameters. Human body is an intricate electrical machine. Its operational flaws can be tweaked by inserting electronic devices. Besides remedying such faults through delivery of electrical impulses, these devices also help in an organized, coordinated release of medication to the body at a predetermined rate. In addition, they assist in defining vital strictures and in sensing abnormal variations to enlighten about the health state of the body.

Keywords Medical device · Invasive device · Noninvasive device · Passive device · Active device · Bioelectricity · Membrane potential · Action potential

2.1 Introduction

Starters/appetizers are food or drink items served preceding a main banquet to stimulate appetite. Similarly, this chapter will take a step further from Chap. 1 to provide a novice's view of the implantable electronic devices. It seeks to kindle reader's interest in the subject. In this chapter, noninvasive and invasive medical procedures and devices are defined. Under the invasive devices, implanted medical devices are introduced. These are grouped into passive and active implants. Under active implants fall the implantable electronic systems [1]. These systems are often called implantable/implanted medical electronic systems. Implantable medical electronics

V. K. Khanna, *Implantable Medical Electronics*,
https://doi.org/10.1007/978-3-031-99868-3_2

constitutes the subject matter of this book. The implantable systems perform biological signal amplification, nerve stimulation, physiological parameter measurement, and drug delivery functions. Historically, progress in the field of implant technologies has closely tracked the developments in electronics. Introspecting deeply, this statement is particularly true for microelectronics, computers, and communications [2]. It is reiterated that a vast majority of implanted devices are used for electrical stimulation of body parts. To understand their operation, it is essential to realize the role of electricity produced and flowing within the human body. With this in mind, the electrical phenomena incessantly taking place in our bodies for its normal functioning are explained.

2.2 Medical Devices and Medicinal Products

A medical device must be clearly set apart from a medicinal product, medicine, or pharmaceutical [3]. The term "medical device" insinuates an appliance, instrument, implant, or similar article together with its associated software. Some of its features are as follows: (1) It is intended for diagnosis, prevention, or treatment of disease. (2) It tries to influence the physiological function or anatomy of the body, for compensation of damages by injury or handicap. (3) It achieves its stipulated purpose by physical means such as electrical, mechanical, thermal, etc. (4) It does not require any chemical action for attaining the aimed objective. Nor does it need to undergo metabolic breakdown and absorption within the body.

The term "medicinal product" is concerned with a chemical preparation. This preparation is formulated for either external or internal use. It acts via chemical route pharmacologically. As another mode of action, it may affect the metabolism. Otherwise, it may work by eliciting an immunological response. Pharmacology is the science of drugs, dealing with their composition, uses, and effects.

Table 2.1 presents the important differences between a medical device and a medicinal product.

Table 2.1 Medical device versus medicinal product

S. No.	Medical device	Medicinal product
1.	An appliance or instrument	A chemical preparation
2.	Acts through physical means	Acts by chemical reaction
3.	Action is mechanical, electrical, thermal, etc.	Influences metabolism or evokes immunological reaction

2.3 Medical Device Classification

Medical devices are subdivided into three classes. These are Class I, II, and III. This subdivision is done according to two considerations: (1) the jeopardies associated with their usage and (2) the authoritarian control level exercised in commercial practices for selling the device [4]. Common examples of Class I devices are tongue depressors, bedpans, and elastic bandages. Also placed in this class are examination gloves, handheld surgical instruments, etc. All these devices manifest extreme design simplicity and pose almost zero risk during use. They are subject to the slightest degree of regulatory control. They also benefit from exemption from premarket notification.

Class II devices, e.g., X-ray machines, acupuncture needles, surgical drapes, etc., are more complex devices. They pose small risks. Besides conformism with general controls, they are consigned under special controls. The special controls include the labeling requirements and performance standards. Surveillance required is very important.

Class III devices are high-risk life-supporting devices such as implanted cardiac pacemakers or cerebral stimulators. These devices must be scientifically reviewed. They ought to be approved by Food and Drug Administration (FDA), USA, prior to marketing. Table 2.2 lists the main ideas of medical device grouping into three classes.

2.4 Noninvasive and Invasive Medical Procedures and Devices

Across the centuries, humankind has been actively engaged in the science and art of healing in one form or another. Through evolutionary processes, medical science has devised several diagnostic and therapeutic procedures to wrestle with diseases. These long-prevalent techniques are either noninvasive or invasive.

A noninvasive medical procedure is one, which does not violate the wholeness or integrity of the body. This violation may take place by necessitating the puncturing of the skin. Making a cut into a body tissue or organ to insert instruments or parts thereof is another way of violation [5]. The doctor recommends a compound or preparation for the mitigation or forestalling of disease. The compound is preeminently a drug taken by mouth or injected through skin. Or the doctor carries out diagnostic tests such as standard eye exam, echocardiography, electrocardiography, CT scan, and MRI.

In the invasive medical procedure of treating illnesses, lacerations, or deformities, the body is intruded or penetrated to access the targeted part. Thus, the impaired physiological function is set right. The doctor cuts into the patient's body, often with the sensation lost, either partially or totally. The loss of sensation can be topical. It may be local, regional, or general. He/she tries to repair or remove the damaged or

Table 2.2 Grouping of medical devices

Class of medical device	Features	Risk	Regulatory requirement	Examples
I	Easily manufacturable device	Lowest (listing with FDA required)	Maximum, if not all general controls: registration of the manufacturer with the FDA, use of good-quality fabrication techniques, scripting brand name properly, labeling the product correctly, notifying FDA before marketing the device, and adhering to general reporting procedures	Oxygen masks, intraoral dental drills
II	More complicated device, albeit not lifesustaining	Medium (clearance by FDA required)	General and special controls. The latter include special labeling obligations, fulfilling compulsory standards of performance, and vigilance by post-marketing surveillance	Syringes, hearing aids, nebulizers, cardiac monitors
III	Life-support or life-sustaining device	Highest (approval by FDA required)	Most stringent control. General controls and in-depth regulatory scrutiny before licensing and vending	Pacemakers, heart valves, implantable urinary continence devices, implantable diaphragmatic/ phrenic nerve stimulators

malfunctioning parts. Examples of surgery span across simple pus collection from abscesses or warts removal to procedures involving opening of large body regions such as abdomen, chest, or skull, e.g., cardiac catheterization performed for the introduction of catheters through blood vessels into the heart.

Non invasive and invasive devices are those that are used by noninvasive and invasive procedures, respectively (Table 2.3). Examples of noninvasive devices are a clinical thermometer used to measure body temperature, a stethoscope used for listening to beats of the heart and lung sounds, a sphygmomanometer for arterial blood pressure measurement, a hearing aid to listen to feeble sounds, or an external splint to restrict a broken arm or leg in a fixed position while it heals. Tout au contraire, catheters inserted into body cavities, ducts, or vessels to allow the passage of fluids or distend passageways are invasive in character. So are the laparoscopic devices used in the surgical treatment of endometriosis in which the tissue lining the uterus or womb grows outside it instead of growing inside.

Table 2.3 Non invasive and invasive medical devices

S. No.	Noninvasive devices	Invasive devices
1.	Those which do not involve skin puncturing or incision	Those whose insertion entails intrusion into the body
2.	Examples: thermometer, stethoscope	Examples: catheter, laparoscopic device

2.5 Implantable Medical Devices

By an invasive procedure, it is possible to cut open the body. Then the doctor can place either a nonelectronic or electronic device inside the body in an invulnerable manner. The device thus secured is called a medical implant. The term "medical implant" refers to a device or a tissue embedded or inserted firmly in the human body. The term is also applicable to an entire system. It represents those devices that are either totally or partially inserted, either surgically or otherwise, into the human body and placed there like a part of the body. It also refers to those devices introduced by medical involvement into a natural orifice of the body. All the above devices are supposed to reside at their placement sites after the medical procedure has been completed [6]. The objectives of medical implanted devices are to substitute a lost biological structure. As another opportunity, they may assist an impaired biological structure. Otherwise, they may augment a biological structure already in existence. These devices are subject to stringent standards (see Sect. 2.3) and definitions. Any failure to comply with the specifications laid out in the standards cannot be tolerated.

The term "procedure" used in the definition of a medical implant needs disambiguation. A procedure means the series of surgical actions conducted in a certain manner for placing the implant inside the body. The procedure also includes the contiguous postoperative care necessary. But it does not broaden in scope to the end of the therapy. If the implant is removed after some period, its withdrawal constitutes another procedure.

2.6 Passive and Active Implantable Devices

Medical implants are of two types: passive and active (Table 2.4). Passive implants do not require an energy source for their working. In this class, mention may be made of artificial joints, vascular grafts, and artificial valves. Active implants need a source of energy for their functioning, e.g., cardiac pacemakers. The implantable electronic system is an active implant containing an electronic circuit. This electronic unit performs one or more of the three tasks: (1) It supplants the function of a diseased part. (2) It measures and observes the physiological parameters of the body, very attentively. (3) It initiates suitable action to remove any fault, if detected. (4) It delivers precisely controlled amounts of drugs.

Table 2.4 Passive and active implants

S. No.	Passive implants	Active implants
1.	Implants requiring no energy source for operation	Implants requiring an energy source such as a primary/rechargeable battery for their working
2.	Examples: artificial joints, heart valves	Examples: cardiac pacemakers, defibrillators

2.7 Active Implantable Devices

2.7.1 Implantable Neural Amplifiers

For untethered recording of simultaneous activity of a vast number of neurons (100–1000) in the human brain, high-performance neural amplifiers are used. These amplifiers are low-noise, energy-efficient, micropower, implantable devices. They operate in the millihertz to kilohertz range. They are endowed with wireless transmission capabilities [7, 8]. Besides possessing acceptably low-noise levels, the bioamplifiers must dissipate as little power as conceivable. This is essential to avoid thermally induced damage to the surrounding tissues. If the power consumption is low, batteries are dispensed with. Then the implant can be powered through energy harvesting strategies. These help to lengthen the implant life. The recordings by neural amplifiers are enabled by the MEMS (microelectromechanical systems) technology-enabled microelectrode arrays.

2.7.2 Implantable Electronic Systems for Electrical Stimulation

These systems are based on the premise that the human body is an electrical machine. The body is made of complex circuits. Therefore, flaws in the operations of the circuits in the body manifest themselves as diseases. The diseases need to be remedied by surgically opening the body. Then electrical stimulators are implanted inside the body. The implanted stimulators deliver pulses of required magnitudes and durations at the intended sites. These electrical stimulators substitute the diseased biological stimulators in the body whose performance has fallen short of expectations. This is an area directly correlated with the electrical system of the human body. It will be elaborated in an ensuing section.

2.7.3 Implantable Electronic Systems for Continuous Health Status Monitoring

Suppose it is required to continuously measure the blood pressure or glucose level of a patient. Further, it is necessary to transmit the data constantly to a doctor living nearby or at a distant location. Also, it is demanded that the freedom of movement or mobility of the patient should not be disturbed by such measurements [9].

Consider blood pressure first. Using a conventional sphygmomanometer machine for this purpose is not feasible. The reason why this is impractical is that the patient will be knotted with one arm inside the cuffs of the machine. He/she will not be able to perform normal duties. Similarly, the blood glucose level measurement requires that the blood samples be taken out at regular intervals. After taking out the samples, they are examined for glucose concentration. Both situations of blood pressure and glucose concentration measurement are highly inconvenient and unviable. The only remedy is to implant suitable sensor systems in the bloodstream, suppose a pressure sensor and a glucose biosensor. The implanted sensor systems must be furnished with signal conditioning and wireless transmission capabilities. Then it is possible to amplify, process, and transmit the signals received from the sensors by implanted electronic unit to the outside world.

2.7.4 Implantable Drug Delivery Systems

The commonest drug delivery route is the oral route. This route is not useful when rapid action is desired. Moreover, many macromolecules are lost. This loss occurs either by digestion in the gastrointestinal tract or through inadequate absorption in the blood. Similarly, pulmonary systems such as inhalers require drug absorption in the blood through lungs. In parenteral drug delivery, the drug is taken through a route other than through the intestinal tract, especially through injection. For receiving the injection, the patient must go to the clinic. Moreover, frequent injections are inconvenient to the patient because of the associated pain. If injections are to be given at defined time intervals, the patient is confined under close attention of the doctor. Such treatment requires extreme care. Portable infusion systems with transcutaneous catheter and external pump constitute another alternative. But consider a situation in which the drugs are to be delivered effectively at selected places at stipulated time intervals without any annoyance to the patient. Then the drug delivery system must be implanted inside the human body. By an implanted system, it is possible to provide site-specific, sustained, faster drug administration to regions that are in the greatest need of drug [10]. This allows lowering of doses. Decrease of doses leads to the reduction of attendant side effects. Besides more rapid delivery and improved targeting of specific organ, patient compliance is also achieved. This is because the method is comparatively less burdensome than either oral pills or injections. By patient compliance is meant the extent to which a patient adheres to a prescribed

Table 2.5 Four roles of active implantable devices

Neural signal amplification	Electrical stimulation	Health monitoring	Drug delivery
Recording of the neuron activity in the brain	Delivery of voltage/ current pulses of correct parameters at the relevant sites	Measurement of the vital parameters of a patient by implanted sensors and transmission of data to the physician/patient for necessary intervention	Provision of drug to specific sites, selectively in proper dose, and at the required rate

diagnostic, preventive, or therapeutic routine. Common applications of implanted drug delivery are in brain tumor or prostate cancer therapy. Table 2.5 defines the principal roles of active implantable devices.

2.8 Brief Historical Background

The historical evolution of different implantable devices will be dealt with in the respective chapters. Nonetheless, an overall commentary on the starting-phase developments in this field can be presented here. The first clinical implantation of cardiac pacemakerwas done on October 8, 1958, in a 43-year-old man. Since 1980, cochlear implants are in widespread use to partly restore the hearing capability of deaf people. The advent of microelectromechanical systems (MEMS) near the beginning to middle part of the 1990–2000 decade uplifted the evolution of retinal implants to treat blindness. In 1997, a deep brain stimulator (DBS) for inhibiting the tremors of Parkinson's disease was approved by the FDA. Today, wirelessly reprogrammable implantable medical devices (IMDs) are finding prolific applications. These are pacemakers, cardioverter defibrillators, and neurostimulators. They use ingrained electronics to observe chronic disorders. These biomedical implants are small and compact. They are neither affected by nor undermine the host biological environment.

Ko [11] takes a view back at the history of the start of implantable electronics during the period 1950–1970. He mentions that the incredibly small size and lower power requirement of the transistor served as an enabling technology. This was particularly so for the construction of practical telemetry systems for implants. Indeed, the development of implantable electronic systems started with telemetry devices. Small-size telemetry transmitters working at low power in microwatt to milliwatt range were developed. These transmitters had small bandwidths (frequency range for transmission). They occupied typically 1 cm^3 in volume. These transmitters were developed for diagnostic and monitoring applications. Their main applications were surveillance of health status of patients from hospitals. The objective was to oversee the response to therapy for early warning and necessary corrective action.

Then came a stage where functional electrical stimulation received attention. Ko [11] cites several examples, notably cardiac pacemakers and defibrillators, pain suppression devices, middle ear and cochlear implants, visual prosthesis, diaphragm pacers, seizure control for epileptics, hand and arm control for patients with spinal cord injuries, leg and foot control, and others.

Further onward came closed-loop electronic control systems. Herein, feedback telemetry circuits were included. Their inclusion served to adjust the parameters of instruments, as in pacemakers. Thus, implantable devices followed the progression: telemetry devices → electrical stimulation → automatic closed-loop systems.

2.9 Electrical System of the Human Body

Implantable electrical stimulators comprise a major chunk of implantable electronics today. These stimulators send electrical pulses through implanted electrodes to the relevant parts of the body. The use of electrical pulses for therapy is not surprising. It becomes obvious when one realizes that low-magnitude electric currents generated within the human body control many activities of the body. Electric currents are always flowing in our bodies. They are responsible for enabling and controlling various activities of the body. Feeling of pain, movements of muscles, secretions from glands, emotions, and thought processes are all triggered by electric currents. These phenomena are clear manifestations of electric signals.

Electroencephalography (EEG) is a neurological diagnostic test. In this diagnostic test, the electrical activity of the human brain is measured using electrodes fixed on the scalp of the patient. The amplitude of EEG signals lies in the range 5–100 μV (Table 2.6). Electromyography (EMG) is a diagnostic technique for recording the electrical signals from skeletal muscles. It seeks to assess the health of muscles and that of the controlling nerve cells. It uses adhesive pad electrodes for signal recording or is done intramuscularly. The amplitude of single action potential lies between 0.05 and 1 mV. The frequency varies from 10 Hz to 3 kHz.

Tingley [12] asserts that from the day of invention of computers, mankind has always contemplated of interfacing them with humans or vice versa. This desire and fancy has arisen because living beings are unsurprisingly electrical in nature. Once in every second, an electrical pulse is generated by a diminutive bunch of cells in the atrium of the human heart. This electrical pulse is responsible for the

Table 2.6 Four types of EEG waves

S. No.	Name of the wave	Frequency (Hz)	Amplitude (μV)
1.	α-wave	8–13	~ 50
2.	β-wave	14–30	< 30
3.	θ-wave	04–07	< 50
4.	δ-wave	0.5–04	10–300

heartbeat. The pulses of the body start from our birth and come to a standstill only on death. It is simply because the human body functions electrically that implantable electronics can influence its faulty operation and bring it back to normalcy. Electric fields comparable in magnitudes to those in lightning have been found inside the cells. At a given instant, the human brain produces sufficient current to energize a 15 W electric bulb. The obvious implication is that a thorough comprehension of the body's electrical system is necessary to appreciate the application of implanted electronic systems for therapeutic use.

2.10 Bioelectricity

It is the electricity produced by a living organism. The source of this electricity is the chemical energy of biological processes [13]. Bioelectric current involves the flow of ions. This is quite unlike the flow of electrons in the currents used for domestic lighting or telecommunication. Bioelectric potentials range in magnitude from one to few hundred millivolts. In contrast, much higher voltages, in the range of a few to hundreds of volts, are used in lighting or communication.

2.10.1 Generation of Bioelectricity by Cells

The basic structural and functional unit of the human body is the cell. Cell counts in the human body are $> 10^{12}$. These cells carry out different tasks. Such tasks are absorbing nutrients from food and converting them into energy. Cells are of different types. According to the tissues formed by them, cells are grouped into several classes, some of which are bone cells, nerve cells, muscle cells, blood cells, gametes or sex cells, etc. The generation of electricity in the body can be understood with reference to the flow of ions across the cellular membrane.

2.10.2 Membrane Potential

It is the difference of potentials, measured in millivolts, between the internal sides of a biological cell with respect to the fluid outside (Fig. 2.1). To clarify, it is the potential difference across the cell membrane between the cell's interior and exterior regions. Hence, it is aptly called membrane potential [14]. Without this potential, human life is not possible. Life of other living creatures is also impossible. In all the living beings, a potential difference is maintained across their cell membranes. This potential difference originates by virtue of the differences in ionic concentrations in the cellular fluids on the opposite sides of the membrane. The fluids on the opposite sides are named as intracellular fluid, which is the fluid inside the cell, and extracellular fluid, which is the fluid outside the cell. The concentrations of four principal ions differ significantly in these fluids. Three of these ions are metal ions: sodium (Na^+), potassium (K^+), and calcium (Ca^{++}) ions$^-$. One ion, the chlorine (Cl^-) ion, is from a nonmetal. Inside the cell, the concentration of K^+ ions is very high. This concentration is about 28 times its value outside. Negatively charged proteins balance this high concentration of positive charges. Outside the cell, Na^+ ions have a high concentration. It is around 14 times more than inside. Also, the concentration of Ca^{++} ions outside the cell is much higher than inside. Thus, there is a very large gradient in terms of Ca^{++} ion concentration. These positive ionic charges are balanced by Cl^- ions. The Cl^- ions are present in the ratio of 25:1 in the cell exterior as compared to its interior. For a resting nerve cell, the ion concentrations generally follow the trend given in Table 2.7.

The transference of the above ions takes place across the cell membrane toward the opposite sides. This transference depends on several factors. Among the forces responsible for ion movement, the important ones are those arising from concentration gradient (diffusion) and charge imbalance (electrical). Besides diffusion and electrical forces, the other contributors to ion movement are membrane permeability and sodium–potassium pump. The membrane has different permeability values for various ions. This means that the membrane is selective in its behavior with regard to passage of different ions across it. It has a high permeability to K^+ and Cl^- ions. On the opposite side, Na^+ ions experience difficulty in crossing the membrane. Further, it is totally impenetrable to the anions. The anions cannot pass through the membrane at all. The differences in ionic permeability arise from the presence of watery pores in the membrane for particular ions. These pores are known as gates or channels. The membrane has 100 times more channels for K^+ ions than Na^+ ions. Thus, the crossing of Na^+ ions across the membrane into the cell is effectively prohibited. Sodium–potassium exchange pump is an energy-consuming mechanism of active transport. It is concerned with transfer of ions from a low concentration to a high concentration region [16]. This process is driven by the energy supplied by the hydrolysis of adenosine triphosphate (ATP). The hydrolysis involves the participation of the enzyme Na^+/K^+-ATPase (sodium–potassium adenosine triphosphatase). The enzyme pumps out Na^+ ions from the cell. It also pumps K^+ ions into the cell. For every 3 sodium ions that are extruded from the cytoplasm into the extracellular fluid, 2 potassium ions are brought into the cytoplasm from the extracellular fluid. Thus, the ratio of number of sodium ions extruded to the number of potassium ions intruded is 3:2.

The end result of the joint action of above forces is that inside the cell, the ionic population is mainly composed of potassium ions. Outside the cell, the ionic population has two ionic components: sodium ions plus potassium ions. This means that the number of positive charges outside exceeds that inside the cell. The consequent charge imbalance leads to a negative potential inside the cell with respect to that outside. This potential has a value of − 70 mV. It is called the resting potential of the cell. It is said to be "resting" because it pertains to the condition when the cell is not disturbed or stimulated. The resting potential is the static membrane potential

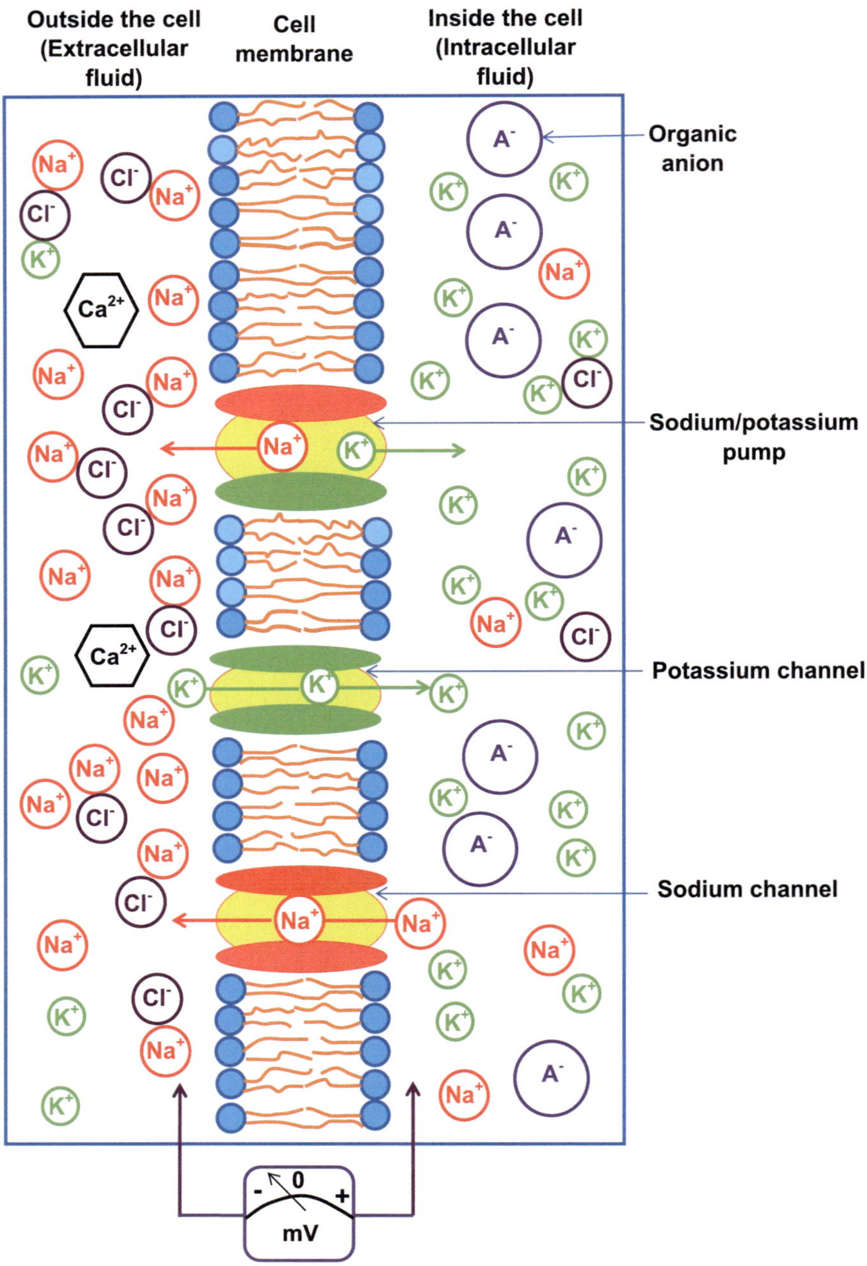

Fig. 2.1 Contributions of different ion fluxes toward the resting membrane potential

Table 2.7 Typical ion concentrations for a resting nerve cell [15]

Name of the ion	Ion concentration inside the cell (mM)	Ion concentration outside the cell (mM)	Ion concentration inside/ ion concentration outside
Sodium (Na^+)	10	140	0.07
Potassium (K^+)	140	5	28
Chlorine (Cl^-)	5	125	0.04
Ca^{++}	0.0001	2	5×10^{-5}

of quiescent cells. There is another important potential related to the cell called the action potential. An action potential occurs under the influence of a stimulus.

2.10.3 *Action Potential*

It is the localized, momentary change in the membrane potential of a cell. It occurs upon stimulation of the cell. Its occurrence leads to the transmission of an electrical impulse [17]. In the span of a few milliseconds, the membrane potential ascends from the resting potential value. This value is characteristically − 70 mV. From this negative value, the membrane potential rises to a positive value, around + 40 mV. Then it tumbles down and marginally overshoots the resting potential value. Ultimately, it sluggishly climbs to the typical resting potentialvalue (Fig. 2.2).

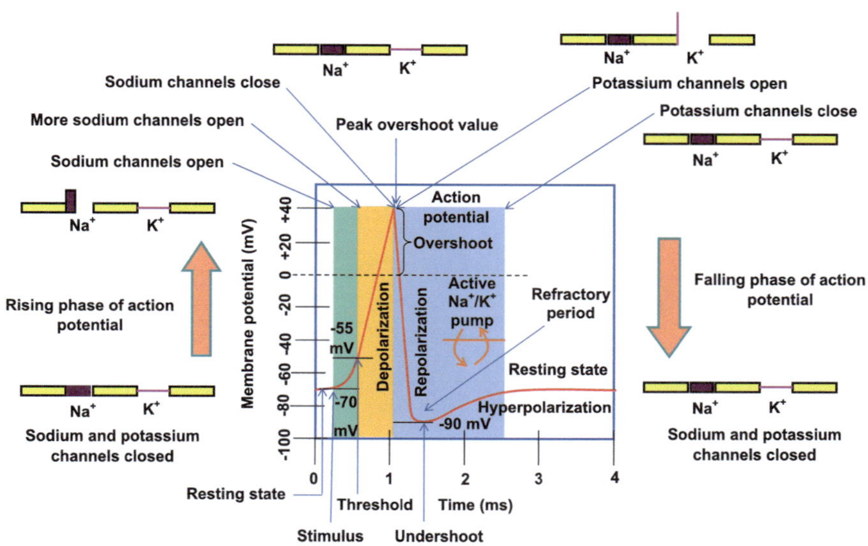

Fig. 2.2 Different phases of the action potential

The sequences of ionic activities in the development of an action potential are as follows: Any disturbance or stimulus (touch, light, sound, etc.) spearheads the opening of a small number of sodium channels in a tiny segment of the membrane. From the sodium channels thus opened, a few sodium ions transfer to the cell. These transferred sodium ions raise and uphold the potential inside the cell. The potential changes from a negative value in the resting condition to slightly less negative value. No sooner than the membrane potential touches a value called the threshold stimulus, a large number of sodium channels are opened. This opening is trailed by the onrush of a large number of sodium ions. It causes the development of an action potential. The membrane potential immediately upsurges to + 40 mV. At this stage, the sodium channels bolt down. So, no further sodium ions can enter the cell. Also, the potassium channels are unsealed. Through the opened potassium channels, potassium ions trickle out from the cell. The loss of positive charge makes the potential inside the cell negative. Its value becomes less than the resting potential. Hence, the membrane potential becomes lower in value than the resting potential. Now the sodium–potassium ion pump leaps into action. It restores the membrane potential to its standard negative value. It may be noticed that the action potential follows an all-or-none law. This law means that there are two possibilities: (1) The threshold stimulus is reached. This triggers an action potential. (2) It is not attained. So, there is no action potential. No intermediate, transitional state of a faint action potential can exist.

In a nerve cell or neuron, the nucleus is located inside the cell body or soma. From the cell body outspread protoplasmic protrusions called dendrites. The dendrites bring impulses toward the soma. Axons carry impulses away from the soma (Fig. 2.3).

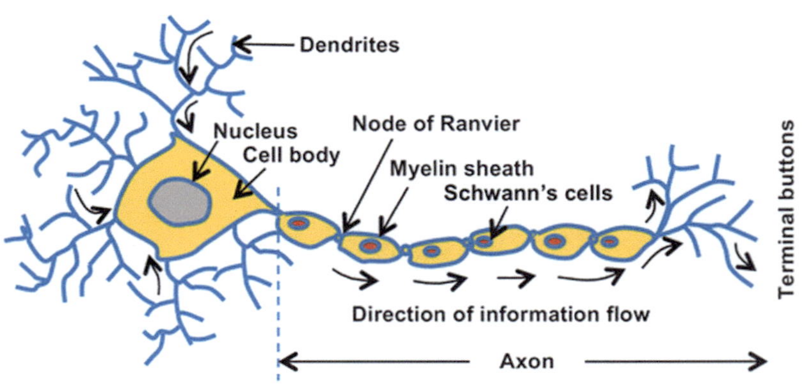

Fig. 2.3 Impulse conduction and information flow in a neuron

2.11 Discussion and Conclusions

In this chapter, the mesmerizing and sensational field of implantable electronic devices was familiarized to the reader. Among the implantable devices, a few devices are meant for neural signal amplification. They assist in measuring the neural potentials accurately. Some devices act as watchdogs of the patient's health. They report immediately any irregularities and aberrations observed. Other devices impart energy pulses to explicit parts of the body to bring back any diseased part to its normal working rhythm of the healthy state. These devices can be better comprehended in the backdrop of the electrical system in the human body. This electrical system was described. The production of electricity in the body was explained. Yet another class of devices includes those that are electronically programmed to administer the prescribed drug at the chosen site at the fixed time to make therapy more effective and reduce its adverse effects.

Review Exercises

2.1 What is the difference between a medical device and a medicinal product?

2.2 Classify medical devices according to the risks inherent in their usage. Explain the characteristics of each class of device. Give examples.

2.3 Under which class will you place the following medical devices: (1) cardiac pacemaker, (2) elastic bandages, and (3) surgical drapes? Which of these devices requires maximum regulatory monitoring and which one needs least control?

2.4 Explain and give examples of the following terms: (1) noninvasive and invasive medical procedures and (2) noninvasive and invasive medical devices.

2.5 Define the term "implantable medical device." Name the two classes into which these devices are subdivided and discuss how they are different?

2.6 What makes us believe that delivering electrical pulses to specific parts of the body can cure diseases? Argue.

2.7 Give your suggestion on how can a doctor continuously know the blood glucose concentration of a patient without any restriction on patient's movement.

2.8 What are the disadvantages of oral delivery of drugs? What is meant by parenteral drug delivery?

2.9 Justify the use of implanted drug delivery systems. What do you mean by patient compliance?

2.10 Trace the historical evolution of implantable electronics following W. H. Ko.

2.11 Comment on the statement, "Human body is an electrical machine."

2.12 What is bioelectricity? What is its source?

2.13 Define membrane potential of a cell. How is it produced?

2.14 Name the four important ions whose concentration differences across the cell membrane are responsible for the origination of membrane potential. Which ion has a high concentration: (1) inside the cell or (2) outside the cell?

2.15 What are the main forces governing the motion of ions across the cell membrane?

2.16 Does the cell membrane show equal permeability for all the ions? If not, which ions can move across the cell membrane with no trouble and which ions face difficulty?
2.17 What is the meaning of gates in the cell membrane?
2.18 Explain the role of sodium–potassium pump in bioelectricity generation. Upon what source of energy does this pump work?
2.19 What is the resting potential of a cell? Mention its typical value. What is action potential of a cell? Describe the phenomena taking place inside the cell that lead to the initiation of an action potential.
2.20 What are dendrites and axons? In what directions do they conduct impulses?

References

1. Bazaka K, Jacob MV (2013) Implantable devices: issues and challenges. Electronics 2:1–34. https://doi.org/10.3390/electronics2010001
2. Greatbatch W, Holmes CF (1991) History of implantable devices. IEEE Eng Med Biol 38–41:49
3. Baker EA (2013) Guidance on legislation: borderlines between medical devices and medicinal products. http://www.mhra.gov.uk/home/groups/dts-bs/documents/publication/con 286964.pdf. Accessed 7 July 2014
4. Syring G (2003) Overview: FDA regulation of medical devices. http://www.qrasupport.com/FDA_MED_DEVICE.html. Accessed 7 July 2014
5. Dictionary.com. Dictionary.com unabridged. Random House. http://dictionary.reference.com/browse/noninvasive. Accessed 5 July 2014
6. Active implantable medical devices directive (90/385/EEC) (1990) http://www.mednet-eurep.com/fileadmin/mednet_ear/download/AIMDD90_385_EEC_2007-en.pdf. Accessed 5 July 2014
7. Rodríguez-Pérez A, Delgado-Restituto M, Medeiro F (2014) A 515 nW, 0–18 dB programmable gain analog-to-digital converter for in-channel neural recording interfaces. IEEE Trans Biomed Circ Syst 8(3):358–370
8. Chae MS, Yang Z, Yuce MR (2009) A 128-channel 6 mW wireless neural recording IC with spike feature extraction and UWB transmitter. IEEE Trans Neural Syst Rehabil Eng 17(4):312–321
9. Theodor M, Fiala J, Ruh D et al (2014) Implantable accelerometer system for the determination of blood pressure using reflected wave transit time. Sens Actuators A Phys 206:151–158
10. Tng DJH, Hu R, Song P et al (2012) Approaches and challenges of engineering implantable microelectromechanical systems (MEMS) drug delivery systems for *in vitro* and *in vivo* applications. Micromachines 3:615–631. https://doi.org/10.3390/mi3040615
11. Ko WH (2012) Early history and challenges of implantable electronics. ACM J Emerg Technol Comput Syst 8(2):1–9. https://doi.org/10.1145/2180878.2180880
12. Tingley K (2013) Annals of medicine: the body electric. New Yorker 25:78–86
13. Jeong Y (2011) Chapter 2: Introduction to bioelectricity. In: Yoo H-J, van Hoof C (eds) Biomedical CMOS ICs. Springer Science + Business Media, New York, pp 13–29
14. Clark JW (1998) The origin of biopotentials. In: Webster J (ed) Medical instrumentation: application and design. Wiley, New York, pp 126–188
15. Purves D, Augustine GJ, Fitzpatrick D et al (eds) (2001) Neuroscience. Sinauer, Sunderland, 680 pp
16. Morth JP, Pedersen BP, Toustrup-Jensen MS et al (2007) Crystal structure of the sodium–potassium pump. Nature 450:1043–1049. https://doi.org/10.1038/nature06419

17. Freudenrich C (2007) How nerves work. HowStuffWorks.com. http://health.howstuffworks.com/human-body/systems/nervous-system/nerve.htm. Accessed 5 July 2014

Chapter 3
Generic Implant Architecture and Organization

Abstract An archetypal implantable microsystem comprises two separate structural units: an external controlling module and a component placed inside the human body (the implant). The internal unit can be partitioned into several sections. These sections include the analog front end, memory, microprocessor (CPU), communication, and power management sections. The analog front end takes care of sensing and stimulation functions. The sensing function decides the magnitude, time, and extent of therapy to be given. It also monitors the therapy of the patient. The stimulation function looks after the output voltage or current pulse delivered by the device to the relevant part of the body. The memory unit stores the computer program and data. The microprocessor is the brain of the implanted device. Communication is established between the implanted device and the external controller for initially setting the device and for its subsequent control. The subsequent control is done to fine-tune the therapeutic parameters. Apart from data telemetry, communication is used for power delivery in rechargeable battery implants. Power management is essential to make best use of the battery life or to minimize the time duration between two consecutive charging steps, if a rechargeable battery is used.

Keywords Induction charger · Resonant coupling · Nonresonant coupling · Voltage regulator · SMPS · Bandgap voltage reference · MICS · MEDS · CPU · RAM · ROM · OP-AMP

3.1 Introduction

An implantable electronic system is capable of performing several functions. Primarily, these functions are:

1. Measurement of patient's physiological parameters and their transmission to the patient/consultant physician. These assignments fall under the head "telemetry"; "tele" means distance and "metry" means measurement. The patient or the physician is monitoring the therapy from outside the body of the patient [1].

V. K. Khanna, *Implantable Medical Electronics*,
https://doi.org/10.1007/978-3-031-99868-3_3

2. Supplying the electrical stimulus of the desired value, as fixed by the patient in dialogue with the physician. This entails actuation from a distance. The physician is advising from a remote place. The patient is controlling the implant inside the body from a switch located outside. Hence, this activity is described as "teleactuation."

The telemetry and teleactuation functions are often carried out in a closed-loop fashion. From these discussions, it is clear that the implantable electronic system must be a two-component system. The component delivering the stimulation or carrying out corrective dosing action resides inside the patient's body, as in a drug delivery device. The other component supervises the operation of the resident component and delivers power to it. This component is placed outside the host body [2]. A continuous exchange of information and also energy takes place between the indwelling and the external modules, whenever necessary. A close examination of the construction and structural constitution of the several available implantable electronic systems will bring home to the readers' mind that the above two-piece arrangement for partitioning of responsibilities has found favor with most designs.

In implantable systems used for sensing action, the sensors inside the indwelling module convert the biological signals into electrical signals. The electrical signals undergo the requisite signal processing. Then they are transmitted to the external module. Herein, they are analyzed. On this basis, suitable advisory is issued to the actuators in the indwelling module [3]. To cite an example, in a patient who has undergone cardiac surgery, a micro-accelerometer is fixed on the surface of the heart wall. So, any abnormality in the motion of the heart wall is noticed. It is conveyed to the surgeon for immediately attending to the patient's discomfort.

In implantable systems meant for stimulating nerves or muscles, the external component transmits the desired commands wirelessly across the patient's skin to the module inside. These commands enable the module inside to supply the pulses of needed voltage or current magnitudes, duration, and frequency in accordance with patient's health parameters. Thus, the treatment is made more efficacious [4]. Both the components of the implantable system operate in a harmonious manner in a closed-loop coupling. Response of the stimuli on the patient is assessed. Based on this evaluation, the stimuli are varied to give the best comfort to the patient.

The block diagram representation of a typical active implantable system is shown in Fig. 3.1. As already said, the system is made up of two components: one external and the other internal. The external component supplies the battery-charging power. It also receives the signals from the internal component and gives commands to control the same. The internal component delivers the therapy.

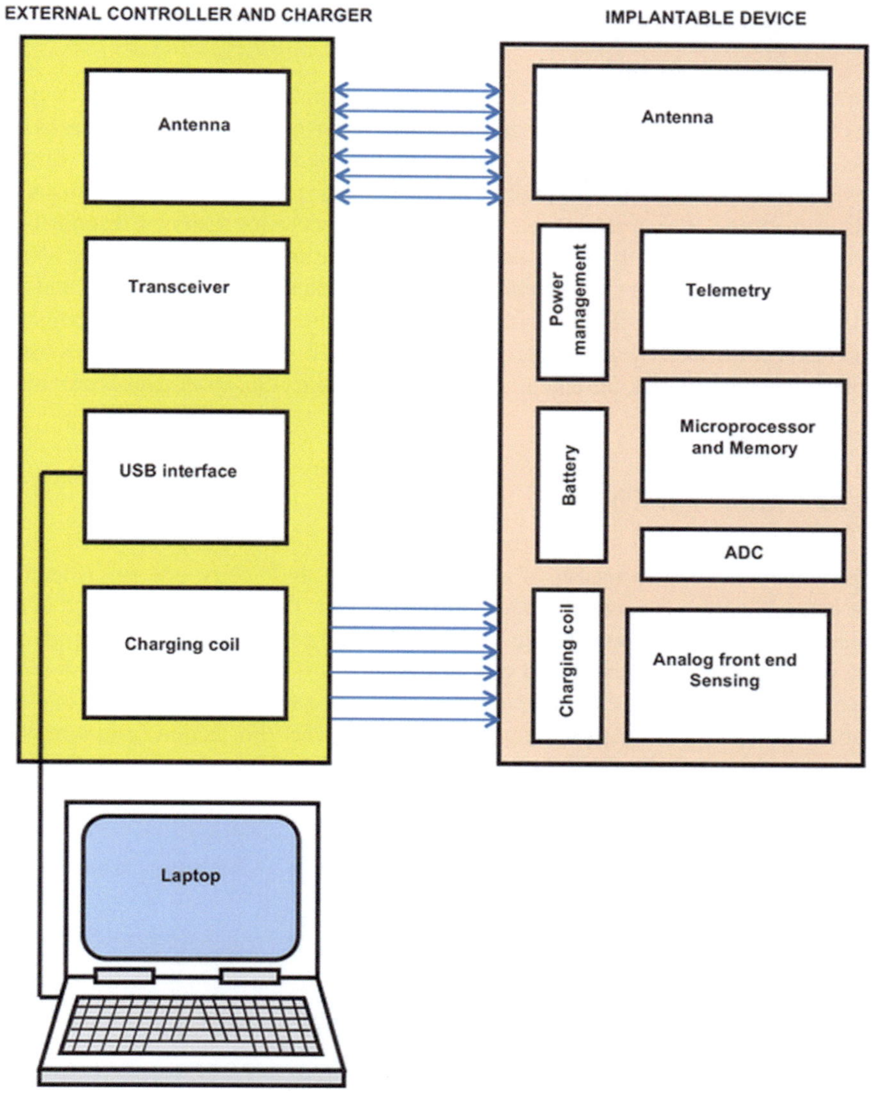

Fig. 3.1 Two-component block diagram representation of an active implantable device system with external controller. Different stages in the working sequence of both the external and internal parts are shown. In many versions, the antenna also performs the function of the charging coil so that a separate coil is not necessary. Here, the antenna is used both for communications and battery charging

3.2 External Part of the Implantable Device

An extensive diversification of the facilities incorporated into implantable devices has taken place. Hence, the external part has to perform a multitude of tasks. In fact, it serves as the part externally controlling the operation of the implant, without any physical contact, and wirelessly too. Sometimes it does so from close distance and sometimes even from remote location. To execute its assigned responsibilities, this part is carefully designed. In general, the external part consists of the following components: (1) induction charger, (2) antenna, (3) transceiver, and (4) universal serial bus (USB) port. In many devices, the function of components (1) and (2) is performed by a single coil, the charging coil. Also, in some devices using nonchargeable batteries, the charging coil is not necessary. So, an antenna is provided.

3.2.1 Induction Charger

The induction charger consists of a pair of coils. Of these coils, one coil is kept in the external part to generate the electromagnetic field. The other coil is placed in the internal part to receive power from this field. Upon receipt of this power, it converts the power into electric current to charge the battery (Fig. 3.2). For action at close distances, the coil in the external part must be in close vicinity of that in the internal part. So, the two coils together constitute a transformer. For action from a remote location, resonant inductive coupling must be established between the two coils.

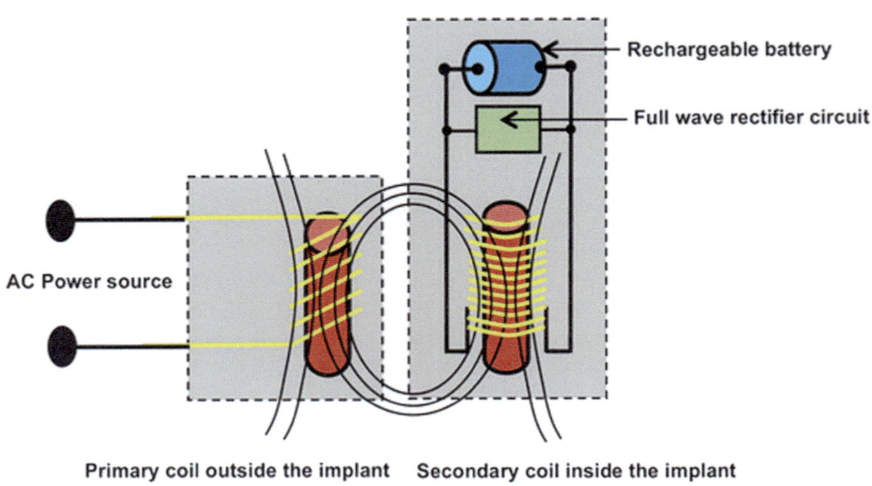

Fig. 3.2 Induction charger based on inductive coupling

3.2.2 Nonresonant and Resonant Coupling

Here, it is necessary to make a clear distinction between nonresonant and resonant coupling. Nonresonant coupling occurs in a transformer. Because the magnetic field generated by the primary coil must embrace the secondary coil, the two coils must be as close as possible. With increasing distance, most of the energy is dissipated through resistive losses in the primary coil.

In resonant coupling of two coils, each of the two coils is capacitively loaded (Fig. 3.3). An LC tuned circuit is thereby formed (*L*-inductance, *C*-capacitance). At a particular excitation frequency called the resonance frequency, a significant amount of energy transference takes place from the primary to the secondary coil. The transference efficiency is appreciable over a distance, which is few times the coil diameter. It is much more than in a transformer.

Inductive coupling is compared with resonantcoupling in Table 3.1.

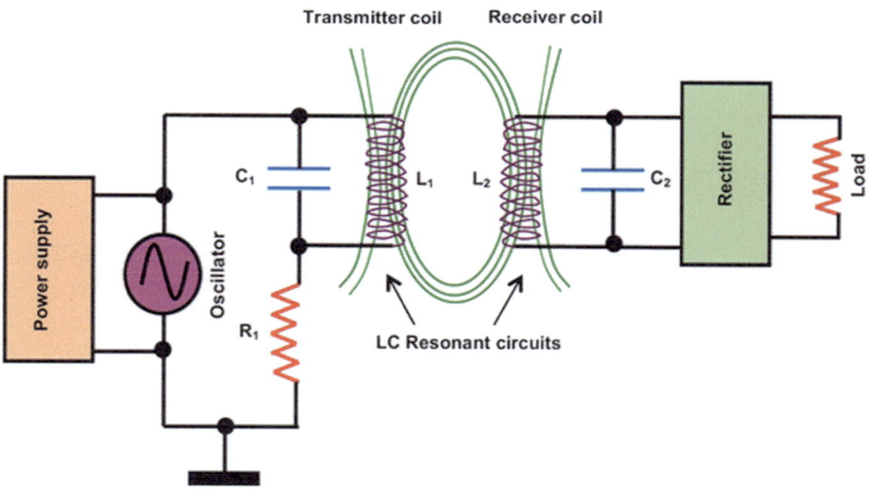

Fig. 3.3 A wireless power transmission system using resonant inductive coupling

Table 3.1 Inductive and resonant inductive coupling

S. No.	Property	Inductive coupling	Resonant inductive coupling
1.	Principle	Mutual inductance; electromagnetic induction	*LC* resonance
2.	Mode of transfer	One-to-one interaction	One-to-many interaction
3.	Range	Small (few cm)	Large (few m)
4.	Efficiency	Low. Energy is lost in air. It may be drawn by any conducting material in vicinity if the range is increased	High. Energy switches back and forth between *L* and *C*. The radiation loss is thereby reduced

3.2.3 Antenna

The antenna is an arrangement of metallic elements connected to the transmitter/receiver. During transmission, an oscillatory current flows through the antenna. The oscillatory current creates time-varying electric and magnetic fields in its neighborhood. These interlinked fields are radiated as electromagnetic waves. During reception, the oscillating electric and magnetic fields in the incoming electromagnetic waves impinge on the antenna. They set up oscillating currents in it.

3.2.4 Transceiver

The transmitter and the receiver are housed together in a single enclosure. They together constitute the transceiver. In the full duplex mode, the operational frequencies of the transmitter and receiver are different. This difference is necessary to avoid interference between the transmitted and received signals. Thus simultaneous to-and-fro communication, i.e., in both directions, is possible.

3.2.5 USB Port

The USB port allows the user to connect a computer to the external port. Along with the computer, mouse, printer, and other accessories are also connected. USB is an industry standard set of specifications for connectivity between computers and peripheral devices.

Thus, the external part is endowed with all capabilities to supply power to the implant. It can also carry out bidirectional exchange of data and information with the implant. These capabilities are further strengthened by interfacing with a computer. The attendant information processing and programmability features of the computer are very helpful. So, the doctor can always get updates on the patient's health status. He/she can program the implant to carry out the intended tasks in the desired sequential scheme.

3.3 The Inner Structural Layout of the Implant

If someone were to break open the implant and look inside to see the constructional units of this box, one would find that any implant contains some or all of the components described in the following subsections.

3.3.1 The Secondary Coil

This coil of the induction charger, located inside the internal part, embraces the magnetic field set up by the primary coil in the external part. It sets up an induced current in the internal part according to the varying current in the primary coil in the external part.

3.3.2 Rectifier, Filter, and Chargeable Battery

Associated with the charging coil is the rectifier circuit. This circuit converts time-varying current into a pulsating direct current. A full-wave rectifier uses four diodes in a bridge configuration. It rectifies both halves of the sinusoidal AC signal (Fig. 3.4). The unidirectional current from the rectifier circuit is smoothened by the filter. The smoothening is done by removing the unwanted AC components of the output waveform. Referring to Fig. 3.4, during the ascending portion of voltage wave, the filter capacitor is charged. During its descending portion, the capacitor discharges through the output resistor. This helps to keep the current constant. The resulting uniform and constant current is fed to the battery. It charges the battery to the rated capacity. So, the battery provides the correct voltage to supply the energy for operation of the implant.

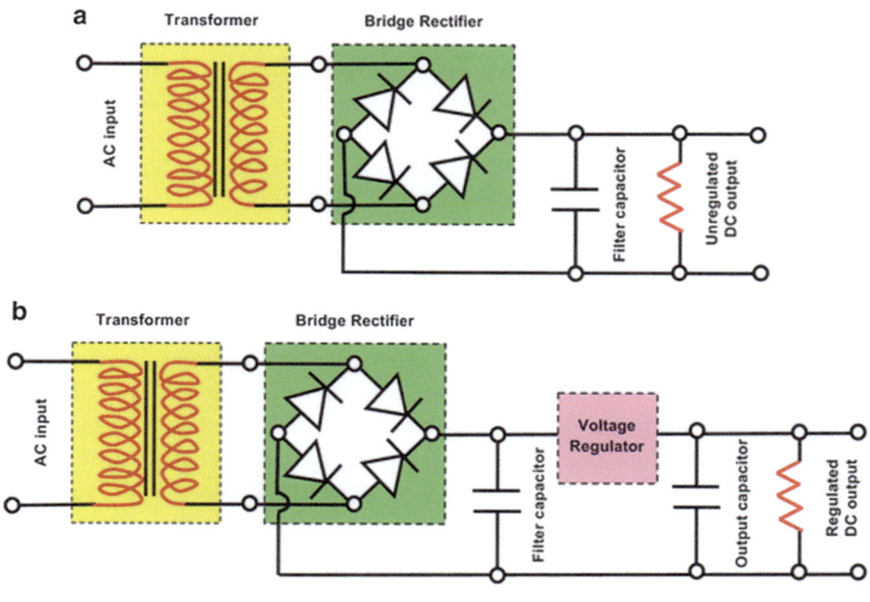

Fig. 3.4 Power supplies: **a** unregulated, **b** regulated

3.3.3 *Voltage Regulator*

The regulator stage furnishes a constant supply voltage. It does so despite variations in input voltage or load current. Linear regulators offer simplicity in design. Electromagnetic interference (EMI) is practically absent in linear regulators. But they are inefficient and waste power. Switching regulators can provide efficiencies exceeding 85% with increased complexity. But switching currents can cause noise problems. Both types of regulators need a steady reference voltage. This voltage is obtained from a specifically designed circuit called the bandgap reference circuit. This circuit is described in Sect. 3.3.3.3.

3.3.3.1 Linear Regulator

In a linear regulator, a power transistor (FET) is operated in its linear region (Fig. 3.5). So, it behaves as a variable resistor connected in series with the load connected across the output terminals [5]. A feedback loop is formed. In this loop, an error amplifier has a most important duty to perform, which is to sense the DC output voltage. A sampling resistor network is employed for this sensing. The network compares the feedback voltage with a reference voltage. The output voltage from the error amplifier acts upon the series power transistor. It propels the base terminal of this transistor via a current amplifier. Let us consider the situation when the input voltage decreases or load current increases. Then the output voltage decreases. So, the feedback voltage also falls. Consequently, more current is produced by the feedback error amplifier and the current amplifier. The current is fed into the base terminal of the transistor. This action lowers the voltage drop between collector and emitter. It forces back the output voltage. This forcing is continued until the feedback voltage equalizes with the reference voltage. By similar reasoning, the working of the circuit is explained when output voltage increases. Excessive power dissipation in the series power transistor is the cause for the low efficiency of this circuit.

3.3.3.2 Switch-Mode Power Supply

The desired output voltage is obtained in a linear regulator by dissipating the excess power as heat in the series transistor. Unlike the linear regulator, an switch-mode power supply (SMPS) regulates the output voltage or current by switching actions. In this switching configuration, storage devices are toggled into and out of various electrical circuits. The storage devices used are inductors and capacitors. An SMPS is based on the switching-mode operation of the transistor instead of the linear-mode operation. The DC supply is chopped at a higher frequency ~ 15–50 kHz. The chopping is done using an active device like the BJT or MOSFET (Fig. 3.6). A chopper or DC–DC converter is a static device which is used to convert a constant DC voltage at the input into a variable DC voltage at the output. The conversion is carried

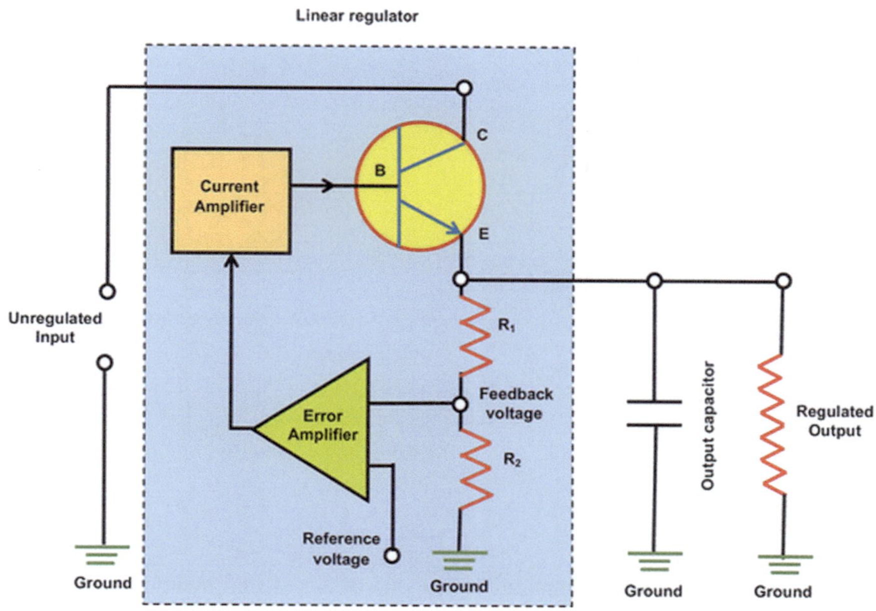

Fig. 3.5 Linear voltage regulator

out by using switching technique. In the switching technique, an electronic switch interrupts one signal under the control of another. One type of switching technique is called pulse-width modulation. In this switching technique, the width of the pulse is varied. The chopping period and frequency are kept constant. The output voltage is altered by changing the on-time. Another type of switching is known as variable frequency control. In variable frequency control, the chopping frequency is varied. The on-time and off-time are maintained at constant values. This technique is also referred to as frequency modulation.

During on-state of the transistor, the voltage dropped by current flowing inside the transistor itself is negligible. The voltage drop is negligible because the transistor operates in the saturation region. In the off-state, a very small current flows through the power path of the transistor. The current is very small because the transistor is working in cutoff mode. Power is only dissipated during the transitions between on- and off-states. The time spent during the transitions must therefore be minimal. By varying the duty cycle or switching frequency, the stored energy in each cycle is adjusted to vary the output voltage. For voltage regulation, a sample of the output voltage is fed back to the drive circuit of the switching transistor. Regulation is provided by controlling the switching action of the chopper circuit in response to the feedback. SMPS is attractive to circuit designers from the viewpoints of obtaining high efficiency at a high density of power (area reduction) and with low power dissipation.

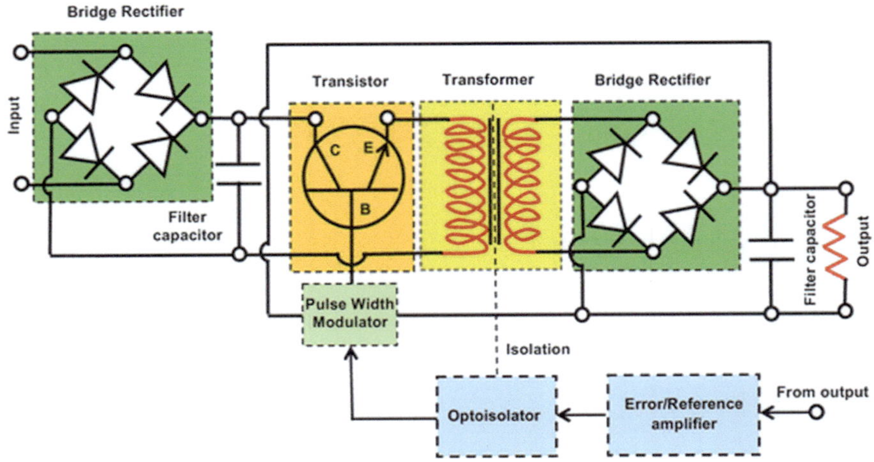

Fig. 3.6 Switch-mode power supply

Table 3.2 brings out a comparative analysis of linear and switch-mode power supplies.

3.3.3.3 Bandgap Voltage Reference Circuit

It is a temperature-independent voltage reference circuit. It supplies a stable output voltage. The output voltage is stable irrespective of manufacturing process and supply voltage variations. The circuit is implemented by canceling out two opposing voltage variations caused by temperature. Therefore, its operation involves summing two voltage sources. One voltage source is producing a voltage V_1. The voltage V_1 has a positive temperature coefficient α. The other voltage source is producing a voltage V_2. The voltage V_2 has a negative temperature coefficient $-\alpha$. Then the sum of the two voltages will yield a voltage $V_{out} = V_1 + V_2$. The voltage V_{out} has a temperature coefficient $= \alpha - \alpha = 0$.

Bandgap circuit implementations are usually done using transistors. The reason is the more nearly ideal characteristics of transistors than ordinary diodes. In a bipolar transistor, the behavior of base-emitter voltage V_{BE} is nearly CTAT (complementary to absolute temperature, i.e., it decreases linearly with temperature). To this voltage V_{BE} is added a voltage that is PTAT (proportional to absolute temperature). The slope of the PTAT term is selected to be equal in magnitude to that of the CTAT term. Then their sum will be independent of temperature. This means that addition of PTAT and CTAT voltages in the proper ratio yields an output voltage whose magnitude = the bandgap voltage extrapolated to 0 K (1.220 V). This output voltage is independent of temperature. Stated otherwise, if the PTAT component is adjusted to make the output voltage equal to V_{G0} at any temperature, the output voltage will be V_{G0} at all temperatures. For obtaining a PTAT voltage, two base-emitter voltages are subtracted, and

Table 3.2 Linear power supply and SMPS [6]

S. No.	Property	Linear power supply	SMPS
1.	Circuit complexity, cost and ease of use	Simple. It uses a regulating IC and noise filtering capacitors. It is low cost and easy to use	Intricate. It uses a controller IC, power transistors, diodes, transformer, inductors, and filtering capacitors. It is an expensive and difficult solution
2.	Size and weight	Small if no transformer is used	Small even if a transformer is used. The size is small because of high operating frequency (50 kHz–1 MHz)
3.	Output voltage	Any voltage if a transformer is used	Any voltage available
4.	Efficiency	30–40% because voltage regulation is accomplished by squandering away the surplus power in the form of thermal energy	60–70% since voltage regulation is performed by varying the duty cycle. Efficiency is lowered by switching losses in transistors, resistive or core losses of inductor, voltage drop across rectifier diodes. But by proper optimization, the efficiency can be raised to 95%
5.	Transient response	Fast	Slow
6.	Radio-frequency interference	Causes meek interference at high frequencies by rectifier diodes during severe current loading and some mains hum induced into unshielded cables	Produces electromagnetic and radio-frequency interference. It requires EMI filters and shields
7.	Electronic noise at input	Introduces harmonic distortion in AC input	Couples switching noise back into mains supply
8.	Electronic noise at output	Can cause audible mains hum or brightness ripples	Noisier
9.	Acoustic noise	Feeble, indistinct, usually muffled mains hum caused by the shuddering windings of the transformer	Inaudible unless provided with a fan
10.	Power factor	Low	Low to medium
11.	Applications	Suitable for use where output and input voltages differ by a small amount	Suitable for applications demanding low power dissipation and high efficiency. It is suitable whenever high power density (small size) is necessary. It is a preferred supply in high current applications

the difference between these voltages is taken. These emitter-base voltages belong to identical transistors. These transistors are operating at two dissimilar values of collector current. More generally, they belong to transistors made in the same fabrication process and operating at two unequal values of collector current density. Then ΔV_{BE} truly is PTAT if the collector current densities are in a fixed ratio. Hence, while each V_{BE} is nearly CTAT, the difference between two V_{BE} values is perfectly PTAT. Thus, a bandgap reference circuit exploits the negative temperature coefficient of emitter-base voltage in combination with positive temperature coefficient of emitter-base voltage differential of two transistors. These transistors are operating at nonidentical current densities. By using these antithetical temperature coefficients, it makes a reference having zero temperature coefficient [7]. The bandgap voltage reference represents the most popular high-performance voltage reference used in IC technology.

Figure 3.7 shows a Widlar bandgap reference circuit. In this classical Widlar bandgap reference circuit, the transistor Q_I is operated at a somewhat high current density. This current density is approximately obtained by multiplying the current density of transistor Q_2 by a factor of 10. The difference ΔV_{BE} in the emitter-base voltages of the two transistors Q_1, Q_2 is measured across the resistor R_3. If the transistors Q_1, Q_2 are high current gain devices, the voltage drop across R_2 is also proportional to ΔV_{BE}. The transistor Q_3 is a stage for providing gain. It controls and adjusts the output signal at a voltage = its emitter-base voltage V_{BE} + the voltage drop $(R_2/R_3)\ \Delta V_{BE}$ across R_2, as shown in the circuit diagram.

3.3.4 Power Saving and Economization Unit

Power must be expended with great restraint. The subdued consumption of power gives a longer battery life, in case of un-chargeable batteries. For the chargeable batteries, it provides a longer time span between successive charging cycles of batteries. Efforts should be made to avoid any wastage of power in any activity, which is superfluous or spurious. The objective is to optimize the consumption of power by different portions of the implant. This optimization depends on whether the concerned portions are active or lazy. Power must be preserved for legitimate use in performing only essential duties of the implant. At any given instant, only those circuits should be fed with energy that are not idling or loafing, but are busy in performing useful work. Other circuits should be denied power supply and should be shutdown. This makes it necessary to choose or select the circuits that are contributing to a given operational condition of the device. Those circuits which are redundant for that particular condition must be avoided. Such capability must be built in the system during its design. Furthermore, at the design level, emphasis must be placed on the inclusion of those circuits, which are able to fulfill the specifications at minimum power consumption. Needless to say, similar results are achievable with different circuits at the expenditure of widely different levels of power. But in implantable devices, even a minute increase in power consumption is intolerable. Even this small increase places an annoying

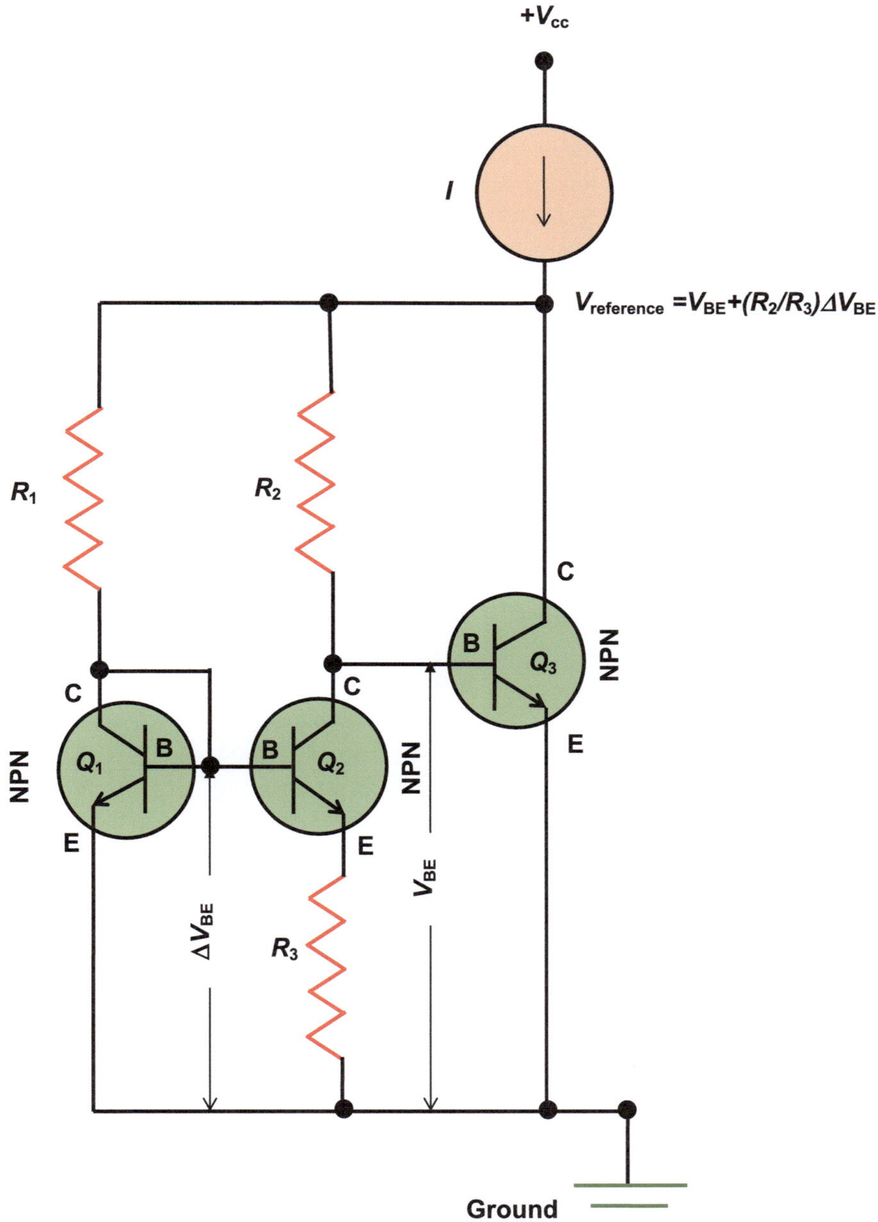

Fig. 3.7 Classical bandgap voltage reference circuit given by Widlar

load on the battery. The designer must note that customary circuits are often designed to execute a broad range of functions. They are more of general-purpose nature than targeted toward a specific aim. So, the designers must propose innovative circuits for their specific requirements. They must keep away from general designs catering to wider needs at an increased power disbursement. The use of off-chip components is a serious impediment. This hindrance is not only from power drainage viewpoint but also from space wastefulness and reliability angles. The necessary long connecting metal lines and larger pads increase the parasitic capacitances unnecessarily. They lead to depletion of power. Hence, these off-chip components must only be used when an integrated on-chip solution is impossible. Apart from the adoption of above power saving measures, the designer has a strong option of framing the design around SMPS concepts. This will help the designer to avail of the benefits of buck and boost converters and charge pumps. An accompanying advantage is reduction in the overhead voltage for the circuit and thereby the extravagance in spending power. To quash the noise in analog circuits, a low power linear regulator is often inserted between the battery and the circuit.

3.3.5 Battery-Less Implant

In a battery-less implant powered by an RF link [3], the energy is harvested from the RF power radiated by the external part. By such energy harvesting, a stable reference and supply voltage is provided for the implant. In these implants, the self-driven synchronous rectifier (SDSR) topology gives the best transaction between efficiency and the input working voltage. It strikes a balance between resistance in the on-state and the current flowing under reverse bias operation. The idea of synchronous or active rectification refers to using active devices such as MOSFETs as replacements for rectifier elements in circuits [8]. P–N junction diodes develop a voltage drop of 0.7 V across their terminals, thereby decreasing the efficiency. Even replacing them with Schottky diodes with lower voltage drop of 0.15–0.45 V does not help much in this regard. MOSFETs have very low on resistance in milliohms range. The low on resistance of MOSFETs leads to a decrease in power loss and hence, increase in efficiency. It thus provides an easy, cost-effective, and reliable method. Generally, one rectification stage is inadequate to produce the required voltage for powering the implant with the RF input received. This insufficiency warrants the use of many charge pump-connected rectification stages.

3.4 Data Telemetry Unit

Let us consider an implant which cannot communicate with the outside world. An implant of this type cannot be controlled without any surgical intervention. The adjustments required could be initial tuning of the parameters. This tuning is done according to the individual patient's needs. The adjustments could also be for subsequently altering the parameters with recovery of the patient's health. The only way of postimplantation control is through surgery. Once an implant has been installed and secured by surgery, any further changes by surgical procedures are painful to the patient. They are therefore grossly inconvenient for the patient. So, for making corrective adjustments to the implant, a reliable wireless communication system is established from the external device to the implant and from the implant back to the external controlling device. The communication system involves bidirectional data exchange. Through this data exchange, necessary manipulations are routinely done by the doctor from outside through the transceiver inside the implant which is similar to the one in the external device.

In today's world of communication revolution, building such capabilities in implantable devices has captivated the designer's alertness. Multipurpose systems have been constructed. Interfacing standards such as the Medical Implant Communication Service (MICS) and Medical Data Services (MEDS) have been formulated. They are used for interfacing between the implant and the external controller (Table 3.3). Short distance data telemetry is done by inductive coupling. It provides service in the ranges < 20 cm. For these short distances, frequencies between 100 and 200 kHz are used. In these systems, load shift keying (LSK) modulation is used for conveying data from the implanted medical device to its external part. Such an arrangement allows the use of the same secondary coil for both data telemetry and battery charging. One coil in the external device along with its associated single coil in the implant is sufficient for this arrangement. This settlement therefore yields a total of two coils. For long-distance data telemetry, higher frequency, shorter wavelength systems working in the MHz region are employed. Then two separate dedicated coils are used. One coil is used for data communication. Another coil is used for charging of the battery. Thus, both the external device and implant have two coils each, in place of one coil each in the previous situation. Perforce, the requirement of coils for long-distance telemetry adds up to a total of four coils. The communication systems will be dealt with comprehensively in a separate chapter.

3.5 Central Processing Unit

The central processing unit (CPU) works as the brain or intellect of the implant. It is also called the microprocessor. It houses one or more arithmetic and logic units (ALUs), the mathematical coprocessor, and the control unit. The ALU is known as the engine of the CPU. It performs arithmetic operations such as addition, subtraction, multiplication, and division. It also executes logic operations, e.g., AND, OR, NOT, NAND, NOR, XOR, etc., including bit-shifting operations. A greater number of ALUs increases the power consumption. So their number needs to be optimized for the particular application. The ALU works together with the register array. A register is a small part of the memory. From this part of the memory, data are retrieved much faster than from other memory locations. The data are loaded into the ALU from the

Table 3.3 Medical device operation in the frequency band (401–406 MHz)

S. No.	Property	Medical Implant Communications Service (MICS)	Medical Data Service (MEDS)
1.	Bandwidth allocation	402–405 MHz	401–402 and 405–406 MHz
2.	Purpose	Facilitating diagnostic and/or therapeutic functions	Facilitating diagnostic and/or therapeutic functions by transferring pieces of physiological information about individual patients which are not critical with respect to time
3.	Communication	Two-way digital communications without voice between an external module serving as the implant programmer or controller and a medical implant or similar communication between medical implants. It includes Medical Implant Telemetry System (MITS) operating in the 403.5–403.8 MHz band. The MITS provides one-way digital communication without voice from an active transmitter of implanted medical device to an external receiving module	Non-voice digital communications from single or multiple sources. One of these devices is an active implanted medical device or a sensor worn by the patient on the body
4.	Channels	Premeditated to access a least number of 9 channels. These channels are evenly spread over the 402–405 MHz band	Planned to ingress a minimum of 18 channels. These channels are evenly and consistently strewn over the 401–402 and 405–406 MHz bands. For each 1 MHz space, a minimum of 9 channels is defined
5.	Maximum channel bandwidth	300 kHz for devices operating in MICS band	150 kHz in 401.85–402 MHz band and 100 kHz in 401–401.85 MHz or 405–406 MHz band
6.	Maximum average effective isotropic radiated power (EIRP)	25 μW for MICS transmitters; 100 nW for MITS transmitters	25 μW for transmitters commissioned in a system using listen before talk (LBT) for channel selection; 250 nW if no LBT

input register. The control unit fetches instructions from the memory. It decodes and executes the instructions according to the program. To cite an example, the instruction tells the ALU about the operation to be executed on the data. The result is transferred to the output register.

The coprocessor enhances the operating speed of computation. It contains the unit capable of performing floating-point computations. Floating-point representation of a number means that the position of radix point, decimal point or binary point, in the number is not fixed. The radix point may be placed at any place with respect to the significant digits of the number. Floating-point operations are those carried out on numbers with floating radix points.

For digitization of the data, the analog-to-digital converter (ADC) is used. It is placed inside the microprocessor as a subcomponent of the analog front end (AFE). The front end collects information in various forms from the user. It processes the collected information to be available in a suitable form usable by the back end. Hence, the front-end interfaces the user with the back end.

Analog-to-digital conversion involves three main steps: sampling, quantization, and coding. Sampling is the process of conversion of signal from continuous-time to discrete-time format. It transforms a continuous-time signal into a discrete-time signal. For sampling, the value of the signal is measured at defined time intervals. Each measurement gives a sample. For faithful reproduction of a signal, the sampling frequency must be twofold the maximum frequency of the signal. This frequency specifying the upper limit is called theNyquist rate. Quantization of a signal is mapping of the sampled values to smaller set of rounded off values. These values can be reckoned. The error introduced during rounding off is called quantization error or distortion. The errors modeled as quantization noise. Encoding or coding is the process of assigning a digital representation to the quantized sample. This assignment is done by representing each quantization level by a unique label such as 0000 for the lowest level and 0001 for the next higher level.

For rigorous data analysis, some microprocessors seek the help of an additional floating-point unit (FPU) or a digital signal processor (DSP). The microprocessor combined with the memory storage is sometimes known as the digital backend.

3.6 Memory Storage

It is used to store the computer programs as well as data. Volatile memory is one which is lost when the computer is switched off. Nonvolatile memory (NVM) is one which is retained even when the computer is not powered. Random access memory (RAM) is a volatile memory. It can be accessed from any physical location. It is also accessible in any order instead of moving sequentially from a given place. There are two versions of RAM devices: dynamic RAM (DRAM) and static RAM (SRAM). (1) DRAM needs to be periodically refreshed. The refreshing is done after every few milliseconds. Without refreshing, data will not be retained, and it will be lost after some time. SRAM needs no such refreshment. To explain the reason for refreshment requirement in case of DRAM, it may be noted that this need arises because the storage cell of the DRAM is a combination of a single transistor and a capacitor. The capacitor tends to lose its charge by leakage. So, it has to be recharged to make amends for the lost charge. Even during reading of memory, some charge may be lost from the storage capacitor. The capacitor loses charge on applying a voltage to the word line. Therefore, information must be rewritten into all the cells of a row following the reading of a single cell from it. (2) SRAM is based on bistable latching circuits called flip-flops using six transistors per unit cell. Hence, it consumes a large area. DRAM uses a single transistor along with a capacitor. Therefore, it requires less area. (3) Because of its larger area, SRAM is expensive. But DRAM is cheaper due to its smaller area. (4) The recurrent refreshment demand slows down DRAM. It makes DRAM slower relative to SRAM. The access times of SRAMs are ~ 10 ns. Those of DRAM are ~ 40–50 ns. (5) At high operating frequencies, SRAM consumes practically identical power to DRAM. But at lower frequencies, its power demand is less than DRAM. In the idling condition, it is almost negligible.

Table 3.4 presents at a glance the main features that tell apart DRAM from SRAM.

As opposed to RAM, read-only memory (ROM) cannot be easily altered or programmed (Table 3.5). Electrically erasable programmable read-only memory (EEPROM)Electrically erasable programmable read-only memory (EEPROM) is a kind of programmable ROM (PROM). It can be erased electrically and programmed. Devices such as metal fuses serve as one-time programmable memories.

A common type of NVM under EEPROM is the flash memory. It is available in two forms: NAND and NOR type. The NAND-type memory has smaller erase and write times. It occupies smaller area per unit cell. By virtue of its smaller area, it allows tighter packing density and reduced price. However, its input/output interface lacks in providing a random excess external bus. So, data reading is possible only in block-wise modes. This makes it inappropriate for a PROM. The cause of inappropriateness is that in PROM, byte-level random access is necessary.

The memory cells of NAND flash memory are made of floating gate MOSFETs (FGMOSFETs). In FGMOSFETs, there are two gates. These are named as the control gate (CG) and the floating gate (FG). The gates are separated by an insulating oxide layer. A cell can be programmed by application of a positive voltage to the CG. By this means, electrons are attracted toward CG. En route their journey toward CG, they are trapped by FG. In FG, they can reside for years. Conversely, a cell is erased by applying a positive voltage on the opposite channel (substrate) side with

Table 3.4 DRAM and SRAM

S. No.	Property	DRAM	SRAM
1.	Refreshment	Yes	No
2.	Area used	Small	Large
3.	Cost	Less	More
4.	Speed	Low	High
5.	Power consumption at low frequencies	More	Less

Table 3.5 RAMand ROM

S. No.	Property	RAM	ROM
1.	Accessibility	Accessible at any time, in whatsoever order and from no matter what physical location	First the ROM information is transferred into RAM and then executed by the processor
2.	Usage	Allows reading and writing	Allows reading only
3.	Role and purpose	Permits the computer to read data fast with great dexterity to run the applications proficiently	Stocks the program required to boot the computer at the time of starting and to act upon diagnostic functions
4..	Storage	Stores temporary information	Stockpiles permanent information
5.	Volatility	Volatile	Nonvolatile
6.	Speed	High	Low
7.	Memory space	Megabytes to gigabytes	Few thousand bytes in a PC
8.	Cost	High	Low

CG grounded. This pushes the electrons away from the FG. The electrons are driven into the channel.

3.7 Analog Front End

It is the set of analog signal processing circuitry comprising operational amplifiers, filters, and application-specific integrated circuits (ASICs). It seeks to provide an electronic functional block that interfaces sensors to analog-to-digital converters.

The operational amplifier or OP-AMP is a direct-coupled voltage amplifier providing a high gain. It is a differential amplifier. Hence, it amplifies the difference of voltages applied between two terminals: (1) a non-inverting input V_+ and (2) an inverting input V_- terminal. This results in an output signal V_{out}. Its power supply terminals are indicated by V_{S+} and V_{S-}. The name OP-AMP is derived from its origin in analog computers. Here, it was used for performing mathematical operations. Examples of these operations are addition, subtraction, multiplication, division, differentiation, integration, etc. It is one of the most widely used integrated circuits in analog systems. An idealOP-AMP has infinite open-loop gain (without positive or negative feedback). It has infinite common-mode rejection ratio (CMRR). The CMRR is the ratio of differential gain to common-mode gain. The input impedance of an OP-AMP is infinite. Its output impedance is zero. It has infinite bandwidth (DC or zero to any AC frequency). Its offset voltage (the output potential difference with equal input voltages) is zero. The slew rate (maximum rate of change of output voltage with time) of OP-AMP imposes high-frequency limitations.

Filters are electronic circuits designed to attenuate certain frequency components and pass the preferred frequency components of a signal. Four types of widely used filters are low-pass, high-pass, bandpass, and bandstop.

Table 3.6 gives a brief account of the main circuits that can be designed usingOP-AMPs.

On the whole, the responsibilities of the AFE are to sense or measure the patient's body parameters. These parameters are heart rate, physical activity level, blood pressure, etc. Accordingly, the AFE delivers the stimulus to the patient. The stimulus is given in the form of an electrical pulse of the correct magnitude and duration. The AFE often involves the use of high voltages. It is therefore more oriented toward analog circuitry unlike digital circuitry. The basic inclination of digital circuitry is toward smaller geometries and lower operating voltages. Digital technologies concentrate more on dimensional shrinkage of device geometry and cost reduction.

3.8 Electronic Block or Feature Grouping

While preparing the organizational plan of an implantable IC, the designer has to study its main features. During this study, he/she should decide the features that can be put together on the same silicon chip to fabricate an IC. Also, he/she must decide as to which circuit components are not amenable to integration. Examples of such components are the inductors and large capacitors for defibrillators, electrostatic discharge (ESD) protection circuits, etc. ESD suppression components clamp the ESD voltage to a value at which the circuit can survive. These components are connected in parallel with the signal line. They shunt away a large proportion of current from the data line to the appropriate reference. This reference is either the power rail or chassis ground. Thus, the protected chip does not see the ESD transient.

Whether in defibrillators or in ESDsuppressors, external discrete devices serve the protection functions better. So, the first step is to classify what features are to be built on the silicon chip. Evidently, those components, which are to be placed outside the chip, have to be connected to the chip at a later stage.

The second consideration is that the chip must be subdivided into different sections in such a way that any future design revision can be undertaken with minimal changes. To exemplify, let us consider a monolithic chip in which microprocessor and memory functions are combined together. Here, one will need to rework the whole design if either the microprocessor or memory part needs to be updated. Had this chip been segmented into two separate microprocessor and memory chips, the design changes would have affected only one chip. The other chip would have remained untouched. Then the designer would be dealing with a less complicated circumstance. Similarly, suppose an RF communications circuitry is embedded into an IC. Later, the communication standards undergo revision. Then an altogether new IC needs to be fabricated. By cautious forethought and brainstorming, the fabrication of new IC is avoidable. Future trends of developments can be visualized. A flexible design can

Table 3.6 OP-AMP circuits

S. No.	Name of the circuit	Function
1.	Non-inverting amplifier	It amplifies a signal without inverting it
2.	Inverting amplifier	It inverts a voltage signal and also amplifies it
3.	Buffer amplifier/voltage follower	It acts as a buffering stage between a high-impedance source and a low resistance load
4.	Summing amplifier/voltage adder	It performs the addition operation by adding several weighted voltages
5.	Difference amplifier/voltage subtractor	It performs a subtraction operation by amplifying the difference between two voltages
6.	Transresistance amplifier/ current-to-voltage converter	It is an amplifying circuit, which converts a current into a proportional output voltage. Its output voltage $\propto$ input current. The constant of proportionality in this relationship is given the name "transimpedance" or "transresistance" from which the circuit is aptly called "transresistance amplifier." This proportionality constant has the unit, "ohm"
7.	Transconductance amplifier/ voltage-to-current converter	It is an amplifying circuit whose output current is directly proportional to its input voltage. The constant of proportionality in this relation is known as transconductance from which the circuit derives its name, "transconductance amplifier." This proportionality constant has the unit, "Siemen"
8.	Integrator/low-pass filter	It is a filter circuit. It integrates the signal with respect to time; hence, it is called integrator. It is used to limit the bandwidth of a given signal
9.	Differentiator/high-pass filter	It is a filter circuit. It differentiates the signal with respect to time; hence, it is termed differentiator. It amplifies AC but blocks DC
10.	Differential amplifier	It is an amplifier used for determining the difference between two voltages. Each voltage is multiplied by some constant, decided by the resistors used in the circuit
11.	Instrumentation amplifier	It amplifies a low-level differential input. It is an amplifier with differential input. Its special features include very high input impedance, very low output impedance, variable gain, high CMRR, and good thermal stability. It is suitable but not limited in use as the input stage of an electronic instrument for making very accurate, low-noise measurements
12.	Logarithmic amplifier	It is an amplifier in which the output signal is proportional to the logarithm of the input signal
13.	Antilogarithmic amplifier	It is an amplifier in which the output signal is proportional to the inverse of logarithm of the input signal

(continued)

Table 3.6 (continued)

S. No.	Name of the circuit	Function
14.	Comparator	It compares two voltages. After comparison, it switches its output to point toward which voltage is larger
15.	Square wave generator	It is an astable multivibrator

be laid out which can adapt well to foreseen or anticipated iterations. In this way, a future-ready chip is designed.

The designer must scrutinize as to what performance levels will be compromised if sensitive analog components are placed along with noisy digital circuits on the same chip. A digital-centric process is aimed at high device densities with least manufacturing steps so that the cost is reduced. So, if a digital process is used, the crucial analog performance characteristics may not be achieved. Hence, the device rating may fall below electrical specifications. Contrarily, even if the digital performance is electrically satisfactory, a larger area might have been used for attaining the rated values. Then both digital and analog functions could have been carried out to fruition. Size reduction of a digital device increases the off-state leakage current. Hence, the digital device requires lower voltage operation. This hampers the normal functioning of the analog device.

Digital ICs tend to follow the latest technologies without waiting for their maturity. Many times these processes go fast into obsolescence. It must be borne in mind that implantable electronic devices are life-supporting devices. They take several years for approval from regulating agencies. Therefore, the hasty practice of device engineers to adopt new technologies must be avoided. Adoption of technologies without observing their sustainability may often result in the fabrication technology becoming obsolete before the device gets its approval. In such situations, the design and the process might need to be completely relooked from the very beginning. Considerable effort and time may be lost if one has to start from a scratch! Had the design and process engineers agreed upon more mature and longer life technologies, such mis-happenings could have been prevented.

3.9 Discussion and Conclusions

Generic implant architecture for wireless communication and wireless charging was presented. The functions of the components in the external and internal modules were discussed. The wireless link is highly desirable for patient's mobility. Moreover, percutaneously connected cables cause skin irritations and infections.

Review Exercises

3.1 "The consensus structural organization of implantable electronic systems is that of a two-component arrangement working in coordination with each other." Explain the meaning of this statement and bring out the principal contractual obligations of the two components of this closed-loop coupling.

3.2 What is meant by telemetry and teleactuation?

3.3 Name the four components of the external part of an implantable device. Underscore the necessity and function of each component.

3.4 Describe the construction of an induction charger. What type of coupling is necessary between its two coils for charging from a: (1) small distance and (2) large distance?

3.5 Mention the differences between nonresonant and resonant coupling of two coils.

3.6 What is the role of antenna in the external part of an implantable device? What is meant by a transceiver? Explain the duplex mode of operation of a transceiver.

3.7 How does the provision of a USB port in the external part of an implantable device help in extending its capabilities?

3.8 Sketch the functional block diagram of the sequential steps through which the oscillating current received in the coil of the implant passes to give a DC current for charging the battery. Explain the function performed by the capacitor in this sequence.

3.9 What is synchronous rectification? How does it increase the efficiency of a rectification system?

3.10 Point out the advantages and disadvantages of linear regulator and switch-mode power supply. Explain with the help of diagrams the operation of both types of regulators.

3.11 How is voltage regulation obtained in a: (1) linear regulator and (2) SMPS?

3.12 What control is exercised by the choice of the duty cycle in the SMPS? What is the role of the feedback in SMPS?

3.13 Explain how voltages with positive and negative temperature coefficients are obtained to implement the function of a bandgap reference circuit? Why is this circuit called by this name?

3.14 The circuit designer of an implantable device is faced with a serious constraint on power consumption. No extravagance is tolerated in spending power. Suggest some steps for power saving that must be kept in mind at the design stage of implantable devices?

3.15 How does the self-driven synchronous rectifier topology help in saving power? Why is it called active rectification?

3.16 For short distance data transmission, what are the typical frequency and distance ranges? What modulation technique is commonly used?

3.17 Is random access memory (RAM) a nonvolatile memory? What are the two types of RAM?

3.18 How do SRAM and DRAM differ in construction? How many transistors are used per unit cell in a flip-flop? Which consumes more area, SRAM or DRAM?

3.19 Why DRAM needs to be periodically refreshed? What typically is the refreshing time of DRAM? Which is costly and why? Which is faster and why?

3.20 What are typical access times of SRAM and DRAM? Compare SRAM and DRAM regarding power consumption.

3.21 Give one example of one-time programmable ROM? Give one example of electrically erasable and programmable ROM (EEPROM). Is it a volatile memory?

3.22 What are the two forms of flash memory? In what respects do these two forms differ?

3.23 Name the device from which the memory cells of NAND flash memory are made. What are the two gates of FGMOSFET called?

3.24 How is a cell of flash memory programmed? How is a cell of flash memory erased?

3.25 What is the role of front-end electronics in implantable devices? Why does this portion of the implantable device use analog circuitry instead of digital circuitry dominating in the remaining device?

3.26 What is an operational amplifier? Why is it called by this name? What are the characteristics of an ideal OP-AMP?

3.27 What is a filter circuit? Name the four types of filters commonly used?

3.28 Prepare a list of important guidelines that the designer of an implantable device must follow from the point of view of: (1) the components that can be put close together on the chip, (2) the use of off-chip devices, (3) future design revisions, and (4) utilization of new unproven technologies?

References

1. Paradiso J, Borriello G, Bonato P (2008) Guest editor's introduction. IEEE Perv Comput 7(1):11–13
2. McDonald J, Dean S, Niewolny D et al (2011) Integrated circuits for implantable medical devices. © 2011 Freescale Semiconductor. Doc. No. ICIMDOVWP REV0, pp 1–8. http://cache.freescale.com/files/32bit/doc/white_paper/ICIMDOVWP.pdf. Accessed 30 July 2015
3. Yakovlev A, Kim S, Poon A (2012) Implantable biomedical devices: wireless powering and communication. IEEE Commun Mag 50:152–159
4. Van der Velde ET, Atsma DE, Foeken H (2013) Remote monitoring of patients with implanted devices: data exchange and integration. Eur J Prev Cardiol 20(S2):8–24
5. Zhang HJ (2013) Linear technology. Application note 140: basic concepts of linear regulator and switching mode power supplies. http://cds.linear.com/docs/en/application-note/AN140fa.pdf. Accessed 27 July 2014
6. SMPS and linear power supply comparison. Copyright © 2014 Sinrace Power Supply Technology (Shenzhen). http://www.sinrace.com.cn/en/knowledge-base/306-smps-and-linear-power-supply-comparison.html. Accessed 27 Sept 2014
7. Widlar RJ (1971) New developments in IC voltage regulators. IEEE J Solid-State Circuits 6(1):2–7
8. Fernández A, Lamar DG, Rodríguez M et al (2009) Self-driven synchronous rectification system with input voltage tracking for converters with a symmetrically driven transformer. IEEE Trans Ind Electron 56(5):1440–1445

Chapter 4
Dilemmas and Enigmas of Implantable IC Design

Abstract Very low power consumption and impeccable reliability are the crucial requirements of implantable electronics. Extremely small power utilization necessitates close attention to power management and budgeting. Together with reliability considerations, it impacts circuit design and fabrication processes, besides influencing the testing methodologies. Reliability physics and failure mechanisms must also be reexamined. An obvious outcome is that the standard designs and processes available from wafer foundries no longer hold; they need to be suitably modified from the viewpoint of power saving and reliability enhancement. The demands become more appalling in the wake of increasing system complexity without concomitantly making more power available.

Keywords CMOS · GIDL · Leakage current · SILC · NBTI · Noise · DIBL · Short-channel effects · ESD · Sense amplifier

4.1 Introduction

Design of implantable ICs is riddled with both challenges and opportunities. So, the designer has to take note of both sides of the story. The designer's confidence is boosted when it is realized that most industrial ICs are designed for operation in the − 40 to + 85 °C temperature range. Contrarily but favorably, implantable ICs operate in the human body environment. For these ICs, the surrounding temperature varies nominally around 37 °C.

Differences and fault tolerances of implants as compared to industrial systems must be always kept in mind. Biological systems are comparatively slower than industrial systems. Hence, the sensors and driver circuits need to work at frequencies ~ 100–200 kHz. Accordingly, the system clock frequencies in implantable devices are in this frequency range. As opposed to this, clocks of 10 MHz or higher frequencies are required in industrial applications. In addition, many (not all) biological systems are more tolerant to deviations from values. Cardiac pacemakers, for example, can tolerate stimulus errors of 0.5%.

V. K. Khanna, *Implantable Medical Electronics*,
https://doi.org/10.1007/978-3-031-99868-3_4

This chapter presents the many puzzling situations that an implantable IC designer has to face frequently.

4.2 CMOSFET: The Digital Workhorse

The trustworthy and ubiquitous structural as well as functional unit of implantable ICs is the complementary metal–oxide–semiconductor (CMOS) device. The CMOS device uses a unification of P-type and N-type metal–oxide–semiconductor field-effect transistors (MOSFETs) for implementation of logic functions. The term "CMOS" refers both to the digital circuit design using symmetrical pairs of transistors as well as the associated technology for circuit fabrication.

4.2.1 CMOS Processes

CMOS processes are of three types (Fig. 4.1):

1. N-well CMOS: The starting substrate is P-type silicon. NMOS devices are formed by N^+ diffusion into this substrate. This diffusion is performed at the source and drain locations. For PMOS devices, diffusion of N-type impurities into the substrate forms an N-type well. P^+-source and drain regions are formed by heavy selective doping of P-type impurities in the source and drain diffusion windows. Compatibility of this process with NMOS process has made it popular.
2. P-well CMOS: Starting from an N-type substrate, P^+ diffusion is done for source and drain regions. A P-type well is formed, and N^+ diffusion is done for source and drain formation. Due to its incompatibility with NMOS process, this process has not gained much popularity.
3. Twin-tub CMOS: This process commences with a lightly doped substrate. In this substrate, N- and P-wells are formed by diffusion. In the wells thus created, the PMOS and NMOS devices are fabricated.

4.2.2 CMOS Combinational Logic

CMOS gates are built from two complementary networks: (1) a pull-up network (PUN) is a network that sets the output to 1. It is built from PMOS transistors. The PMOS transistors are arranged between the output terminal and the power supply rail at higher voltage (V_{dd}). (2) A pull-down network (PDN) is a network that sets the output to 0. It is made from NMOS transistors. The NMOS transistors are arranged between the output terminal and the power supply rail at lower voltage (V_{ss}).

The pull-up and pull-down networks are interrelated as duals of each other. They are configured in a complementary topology. At a particular instant, only one network

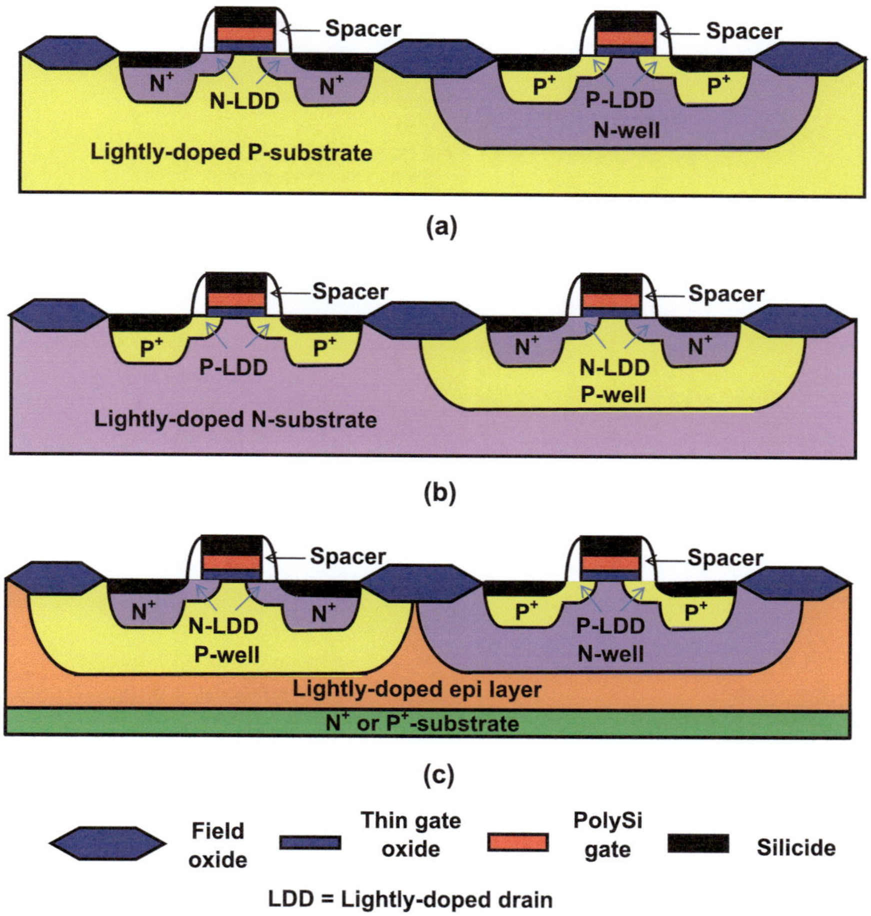

Fig. 4.1 CMOS processes: **a** N-well, **b** P-well, and **c** twin well

is on and the other is off. So, the output is either connected to ground or to drain supply terminal. Suppose both networks are simultaneously on. Then the drain path between drain and ground terminals will draw excessively high current. The high current drawn will damage the circuit.

The simplest CMOS circuit is the NOT gate or inverter. It is a combination of a pull-up network (PUN) and a pull-down network (PDN), as depicted in Fig. 4.2. When the input voltage is high, the PMOS transistor is off, and the NMOS transistor is on. This sets up a connection between output and V_{ss}. The connection pulls the output to low. Conversely, when the input voltage is low, the PMOS transistor is on and NMOS transistor is off. This creates a connection between the output and V_{dd}. The connection pulls the output to high.

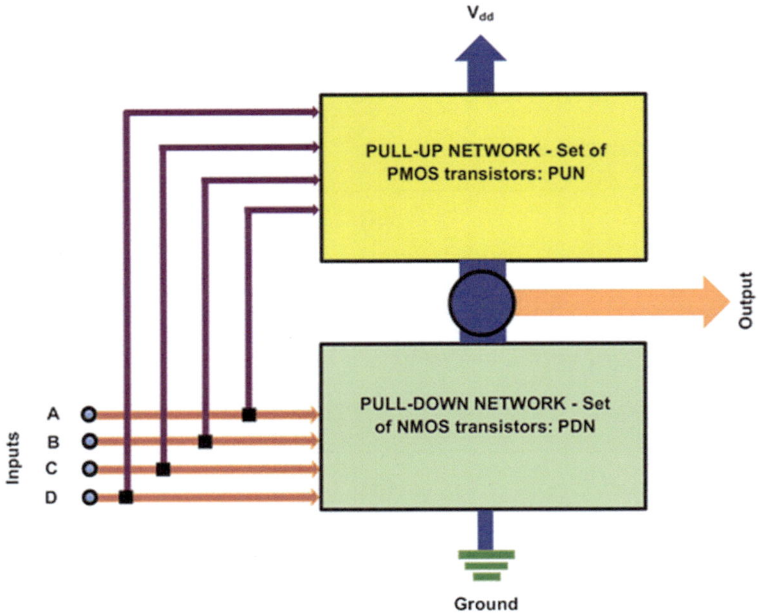

Fig. 4.2 Generic CMOS gate

A two-input CMOS NOR gate comprises: (1) two pull-down N-channel MOS transistors connected in parallel, and (2) two pull-up P-channel MOS transistors connected in series (Fig. 4.3a).

The two-input CMOS NAND gate consists of: (1) two pull-down N-channel MOS transistors connected in series and fixed to V_{ss}, and (2) two pull-up P-channel MOS transistors connected in parallel and tied to V_{dd} (Fig. 4.3).

4.2.3 CMOS Advantages

CMOS circuit benefits from the fact that one transistor of the pair is always off. So negligible power is wasted in the series combination of P- and N-transistors in the static mode. Power dissipation only occurs during the short periods that the device switches between on- and off-states. Thus CMOS circuit has very low static power dissipation. The low power consumption makes it suitable for ultralow power implantable ICs. CMOS has an advantage over NMOS due to its fast transition from 0 to 1 and 1 to 0 logic states. NMOS switches slowly from 0 to 1 state. The switching is slow because NMOS uses a resistor in place of the pull-up network.

Continued shrinkage in geometries of CMOS ICs without increasing power consumption is making it possible to design implantable circuits with enhanced functionalities and features.

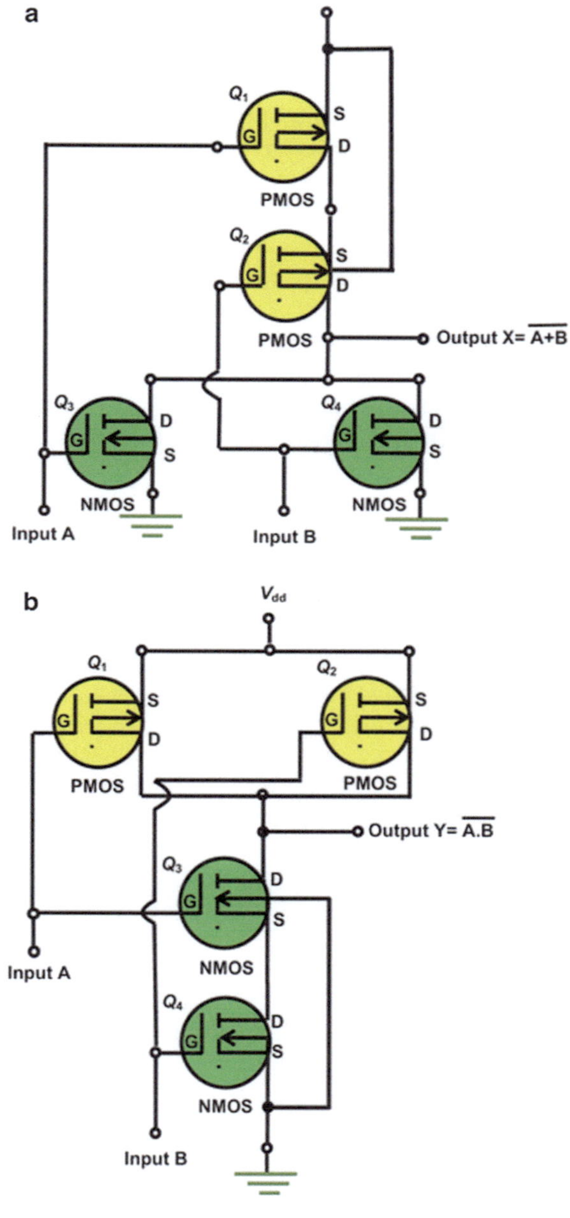

Fig. 4.3 Two-input CMOS gates: **a** NOR gate and **b** NAND gate

4.3 Single-Chip Versus Multiple-Chip Design

Because more dynamic power is consumed at the crossing points across chips, the system-on-chip (SoC) concept is gaining popularity. The SoC is an integrated circuit that assimilates all the miscellaneous functions consummated by an electronic system on a single small chip. These functions may be digital, analog, mixed-signal, or RF signal tasks. The components included in the SoC are CPU, memory, power management, and wireless communication circuits. The SoC concept decreases the overall size of the circuit. The size is smaller because SoC is only slightly larger than CPU. Nevertheless, it provides much greater functionality. Also, it employs much higher level of integration. Together with more integration, it also uses shorter interconnects. So, the power consumption decreases. Let us count the number of chips that will be required to assemble a system. Then lesser number of physical SoC chips will be needed. The smaller number of SoC chips will enable easier and cheaper construction of system. The main disadvantage of using SoCs is the loss of design flexibility. This flexibility is lost because it may be economical to replace a CPU only if it gets faulty. But with the SoCs, the integrated memory and other functions come along with the chip. It is therefore both wasteful and costly to replace the entire range of functions when only a small part is unserviceable.

4.4 Speed and Threshold Voltage Trade-Off

Human body responses are measured in milliseconds, not in nanoseconds. Hence, implantable ICs need not be as fast as consumer electronics ICs. Fortunately, the slower device requires less power: low power → slow device; high power → faster device.

On the opposite side, for smaller geometries, the maximum supply voltage is lowered. Therefore, the transistor threshold voltage decreases. Regrettably for low power systems, lower threshold voltages lead to higher leakage currents: low threshold voltage → high leakage current. In turn, the high leakage current raises standby power dissipation: high leakage current → high standby power dissipation.

4.5 Matching the Threshold Voltages of N- and P-Channel Devices

At large process geometries ~ 1 μm, the threshold voltages of P-channel transistors are kept intentionally high. The reason for keeping the threshold voltages high is that it is difficult to keep the leakage current of these transistors low in the buried channel architecture with N^+ polysilicon gate [1]. Consequently, appreciable differences arise in drive capabilities of P- and N-channel transistors. These differences lead to concerns with common transistor sizing in standard cells. At the other extreme of small process geometries in the deep submicron region, surface P-channel devices are fabricated with P^+ polysilicon gates. Hence, the mismatching of threshold voltages of N- and P-channel transistors is a nonissue. Low threshold voltages of P-channel transistors are achieved with acceptable leakage currents like N-channel transistors. From these arguments, a successful strategy for low power operation is evolved. This strategy utilizes all the advantages of small-geometry processes but keeps the threshold voltages high. This will reduce the leakage currents to negligible levels. The expense is a decrease in operational speed. Thus by sacrificing speed, less leaky, low power operation will be ensured. This calls for ion implantation step at appropriate dose to adjust the threshold voltages. Hence, there is a departure from the regular foundry process.

4.6 Rise of Leakage Currents in Deep Submicron Transistors

For downscaling toward 150 nm and below, two additional forms of leakage current become significant. These leakage currents were effectively nonentities at larger geometries

4.6.1 Gate-Induced Drain Leakage

Leakage current minimization is an issue at the forefront for low power implantable devices. A major contributor to the leakage current is the gate-induced leakage current.

At small geometries, the doping levels of the transistor well and channel increase. Also, the gate oxide thickness decreases. Without similar reduction in supply voltage, the electric field present at the gate edge of the drain junction rises. The rise of electric field takes place to the level at which a Zener-type leakage (band-to-band tunneling) in the drain region below the gate becomes meaningful at high gate–drain voltages [2]. This happens because there is appreciable band bending near the interface between silicon and the gate dielectric. So, electrons in the valence band can tunnel into the conduction band. The problem is further worsened by the fact that the adjustment of transistor threshold voltages at higher values, as recommended for lowering the leakage current, now has a derogatory effect. This depreciating effect occurs because the threshold voltage adjustment raises the electric field at the gate edge of the drain junction. Thereby, it elevates gate-induced drain leakage (GIDL). Thus as the drain supply is increased, at a particular drain voltage, the GIDL exceeds the subthreshold leakage current. This is the crossover point. With further downscaling of dimensions, the GIDL issue is further magnified.

For all intents and purposes, the leakage in the drain is a bottleneck for dimensional reduction of CMOS toward the deep submicrometer dominion. On reduction of channel length below 50 nm, leakage current control becomes a real challenge to IC engineers because: (1) the subthreshold conduction current increases in an exponential manner with falling threshold voltage, (2) the band-to-band tunneling current at the surface or GIDL increases as an exponential function with decreasing gate oxide thickness, and (3) the band-to-band tunneling current in the bulk increases exponentially with rise in doping concentrations in the bulk and well regions. The GIDL current is an exclusive function of the conditions prevailing in the gate-to-drain overlap region. It is extremely responsive to the oxide geometry under the boundary of the gate. The upper limiting value of the electric field in drain region is bumped up by the bird's beak of the gate oxide. It is also influenced by a flawed optimization of the drain structure and the gate–drain overlapping [3].

4.6.2 Leakage Current Flow Through the Gate Oxide

The gate oxide thickness for 130 nm process is 200 nm. For the 90 nm process, it is 160 nm. For every 20 nm reduction in oxide thickness, the leakage current due to tunneling increases by an order of magnitude. It thus attains overly serious proportions.

4.7 Reliability Degradation of Deep Submicron Transistors

Among the reliability concerns, mention may be made of many issues. These issues are explained below.

4.7.1 Stress-Induced Leakage Current and Soft Breakdown

For deep submicron geometries, oxide does not undergo immediate rupturing. It does so in gradual stages through a special mechanism. This mechanism is known as *quasi-* or *soft breakdown*. Stress-induced leakage current (SILC) flowing through the oxide causes this breakdown. Battery depletion and infant mortality failures may rise due to high initial leakage currents.

Theunderstanding of microscopic level physics of SILC is deficient. However, there is a wide-ranging agreement that it is caused by trap-assisted conduction. In all likelihood, it involves the tunneling mechanism [4]. As is well known, thick oxides obey the feedback runaway model. The thin oxides of thickness $\leq$ 10 nm conform to a wear-out model in place of the feedback runaway model. In the wear-out model, defects are produced in the oxide network when it is stressed electrically. These defects are known as neutral traps. This has been done to indicate that the traps causing breakdown need not be related with any charge. Also, it has been done to demarcate the historical models on charged defects activating breakdown. By capturing a hole, the neutral trap becomes positively charged. By capturing an electron, it is negatively charged.

The stress-induced traps in the oxide produce conductive percolation paths through it. No sooner than the density of neutral traps attains a critical value, a conduction path is formed between the cathode and anode. This path sets off a breakdown. The physical phenomena underlying traps and the kinetics of trap creation must be clearly comprehended. Then it is possible to predict the reliability of thin SiO_2 gate dielectric under operational conditions [5].

4.7.2 Negative-Bias Temperature Instability

This is a prime issue determining the reliability. It impacts mainly P-channel transistors. This is because they normally have negative gate-to-source/body voltages in on states [6]. NBTI appears in the form of a rise in P-channel threshold voltage on transistors that have a bias applied across them. The appearance of negative-bias temperature instability (NBTI) takes place irrespective of activity, particularly at elevated temperatures. NBTI results in a fall of drain current and decrease in transconductance of the device. Temporally, the degradation takes place logarithmically. It must be distinguished from the hot electron effects. It produces shifts that are dependent on the state of the transistor and its period in that state.

Figure 4.4 shows the phenomena responsible for NBTI. Two simultaneous mechanisms have been proposed for this instability. NBTI is a consequence of, firstly, production of interface traps and, secondly, the filling of preexisting traps by holes from the P-MOSFET channel. The SiH bonds at the interface between SiO_2 and Si substrate are broken. This bond breakage occurs by the joint effect of electric field and temperature. Holes also participate in the process. Consequently, dangling bonds or interface traps are produced at the aforesaid interface. Also, positive oxide charge results from either H^+ or trapped holes. Hence, NBTI is a combined effect of interface state generation and positive trapped charge. For low-supply voltage systems, the shift occurs slowly. This is an advantage. But for a given value of shift, the effect on performance is more conspicuous due to the lower clearance above the threshold for the transistors.

The most prevalent model of NBTI is the reaction–diffusion (R–D) model. In this model, there are two phases. The first phase is the reaction phase in which interface traps are produced at the SiO_2/Si interface. An interface trap consists of any interfacial silicon atom with an unsaturated or unpaired valence electron at the SiO_2/Si interface. This silicon atom has a dangling bond. Such a silicon atom is produced by the reaction between a hole h^+ and SiH. The hole tunnels through the oxide from the inversion region. It is captured by SiH. The products of this reaction are a positive silicon ion and free hydrogen: $h^+ + \text{SiH} \leftrightarrow \text{Si}^+ + \text{H}$. Hydrogen is liberated during this reaction phase. The hydrogen thus released spreads by diffusion and moves away from the Si/SiO_2 interface. Sometimes, it reacts again with silicon forming Si–H. Some hydrogen atoms interact among themselves forming H_2. Besides the Si^+ ions created by the above process, more unsaturated Si atoms occur in SiO_2 itself. These unsaturated Si atoms together with other oxide defects increase the number of interface traps. With

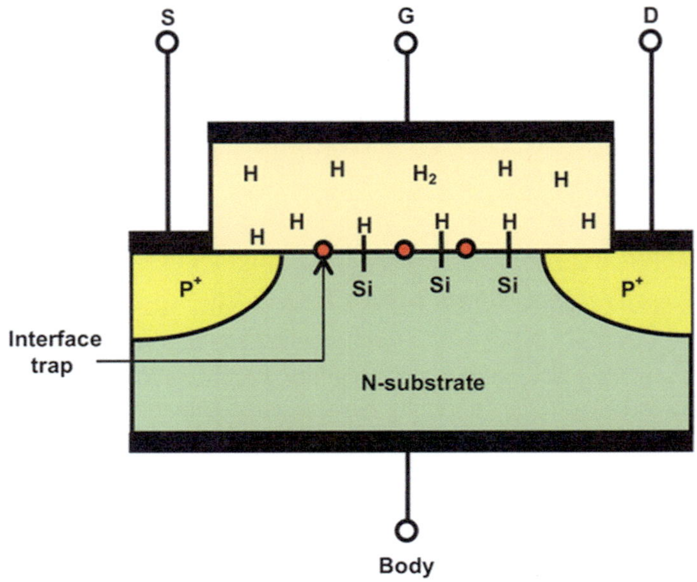

Fig. 4.4 NBTI effects in P-channel MOSFET

respect to stress time, the interface traps have a linear dependence on stress time. The shift in threshold voltage arises from the building up of interface traps. It also arises from oxide-trapped charge.

The next phase is thediffusion phase. In this phase, the hydrogen disperses away from the interface and disseminates into the oxide. The time dependence of this phase has the form t^n, where $n = 2.5$ for neutral hydrogen species. Hydrogen may as well diffuse to the interior of the substrate. This diffusion-limited regime has been corroborated on several instances. However, the reaction-limiting regime has not been noticed that much frequently. The nonobservance is because this regime lasts only for a very short time [7].

4.7.3 CMOSFET Noise Sources

Most signals are noise-contaminated. Noise is an unprompted variability observed in current or in voltage. It is noticed as a spontaneous, random fluctuation in the electrical signal. It may be described as the unwanted signal that degrades the main signal.

The smallest signal that the transistor can process is limited by the internal noise produced by it. The comportment of CMOS devices is controlled mainly by two sources of noise, thermal noise and flicker ($1/f$) noise. Occasionally present in the noise spectrum are three other types of noise: shot noise, generation/recombination noise, and popcorn noise [8].

4.7.3.1 Thermal Noise

Thermal or Johnson–Nyquist noise is the electronic noise produced by the random movement of charge carriers inside a conductor of electricity. Usually these charge carriers are electrons. The underlying reason for the production of this noise is the thermal excitation of charge carriers inside the conductor in equilibrium condition. The thermal excitation and hence noise take place in any case, regardless of whether a voltage has been applied to the conductor or not.

4.7.3.2 Flicker Noise

$1/f$ noise is the preponderant noise in the range of low frequencies. It has a spectral density function, which is inversely proportional to frequency. Being an electronic noise characterized by $1/f$ or pink power density spectrum, it is often called by the name, "pink noise." This kind of noise is habitually present in all semiconductor devices under biasing conditions. It is usually correlated with defects in the semiconductor material and with imperfections in the fabrication process. It perpetuates even at very low frequencies down to 10^{-6} Hz, corresponding to a long period around a month. Frequently, it is used in modeling the temporal fluctuations of operational parameters of semiconductor devices.

Two important models have been propounded to elucidate the origination of flicker noise: (1) The surface model or carrier number fluctuation model [9] is primarily used for MOSFET devices. It is proposed under the assumption that there are uniformly distributed trap centers in the silicon dioxide adjacent to the silicon surface. Further, it is supposed that the likelihood of the carrier intrusion to trapping centers decreases exponentially as the remoteness from the surface increases. Going further, it is believed that the time constants of trapping centers increase with the separation from the surface. Finally, it is assumed that trapping mechanisms by distinct centers are self-determining and not interconnected in any way. The cause of the flicker noise is given as the random trapping and de-trapping of charges in the oxide traps neighboring the Si–SiO_2 interface. The charge variations produce instabilities in the potential of the surface. Consequently, the mobile charge density is modulated. Charge exchange is postulated between the channel and the interfacial oxide traps by tunneling. The $1/f$ characteristics are obtained by the superposition of multiple dissimilar spectra of generation recombination noise. In these spectra, free carriers are haphazardly trapped and liberated by centers with assorted lifetimes. (2) Bulk model or mobility fluctuation model [10] is more appropriate for the description of bipolar transistors. The model is based on the presupposition that the flicker noise occurs by way of the fluctuations in bulk mobility. A clear dissonance exists in views regarding the various physical mechanisms that have been proposed for the same [11,

12]. In this noise model, two main mechanisms of carrier scattering are considered during the carrier transport. The mechanisms included are scattering from lattice atoms and scattering due to ionized impurity ions. It is postulated that scattering from the crystal lattice alone is the originator of 1/*f* noise. Scattering from the impurity ions does not affect the noise level. All inadequacies of the crystal lattice lead to pronounced 1/*f* noise.

MOSFETs/CMOSFETs are more noisy devices than their bipolar counterparts. This is especially true at low frequencies. At these frequencies, flicker (1/*f* noise) is high [13]. The current flowing through the gate oxide influences transistor noise. This current upsets the Si–SiO_2-interface passivation that was done during final annealing. As a result, the number of interface traps increases. This is not always a foreseeable change. The unpredictability crops up because modest alterations in geometrical layouts steer differences in the number of interface traps among transistors. These differences result in threshold voltage mismatching. Hence, variances are seen in the noise behavior among devices. Analog design in the deep submicron devices is a formidable task due to the very low noise required. Also precisely matched-pair transistors present difficulties. In addition, the noise and likely changes in noise over a lifetime make the design more unmanageable.

Downscaling has a great impact on 1/*f* noise in CMOS transistors. Dimensional reduction of CMOS transistors is associated with upsurge of 1/*f* noise in low-frequency operation. 1/*f* noise scales inversely with chip area. Hence, it becomes more prominent for smaller devices [14]. Newer generations of submicron CMOS transistors have higher 1/*f* corner frequencies. The high 1/*f* corner frequencies are more troublesome. Corner frequency marks the boundary between the flicker noise at low frequencies and the flat-band noise at high frequencies. The magnitude of the low-frequency noise depends on the fabrication process used, and the design and the size of the transistor. Using larger area input transistors is one approach used in amplifiers/oscillators to lessen noise. But it leads to area wastage.

4.7.3.3 Shot Noise

Shot noise is also called Poisson noise. It is a kind of electronic noise whose origin lies in the discretized or quantized structure of electricity. The discrete nature of electricity arises from the fact that electrical charge is found only in integral multiples of the elementary electronic charge. In shot noise, time-dependent fluctuations are found to take place in electric current. These fluctuations occur when the number of energy-carrying particles is very small. It is so infinitesimally small that statistical fluctuations are clearly revealed in a measurement. In an electronic circuit, these particles are electrons. In an optical device or assemblage of optical components, the relevantparticles are photons.

4.7.3.4 Generation–Recombination Noise

Generation–recombination noise is a type of noise in which current or voltage fluctuations are produced. The fluctuations are due to variations in the number of charge carriers. These variations are caused by the statistical generation and recombination of charge carriers through the generation–recombination centers in the semiconductor. To explain this type of noise, let us consider a semiconductor material containing several trapping centers. Due to the presence of these centers, at any given instant of time, the situation inside the semiconductor is visualized as a nonstop trapping and de-trapping of the charge carriers taking place within the semiconductor. Due to the constant capture and release of carriers from the traps, there is a fluctuation in the total number of carriers available in the conduction and valence bands. The fluctuation leads to changes in electrical conductivity of the material and thereby the conductance of the device made from the semiconductor. Such generation–recombination noise resulting from endless free carrier generation (de-trapping) and recombination (trapping) depends on the temperature of the device. It also varies with the biasing conditions.

4.7.3.5 Popcorn Noise

Popcorn or burst noise is an important concern in low-frequency, high-gain applications. It is primarily process-related. It is caused by semiconductor defects and/or processing issues.

4.8 Input DC Offset

A common parameter of an amplifier is its DC offset voltage. It is the DC voltage that must be impressed across the input terminals of the amplifier to nullify any deviation of the output voltage from zero value and thereby give an output voltage of zero volts [15]. Symbolically, the DC offset is represented by a voltage source in series with input. This source can have positive or negative polarity. It varies from device to device.

In the ideal situation, when the input terminals of an amplifier are shorted together, the output voltage should be zero. But in practice, this is never so. The nonzero value comes up because of the inherent mismatching in the characteristics of input transistor and other components of the silicon chip. The characteristics mismatch because of differences arising during fabrication. Along with the mismatching, the stresses produced during packaging constitute another likely reason. These two effects collectively cause a mismatching of currents. They lead to a voltage differential between input terminals.

4.9 Drain-Induced Barrier Lowering

Here, the barrier talked about is the potential barrier that a charge carrier has to overcome in moving from source to the drain terminal (Fig. 4.5). Drain-induced barrier lowering (DIBL) is one of the several phenomena that come under the theme of "short-channel effects" because they do not occur in a long-channel MOSFET device. The question may be raised as to when a channel should be branded as a "short channel." A channel is considered to be short when its length becomes comparable to the depletion region widths of source and drain junctions.

DIBL determines the ultimate propinquity of diffusions at the surface of a semiconductor. It is acknowledged as one of the deep-seated restrictions on the precincts of VLSI from electrical viewpoint. There is a plain electrical constraint in VLSI regarding the ultimate distance of separation between the surface diffusions that constitute P–N junctions [16]. This limitation arises because the application of reverse bias on a diffused P–N junction establishes a pattern of electric field surrounding it. This field pattern lowers the potential barrier isolating the reverse-biased junction from a nearby diffused junction. As soon as this lowering of barrier is sufficiently sizeable, the neighboring diffusion works as a source. An unnecessary current flow path is created.

In DIBL, the reverse-biased depletion region produced at the drain junction by the applied drain voltage intrudes into the channel. It thus offers some of the bulk depletion effects necessary for channel inversion. Under normal conditions, these are essentially gate-induced effects. Elaborating this explanation further, it may be stated that with rising drain voltage, the depletion region of the P–N junction between the drain and the substrate enlarges. It stretches underneath the gate. Hence, the drain

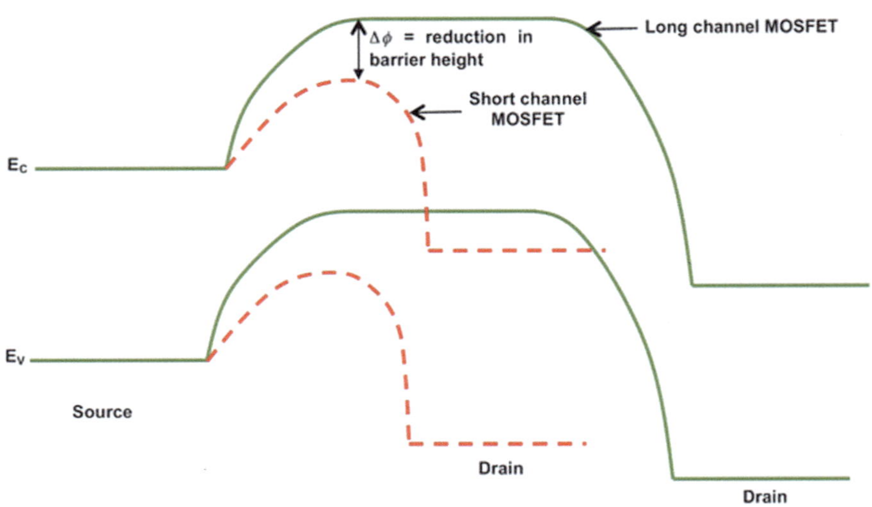

Fig. 4.5 Energy-band diagrams for a long-channel and a short-channel MOSFET illustrating the DIBL effect

is more burdened with the onus of balancing the charge of the depletion region. So, it relieves and allays the gate of some of its responsibility. Consequently, the gate voltage attracts more carriers into the channel for maintaining the charge balance. This has the same effect as a decrease in threshold voltage will cause at high drain voltages. The statement may be affirmed in another form as a fall in potential energy barrier for the carriers. Hence, it is called barrier lowering.

In DIBL, the drain voltage depresses the potential barrier. The transistor turns on at a lower threshold voltage. Therefore, a major chunk of the gate voltage is dedicated toward inversion layer creation. This in turn raises the subthreshold leakage current.

It has been conclusively proven that DIBL depends on all those parameters which can be foreseen to influence the two-dimensional pattern of electric field in a MOSFET. To name a few of these parameters, mention may be made of thickness of gate oxide, bulk doping of the substrate, length of the channel, depth of the junction, lateral spreading of drain–source diffusions, and the biasing voltage applied to the MOSFET body or substrate, i.e., the substrate bias. The above results have been established through two-dimensional modeling and simulation studies.

Table 4.1 explains some phenomena that are exclusive and peculiar to short-channel MOSFETs. These phenomena are known as short-channel effects. As already briefly mentioned, the short-channel MOSFETs are defined as those devices in which the channel length is comparative in dimensions, i.e., of the same approximate size as the widths of the depletion regions formed at the source and drain junctions.

4.10 Channel Lengthening

With shortening of the channel, the barrier lowering effect is aggravated further. The aggravation happens even when the drain voltage is zero. This aggravation occurs because the depletion regions formed between source-substrate and drain-substrate junctions participate actively in charge balancing, even in absence of reverse bias. On the opposite side, the increase of the minimum channel length for a particular CMOS technology is beneficial. The advantage obtained is the resultant uplift of drain-induced barrier. If the channel is lengthened, the threshold voltage of the transistor increases. Correspondingly, the leakage current is reduced. Because of an exponential decrease in subthreshold leakage current for a linear change in threshold voltage, this effect attains significant scale. The convenience accrued to device engineer is that no extra process requirements are obligatory on the semiconductor fabrication facility. Also, the general variation in threshold voltage is negligible. The negligible variation is due to the feebler influence of channel length variation on this parameter.

Table 4.1 Short-channel effects

S. No.	Effect	Explanation
1.	Drain-induced barrier lowering, threshold voltage reduction, and punchthrough	As the drain depletion region continues to increase in thickness with bias, it eventually interacts with the source-to-channel junction. A lowering of the potential barrier takes place. Then free carriers are effortlessly introduced into the channel. The gate voltage has only fractional influence or authority over the drain current. This is because part of the depletion is accomplished by drain–source bias. Since less gate voltage is required to deplete the semiconductor charge, threshold voltage falls with decreasing channel length. Thus the drain current at a given gate voltage increases when the drain–source voltage is amply high. In this condition, the depletion layer in the region of the drain may extend to the source. It may cause the current to flow even if gate voltage is zero. This is drain punchthrough condition
2.	Mobility degradation	In short channels, the lateral field increases. This makes the mobility field-dependent. Also, the vertical field increases. The increase in vertical field enhances surface scattering. It reduces surface mobility
3.	Velocity saturation	When the electric field reaches a critical value E_C, the velocity of the carriers tends to saturate. This saturation is due to scattering effects. $E_C = 1 \times 10^6$ V/cm for electrons and 5×10^6 V/m for holes. The saturation velocity $v_s = 10^5$ m/s for silicon. After velocity saturation, the drain current no longer increases with drain–source voltage
4.	Hot carrier effects	With the shortening of channel length, there is an increase in the electric field inside the channel near the drain end. In this high electric field region, the free carriers acquire sufficient kinetic energy above the average energy of those carriers, which are in thermal equilibrium with the semiconductor material. These free carriers are known as hot carriers. Naturally, hot carrier effects follow. These hot carriers can overcome the oxide–silicon barrier. They can cause gate current flow. Trapping of the charge of hot carriers in the oxide causes permanent changes in threshold voltage. Avalanche action expedites electron–hole pair generation. The reverse-biased drain seizes some of these carriers. A parasitic bipolar transistor action results. It hinders the control of drain current by gate action. Using the lightly doped drain (LDD) structure minimizes the hot carrier effect

4.11 Revision of Transistor Models for Implantable Electronics

Many effects such as GIDL, DIBL, NBTI, etc., come into play during implantable device simulations. Consideration of these effects calls for building new models. These models must also include the accompanying parasitic capacitance and resistance effects. Additionally, in routine design, the emphasis is on speed. ICs are being made faster. So, the transistors are wider in breadth and shorter in length. But implantable electronics is speed-tolerant. So, it calls for shift in focus on design strategy. As seen above, favoring longer transistors in implantable devices shifts the methodology toward a different approach. Thus the transistor models must be reworked. New models must be crafted. These models must be capable of correctly predicting the parameters of implantable devices

4.12 Analog Signal Processing

The analog nature of the real world is well recognized. According to the consensus opinion [17], the participation of analog circuits in VLSI systems is chiefly curbed to that of an interfacial region where two types of systems meet and interact mutually. This interface is the very slender "analog crust." The crust separates the completely analog exterior world from the entirely digital core of the swelling egg called "digital signal processing." Analog circuit designers have been imaginative ad infinitum. They are working zealously, notwithstanding the fears that they may be wiped away some day by the aggrandizement of the digital sovereignty into the analog kingdom.

All-digital computation is marked by its low price. Potentially boundless precision and dynamic range are achieved. This is owed to the methodical rejuvenation of the binary states at all stages of processing. Supplemental recompenses are the easy design and testing of the circuits. By running sophisticated synthesis softwares, computer engineers are able to design digital circuits for the implementation of detailed programmable algorithms.

But at larger geometries, digital signal processing is disfavored. This is because of the wastage of power. Then analog signal processing is used. This is particularly done in the subthreshold voltage regime. This regime involves lower currents and higher gains. To avoid power wastage, the matching of the output drive to input load should be ideal.

4.13 Electrostatic Discharge Failure Limit and Protection

Electrostatic discharge (ESD) is the transference of electrical charge between two bodies at different potentials. This transference takes place, either through direct contact (contact discharge) or through an electrical field. It occurs when the bodies are in close vicinity (air discharge).

This discharge is lethal to the sensitive microelectronic devices. It triggers several failure mechanisms. Gate oxide breakdown may result in excessive leakage. Junction spiking by a migration of the metal to the drain–source junction of MOS transistors causes failure. Latch-up occurs through an internal feedback mechanism. It is associated with temporary or permanent loss of circuit function. To eliminate electricity from the workplace, the operators, equipment, and devices are grounded. Electric field damage is avoided by transporting devices and circuits in protective, electrostatically shielded packages [18].

In implantable electronics, the post-stress leakage current must be in nA–pA range. This is opposed to the usual 1–10 μA range in commercial ICs. In general-purpose ICs, ESD protection is achieved by increasing the capacitive load. The associated decrease in speed is confronted by using larger driver transistors in input/output cells. These measures are not preferred in implantable electronics because the switching energies and related losses are to be minimized.

4.14 Digital Signal Processing

The general trend in digital design is to aim first at reaching the required speed. Chip area comes next. Power consumption is placed at the end. Implantable electronics follows the reverse direction of prioritization. The main cause is that power consumption has to be minimized. Then only the battery will last longer. The second priority is the area. This is because the device has to be implanted in the human body. The last priority is the speed. Its placement at the end follows from the grounds that human body responses work slower.

Because speed is at low priority in implantable electronics, smaller gate sizes and drives are usable. This practice is in sharp contrast to that of using larger size gates and drives in commercial ICs. A feature of low power consumption geometries is that they involve smaller routes. Lower parasitic capacitances are found within and between cells. From this viewpoint, they are more complex than normal geometries. Such gate topologies must be invariably adopted.

For minimizing the power consumption, many circuit blocks/clocks of the system are switched off whenever not required. Power supply voltages must be kept at the minimum level. This minimum level equals the sum of N- and PMOS threshold voltages. If necessary, some circuits or functions can be replicated for decreasing the power demand. This may be done by utilization of more area. At the IC level, power consumed by the clock network ought to be reduced.

4.15 Memory Design Artifices

4.15.1 Sense Amplifiers

With increasing memory capacity of the computer, the bit line capacitances have risen high. These high capacitances have made the memory slower. They have also increased the power consumption. The sense amplifier is an analog circuit used in the periphery of the memory, in a subsystem known as the read circuitry. Its purpose is to feel or discern data from the memory cell chosen by the read circuit. It executes this task by sensing the low power signal from a bit line. The bit line symbolizes a bit of data, either 1 or 0 piled up in a memory cell. The sense amplifier reinforces the petite voltage sway to decipherable logic levels. The amplification is necessary for accurate data interpretation by logic circuits external to the memory. The performance of the sense amplifier is crucial to memory function. This cruciality arises because it affects both the memory access time and its power dissipation. Only one sense amplifier handles each column of memory cells. Hence, there are usually hundreds or thousands of indistinguishable sense amplifiers on a memory chip. The task assigned to each sense amplifier is to read data from its respective column.

Reading of a memory cell that produces a current corresponding to logic 1 involves the removal of a small element of charge ($\mathrm{d}Q$) stocked on the previously charged bit lines. It may be remarked that the bit lines are extremely lengthy. Many similar cells share them. Hence, the parasitic resistance R and capacitance C due to the long bit lines are both very high. Due to this reason, the oscillation $\mathrm{d}V$ of the bit line voltage taking place when a small quantity of charge $\mathrm{d}Q$ is withdrawn from the bit line has a very small value $\mathrm{d}V = \mathrm{d}Q/C$. Sense amplifiers are assigned the task of upgrading this small voltage signal to a complete logic signal. The full logic signal should be further utilizable by digital logic. Thus the sense amplifiers convert the hazy logic levels produced on a bit line to the distinct logic levels to operate the fringe-lying Boolean circuits. Understanding and analyzing the operation and design of different sense amplifier circuits are imperative. It is essential in order to make available signals that comply with the requirements of driving the outlying logic circuits. The performance of memory is thereby improved.

According to the circuit type, sense amplifiers are classified into differential amplifiers and nondifferential amplifiers. By operation modes, they are classified as voltage amplifiers, current amplifiers, and charge amplifiers.

A differential sense amplifier can discern signals of smaller magnitudes from noise than its nondifferential amplifier counterpart. Some silicon area is negotiated in building the differential amplifier. Still in many situations, the use of differential amplifier allows to fabricate intensely close-packed circuits. Together with the high density of circuits squeezed into tiny packages, a realistic time of accessing memory, and low power consumption are obtained.

4.15.1.1 Voltage-Mode Sense Amplifier

A voltage-mode sense amplifier is an amplifier performing the job of detection of the voltage difference on the bit lines. The cushy approach to building a voltage sense amplifier is based on the principle of differential voltage amplification. The process of reading a cell is accompanied by the appearance of a small voltage swing on the bit line. The differential couple amplifies this swing to drive the digital logic circuit.

Figure 4.6 shows cross-coupled voltage sense amplifier. A main feature of this full complementary positive feedback differential sense amplifier is its very large differential gain. It can also rewrite destructively read data automatically. By virtue of its high gain, it can easily detect diminutive voltage sways on the bit line. Hence, the sensing time is considerably reduced. The circuit of this amplifier contains two data nodes $V_{\text{IN/OUT1}}$ and $V_{\text{IN/OUT2}}$. The signals at the data nodes $V_{\text{IN/OUT1}}$ and $V_{\text{IN/OUT2}}$ serve the dual functions of input and output signals to the sense amplifier. Besides these data nodes, the circuit has three control nodes SAN_{EN}, SAP_{EN}, and PRE. The symbols SAN_{EN} and SAP_{EN} represent signals that are externally applied to enable data sensing. PRE stands for pre-charge. During a read operation, the input nodes are pre-charged to voltage V_{PRE}. This pre-charging is done through the PRE node. It causes the equalization of output nodes to the same voltage level. Then immediately, on selecting and addressing a particular memory cell, a trivial difference of voltages appears on the nodes $V_{\text{IN/OUT1}}$ and $V_{\text{IN/OUT2}}$. During this period, the biasing of transistors M_1 and M_2 is such that they lie in the saturation region. The transistor M_6 is switched on by the SAN_{EN} signal. Both $V_{\text{IN/OUT1}}$ and $V_{\text{IN/OUT2}}$ start to decrease in voltage. Accordingly, the difference between them narrows down. One of the two signals, either $V_{\text{IN/OUT1}}$ or $V_{\text{IN/OUT2}}$, decreases faster than the other. This causes one of the transistors, either M_1 or M_2, to be cut off. At the same time, the other transistor is operating in linear region. At this moment, the transistor M_5 is turned on by SAP_{EN} signal. This pulls the signal levels rapidly apart. The data is written to the memory cell that has been destructively read. This happens because $V_{\text{IN/OUT1}}$ and $V_{\text{IN/OUT2}}$ are directly connected to the bit lines.

With technology scaling in memories, increasing numbers of cells are attached to the columns. Hence, the capacitances of the bit lines are large [19]. In such advanced memories, voltage mode falls short of expected performance. This creates the need for faster sensing techniques. These sensing techniques must remain unaffected by the bit line capacitance. To overcome these deficiencies, the current-mode approach is followed. This approach does not depend on bit line capacitance.

4.15.1.2 Current-Mode Sense Amplifier

In the current sense amplifier, a comparison of currents is implemented. A current mirror directly obtains the current of a chosen memory cell. Then the mirrored current and the reference current are mutually compared [20].

Current-mode sense amplifiers are able to decrease the delays encountered in the sense circuit. This is because of the provision of low common input/output

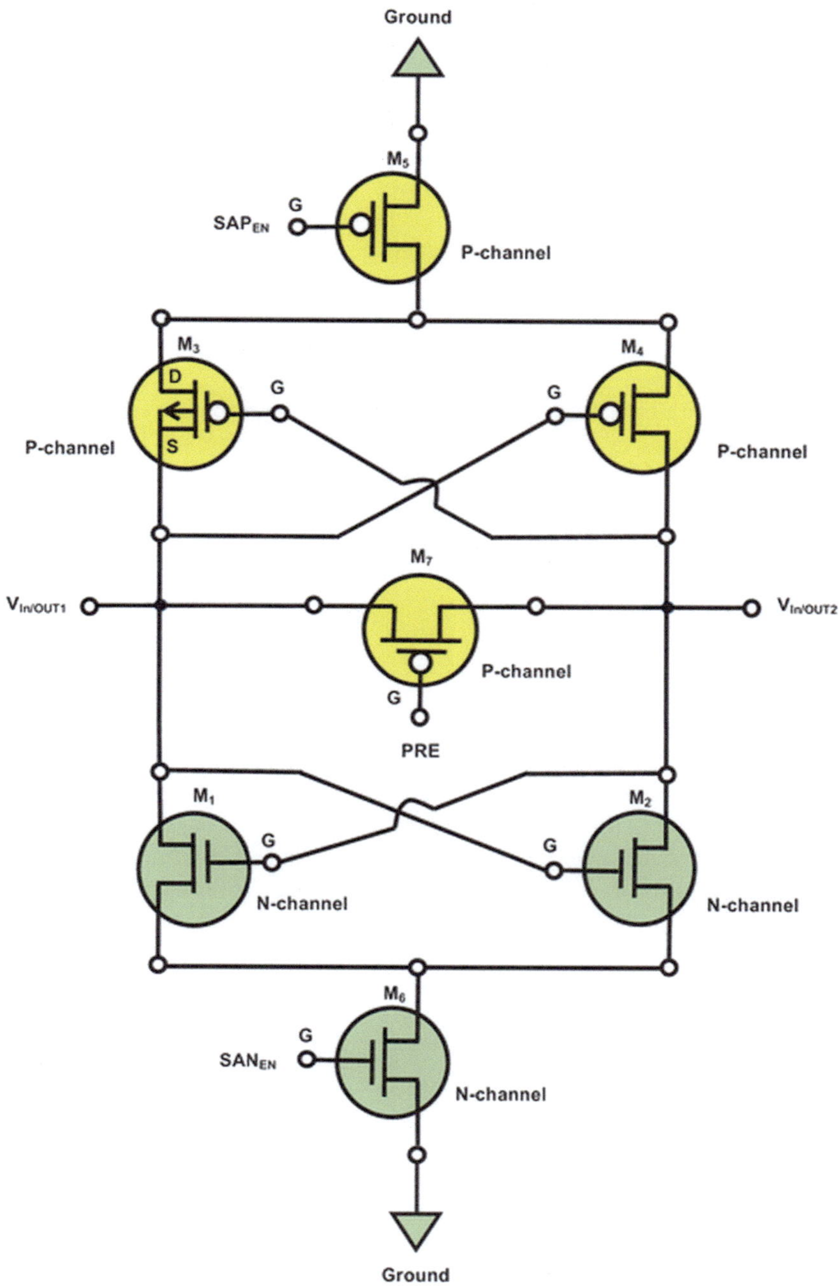

Fig. 4.6 Differential voltage sense amplifier

impedances by them. The voltage swings are lessened by their small input impedance to the bit lines. It also decreases the cross-talk. Moreover, substrate currents and modulations of substrate voltage are lowered [21].

4.15.1.3 Charge Transfer Sense Amplifier

This amplifier works on the mechanism of charge redistribution. The charge is redistributed between the exceedingly high bit line capacitance and low capacitance output nodes of the amplifier. The increased bit line capacitance proves to be advantageous to a differential charge transfer amplifier. This amplifier offers operation at a low consumption of power. The circuit speed is not forfeited.

In implantable electronics, the events take place at a slower gait. The activity and duty cycles for memory access and utilization are low. Therefore, simple CMOS sense amplifiers for voltage transitions suffice. Dividing the memory array into sub-arrays further decreases power dissipation. Furthermore, it must be confirmed that only the sub-arrays targeted for a given task are active. The remaining sub-arrays must be docile.

4.15.2 Soft Errors

In implantable electronics, the probability of soft errors is very high. Examples of soft error are the radiation-induced bit flips in memories. The high probability of soft errors arises from the low operating voltages required. For error detection and fixing, error-correcting codes (ECCs) must be used. An ECC is an algorithm, which picks out and sets the errors right. Within limits, any errors introduced are spotted. They are repaired on the basis of the available remaining numbers.

4.16 IC Testing and Evaluation

The ICs of implantable electronic systems are life-supporting tools. Hence, any error in their functioning may be often de*I*cisive for a person's life or death. In view of this, the testing scheme must be carefully planned and executed. Voltage overstressing for enhancing infant mortality rates may actually consume and hence decrease the lifetime of thin insulating oxide films. It may be more harmful than beneficial for the patient. Physics- and statistics-based approaches may be superior to chip functionality examination.

I_{DQQ} denotes the quiescent power supply current in a CMOS IC. $_{DQQ}$ testing is based on the principle that in a properly operating CMOS digital circuit operating in static state, the only current flowing from power supply to the ground terminal is the reverse bias leakage current. The presence of any manufacturing defects or faults can increase this current by a large amount. Hence, an increase in I_{DQQ} indicates the presence of defects.

4.17 Discussion and Conclusions

An implantable IC designer should leave no stone unturned for chip area and weight reduction [22]. Power consumed by the output load, i.e., the nerve or tissue to be excited, is governed biologically. Stimulus thresholds determine it. Therefore, this component of power is out of designer's hand. But the IC itself during its normal functioning must expend power frugally. By this approach, the battery life is prolonged. Harmful heat dissipation in the body tissues surrounding the implant is also avoided. Power is also related to implant size. Battery capacity and therefore battery volume increase with power. Hence, if implant size is to be reduced, power must be decreased. Short-channel effects in CMOSFETs must be paid due attention. Reverse leakage currents should be restricted to infinitesimally small values by careful design and process techniques.

Review Exercises

4.1 Compare the operational frequency and temperatures ranges of implantable devices with those used in industry. Comment on the answer.

4.2 Explain the following CMOS processes: (1) N-well CMOS, (2) P-well CMOS, and (3) Twin-tub CMOS.

4.3 What are pull-up and pull-down networks? Using these networks, how do you construct the following logic gates: (1) NOT circuit, (2) NAND circuit, and (3) NOR circuit?

4.4 Explain the idea of a system-on-chip. In what ways does it help to cram more functions in a single chip? What benefits are lost in making such chips?

4.5 (1) Which CMOS device consumes more power: faster device or slower device? (2) Which CMOS device has a larger leakage current: lower threshold voltage or higher threshold voltage device?

4.6 How does the electric field present at the gate edge of the drain junction induce leakage current at small geometries? What is this current called?

4.7 By what factor does the leakage current due to tunneling increase for every 20 nm decrease in gate oxide thickness?

4.8 What is meant by soft breakdown of oxide film? What is stress-induced leakage current? Explain the basic mechanism of its origin.

4.9 Why does NBTI affect only P-channel transistors? What MOSFET parameters are altered by NBTI? What is the most common model for NBTI? What does it enlighten about this instability?

4.10 What is the meaning of noise in a transistor? What are the prominent types of noise found in CMOS devices? What kinds of noise are sporadically present?

4.11 What is flicker noise? Describe the two major models for explanation of origin of flicker noise? Which of the models is used for MOS devices? Which one is used for the bipolars?

4.12 How does $1/f$ noise of CMOS transistors depend on chip area? How can this noise be decreased?

4.13 What do you mean by the input DC offset of an amplifier? How is this offset produced? How is this offset represented in a circuit diagram?

4.14 Name the kind of noise arising from the quantization of electric charge. What type of noise is related to the generation–recombination centers in a semiconductor?

4.15 When is the channel of a MOSFET said to be short? Is DIBL a short-channel effect? How does the threshold voltage decrease at high drain voltages? Upon what structural and electrical parameters does DIBL depend?

4.16 Explain the statement, "DIBL settles the final closeness of surface diffusions."

4.17 What advantage is derived by increasing the minimum channel length for a particular CMOS technology? How is this advantage gained?

4.18 Argue in favor of revising the existing transistor models for application to implantable electronics design.

4.19 Define the roles played by analog and digital signal processing in implantable electronics. Give reasons.

4.20 What is ESD? What kind of failure mechanisms does it trigger? How can ESD be prevented?

4.21 To what limiting boundaries the post-stress leakage current must be confined in implantable electronics? What is the tolerable limit in general-purpose electronics?

4.22 How does digital design for implantable electronics differ from the general trend? Explain giving reasons.

4.23 What is the task performed by the sense amplifier in a computer memory? How many such amplifiers can be found on a memory chip?

4.24 What is a voltage-mode sense amplifier? What problem is encountered in the use of this amplifier with technology scaling? How does a current-mode sense amplifier provide the solution?

4.25 What are the chances of soft errors in the memory section of implantable devices? What is an ECC? How does it function?

4.26 Why should techniques like voltage overstressing be avoided in testing an implantable device? How does I_{DQQ} testing help in distinguishing a defective IC from a good one?

References

1. Gerrish P, Herrmann E, Tyler L et al (2005) Challenges and constraints in designing implantable medical ICs. IEEE Trans Device Mater Reliab 5(3):435–444
2. Semenov O, Pradzynski A, Sachdev M (2002) Impact of gate induced drain leakage on overall leakage of submicrometer CMOS VLSI circuits. IEEE Trans Semicond Manuf 15(1):9–18
3. Kurimoto K, Odake Y, Odanaka S (1989) Drain leakage current characteristics due to the band-to-band tunneling in LDD MOS devices. In: Technical digest international electron devices meeting (IEDM '89), Washington, DC, 3–6 Dec 1989, pp 621–624
4. Riccò B, Gozzi G, Lanzoni M (1998) Modeling and simulation of stress-induced leakage current in ultrathin SiO_2 films. IEEE Trans Electron Devices 45(7):1554–1560
5. Pantisano L, Cheung KP (2001) Stress-induced leakage current (SILC) and oxide breakdown: are they from the same oxide traps? IEEE Trans Device Mater Reliab 1(2):109–112
6. Stathis JH, Zafar S (2006) The negative bias temperature instability in MOS devices: a review. Microelectron Reliab 46:270–286
7. Schroder DK (2007) Negative bias temperature instability: what do we understand? Microelectron Reliab 47:841–852
8. Lundberg KH (2002) Noise sources in bulk CMOS, pp 1–12. http://web.mit.edu/klund/www/papers/UNP_noise.pdf. Accessed 30 July 2015
9. McWhorter AL (1957) *1/f* noise and germanium surface properties. In: Kingdton RH (ed) Semiconductor surface physics. University of Pennsylvania Press, Philadelphia, pp 207–228
10. Hooge FN (1969) *1/f* noise is no surface effect. Phys Lett A 29(3):139–140
11. Hooge FN, Vandamme LKJ (1978) Lattice scattering causes *1/f* noise. Phys Lett A 66(4):315–316
12. Jindal RP, van der Ziel A (1981) Phonon fluctuation model for flicker noise in elemental semiconductors. J Appl Phys 52(4):2884–2888
13. Hung KK, Ko PK, Hu C et al (1990) A physics-based MOSFET noise model for circuit simulators. IEEE Trans Electron Devices 37(5):1323–1333
14. van der Wel AP, Klumperink EAM, Nauta B (2001) Effect of switched biasing on *l/f* noise and random telegraph signals in deep-submicron MOSFETs. Electron Lett 37(1):55–56
15. Palmer R (2001) DC parameters: input offset voltage (VIO). Application report SLOA059, Mar 2001. Texas Instruments, pp 1–24
16. Troutman RR (1989) VLSI limitations from drain-induced barrier lowering. IEEE J Solid-State Circuits 14(2):383–391
17. Vittoz EA (1994) Analog VLSI signal processing: why, where and how? Analog Integr Circuits Signal Process 8:27–44
18. Intel (2000) ESD/EOS. 2000 packaging databook, pp 6-1–6-8
19. Patel P, Zafar S, Soni H (2014) Performance of various sense amplifier topologies in sub100 nm planar MOSFET technology. Int J Emerg Trends Technol Comput Sci 3(2):42–49
20. Jiang L, Xueqiang W, Qin W et al (2010) A low-voltage sense amplifier for high-performance embedded flash memory. J Semicond 31(10):105001-1–105001-5
21. Kumar S, Singh SK, Noor A et al (2011) Comparative study of different sense amplifiers in submicron CMOS technology. Int J Adv Eng Technol 1(5):342–350
22. Kelly A (2012) Integrated circuit design for miniature implantable medical devices. In: Embedded systems conference, Design East, Boston, 18 Sept 2012, pp 1–14

Chapter 5
Neural Stimulation and Charge Balancing Approaches

Abstract The stimulating electrodes are generally configured in monopolar and bipolar configurations. Common simulation modes are the current mode and voltage mode. The usual stimulation waveforms are either monophasic or biphasic. Charge imbalance occurs by semiconductor failure. Such imbalance may also arise from leakage currents. The main cause is cross-talk between adjacent stimulating channels (sites) as well as cable failure. Positive charge balance is provided by a blocking capacitor connected in series with each electrode. This protective mechanism is used for electrical safety against fault conditions. The large capacitance value required for the blocking capacitors (sometimes a few microfarads) is realized through off-chip surface-mount components. In applications, e.g., retinal implants, the large-size blocking capacitors cannot be used. This inability is due to physical size limitations. Then, other methods for active charge balancing are resorted to. A stimulator circuit that is foolproof without off-chip blocking capacitors produces an active stimulation phase by high-frequency current switching. This phase is followed by a succeeding passive discharge phase.

Keywords Monopolar electrode · Bipolar electrode · DAC · ADC · VIC · Voltage multiplier · CCS · VCS · ChCS · Charge balancing · Blocking capacitor

5.1 Introduction

In electrostimulation, different types of stimulating electrodes, modes, and pulse waveforms are used.

V. K. Khanna, *Implantable Medical Electronics*,
https://doi.org/10.1007/978-3-031-99868-3_5

5.2 Monopolar and Bipolar Electrodes

A monopolar electrode has a single pole. Hence, it is a single electrode starting from the stimulator. It consists of an insulated wire whose tip is uncovered to deliver the current to the site to be stimulated (Fig. 5.1). The return path followed by the current is through the body of the patient to the earth. Abipolar electrode has two poles. Hence, it comprises two electrodes. One electrode is for entry of current from the stimulator into the patient tissue. Another electrode is for its exit from the tissue back to the stimulator.

In monopolar mode, the current begins its journey from the first terminal of the stimulator and traverses all through the stimulated site. Then, the current flows through the body of the patient to the ground to complete the circuit. In bipolar mode, the current only passes through the tissue between the two electrodes of the stimulator. The active and return functions take place at the targeted site. Thus, in the monopolar mode, the region of confinement of current in the body is larger. In the bipolar mode, it is smaller; in effect, the current is localized around the site. Concentric bipolar electrode and twisted wire bipolar electrode are the two popular

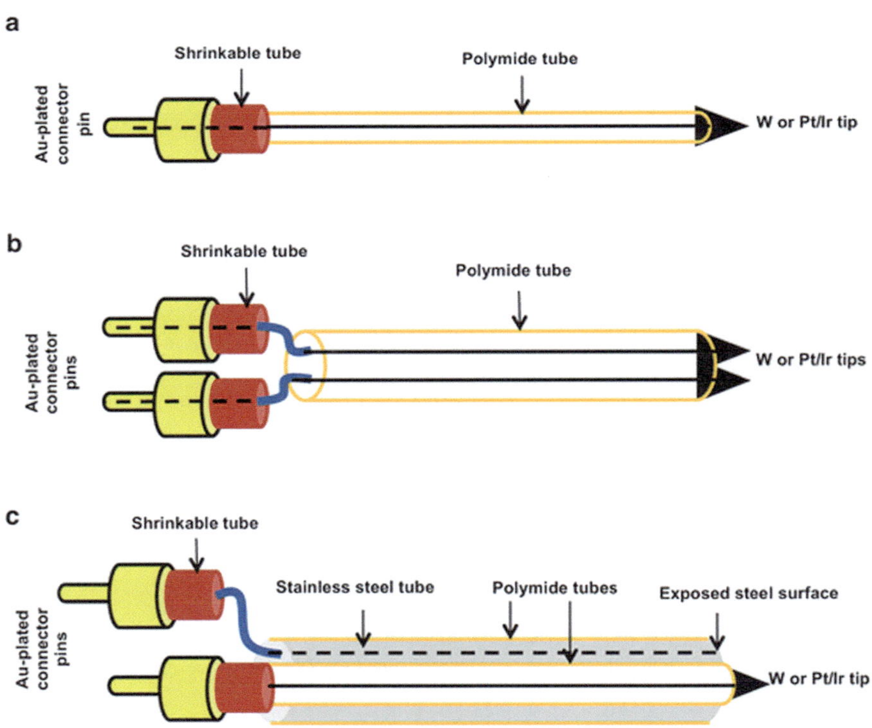

Fig. 5.1 Stimulation electrodes: **a** monopolar, **b**, **c** bipolar, **d** concentric bipolar electrode

Table 5.1 Monopolar and bipolar stimulation

S. No.	Monopolar stimulation	Bipolar stimulation
1.	Current flow path: starting from the pulse generator, it passes though the tissue, moves through the subject's body, and then goes back to the ground	Current flows between the regions separating the tips of the two electrodes: active and return electrodes that are incorporated into a single device
2.	Current flows across a larger chunk of the body of the patient receiving the implant	Current flow is confined to a limited region of the body of the implant receiver
3.	Stimulation effect is spread over a larger region	Stimulation effect is restricted to a smaller area

types of bipolar electrodes. Table 5.1 explains the manner in which stimulations are carried out using the two types of electrodes and their resulting effects.

5.3 Monophasic and Biphasic Waveforms

Considering square-shaped stimulating pulses, a monophasic wave is a unidirectional wave. It consists of only one type of pulses, either positive pulses or negative pulses (Fig. 5.2).

A biphasic wave is a bidirectional wave. It comprises a positive pulse (anodic pulse) immediately followed by a negative pulse (cathodic pulse) or vice versa (Fig. 5.3). A train of pulses is a sequence of pulses of the intended type at repeated definite intervals of time.

To visualize the relative impact and effectiveness of these waveforms, defibrillators using the two waveforms are considered in Table 5.2.

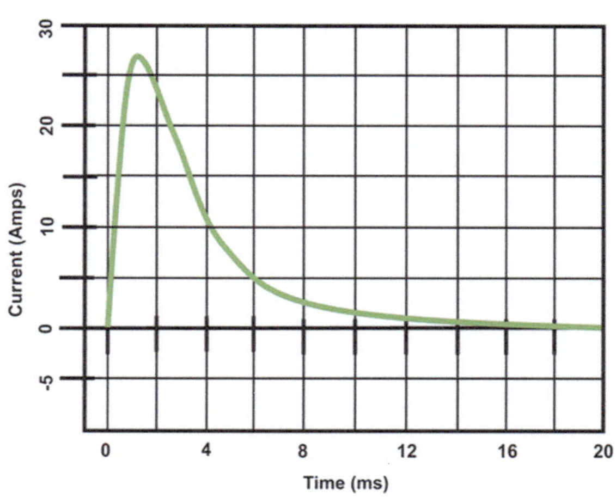

Fig. 5.2 Monophasic waveform

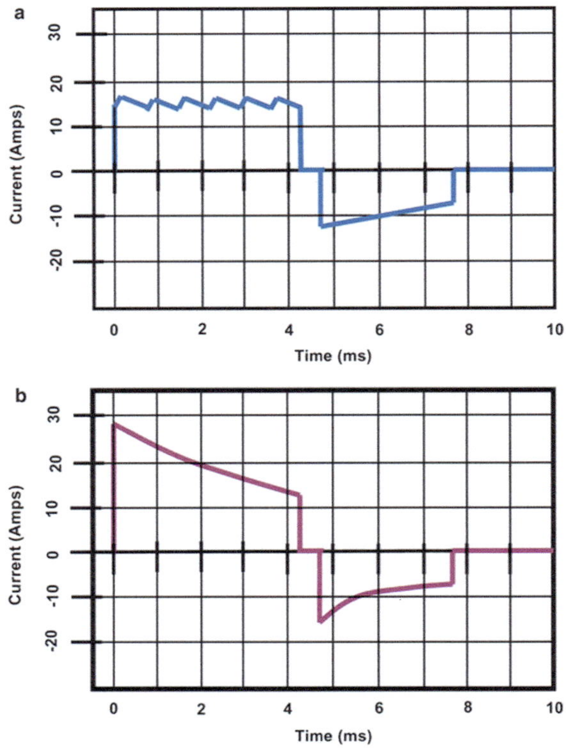

Fig. 5.3 Biphasic waveforms: **a** rectilinear biphasic and **b** biphasic truncated exponential

Table 5.2 Monophasic and biphasic defibrillators [1]

S. No.	Monophasic defibrillator	Biphasic defibrillator
1.	The shock is given in one direction only from a given electrode to another electrode	The direction is reversed in the latter part of the shock, and magnitude of voltage in this latter part is generally lower than in the initial part. Biphasic truncated exponential and rectilinear biphasic waveforms are commonly used
2.	Less effective	More effective
3.	Need more energy ~ 360 J	Require less energy ~ 200 J. A biphasic protocol using 120–200 J has the same efficacy as a monophasic protocol having 200–360 J
4.	Chances of more potential damage to the heart by the energy shock	Less likely damage to the heart by the lower energy shock

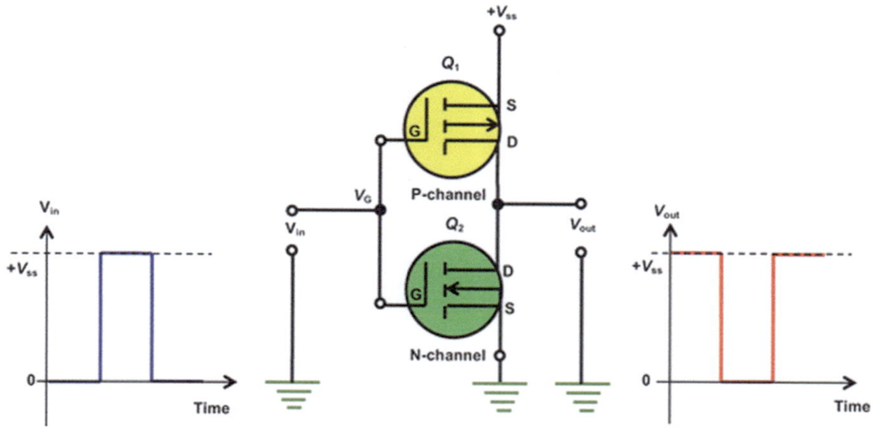

Fig. 5.4 The CMOS switch

5.4 Functional Circuit Blocks

Many elementary circuits occur repeatedly in discussions on stimulators. It is worthwhile to recapitulate some of these circuits to facilitate understanding.

5.4.1 CMOS Switch

A CMOSswitch (Fig. 5.4) is an electronic switch. It consists of two MOS transistors that are always in opposite operational conditions. When $V_G = 0$ V, the P-channel transistor Q_1 is on. At this time, the N-channel transistor Q_2 is off. An output approximating V_{ss} is obtained. This is the open switch state. When $V_G = V_{ss}$, the output ~ 0. In this condition, the P-channel transistor Q_1 is off. The N-channel transistor Q_2 is on. This is the closed switch state. Thus, the transistor combination is driven back and forth between V_{ss} and 0.

5.4.2 Digital-to-Analog Converter

This circuit lies on the borderline between the notional digital world and the realistic analog life. It is a circuit which converts a signal having a few, usually two (binary) defined levels/states into a signal (current, voltage, or charges) having theoretically infinite number of levels/states.

The main kinds of digital-to-analog converters (DACs) are:

1. *The Pulse-Width Modulator*: It is an ingenious DAC of preliminary nature. It involves the switching of a steady current or voltage into an analog filter of low-pass type (Fig. 5.5). The digital input code determines the duration of switching. By analog filtering, the high-frequency components of the signal are removed. Only the DC component is left behind. The bandwidth of the DAC depends on that of the low-pass filter used.
2. *The Binary Weighted DAC*: It contains *n* switches. In this arrangement of switches, one switch is reserved for every single bit fed to the input. The DAC also

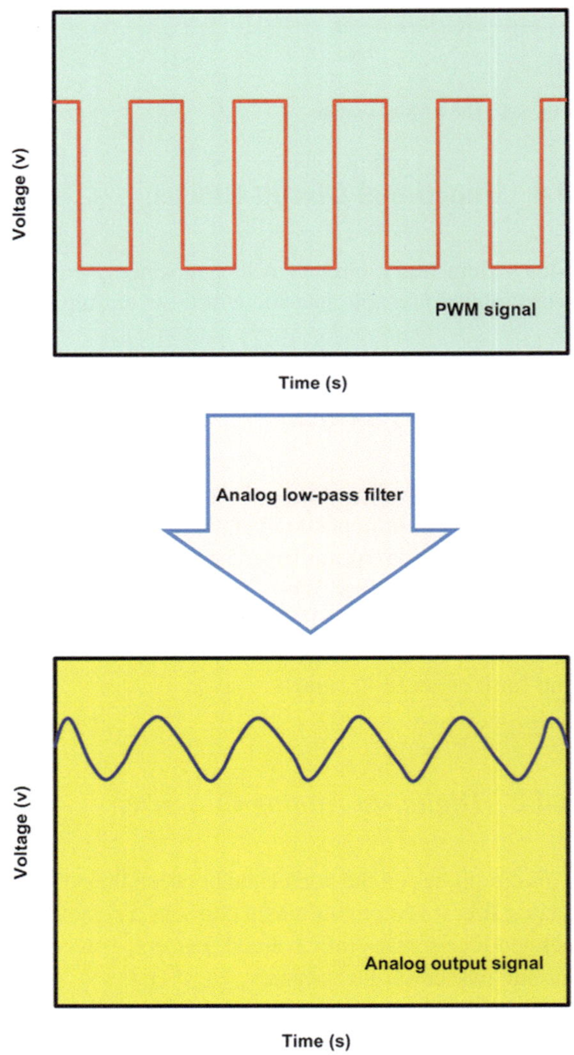

Fig. 5.5 Digital-to-analog conversion by analog low-pass filtering

has a weighted resistor ladder network. In this network, the values of resistances are selected in such a manner that they bear inverse proportionality relationship with binary weights of the input bits. Besides, the circuit has a reference voltage V_{Ref}. A summing amplifier performs addition of all the currents flowing in the resistor ladder network. By summing, it produces a signal proportional to the digital input (Fig. 5.6). In voltage scaling, the digital inputs are decoded. The corresponding switches are turned on to produce the analog output. This output is generated by dividing the reference voltage V_{Ref} by the series resistor network. The discrete voltages appearing on the nodes are smoothened by interpolation circuit.

It is a fast conversion method. However, it suffers from poor accuracy. The reason for inability to provide adequate accuracy is that the circuit requires very exact values of every individual voltage or current. To meet this demand, high-precision resistors or current sources are necessary. These are expensive, so often one has to manage with components of average precision. Hence, application of this type of converter is seldom done beyond 8-bit resolution. For an 8-bit converter, the eight resistor values are distributed in the range between R and $128R$ in binary weighted steps. Large-scale manufacturing of this type of DAC is problematical. This difficulty arises due to the range of resistor values needed with desired tolerances. The tolerance required is $< 0.5\%$ to accurately convert the input. Further, it is very difficult to match the temperature coefficients of all the resistors.

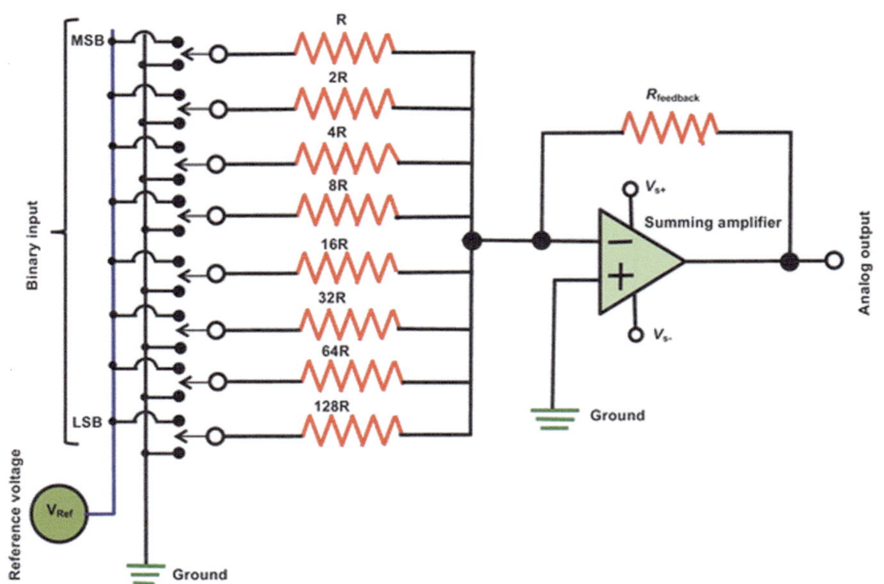

Fig. 5.6 An 8-bit binary weighted DAC constructed from weighted resistors and a summing amplifier

3. *The R/2R Ladder DAC*: This DAC employs repeating resistors of only two different values, namely, *R* and 2*R* (Fig. 5.7). Hence, 2 *N* resistors are required to assemble an *N*-bit DAC. These resistors are quite easily laser trimmed to obtain correct values because only few resistors have to be trimmed. In comparison, for high-accuracy conversion, a weighted resistor DAC necessitated a much expansive range of resistance values and switches for its different bit positions. An *R*/2*R* ladder DAC eradicates the snags of weighted resistor DAC at the outlay of an extra resistor for each bit. DAC precision is thereby increased considerably. The chief cause is the comparative ready availability of equal matched values of resistors or current sources. But conversion speed is lowered owing to parasitic capacitance.

There are two possible modes of operation of an*R*/2*R* ladder network. These are the voltage mode and the current mode. The voltage-mode *R*/2*R* ladder DAC is implemented by switching the arms of the ladder between V_{Ref} and ground. The output signal is obtained from the end of the ladder.

The function of current-mode *R*/2*R* ladder DAC is based on the adjustment of the gain of the DAC. This adjustment is made by connecting a series resistor at the V_{Ref} terminal. The output is connected to an OP-AMP. This OP-AMP is configured as a current-to-voltage (*I*/*V*) converter.

DACs are prone to switching glitches. A glitch means a sudden, temporary malfunction. Current-mode operation is more liable to the hazard of switching

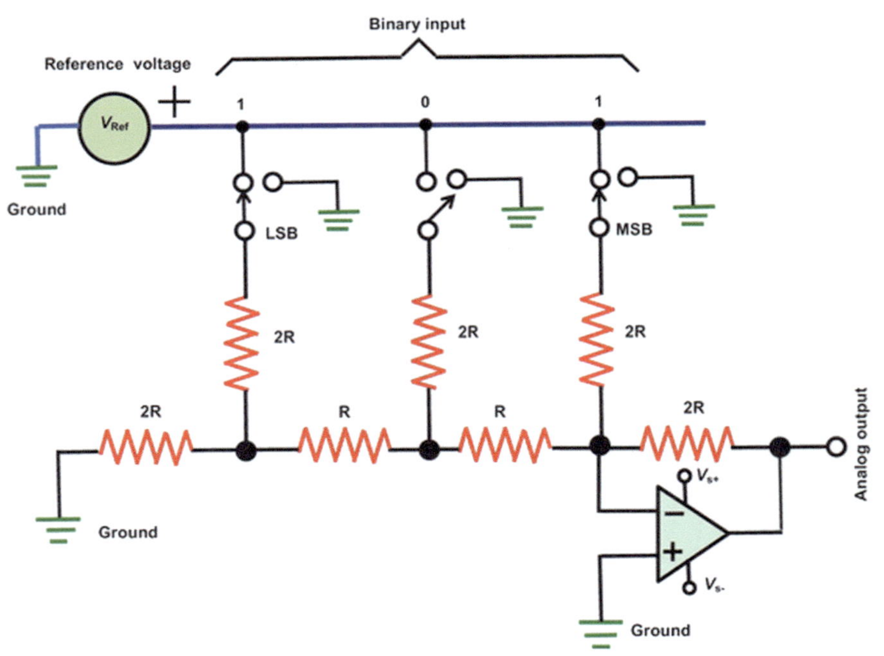

Fig. 5.7 R/2R ladderDAC

glitch than voltage mode. The glitch encountered with current-mode switching arises from the manner of connection of switches. In this mode, the switches are directly connected to the output line(s). But the design of switches of a current-mode ladder network is less tricky. This is because they are always at ground potential. On the whole, their voltage specification does not impinge on the V_{Ref} rating.

5.4.3 Analog-to-Digital Converter

It is a circuit which converts a continuously variable signal having theoretically infinite number of levels/states into a signal having a few, usually two (binary) defined levels/states. An A/D converter translates analog electrical signals portraying actual world events, e.g., mechanical, optical, acoustic, or thermal signals, into digital form for handling and manipulating data. The analog-to-digital converters (ADCs) are at the front end of any digital circuit used to process signals from the exterior world.

ADCs are of the following types: flash, successive approximation register (SAR), pipelined, integrating or dual slope, and sigma delta ($\Sigma\Delta$). All the ADCs work on the same principle. They produce a signal consisting of a certain number of bits N. The sequence of bits represents the signal. Beginning from the most significant bit (MSB) and moving to the least significant bit (LSB), each bit has doubled the weight of the next. The different ADCs to be discussed have their own strengths and weaknesses. The choice of the ADC for a given application is defined by the priority in requirements of speed, precision, compactness, power consumption, etc.

Also called the parallel A/D converter, aflash converter uses a resistive ladder. This ladder divides the reference voltage linearly into 2^N equal parts (Fig. 5.8). OP-AMP comparators at different rungs of the ladder compare the input signal to successive unique reference voltages supplied by the corresponding part of the resistive ladder. One comparator is provided for each reference level. Starting with $V_{Ref} = 1/2_{LSB}$, each comparator compares V_{in} to a different reference voltage. If $V_{in} > V_{Ref}$, the output signal is high logic level. If $V_{in} < V_{Ref}$, the output signal is low logic level. The output signals from these comparators are fed into a priority encoder circuit. The output signal from the priority encoder is a binary number. This binary number is based on the highest-order active input signal, paying no heed to any other active input signal. The flash converter is a very fast ADC. Hence, it is aptly named flash converter. But it needs many parts (255 comparators for 8-bit ADC), doubling in size for each bit added to the representation. Resolution is low and power consumption is high. The ADC is expensive.

Successive-approximation-register (SAR) ADC is the preferred structural design for achieving average-to-large resolution and applications demanding accuracy. It provides low power consumption. Besides, it has a small form factor. It offers a worthy compromising solution between speed and cost. Its speed is limited to ~ 5 Msps. The SAR ADC (Fig. 5.9) consists of a successive-approximation-register subcircuit. Along with it a comparator, a DAC, and digital control logic are included. As obvious from its very name, the SAR ADC implements a fundamental algorithm

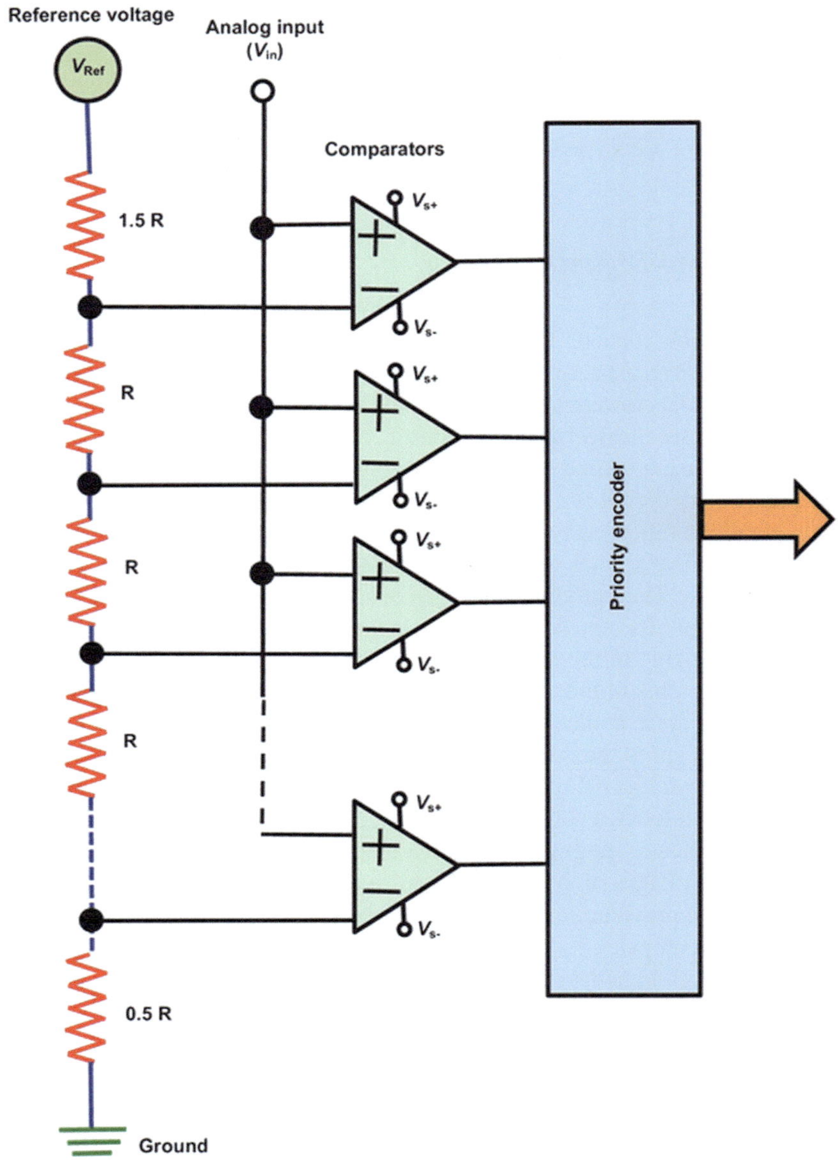

Fig. 5.8 Flash ADC based on direct conversion architecture including comparator banks and reference resistor divider networks

in computer science called the binary search algorithm to look for the position of a targeted value in an organized collection. In this algorithm, the *N*-bit register is initialized and placed at midscale. This register is a counter circuit. It counts by attempting all values of bits. During counting, it begins with the MSB and terminates at the LSB. The ADC assumes the MSB to be 1 and the remaining bits to be 0. So, the DAC produces an output value of 0.5 V_{Ref} where V_{Ref} is the reference voltage furnished to the ADC. Then, a comparison is made to find whether V_{in} is $< V_{DAC}$ or V_{in} is $> V_{DAC}$. If V_{in} is $> V_{DAC}$, the output signal of the comparator is a high logic level. In this situation, the MSB of the *N*-bit register remains at 1. In opposition, if V_{in} is $< V_{DAC}$, the output signal of the comparator is a low logic level. The MSB of the register is reset to logic 0.

The SAR control logic budges to the succeeding bit downwards by following a similar procedure. It compels that bit to ascend to high logic level and performs one more comparative assessment. The above sequence of steps carries on, treading along the path toward the LSB. When the sequence reaches LSB, the conversion has been completed. The register shows the *N*-bit digital word.

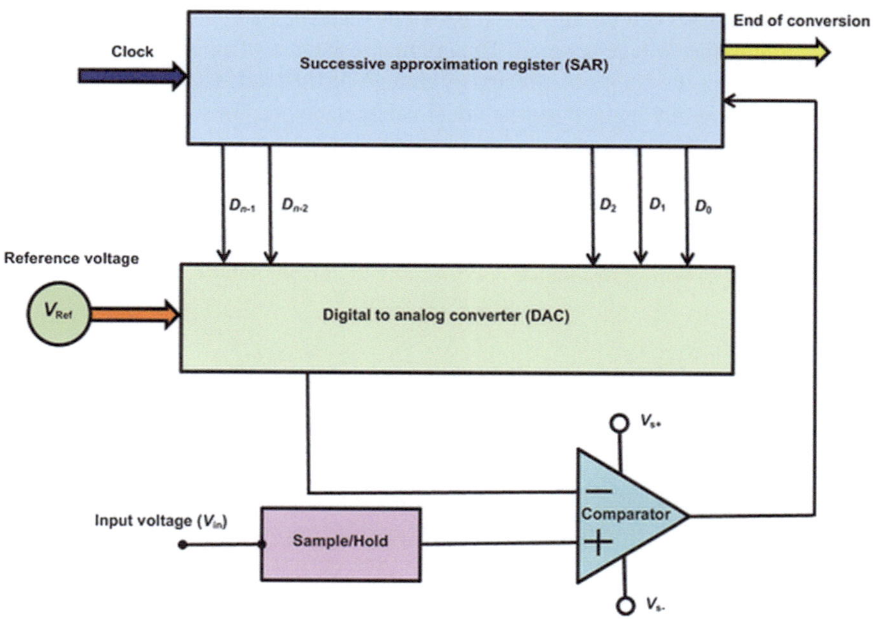

Fig. 5.9 Successive-approximation-register ADC

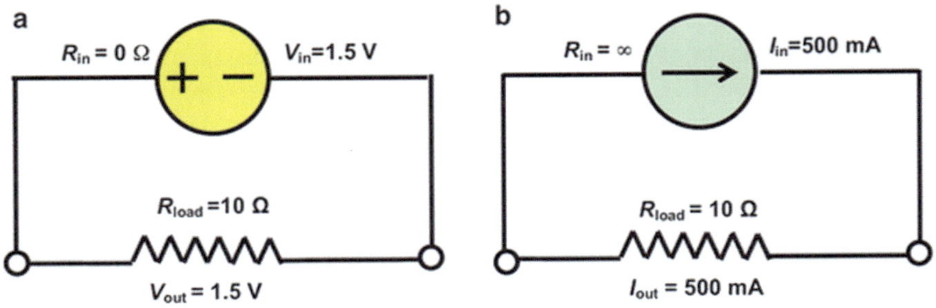

Fig. 5.10 Ideal sources: **a** voltage source and **b** current source

5.4.4 *Voltage and Current Sources*

A voltage source is a device having two terminals, which maintains a fixed output voltage across it, independent of load resistance variations. The internal resistance of an ideal voltage source is zero. So, load resistance changes do not affect the voltage.

A current source is a circuit that supplies or receives fixed electric current in it, independent of the voltage across it. The internal resistance of an ideal current source is infinite. So, the current is not affected by changes in the load resistance. The current source is the dual of a voltage source. Ideal voltage and current sources are depicted in Fig. 5.10.

In a circuit diagram, a real voltage source is shown as an ideal voltage source in series with a resistance r. For the ideal voltage source, resistance $= 0$. The circuit diagram representation of a real current source is in the form of an ideal current source in parallel with a resistance r. For the ideal current source, resistance $= \infty$.

5.4.5 *Current Source Versus Current Sink*

The term "current source" is used when current flows from the invisible region such as the battery to the visible region, making the electric circuit to do some work. This is analogous to water current ejected from a mountain spring (invisible underground) to the surrounding areas. In a current source, the current flows toward the output, i.e., the load.

The term "current sink" is used when the current flows from the visible region such as a circuit to the invisible region inside the battery for charging the same. This flow is similar to the flow of water current from the tap to the drain where it cannot be seen. In the case of a current sink, the current flow occurs toward the input, i.e., away from the load. Figure 5.11 shows ideas of current source and current sink.

Sources and sinks are analysis formalisms. They are applied for distinguishing between directions in which current enters or exits a system.

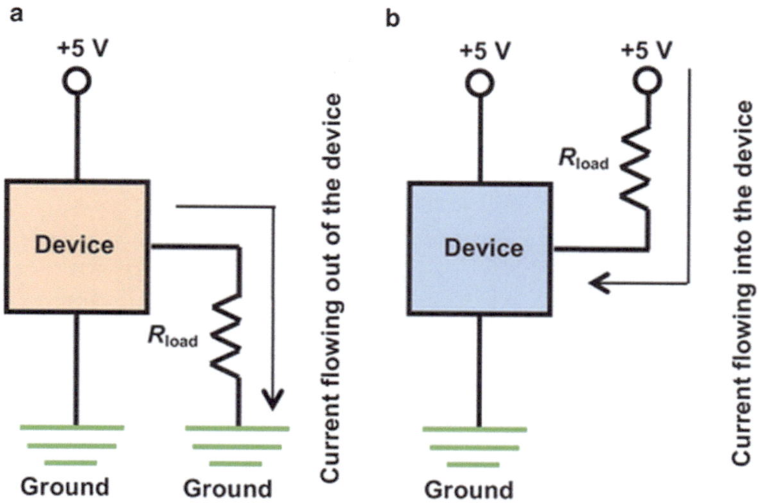

Fig. 5.11 Source and sink: **a** current source and **b** current sink

5.4.6 Current Mirror

It is also called current-controlled current source (CCCS). The current mirror circuit reads a current entering a read node. It copies the current to an output node(s). It does so with a suitable gain. In other words, it is a replicating circuit. It copies the current flowing through one active device in the circuit. It does so by controlling the current passing through another active device of the circuit. The current is maintained constant irrespective of any loading effect. Hence, a high output resistance is shown. The current mirror also has a low input resistance. This low value seeks to keep the input current constant notwithstanding the drive conditions. Current mirrors form the fundamental building blocks of analog circuit design.

Figure 5.12 shows the implementation of a current mirror using MOS transistors. The current mirror consists of two parallel branches. These branches carry approximately equal currents; hence, the name "mirror." The circuit has a compliance voltage. The compliance voltage is the minimum output voltage required for correct circuit operation. For this circuit, the MOSFET Q_2 on the output side (right) should be in the saturation region. For the left MOSFET Q_1, the gate being shorted to drain, $V_{DG} = 0$ and $I_{in} = I_{DS1}$. Since the circuit in the diagram forces the same gate-to-source voltage to be applied to transistor Q_2, $V_{GS1} = V_{GS2}$. Hence, the current flowing in the right MOSFET must be the same, i.e., $I_{DS2} = I_{DS1}$, so that $I_{out} = I_{in}$, provided Q_2 is operating in saturation mode, and transistors Q_1 and Q_2 match well in their properties, such as channel length, width, threshold voltage, etc. Temperature should also be same for both Q_1 and Q_2.

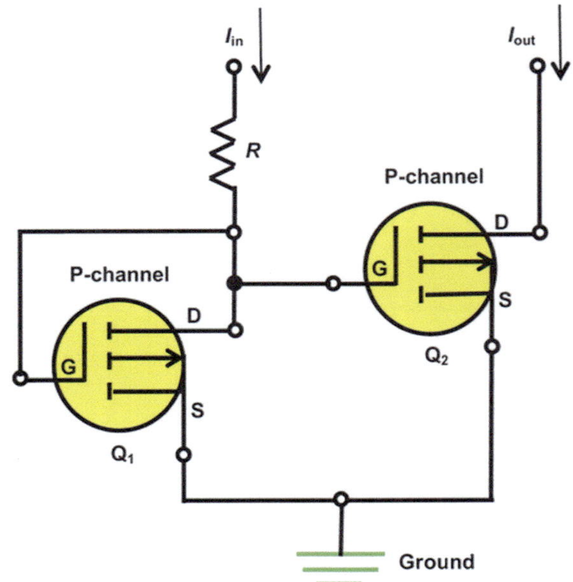

Fig. 5.12 MOS current mirror

5.4.7 Voltage-to-Current Converter

Voltage-to-current converter (VIC) is also called a voltage-controlled current source (VCCS) or transconductance amplifier. Transconductance (transfer conductance) or mutual conductance (g_m) is the ratio of a change in current (ΔI) to change in voltage (ΔV) which produced the current change; hence, $g_m = \Delta I/\Delta V$. The unit of transconductance is siemens or Ω^{-1}.

The VIC circuit (Fig. 5.13) is actually an amplifying arrangement. It generates an output current signal proportional to the input voltage signal. The constant in this proportionality relation is referred to as transconductance. The circuit employs an OP-AMP configuration furnished with negative feedback. The load resistor R_L is left floating without any connection to ground. The input voltage V_{in} is applied to the noninverting input terminal of the OP-AMP. The inverting input terminal of OP-AMP is driven by the feedback voltage across the load resistor R_L. Then, the output current flowing through the load resistor R_L is proportional to the input voltage V_{in}. This statement is valid on the presupposition that the OP-AMP has infinite input impedance. The circuit keeps the current fixed at a prescribed value by applying the required voltage of sufficient magnitude to the load resistor to maintain the current constant. Any perturbation of the circuit from its normal functioning, as induced by outside influences, is very small because the operational amplifier has very high impedance.

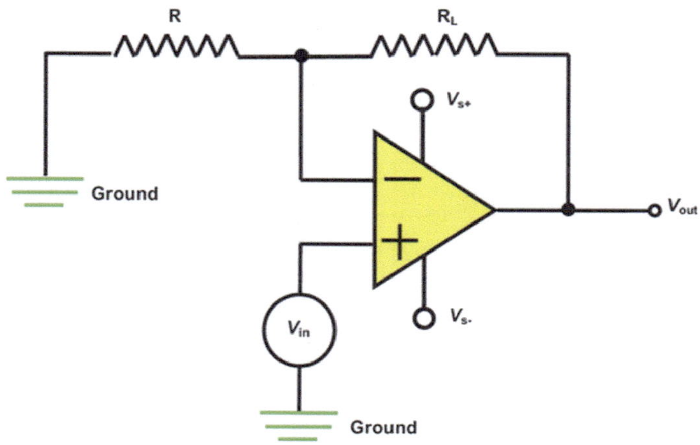

Fig. 5.13 Voltage-to-current converter (transconductance amplifier)

5.4.8 Voltage Multiplier

It is a specialized rectifier circuit. It is made of combinations of diodes and capacitors. It serves to produce an output DC voltage (V_{dc}) having magnitude equal to either an odd or even multiple of the peak or crest value of alternating input voltage (V_{in}), e.g., 2, 3, 4 … times the peak AC input voltage (V_{in}). A voltage multiplier employs charge pumps. The charge pumps are DC-to-DC converters using capacitors as storage devices to create voltage sources supplying higher voltages without using a transformer. On theoretical grounds, the voltage produced by a voltage multiplier circuit can be indefinitely large. But practically, this is hardly feasible. A major cause is the resulting low current capability. Another factor is the poor voltage regulation achieved. Due to such types of performance degradations accompanying the voltage multiplication, the design of voltage multipliers is not encouraged up to multiplication factors beyond 10 [2].

The full-wave voltage doubler is a symmetrical voltage multiplier circuit (Fig. 5.14). It is assembled from two half-wave rectifier circuits. Its operation can be described with reference to the successive time durations in which the sinusoidal input voltage acquires positive and negative values.

During the period of time in which the input voltage has a positive value, capacitor C_1 charges through diode D_1. This charging proceeds to the peak value of the input voltage (V_p).

Afterward, during the time interval in which the input voltage acquires a negative value, capacitor C_2 charges through diode D_2 to V_p.

Thus, the total output voltage developed across the capacitors C_1 and C_2 connected in series = voltage across capacitor C_1 (V_p) + voltage across capacitor C_2 (V_p)—the sum of voltage drops across the diodes = $2V_p$, neglecting the forward voltage drops of the diodes.

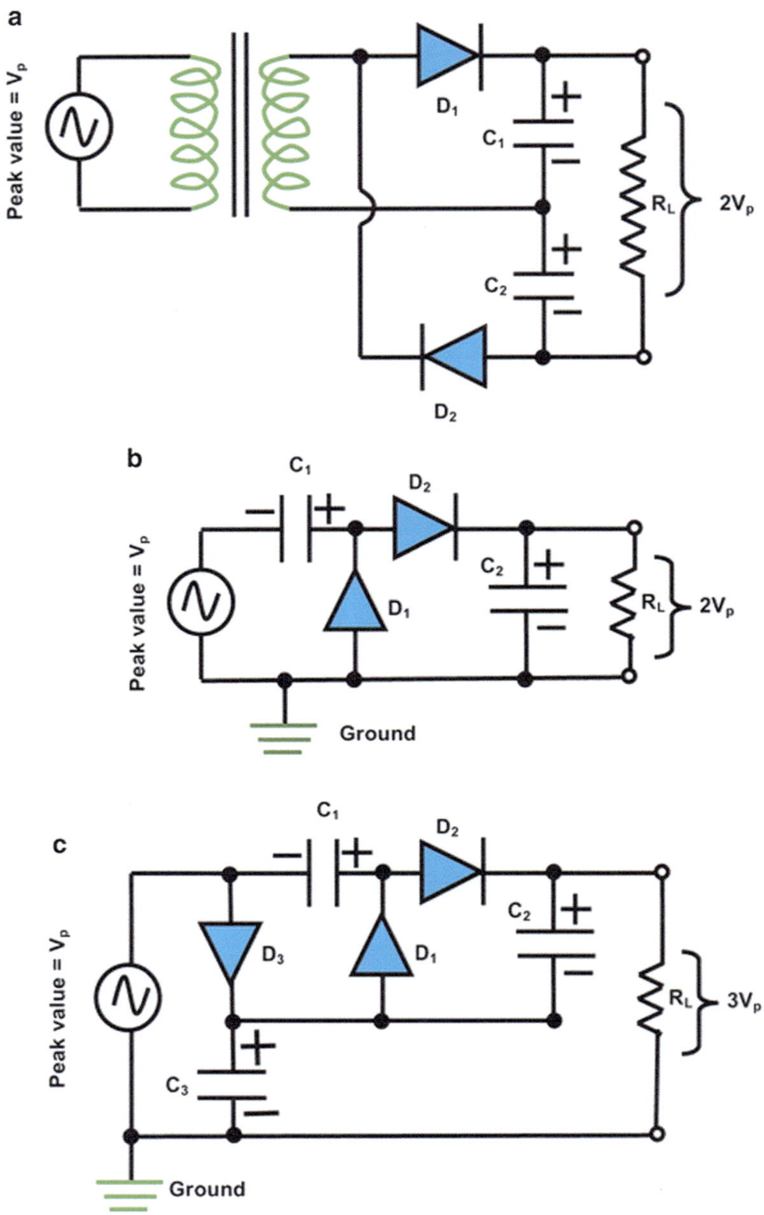

Fig. 5.14 Voltage multipliers: **a**, **b** full-wave and half-wave doubler circuits, **c** voltage tripler circuit

Referring to the negative and positive half cycles of the sinusoidal input voltage, the half-wave voltage doubler circuit works as follows: For the period in which the input voltage is negative, the diode D_1 operates in a forward-biased mode. It conducts current. Consequently, the pump capacitor C_1 is charged to the maximum voltage = peak voltage V_p. With no path left for capacitor C_1 to lose its charge, it has no other option but to retain its charge. It remains in this fully charged condition. Therefore, it acts as a charge storage device in series with the voltage supply. At the same time, diode D_2 conducts via D_1. The resulting current charges the capacitor C_2. During the time interval in which the input voltage is positive, diode D_1 is reverse biased and nonconducting. It blocks the discharging of capacitor C_1. However, the forward-biased diode D_2 allows current to flow. This current charges the capacitor C_2. But a previous voltage $= V_p$ persists across the capacitor C_1. So capacitor C_2 charges to the voltage $2V_p$, twice the peak voltage of input signal.

A voltage tripler circuit contains an ancillary single diode–capacitor stage. This stage is connected to the half-wave voltage doubler circuit explained above. A voltage quadrupler circuit is assembled by connecting together two full-wave voltage doubler circuits in succession. Thus, it is a cascade of two such voltage doubler circuits.

5.4.9 Boost Converter

It is also called a step-up converter. It is basically a DC–DC converter circuit providing an output voltage > the input voltage, albeit it supplies a higher voltage at a reduced current because power = voltage × current. So, an increase in voltage irrefutably goes hand in hand with a decrease in available current. Essentially, it is a type of switch-mode power supply (SMPS) used with battery-powered implant circuits. It is used when the battery cannot supply the required high voltage for circuit operation. Moreover, the use of extra batteries is disallowed from weight and volume restrictions [3]. It also takes care of the dropping battery voltage with time. It does so by increasing the battery voltage, thereby extending its service life.

Referring to Fig. 5.15, the circuit operation is explained by examining the incidents that take place during the positive and negative logic levels of the square wave. When the square wave acquires high logic level, the switching power MOSFET turns on. Hence, current starts to flow from the positive terminal of the supply. On the way, it passes through the inductance L_1. It then moves back to the negative terminal of the supply. During the current flow through the inductance, energy is accumulated in the magnetic field surrounding the inductance. Practically, no current flows through the high-impedance path formed by diode D_1, capacitor C_1, and the load.

When the square wave attains the low logic level, the MOSFET is turned off. As a result, current flow in the inductance stops. Its magnetic field begins to collapse. A back electromotive force (EMF) is generated in the opposite direction to the voltage across L_1 in the on-state to maintain the flow of current. This back EMF (V_L) is in series with the supply voltage (V_{in}). Therefore, the higher voltage $= V_{in} + V_L$ forward biases the diode D_1, as the MOSFET is not conducting. This higher voltage

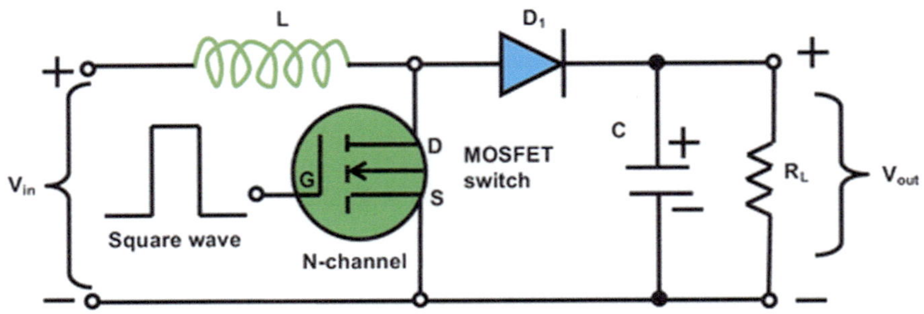

Fig. 5.15 Boost converter circuit

minus the voltage dropped across the diode charges the capacitor C_1. It also feeds the load.

The scenario during the next high logic level of the square wave is as follows: By virtue of the charge on C_1, the cathode of D_1 is more positive than its anode. The diode is reverse biased. However, current is still supplied to the load by the charged capacitor C_1. The capacitor C_1 being discharged, it is recharged when the MOSFET again turns off. Thus, a constant voltage is produced across the load. In terms of the input voltage V_{in} and the duty cycle D of the switching square waveform, the DC output voltage is expressed as

$$V_{\text{out}} = {}^{V_{\text{in}}}\!/_{(1-D)} \tag{5.1}$$

In an IC boost converter, a suitable fraction of V_{out}, dependent on the resistance ratio R_2/R_3 is compared with a reference voltage within the IC. The aim is to produce an error voltage. This error voltage is the key controlling variable. It is used to vary the duty cycle of the switching oscillator. Therefore, a series of boost voltages can be obtained. Also, these voltages are automatically regulated. The IC contains an internal oscillator and an FET switching transistor. To minimize conduction losses and achieve higher switching speeds, a Schottky diode is used in place of D_1. In order to save power, the IC also has a shutdown (SHDN) facility. This facility disables the boost converter when not required.

5.4.10 Timer Circuit

It is a circuit made from resistors, capacitors, and transistors/timer ICs. It is used for introducing a time delay to switch a circuit on or off. Usually, the user has the freedom to adjust the time delay to trigger the load as desired in the application. This delay is determined by the values of resistive and capacitive components.

5.4.11 Driver Circuit

It is a circuit used to control another circuit or component, e.g., the output driver circuit of an implant serves to source/sink a current at programmed level through the resistive load offered by the body tissue.

5.5 Current-, Voltage-, and Charge-Mode Stimulation

The front-end circuit selects the active electrodes. It either sends and controls current between them or applies and controls voltage across them. The power consumption in any arrangement consists of two parts. The first part is the power utilized by the electronic circuit inside the stimulator for its operation. The second part is the power utilized by the stimulator in delivering the required current or voltage pulse to the tissue for treatment of the malady. By lowering the operating voltage, the internal power consumption by the circuit is decreased. This is possible through careful circuit design. For bipolar electrodes, the instantaneous power consumption is found by multiplication of the voltage V applied between the two electrodes and the resulting current I flowing between the electrodes. In the case of monopolar electrodes, the voltage of interest is that between the active monopolar electrode and the distant ground electrode.

5.5.1 Current-Mode Stimulation

Figure 5.16 shows the basic idea of a current-mode stimulation circuit. In current-mode stimulation, or current-controlled stimulation (CCS), the voltage signal from the microprocessor is fed to a DAC [4]. The output signal from the DAC flows to the VIC. Thus, the stimulator output is a current pulse. The amplitude of this current pulse is controlled by the DAC. The current pulse is applied to the tissue.

In another version [5, 6], the voltage signal from the microprocessor is transformed into a current signal. A VIC is used for this transformation. The obtained current

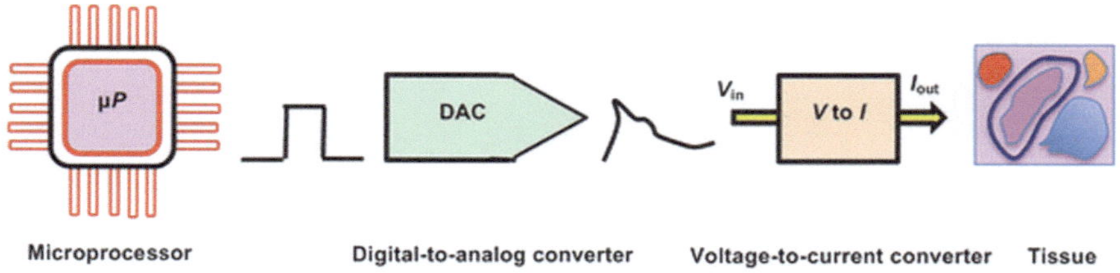

Fig. 5.16 Steps in current-mode stimulation

signal is changed into an analog form by the DAC. The analog signal is fed to a current mirror. The output signal obtained from the current mirror is applied to the tissue.

In a current source with high output impedance, any changes in load impedance, i.e., the tissue–electrode impedance, cannot affect the amplitude of current. So, the amount of charge supplied per stimulus is easily maneuvered. If current pulse has magnitude I and it lasts for a duration T, the charge injected into the tissue is $Q = I \times T$. Knowledge of the charge injected and its controllability makes this mode safe in implementation.

In reality, the medic programs the current I. In an automatic fashion, the front-end circuit adjusts the voltage V between the electrodes. So, the current of programmed value I flows athwart the electrodes. The current flows all the way through the tissue having impedance Z. The value of Z is composed of resistive and capacitive components. It is a nonlinear parameter. The parameter varies with time and frequency. It also changes according to any reactions that take place in the tissue with passage of current. However, despite all these complications, the stimulator circuit maintains a constant current flow I. The only exceptional case occurs when the maximum voltage available from the stimulator is not able to keep the current constant. In that situation, the current value cannot be kept constant.

5.5.2 *Voltage-Mode Stimulation*

Figure 5.17 illustrates the principle of voltage-mode stimulation. In voltage-mode stimulation, or voltage-controlled stimulation (VCS), the stimulator is an adjustable-gain amplifier. The input voltage of this amplifier is set by the DAC [4]. The load resistance is the sum of the resistances of the conducting lead and the tissue.

In another version of voltage-mode stimulation [7, 8], a rectifier chip is used to charge several capacitors to certain voltages. In this chip, the capacitive voltage repositories are derived from a solo AC voltage on a secondary coil. The capacitors discharge and deliver the charge to the relevant tissue. Controlled synchronous rectification is used in the above system. In synchronous or active rectification, active devices such as power MOSFETs are used in place of junction diodes to increase the efficiency.

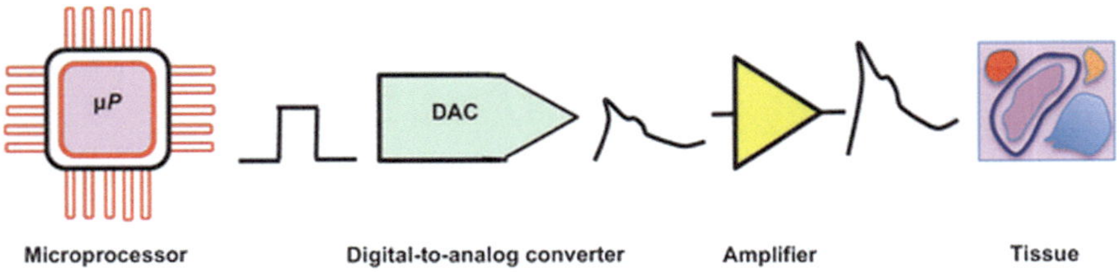

Fig. 5.17 Steps in voltage-mode stimulation

The magnitude of current supplied to the tissue varies with interelectrode impedance. Hence, the load impedance variations do not allow easy controllability of the charge supplied to the load. Within safety limits, the voltage can be set by the clinician to obtain the desired result. The clinician has limited options to vary the current *I*. The current must be always restricted within safe limits. Irreversible or damaging effects to the tissue must be always avoided. Naturally, therefore, VCS is less safe than CCS. It must be used with precaution. Safety limits must be clearly mentioned by the manufacturing companies. Nevertheless, its design simplicity and power efficiency make it more appealing than CCS. The power consumption is decreased at lower supply voltages. For this reason, cardiac pacemakers and deep brain stimulators use this mode [9].

In actual practice, the clinician chooses a particular voltage *V* between two electrodes. Then, the current drawn by the tissue is governed by the impedance *Z* of the tissue. Application of this approach is therefore restricted to limited cases. These are the cases where the temporal *Z* variations across different electrode pairs are small. These variations must be invariably already known. The efficiency is maximum when the applied voltage *V* equals the battery voltage. But efficiency falls above or below this value.

5.5.3 Charge-Mode Stimulation

In charge-mode stimulation or switched-capacitor-based stimulation, the stimulator consists of a bank of capacitors (Fig. 5.18). This bank of capacitors is charged sequentially from a DC–DC converter. The supply voltage of DC–DC converter is adjustable. So, the total charge on a capacitor is alterable [4]. The capacitors are discharged into the tissue. This discharging action is carried out through several electrodes by digitally controlled switches. Every capacitor can be discharged into the load consecutively. In this way, quantized amounts of charge are impelled into the tissue. The charge injected over each stimulation pulse is measured. For this measurement, the current is integrated over the period of the pulse. But the main issue is that the circuit requires large capacitors. The large capacitors can only be provided in off-chip form. Difficulty in provision of large-value on-chip capacitors is the main disadvantage of this technique. Otherwise, it serves as a unification of advantages of VCS and CCS methods. It combines the improved power efficiency and safety of VCS with the superior charge-controlling ability of CCS method.

Table 5.3 gives an overall view of CCS, VCS, and ChCS.

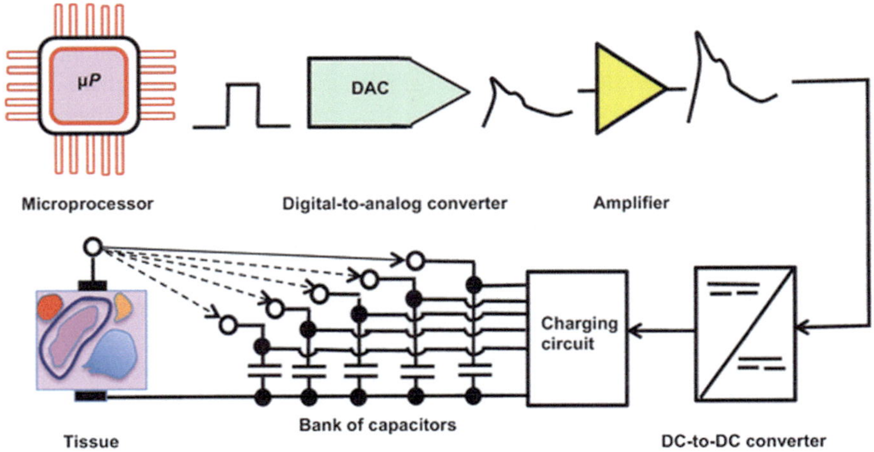

Fig. 5.18 Steps in charge-mode stimulation

Table 5.3 Current-controlled stimulation (CCS), voltage-controlled stimulation (VCS), and charge-controlled stimulation (ChCS) [4]

S. No.	Property	CCS	VCS	ChCS
1.	Design	Complex	Simplest	Simple
2.	Safety	High	Poor	Good
3.	Power efficiency	Poor	Highest	Between CCS and VCS

5.6 Charge Balancing

In electrical stimulation, charge is transferred from the electrode to the tissue. If a large potential is applied for a long period of time, enormous charge exchange occurs between the electrode and the tissue. The enormous charge exchange leads to flow of a strong faradaic current. The outcome is the inception of irreversible electrochemical processes. These processes are electrolysis, pH changes, electrode dissolution, and tissue destruction. To avoid the onset of these untoward incidents, net charge in each stimulation cycle must be zero. This requires that appearance of any direct currents during stimulation be prevented. From here arises the notion of charge balancing. On first thought, it seems that use of biphasic pulses can straightaway stop the charge unbalance. In biphasic pulses, the first phase is the stimulating phase for evoking the physiological response. The second phase is the reversal phase to quash any electrochemical reactions that are initiated during the first phase. So each stimulation phase is free of DC currents. But the nonuniformities and variations in IC fabrication processes are unavoidable. These result in mismatches between current pulses. The mismatches upset the charge balance. Adoption of preventive means for charge upsetting is therefore essential. Table 5.4 outlines the key features of active and passive charge balancing techniques.

Table 5.4 Passive and active charge balancing

S. No.	Active charge balancing	Passive charge balancing
1.	A monitoring system is engaged to perceive the remnant voltage on electrode–tissue interface when no current is flowing through the tissue. It also equalizes the charges of the two phases: cathodic and anodic	No monitoring system is used. Balancing is done by charge storing and discharging phases. The connection of capacitors between the output circuits of the pulse generator and the electrodes aids in charge balancing. The capacitors block errant continuous direct current. They are discharged typically after delivery of an individual pulse
2.	No requirement of large off-chip components	Due to restriction in the voltage tolerance, the blocking capacitor must be sufficiently large. Its value is generally approximately several μF. Several off-chip blocking capacitors may be necessary. The area consumption by these capacitors may cause a space crisis in very small implants like those for multichannel retinal stimulation
3.	Usable for different stimulation parameters	Anodic duration is uncontrollable for different stimulation parameters. The large blocking capacitor value lengthens the time constant of discharge. Therefore, it is possible that the discharging phase may terminate prematurely. It may reach the stimulation phase

5.6.1 Passive Charge Balancing

5.6.1.1 Blocking Capacitor

A large off-chip DC capacitor in μF range is connected in series with the electrode. By connecting this capacitor, DC currents are stopped from reaching the tissue. Protection against any accidental semiconductor chip failure is also assured. This is because any resulting excessively harmful current cannot flow in the body. It must be remembered that DC current integration will cause voltage buildup on the capacitor. Hence, the capacitor must be regularly discharged.

Among the disadvantages of this technique, it must be mentioned that the value of blocking capacitor has to be the electrode–electrolyte capacitance. Generally, this large capacitor cannot be integrated. It is externally connected with the chip. In multichannel implants, several such capacitors are required. One capacitor is needed for each electrode. A large amount of space is therefore wasted. Lastly, the discharging step for charge balancing is an uncontrolled process.

5.6.1.2 Short-Circuiting of Electrodes

The method is commonly used for a biphasic pulse. It removes any leftover charges present due to mismatching effects. Following each stimulation cycle, short-circuiting of the electrodes helps to discharge them. However, a perplexing situation often arises. It appears because the discharging time is not exactly known. The reason for the inability to know discharging time accurately is the wide variation in current matching and electrode impedances. Further, changes take place in their behavior over a period of time. Nonetheless, the blockingcapacitor issues are steered clear off.

5.6.2 *Active Charge Balancing*

5.6.2.1 Charge Surveillance

This scheme is based on calculation or measurement of charge in the stimulation phase. This knowledge is used in the discharge phase to control its intensity and duration. Thus, the charges in cathodic and anodic phases are compensated.

5.6.2.2 Pulse Insertion

Suppose, after each stimulation phase, the remnant charge on the electrode exceeds a defined safety limit. Then, a corrective short-duration discharging pulse is inserted to remove this charge. This routine is repeated until the electrode charge falls sufficiently to lie within safe bounds.

5.7 Discussion and Conclusions

The criteria for choosing among the three types of stimulation, CCS, VCS, and ChCS, depend on their relative merits and demerits. VCS has the highest power efficiency but less safety. CCS is poor in power efficiency but relatively safe. ChCS has power efficiency between CCS and VCS but is safe like CCS. In charge balancing, human safety is the prime consideration. The main aim of charge balancing is to keep the electrode voltage below the safe limit. It also safeguards against failures.

Review Exercises

5.1 What are monopolar and bipolar electrodes? Describe the paths of flow of current in the two types of electrodes. In which type of electrode does the current flow over a greater region of the body?

5.2 What are monophasic and biphasic waveforms? Which kind of waveform is less effective? Which one requires more energy? Which one can be more damaging to the heart?

5.3 Draw the circuit diagram and explain the operation of a CMOS switch. What is meant by "open switch" and "closed switch" positions in this circuit?

5.4 Name three types of DACs. Describe the working of a binary weighted DAC. What is the reason of its poor accuracy?

5.5 How many resistors are required to build an N-bit $R/2R$ ladder DAC? Discuss the relative requirements of binary weighted DAC and N-bit $R/2R$ ladder DAC for high-achieving accuracy.

5.6 What are the two principal modes in which an R/2R ladder network may be operated as a DAC? In what respects do these modes differ?

5.7 What is an analog-to-digital (ADC) converter? Name two types of ADC circuits.

5.8 What does a flash converter circuit do? Why is it called by this name?

5.9 What is the full form of SAR ADC circuit? Draw its diagram and explain its operation.

5.10 What is a voltage source? Name the dual circuit of the voltage source.

5.11 What are the requirements of the following circuits: (1) a perfect voltage source and (2) a perfect current source?

5.12 How does a current source differ from a current sink?

5.13 What is a current mirror? By what other name is this circuit called?

5.14 Draw the circuit diagram of a MOSFET-based current mirror circuit and explain its working.

5.15 Write two alternative names of a current-to-voltage converter circuit. Explain the working of a VIC circuit using OP-AMP.

5.16 What is the task performed by a voltage multiplier circuit? Is it theoretically possible to increase the voltage indefinitely? What is its practical limitation? Up to what limiting factor is voltage multiplication generally done?

5.17 How is a full-wave voltage doubler circuit constructed from two half-wave rectifier circuits? Explain the working of a half-wave voltage doubler circuit.

5.18 A boost converter provides a higher voltage than the input voltage. What about the output current?

5.19 Draw the circuit diagram of a boost converter circuit and describe its operation. How does an IC boost converter work?

5.20 Describe two approaches used for making a current-mode stimulation circuit. What is the reason for safety in implementation of a current-mode stimulation circuit?

5.21 Describe two schemes for making a voltage-mode stimulation circuit. Why is this mode said to be less safe as compared to the current-mode circuit?

5.22 How does charge-mode stimulation combine the beneficial features of current-controlled and voltage-controlled stimulation circuits? What is the main shortcoming preventing its extensive utilization?

5.23 Why should any charge unbalance be stopped during electrical stimulation? How does the use of a biphasic pulse help in this regard?

5.24 Explain the concept of charge balancing in electrical stimulation. Discuss the main features of active and passive charge balancing. Point out any similarities and dissimilarities between the two methods.

5.25 How is a blocking capacitor used for charge balancing? What protection does it provide against semiconductor chip failure? What is the main limitation of this method?

5.26 How is charge balancing done by short-circuiting electrodes? What is the main difficulty of using this method?

5.27 Describe the charge surveillance and pulse insertion methods of charge balancing.

References

1. Scholten M, Szili-Torok T, Klootwijk P et al (2003) Comparison of monophasic and biphasic shocks for transthoracic cardioversion of atrial fibrillation. Heart 89:1032–1034
2. Storr W (2014) Voltage multiplier—basic electronics tutorials. http://www.learnabout-electronics.org/PSU/psu32.php. Accessed 26 Apr 2015
3. Coates E (2014) Learn about electronics: power supplies: module 3: switched mode power supplies. http://www.learnabout-electronics.org/PSU/psu30.php. Accessed 14 Sept 2014
4. Simpson J, Ghovanloo M (2007) An experimental study of voltage, current, and charge controlled stimulation front-end circuitry. In: IEEE international symposium on circuits & systems, ISCAS 2007, New Orleans, LA, 27–30 May 2007, pp 325–328
5. Valente V, Demosthenous A, Bayford R (2012) A tripolar current-steering stimulator ASIC for field shaping in deep brain stimulation. IEEE Trans Biomed Circuits Syst 6(3):197–207
6. Liu X, Demosthenous A, Donaldson N (2008) An integrated implantable stimulator that is fail-safe without off-chip blocking-capacitors. IEEE Trans Biomed Circuits Syst 2(3):231–244
7. Kelly SK, Wyatt J (2004) A power-efficient voltage-based neural tissue stimulator with energy recovery. In: IEEE international solid-state circuits conference ISSCC 2004, 17 Feb 2004, 10 pp
8. Kelly SK, Wyatt JL Jr (2011) A power-efficient neural tissue stimulator with energy recovery. IEEE Trans Biomed Circuits Syst 5(1):20–29
9. Ghovanloo M (2006) Switched-capacitor based implantable low-power wireless microstimulating systems. In: Proceedings of IEEE international symposium on circuits and systems, ISCAS 2006, pp 2197–2200

Chapter 6
Implant Clocking and Timing Circuits

Abstract A clock circuit is a redstone circuit (a structure built to activate or control mechanisms). It produces a repetitive pattern of pulses referred to as a clock signal. A clock generator IC can perform various functions. The functions include the generation, conditioning, manipulation, distribution, or control of a timing signal in an electronic system. The output of clock generators is toggling on/off constantly. To build clock generator circuits, a number of oscillator designs exist. These circuits produce sinusoidal or square waves. The frequencies of the waves range from < 1 Hz to several MHz. A linear or sinusoidal wave RC oscillator consists of a tuned RC network connected to regenerative feedback amplifier. Sustenance of steady oscillations requires the fulfillment of Barkhausen's criterion stating that the loop gain at the oscillation frequency must be unity in absolute magnitude and the total phase shift around the loop should be zero or an integral multiple of 180°. RC oscillators for producing waveforms of square, triangular, or sawtooth shapes are often called relaxation oscillators. The term "relaxation" is related to the charging and discharging states of a capacitor. Many types of relaxation oscillators have been built, starting from simple two-transistor multivibrator circuits and advancing to more intricate circuit topologies. All these oscillators work on the same operating principle. The generated square waves are used to control the timing of operations in digital systems, such as clock generators for microprocessors. These different circuits are known by a variety of names, such as monostable, astable, and bistable multivibrators. The multivibrator functions can be easily implemented with timer integrated circuits. These circuits are therefore extensively used in timing applications.

Keywords Linear oscillator · Nonlinear oscillator · Crystal oscillator · Monostable/astable/bistable multivibrator · Timer IC · Time delay · Missing pulse detection · Frequency division · PWM · PAM

V. K. Khanna, *Implantable Medical Electronics*,
https://doi.org/10.1007/978-3-031-99868-3_6

Table 6.1 Linear and nonlinear oscillators

S. No.	Linear oscillator	Nonlinear oscillator
1.	It is also called sinusoidal or harmonic oscillator	It is also known as nonsinusoidal or relaxation oscillator
2.	It produces sine wave signals	It produces rectangular, square, triangular, or sawtooth wave signals
3.	It consists of an amplifier and a circuit providing positive feedback to produce phase shift	It consists of an RC timing circuit and a device such as transistor or OP-AMP. The transistor/OP-AMP changes states to charge and discharge a capacitor through a resistor, in alternation

6.1 Introduction

Oscillators are electronic circuits whose distinctive trait is that they provide periodic waveforms on their output terminals. But they use only the DC supply voltage at their input terminals. Depending on the type of oscillator, the output voltage can be shaped like a sine wave, hence said to be sinusoidal oscillator, or a shape different from it, hence known as a nonsinusoidal oscillator (Table 6.1). A kind of oscillator termed the feedback oscillator consists of an amplifier for obtaining gain. A second type of oscillator is the relaxation oscillator. It uses an RC timing circuit to produce a waveform. This waveform is normally a square wave or some other waveform falling under the nonsinusoidal category. A timer is an automatic mechanism for controlling the sequence of an event. In its simplest form called a register, the timer waits until a certain time interval has elapsed. Then it fires and sends a specified message to an object. When a timer is signaled, the processor must run to process the associated instructions. Timer ICs are widely deployed for making oscillator and pulse generator circuits.

6.2 Clock Generators

The clock is the vital unifying force of any processor. It is an oscillator circuit producing a waveform as under: high voltage, low voltage, high voltage, low voltage, etc., or one, zero, one, zero, etc. The signal produced by the clock is usually a constant-frequency square wave with 50% duty cycle. Its speed is given in megahertz (MHz) or gigahertz (GHz). The clock signal coordinates and synchronizes various parts of the processor circuit. It regulates the rate at which the processor executes the instructions. Every time the clock changes level, the processor executes one instruction. This coordinating action is similar to a metronome. The metronome is a practice tool which plays steady beats to help musicians in playing their instruments rhythmically.

Clock sources for microprocessors/microcontrollers are of two types: (1) those based on mechanical resonance devices like crystals, showing a maximum amplitude of oscillation at a resonant frequency and (2) those using electrical phase-shift circuits, e.g., resistor–capacitor (RC) networks. Crystal-based oscillators usually make available extremely high starting accuracy along with reasonably low thermal sensitivity. RC oscillators are less costly. But they offer reduced accuracy with varying temperatures and supply voltages. They also suffer from frequency changes in the range of 0.05–0.5 fraction of the supposed output frequency [1].

6.3 Oscillator Circuits

6.3.1 Crystal Oscillator (XO)

This is an electronic oscillator based on inverse piezoelectric effect, viz., deformation of a crystal by a voltage applied across it (Table 6.2). The frequency-determining component of this oscillator is an appropriately cut piezoelectric crystal of desired shape, size, and thickness. The oscillator utilizes the electromechanical resonance of the vibrating crystal to produce an electrical signal of precise frequency with a high stability and quality factor. The stability and Q-factor are independent of temperature or load variations and DC supply fluctuations. Electrically, the crystal is modeled as a series-connected inductance–capacitance–resistance (L–C–R) branch in parallel with a capacitor (Fig. 6.1). The L–C–R segment represents the inertia, friction, and stiffness of the crystal. The shunt capacitor stands for the self-capacitance of the crystal. A crystal oscillator uses the same circuit as an *LC* (inductance–capacitance) oscillator. The crystal replaces the *LC* tuned circuit. The oscillations are sustained by amplifying the voltage signal from the crystal resonator. The signal is fed back to the resonator to provide positive feedback.

There are many types of piezoelectric materials that are used in oscillator circuits. Among these materials, quartz minerals occupy an unmatched position. The uniqueness is because of their high mechanical strength. The quartz crystal used in a crystal oscillator is a very small, thin piece or wafer of quartz. The two parallel surfaces of this wafer are metalized for the electrical connections.

The quality factor Q for a quartz oscillator (10^4–10^6) is much higher than that for an LC oscillator (10^2). The oscillation frequencies range from a few kilohertz up to several hundred megahertz.

Square wave clock oscillators are of Pierce type. They are made by using a crystal with a digital inverter circuit. Two resistors along with two capacitors are included in the circuit. The quartz crystal acts as an exceptionally selective filtering component. These oscillators will be described in Sect. 6.3.3.

Table 6.2 Piezoelectric and inverse piezoelectric effects

S. No.	Piezoelectric effect	Inverse piezoelectric effect
1.	On applying a stress along any pair of opposite faces of a crystal of certain substances called piezoelectric materials, electric charges of opposite polarity are induced on the opposite faces of the crystal perpendicular to the direction of applied stress. Hence, an electrical potential difference is generated across the opposite faces of the crystal by the applied stress	When a potential difference is applied to the opposite faces of a crystal of certain substances known as piezoelectric materials, the crystal experiences mechanical strain in the direction perpendicular to the applied potential difference. Hence, a strain or deformation is induced in the crystal by the application of an alternating potential difference
2.	It involves conversion of mechanical energy into electrical energy	It involves conversion of electrical energy into mechanical energy
3.	Examples: quartz (SiO_2), Rochelle salt ($KNaC_4H_4O_6$, $4H_2O$), tourmaline, and lead titanate zirconate (PZT), $Pb(Zr_xTi_{1-x})O_3$ ($0 \le x \le 1$) ceramics	Examples: quartz (SiO_2), Rochelle salt ($KNaC_4H_4O_6$, $4H_2O$), tourmaline, and lead titanate zirconate (PZT), $Pb(Zr_xTi_{1-x})O_3$ ($0 \le x \le 1$) ceramics

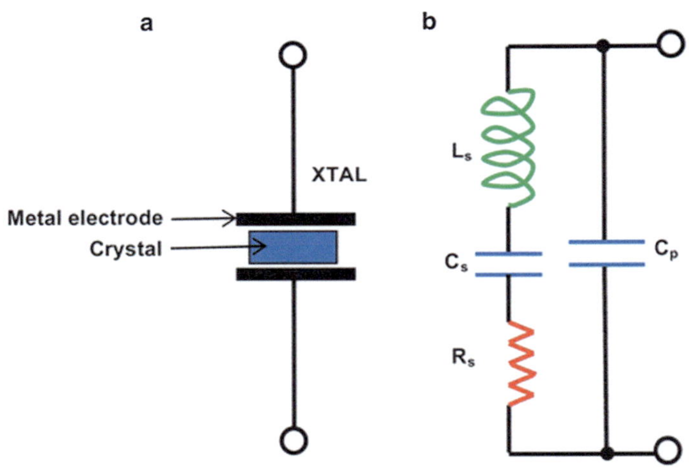

Fig. 6.1 Crystal: **a** schematic symbol and **b** series–parallel LCR equivalent circuit

6.3.2 *Resistance–Capacitance Oscillator*

It is an oscillator using a resistance–capacitance (RC) network for frequency selection. An operational amplifier connected in "inverting amplifier" configuration with three-section RC network may be used. An alternative name for RC oscillator is "phase-shift oscillator." Its output signal has the shape of a sine wave. The phase difference between each RC section and the preceding RC section is 60°. Therefore, the phase difference between the output of three-section RC network and its input

$= 3 \times 60° = 180$. When the 180° phase shift produced by the amplifier is added to 180° phase shift due to the RC circuit, a total of $180° + 180° = 360°$ phase shift is obtained. The 360° phase shift causes positive or regenerative feedback for oscillation.

The operation of RC oscillator is dependent on the provision of positive feedback by the RC network. This positive feedback is derived from the rudimentary property of a capacitor to act as a storehouse of electric charge. The values of resistive and capacitive components determine the frequency of the output waveform of the RC oscillator. This oscillator is suitable for and finds applications in circuits designed for low-frequency (audio range) and moderate-frequency (5 Hz–1 MHz) applications.

To generate a square, rectangular, or any other waveform of nonsinusoidal shape, a relaxation oscillator employs an *RC* timing circuit and a variable-state device, i.e., one which changes states. This device could be an OP-AMP. A relaxation oscillator is a circuit performing a repetitive operation. It acquires its repetitive behavior from the acquisition of charge by the conductive plates of a capacitor to reach a threshold limit. After reaching the limit, the capacitor discharges. Its recharging time determines the repetition time. The square wave generator shown in Fig. 6.2 is a relaxation oscillator. It essentially functions on the phenomena of successive charging and discharging of a capacitor. The potential produced on the capacitor is connected to inverting input of the OP-AMP. A fraction of the output voltage is supplied back via the resistors R_2 and R_3 to the noninverting input. To begin with, at the time of turning the circuit on, the capacitor is in uncharged state. The inverting input of the OP-AMP is at 0 V. This pushes its output at maximum positive voltage, i.e., high-logic level. So, the capacitor is charged to the voltage V_{out} through the resistor R_1. When the capacitor attains a voltage V_{c} = feedback voltage V_{f} on the noninverting input, the OP-AMP output reaches the utmost negative voltage, i.e., low-logic level. Now the capacitor discharges from voltage $+ V_{\text{f}}$ to $- V_{\text{f}}$. With the capacitor voltage attaining the value $= - V_{\text{f}}$, the OP-AMP again changes to high-logic state. By the continuous rhythmic sequence of the above cycles, a square wave output voltage waveform is generated.

6.3.3 Crystal-Based CMOS Square Wave Oscillators

Square wave oscillators made from CMOS logic components are highly stable against power supply variations. They perform reliably over a wide range of supply voltages (3–15 V) and a broad range of frequencies (< 1 Hz–15 MHz). Besides, they offer low power consumption and easy interfacing with other logic families [2].

The stability of the CMOS RC oscillator is sufficient for the major chunk of applications. However, for more demanding precision applications, crystal oscillators are essential (Table 6.3). A large number of applications in several diverse designs are implemented with these fairly simple, easy to build, reliable, inexpensive, and adaptable RC or crystal-based CMOS oscillators.

Crystal oscillator circuits apply the gated Pierce design [3]. In this design, the oscillator is built around one/more CMOS inverting gate(s). The Pierce design has

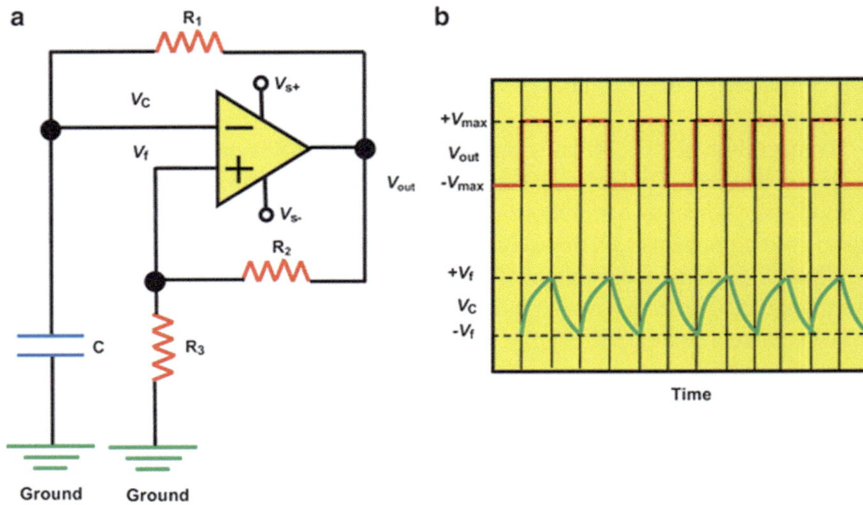

Fig. 6.2 Square wave relaxation oscillator: **a** circuit and **b** waveforms

Table 6.3 Crystal and RC oscillator [1]

S. No.	Property	Crystal oscillator	RC oscillator
1.	Precision	High	Low
2.	Temperature stability	High	Low
3.	Cost	High	Low
4.	EMI, dampness	Sensitive but insensitive in an oscillator module	Sensitive
5.	Vibration	Sensitive	Sensitive
6.	Applications	For precision applications, no other material/technology has been able to replace the quartz crystal so far	Suitable for less critical applications

many desirable features [4]. It steers the crystal at a low power level. Hence, less power is dissipated especially at high frequencies. Nevertheless, it provides a large output signal at a constant frequency. The signal is relatively independent of power supply and temperature changes. Apart from its low cost, the oscillator is usable at any frequency. Its only disadvantage is the mandatory requirement of a high-gain amplifier in the circuit. This is necessary to make up for the appreciably large losses of gain incurred in the circuit adjoining the crystal.

A crystal oscillator may be constructed using only one CMOS inverter as the active component. An inverter comprising one P channel and one N channel enhancement-mode MOSFET is usually referred to as an unbuffered inverter. It may be noted that it is not essential to use only a single gate. Any odd number of inverting logic gates will oscillate if they are knotted together in the form of a ring. It is also feasible to use a buffered inverter made of three P–N MOSFET pairs connected in series. But the associated gain, typically approximately many thousands, provides a less stable oscillator.

A practical oscillator circuit comprises the unbuffered inverter, two capacitors, two resistors, and the quartz crystal (Fig. 6.3). To understand the operation of this oscillator, the CMOS inverting gate must not be considered as a logic device working with 1s and 0s. Instead, it must be looked upon as a linear amplifier. It has gain, phase, and propagation delay limitations. For the inverter biased in its linear region, a small change of the input voltage is amplified by the gain factor. It appears as a larger change in the output voltage. In order that this biased inverting gate is used as an oscillator, it must have adequate gain to surmount the losses of the feedback network including the capacitors, the gain limiting resistor, and the quartz crystal. It must also have ample negative resistance at the frequency of oscillation to prevail over the equivalent series resistance of the crystal. Further, the net phase shift around the circuit must be 360°. All CMOS inverting gates are characterized by parameters such as an input capacitance, an output capacitance, an output resistance, and a propagation delay. These parameters affect the choice of capacitors and the gain limiting resistor. They eventually affect the upper operational frequency of the oscillator.

A useful oscillator is made with three inverters (Fig.6.4). The duty cycle is ~ 50%. The frequency of oscillation is determined by the total propagation delay through the ring. The lesser the number of inverters used, the higher is the frequency. The oscillator is nearly insensitive to power supply variations. This is mainly due to the threshold tracking ~ 50% of the supply voltage. Its stability is determined by the frequency of oscillation. The lower the frequency, the more stable is the oscillator

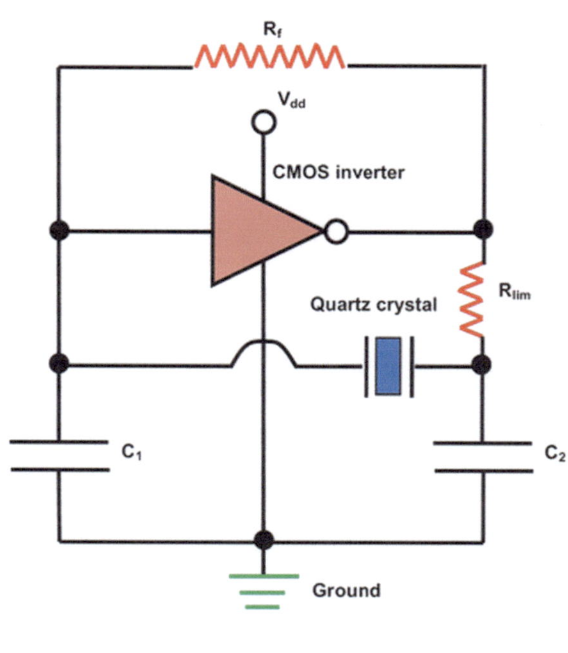

Fig. 6.3 Practical oscillator circuit using CMOS inverting gate

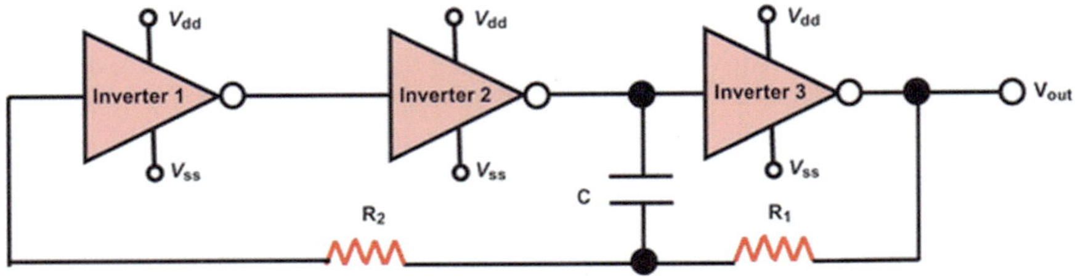

Fig. 6.4 Three-inverting gate RC oscillator

and conversely. This is because propagation delay and the effect of threshold changes constitute a minor portion of the total period.

6.3.4 Multivibrator Circuits Using Logic Gates

Multivibrators (Table 6.4) are called by various names, such as nonlinear oscillators or function generators. Their output waveforms are square or rectangular in shape. These waveforms look like pulse trains [5, 6]. From the circuit viewpoint, the multivibrator circuits basically consist of two amplifier sections. These sections are arranged with regenerative feedback. Multivibrators are of three types: monostable, astable, and bistable. These types are described below.

Table 6.4 Multivibrators

S. No.	Property	Monostable multivibrator	Astable multivibrator	Bistable multivibrator
1.	Input	One triggering pulse required	No triggering pulse required	Two triggering pulses required for one complete cycle
2.	Number of stable states	One; hence called single shot multivibrator	None; hence called free-running multivibrator	Two
3.	Number of quasi-stable states	One	Two	None
4.	Output states	One stable, one quasi-stable	Two, both quasi-stable	Two, both stable
5.	Applications	Timing circuits. These circuits are used for producing a timing interval of fixed length in retaliation to some external occurrence, pulse train generation, etc.	Clocking, e.g., square wave generators	Flip-flop or latch, memory element, divide-by-2 counters

6.3.4.1 Monostable Multivibrator

A monostable multivibrator (Fig. 6.5) is also called by another name, "one-shot multivibrator." A common name is "flip-flop." It is a circuit in which one state of the output signal is stable. But the other state of this signal is unstable; hence, it is said to be a monostable multivibrator. On applying a proper trigger signal from outside or a pulse signal T, a timing cycle is kicked off. The induction of this cycle brings about a change of state of the output signal of the multivibrator. The change of state occurs at the start of the timing cycle, at an instant of time t_1. The multivibrator stays in this second state for a period of time determined by the timing cycle. It changes its state only after the termination of the cycle, at time instant t_2. The period of time $(t_2 - t_1)$ is determined by the time constant $R_T\ C_T$ of the resistor–capacitor combination, which is the product of the timing capacitor C_T and the resistor R_T.

The second state is maintained by the multivibrator until the end of a period of time $= R_T\ C_T$ time constant. Thereafter, the multivibrator resets or reverts to its stable state at the beginning without any human intervention. This kind of multivibrator finds application in the conversion of short sharp pulses into wider ones for timer circuits.

Initial State

- To begin with, let the resistor R_1 clasp the trigger input signal T at high-logic level. Then the output signal from the NAND gate G_1 is an inverted form of trigger signal, i.e., low-logic level. It may be noted that the timing resistor R_T is joined to a voltage level = low-logic level. This causes the capacitor C_T to discharge through the resistor R_T. The output of G_1 being low, the discharging of the timing capacitor C_T continues until it is completely discharged. The junction J is at a voltage level = low-logic level, wherefore the output from the NAND gate G_2, connected as an inverting NOT gate, is at high-logic level. This output is applied to one input of gate G_1. The needful positive feedback is thus provided. The junction J and the output of G_1 are both at low-logic level. For this reason,

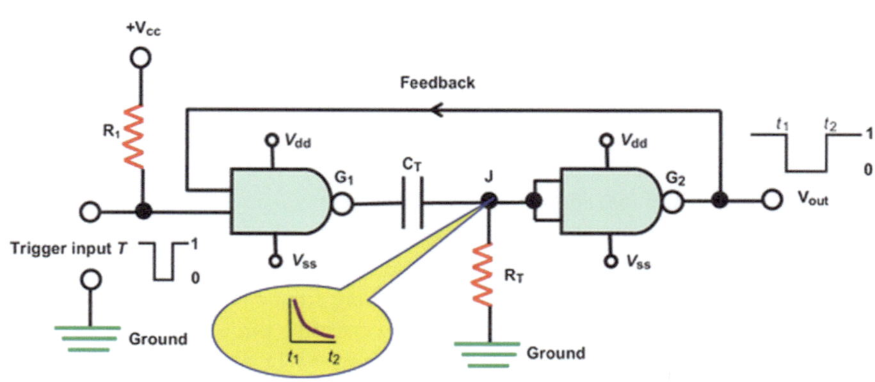

Fig. 6.5 NAND gate monostable multivibrator circuit

current cannot flow to the capacitor C_T. Consequently, the circuit maintains its status quo. It prolongs in this state placidly, pending any change in the trigger input signal T. It may thus be said to be a waiting state.

Triggered State

1. *Metastable State*: This state commences with the application of a negative pulse to the trigger input signal of the NAND gate G_1. As a result, the output of G_1 goes to high-logic level. An instantaneous change in voltage across the capacitor is not possible. That being so, the junction J and the input to G_2 both climb to high-logic level. This makes the output of the NAND gate G_2 to shift to low-logic level. Even when the trigger input pulse T is disconnected, the circuit remains in this second state. This second state is known as the metastable state. It is a seemingly equilibrium state of short duration, which can rapidly change to a more stable state when slightly disturbed. Theoretically, it is an unstable state but is sufficiently long-lived practically to be called stable.
2. *Stable State*: The capacitor C_T begins to charge up from the output of G_1. This charging takes place at a time constant determined by the resistor–capacitor combination. Due to this charging, the voltage across the capacitor increases. This charging process extends until the charging current is able to bring the input of G_2 at low-logic level. Therefore, it continues until the junction J is at low-logic level. No sooner than this happens, the output of G_2 switches again to high-logic level. On this account, the output of G_1 goes low. The capacitor discharges through resistor R_T into the output of G_1. The circuit has now switched backwards to its original stable state.

Effectively, the multivibrator circuit produces an output pulse at low-logic level corresponding to each negative-going trigger pulse. The values of capacitance C and resistance R of capacitor–resistor network dictate the time period of output signal. The time period is given as the time constant $\tau = 0.69\ RC$ of the circuit in seconds. The input impedance of a NAND gate is very large. In consequence, high values of timing periods are achievable.

6.3.4.2 Astable Multivibrator

Astable multivibrator (Fig. 6.6) is a circuit, which does not have any stable state. It is unceasingly oscillating from one state to the other. It is a type of regenerative oscillator. It consists of two amplifying devices. These two devices are cross-coupled by resistors and capacitors. No external triggering is necessary for this type of multivibrator. In the absence of any external signal, it switches automatically between the two states at a certain frequency. The switching frequency is controlled by the (resistance $\times$ capacitance) time constants of the coupling circuits.

In the beginning, let the output signal from the NAND gate G_2 be at high-logic level. In this condition, one can easily surmise that the input signal of G_2 will be at low-logic level. The output from the NAND gate G_1 will also be at low-logic level.

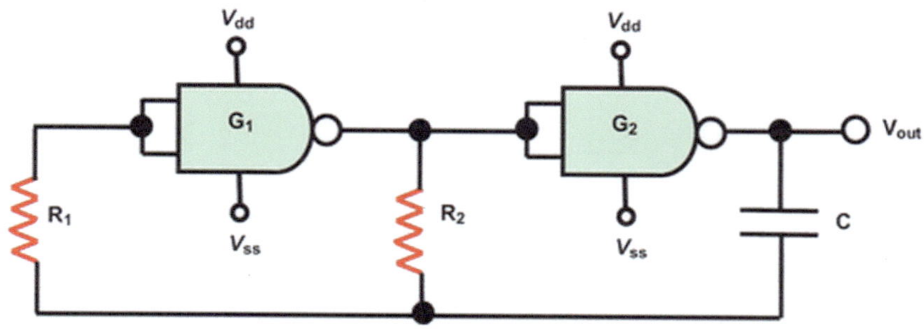

Fig. 6.6 NAND gate astable multivibrator circuit

Looking at the circuit diagram, one finds that capacitor C is connected between the output terminal of the NAND gate G_2 and its input terminal through the timing resistor R_2. Hence, this capacitor starts to store charge across its plates. The charging rate is decided by the time constant R_2C. With the progress in charging of capacitor C, there is a relentless decrease in potential of the meeting or convergence point between the resistor R_2 and the capacitor C. This meeting point is also joined to the input of the NAND gate G_1 through the stabilizing resistor R_2. The above decrease in potential cannot proceed indefinitely. It lasts as long as the lower threshold value of G_1 is arrived at. On reaching this threshold point, the gate G_1 alters its state. Its output becomes high-logic level. This causes NAND gate G_2 to change state. The change comes about because its input has now changed from low-logic level to high-logic level. A shift of the output of NAND gate G_2 to low-logic level takes place. Capacitor C is now biased in the opposite direction. The capacitor loses charge by discharging through the input terminal of NAND gate G_1.

Once again, the capacitor C begins to charge, but now it charges in the reverse direction. The rate of charging is determined by the time constant R_2C, as before. The charging of capacitor is continued until it attains the upper threshold value of NAND gate G_1. Consequently, gate G_1 changes state. The full cycle of events replicates all over again. The time constant for the astable multivibrator is $\tau = 2.2\,R_2C$ in seconds.

6.3.4.3 Bistable Multivibrator

A bistable multivibrator (Fig. 6.7) is a circuit having two stable states. These states are designated as high and low. This multivibrator is stable in both the states. It remains in either state for an unlimited period. For toggling or flipping between the high and low states of the output, an outside trigger pulse signal T is required. To undergo a complete SET–RESET sequence, two triggering pulse signals must be supplied. It finds application in flip-flops.

The circuit uses Schmitt trigger NAND gates represented by the symbols shown in the diagram. The special feature of the Schmitt NAND gate is the provision of two different input logic thresholds. It has two input voltage levels. One is the input

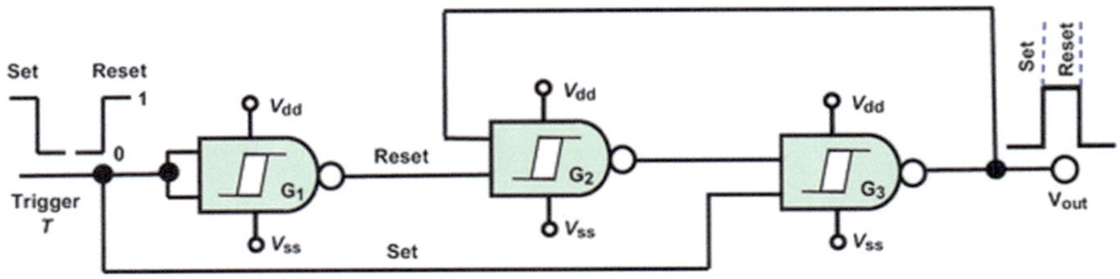

Fig. 6.7 NAND gate bistable multivibrator circuit

voltage level causing a positive-going change from logic 0 to 1 at the output of the gate. The other is the input voltage level causing a negative-going change from logic 1 to 0. The two input voltage levels are different.

To start with, let the input trigger signal be at a low-logic level. Clearly, the output signal is at a high-logic level. This is the SET state. During this state, the output signal of NAND gate G_1 lies at high-logic level. Since both the input signals of G_2 are at high-logic level, the output signal is obviously at low-logic level. Inasmuch as both the input signals of G_3 are at a low-logic level, the output signal is at high-logic level.

Considering the input trigger signal as high-logic signal, for the RESET state, the same operation manifests itself. The circuit alters state with its output signal at low-logic level. The output signal sojourns in this RESET state. Application of another input trigger pulse changes its state. Then the complete cycle starts another time passing through the aforementioned series of rungs.

6.4 Timer ICs and Timing Circuits

Implantable circuits are preferably built-in monolithic IC form. However, for the purpose of convenience, their operation can be easily understood by following a building block approach using timer ICs. The 555 timer IC was designed in 1971 by Hans R. Camenzind. The design was completed under contract to Signetics. The timer IC is a highly stable device. It is used in a variety of accurate timer, pulse generator, and oscillator applications [7, 8]. Its output is one of the three types: (1) monostable mode, which functions as a one-shot multivibrator and is used in timers and for missing pulse detection; (2) astable mode, which functions as a free-running mode and is used as an oscillator for pulse generation; and (3) bistable mode, which is used as a flip-flop. Applications of timer IC in controlling and generation of pulses are listed in Table 6.5.

Table 6.5 Control and generation of pulses by timer IC

S. No.	Task	Mode	Function
1.	Control	Monostable multivibrator	Time delay inserter
2.	Control	Monostable multivibrator	Frequency divider
3.	Control	Monostable multivibrator	Missing pulse detector
4.	Control	Monostable multivibrator	Pulse-width modulator
5.	Generation	Astable multivibrator	Pulse generator
6.	Control	Astable multivibrator	Pulse amplitude modulator
7.	Control	Astable multivibrator	Pulse position modulator

6.4.1 Block Diagram

The functional block diagram of 555 IC is shown in Fig. 6.8. It includes two comparators (threshold and trigger), an R-S flip-flop, two transistors, and a power output stage. Also seen are three 5 kΩ resistors in series connection. The use of trio 5 kΩ resistors is a likely reason for the code number 555. These resistors produce 1/3 and 2/3 V_{cc} levels (V_{cc}, supply voltage). These levels are used for monitoring the function of trigger and threshold comparators. The two comparators run the flip-flop. The flip-flop controls the output states, on or off.

6.4.2 Pin Diagram

A standard 555 IC is housed in 8-lead dual-in-line package (DIL-8 or DIP-8). It contains more than 20 transistors. In addition, there are 2 diodes and 15 resistors. These components are fabricated on a silicon chip. The pins (Fig. 6.8) are named as follows:

1. Pin 1 (Ground): It is connected to the negative power supply or battery terminal.
2. Pin 2 (Trigger): It triggers when the voltage at this pin is < 1/3 V_{cc}. Triggering voltage = 1.67 V, 5 V for V_{cc} = 5 V, 15 V, respectively; triggering current = 0.5 μA for 0.1 μS.
3. Pin 3 (Output): It sources or sinks a maximum current of 200 mA. Sourcing means that the output signal of the IC is at high-logic level. So, it can give out current. Sinking means that the output signal of the IC is at low-logic level. So, it can absorb current.
4. Pin 4 (Reset): It is connected to positive terminal of the supply for normal operation of the IC. When it is grounded, the IC stops working. Required rest voltage is typically 0.7 V at 0.1 mA.
5. Pin 5 (Control voltage): To this pin, the 2/3 V_{cc} point on the terminal voltage divider is brought out. Use of this terminal gives an option to the user for the

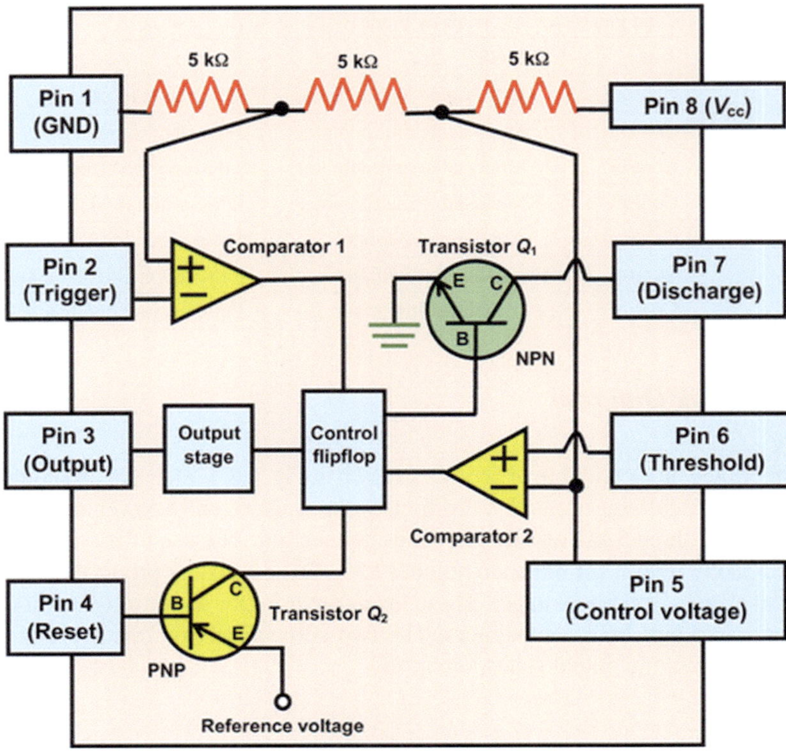

Fig. 6.8 Block and pin diagram of 555 timer along with internal circuit

connection of an external DC control voltage. The connection is able to modify the timing cycle.

6. Pin 6 (Threshold): It accepts positive-going pulse to end the timing cycle. The threshold current is typically 0.1 mA.
7. Pin 7 (Discharge): It discharges the external capacitor into itself. For the safety of internal transistor, the discharging current must be < 50 mA.
8. Pin 8 (+ V_{cc}): It is the positive supply terminal of the IC.

6.4.3 Monostable Mode for Timer or Time Delay Function

The timer is an electrical circuit used for producing a delay interval after which an electrical load is set going. Generally, this delay interval is easily changeable at the discretion of the user.

In Fig. 6.9, let us start with the condition in which the external capacitor has been discharged by the transistor inside the timer IC. So, the output terminal is at a low-logic level. Suppose a negative pulse < 1/3 V_{cc} is applied to Pin 2, the trigger pin. By this action, the internal flip-flop is set. This happens because the threshold on the lower internal comparator is surpassed. As a result, the internal transistor T_1 is cut off. The short circuiting across the external capacitor C is released. The release of short circuiting gives rise to a positive-going output pulse. This pulse drives the output to high-logic level.

The capacitor C charges through the external resistor R for a period of time $T = 1.1\ RC$. That being the case, the voltage across this capacitor increases at an exponential rate. When the capacitor voltage reaches a potential of 2/3 V_{cc}, the upper internal comparator is triggered. This triggering resets the flip-flop. Now, the internal transistor T_1 becomes conducting. This discharges the capacitor C. Consequently, the output is driven to low-logic level.

It is worth noting that for the condition in which the output signal is at high-logic level, any further triggering pulse does not influence the circuit. If left unused, the reset pin should be connected to V_{cc} terminal. This is required for avoidance of false triggering. Further, during the period that the output signal is at high-logic level, the circuit is resettable by means of a negative pulse applied to Pin 4, the reset pin. Then the output remains at low-logic level. This level lasts until the application of another trigger pulse. It must also be emphasized that the timing interval is independent of supply voltage. This is because of the direct variation of charge and the threshold level of the comparator with this voltage. The output waveforms are illustrated in the figure. To use this circuit for introducing various time delays, the R and C values must be changed. The timing can be altered as desired by changing either the value of the capacitor or the resistor. The timing changes according to the formula $T = 1.1$ R (ohms) $\times$ C (farads), mentioned above. The lower limit is 10 μs.

6.4.4 Monostable Mode for Frequency Division

The monostable circuit described in the preceding subsection can be used as a frequency divider. For operation as a frequency divider, the extent of the timing cycle is changed. The waveforms produced by a divide-by-two circuit are shown in Fig. 6.10.

6.4.5 Monostable Mode for Missing Pulse Detection

In a regular train of pulses, it may sometimes turn out that a pulse is missing. Then the time gap between two consecutive pulses will be unusually long. Such an abnormal behavior in a pulse train may occur in the ECG of a heart patient. It must be carefully detected. To detect such an abnormality, the resistor and/or capacitor values in a monostable multivibrator circuit are chosen according to the situation. The component values are selected such that the spacing between successive pulses is 30–40% less than the delay or timing interval of the multivibrator. Then the train of pulses being examined is fed to the Pin 2, the trigger pin of the multivibrator whose output

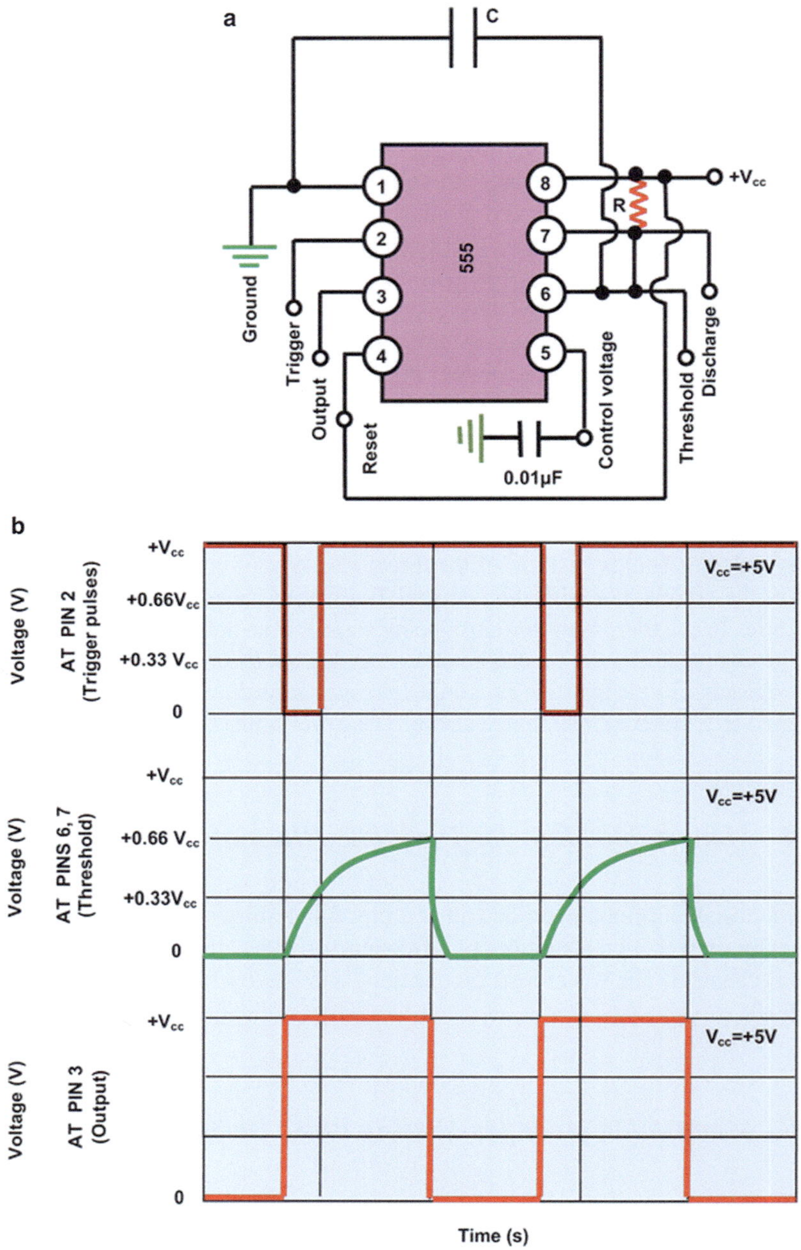

Fig. 6.9 Monostable multivibrator: **a** circuit using 555 timer IC and **b** waveforms

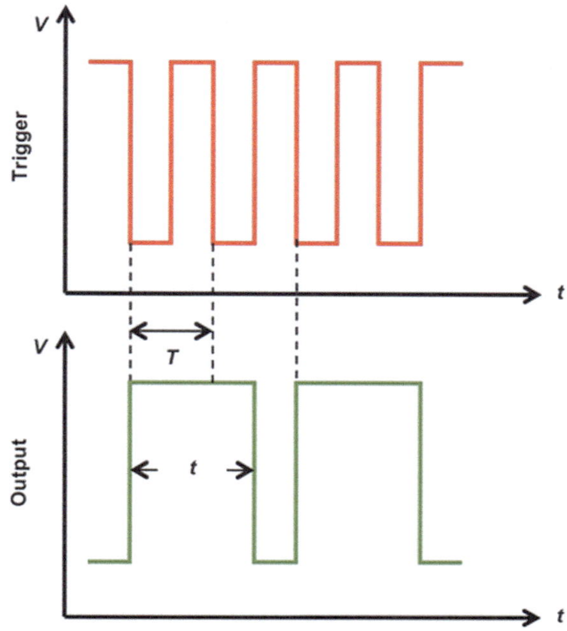

Fig. 6.10 Waveforms of a divide-by-two circuit in which the timing interval *t* is kept slightly longer than the time period *T* of trigger signal

is at a low-logic level (suppose). The circuit is shown in Fig. 6.11. As long as the spacing between the pulses is less than the timing interval, the multivibrator is continuously reset by the input pulses. This resetting is accompanied by the output rising to a high-logic level. Thus these incoming pulses are continuously retriggering the multivibrator. The output continues to show a high-logic level. However, when a pulse is missing in this train, the triggering that was so far taking place will not be possible. Therefore, the absence of a pulse will allow the completion of time interval without retriggering. As there was no retriggering during this period, the output will remain in low-logic level. This anomaly in output will enable the detection of the missing pulse.

6.4.6 Pulse-Width Modulation

The pulse-width modulation (PWM) is often termed pulse duration modulation (PDM). It is a digital modulation technique. In this technique, the width of pulses is made to vary in concordance with the modulating voltage. In this modulation, the leading edge of the pulses is kept fixed. The occurrence of the trailing edge is variable. Suppose the timer IC is connected in monostable mode. Let it be subjected to triggering under a train of pulses (Fig. 6.12). A triangular wave oscillator is connected to the control voltage Pin 5 of the IC. This wave controls the voltage which ends the timing intervals. Internally the control voltage changes the reference voltage of

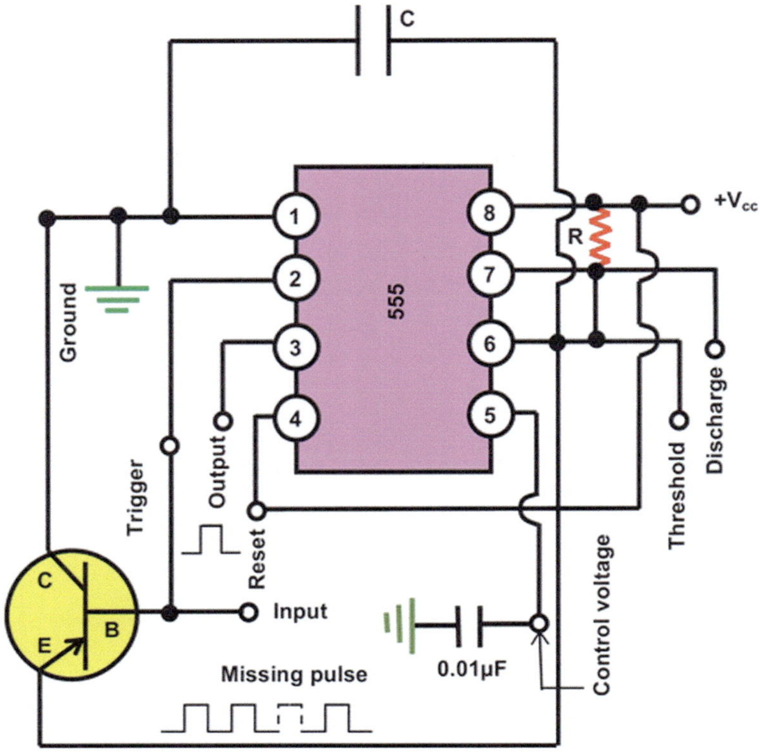

Fig. 6.11 Missing pulse detection circuit. A missing pulse prevents retriggering until a cycle is completed. This causes Pin 3 to go low until the arrival of a new pulse

comparator. So, charging time of capacitor changes every time. Hence, the widths of output pulses also change. The output waveforms are shown in the figure.

6.4.7 Astable Mode for Pulse Generation

For operation as an astable multivibrator, the Pins 2 and 6, the trigger and the threshold pins of the IC, are connected together. They are also connected to the external capacitor, as shown in Fig. 6.13. The external capacitor charges through the resistor combination ($R_A + R_B$). As the discharge Pin 7 connected to the internal transistor is joined to the junction of R_A and R_B, the external capacitor discharges through resistor R_B.

When the circuit is switched on, the capacitor has no charge over its plates. So, the potential of the trigger and threshold pins is 0 V. The flip-flop is set by the lower comparator. Hence, the output becomes high. The internal transistor T_1 is turned on. Charging of the external capacitor starts via ($R_A + R_B$). When the capacitor potential reaches a potential 2/3 V_{cc}, the upper comparator triggers. This triggering

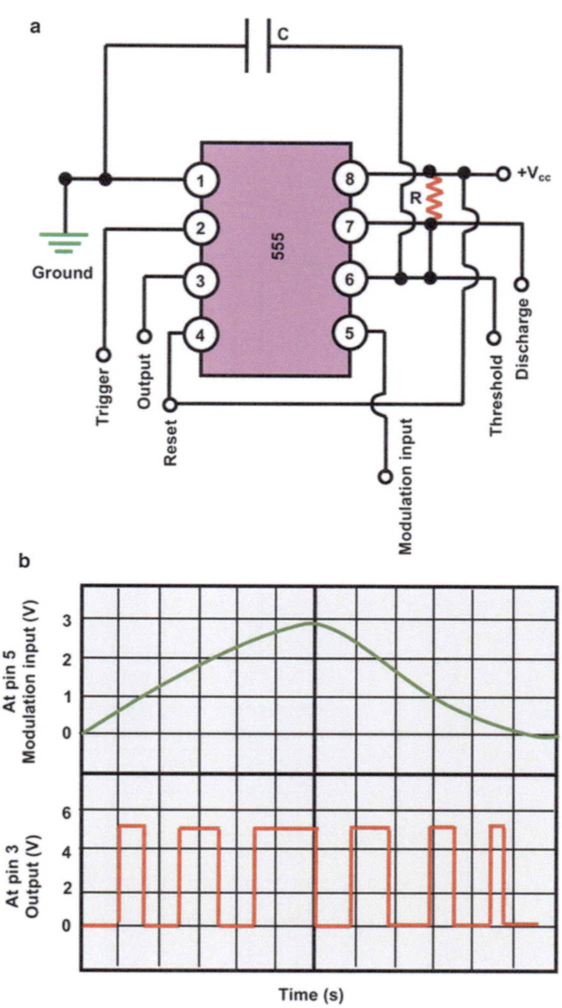

Fig. 6.12 Pulse-width modulation: **a** circuit and **b** waveforms

is followed by resetting of the flip-flop. The output becomes low. Internal transistor T_1 is turned on. Thereby, the resistor R_B is connected with the external capacitor. The capacitor discharges through resistor R_B. In turn, R_B is grounded through T_1. During discharging, at the instant the capacitor voltage falls to 1/3 V_{cc}, the lower comparator is triggered. Then the flip-flop is set. The output goes high. The internal transistor T_1 is cut off. The capacitor starts charging again through $(R_A + R_B)$. Thus a cyclic repetition of events takes place. In this repetition, the flip-flop is alternately set and reset. The external capacitor continues to charge and discharge between two voltage levels, viz., 2/3 V_{cc} and 1/3 V_{cc}. The output of this sequence of events is a succession of square or rectangular pulses. The times of charging and discharging of the externally placed capacitor are independent of the supply voltage. Therefore,

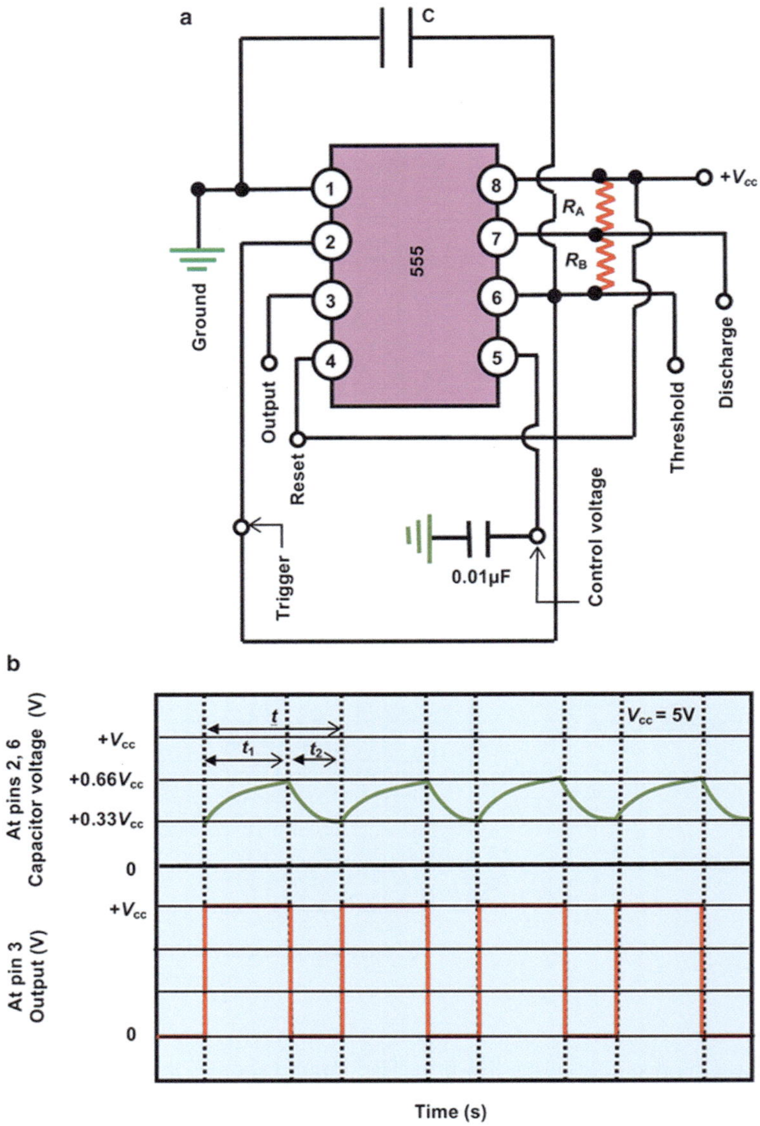

Fig. 6.13 Astable multivibrator: **a** circuit and **b** associated waveforms

the frequency of oscillation of the free-running multivibrator also does not depend on the applied voltage.

Let us explore how the oscillation frequency and duty cycle vary with the values of two external components: two resistors R_A, R_B and one capacitor C. The charging time of the external capacitor with output high is the on-time of the pulse. It is expressed as

$$t_1 = 0.693(R_A + R_B)C \tag{6.1}$$

The discharging time of the external capacitor with output low is off-time of the pulse. It is given by

$$t_2 = 0.693 R_B C \tag{6.2}$$

So, the total period is the sum of t_1 and t_2 written as

$$T = t_1 + t_2 = 0.693(R_A + R_B)C + 0.693 R_B C = 0.693(R_A + 2R_B)C \tag{6.3}$$

The free-running oscillation frequency is

$$f = \frac{1}{T} = \frac{1}{0.693(R_A + 2R_B)C} = \frac{1.44}{(R_A + 2R_B)C} \tag{6.4}$$

The duty cycle is the proportion of time period T during which the signal is active. It is the ratio of time interval during which the output is on or high to the time period T:

$$D = \frac{t_1}{T} = \frac{0.693(R_A + R_B)C}{0.693(R_A + 2R_B)C} = \frac{R_A + R_B}{R_A + 2R_B} = \frac{\left(R_A/R_B\right) + 1}{\left(R_A/R_B\right) + 2} \tag{6.5}$$

Hence, the duty cycle is precisely fixed by the ratio of resistors R_A, R_B.

6.4.8 *Pulse Amplitude Modulation*

Pulse amplitude modulation (PAM) is the simplest of all pulse modulation techniques. In PAM, the amplitude of the modulating signal is mapped to a series of pulses. This means that the pulse amplitudes are adapted according to the signal used for modulation. As a consequence, the output is a series of pulses. The amplitudes of this series of pulses vary in proportion to the modulating signal.

The principle of PAM is explained with reference to the block diagram shown in Fig. 6.14. The sampling of the analog modulating signal is performed with high-frequency pulses. After completion of the sampling process, the output of the circuit is a series of pulses having amplitudes according to the shape of analog signal.

A separate circuit produces the modulating signal. Fixed frequency square waves from the astable multivibrator are used to drive the monostable multivibrator. The output of the multivibrator is in the form of a series of pulses of fixed amplitude. This output is fed to a switching transistor. The transistor opens and closes according to the incoming pulses from monostable multivibrator. In this way, the modulating signal is sampled. The resultant output is a series of pulses with changing amplitudes and

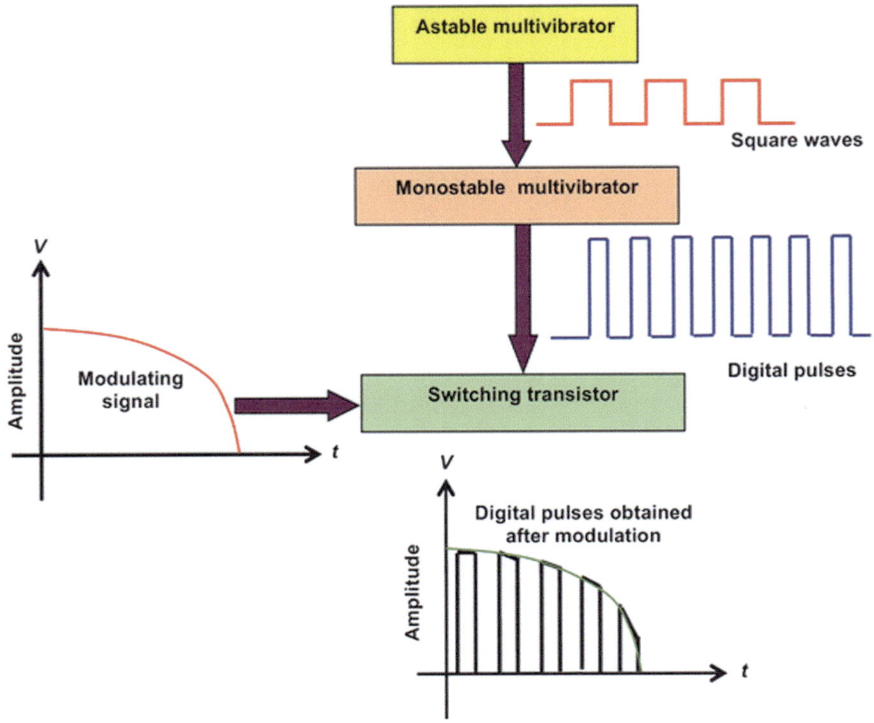

Fig. 6.14 Block diagram and waveforms of pulse amplitude modulation system

shapes replicating the contour of modulating signal. In other words, the amplitude of pulses generated by timer IC varies in accordance with the instantaneous amplitudes of the modulating signal.

6.4.9 Pulse Position Modulation

The timer IC is connected in astable mode as before. By application of a triangular wave modulating signal to the control voltage Pin 5, the position of the pulse is changed with the modulating signal. The change takes place because the threshold voltage and time delay vary accordingly. The output waveform is shown in Fig. 6.15.

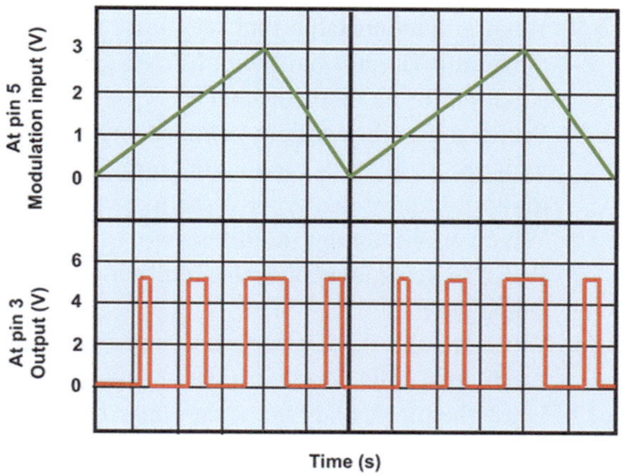

Fig. 6.15 Pulse position modulation: output waveform produced by the modulating input shown

6.5 Discussion and Conclusions

Circuit blocks and structures for implantable devices capable of operating at low voltages and consuming trifling amounts of power were described. Crystal-based and RC oscillators are two varieties of clock sources suited for use with implanted devices. The optimum type of clock source for a particular solicitation is decided by cost and accuracy considerations. Multivibrators can be built using NAND gates or timer ICs. Modulation circuits made from timer ICs were discussed.

Review Exercises

6.1 What is the function of the clock signal in a processor? Name the two types of oscillator circuits used in making clock generators for microcomputers. Compare their relative performances.

6.2 What is inverse piezoelectric effect? How is it utilized in making an oscillator circuit? Name the piezoelectric material most commonly used for making crystal oscillators. Why?

6.3 Draw and explain the electrical equivalent circuit of a quartz crystal. Write the *Q*-factor values obtainable from a quartz crystal and an *LC* circuit.

6.4 How does an *RC* oscillator work? How is 360° phase shift achieved for sustained oscillation? To what frequency ranges is its usage restricted?

6.5 Name the design approach followed by CMOS square wave generator circuits. What are the advantages of this approach? Can you build a crystal oscillator using a single CMOS inverter? How? Explain using a circuit diagram. What is the problem with using a buffered inverter?

6.6 Compare the oscillation frequency and stability of CMOS crystal oscillators made using: (1) a small number of inverters and (2) a large number of inverters.

6.7 What is a multivibrator? Name the types of multivibrators you know, and distinguish between their characteristics.

6.8 What is a monostable multivibrator? Draw the circuit diagram of a monostable multivibrator built with NAND gates and explain its working. Write the equation for its time constant.

6.9 Which multivibrator does not need any external signal for triggering? Explain with the help of a circuit diagram the operation of a NAND gate-based multivibrator of this type. Write the equation for its time constant.

6.10 Which multivibrator requires two triggering pulses for a complete SET–RESET cycle? How does this multivibrator change state when a trigger signal is applied?

6.11 What are the three categories into which the outputs from a timer IC are subdivided? Give one example of each category.

6.12 Draw the block diagram of the timer IC and label the different parts. Why is it called 555 IC?

6.13 Draw the pin diagram of a timer IC and briefly describe the connections of its pins.

6.14 What is meant by a timer circuit? Explain with the help of a diagram the use of a timer IC for implementing a time delay function. How many external components do you need for this implementation? Write the formula for the timing interval in terms of the values of components used.

6.15 Explain the use of the monostable mode of timer IC as a detector of missing pulse. How should the values of the external components be selected in order that a missing pulse is noticed?

6.16 The trigger and threshold pins of a timer IC are connected together? What is the resultant operating mode called? Explain in detail giving the sequence of events that repeatedly occur in this mode. Write the equations for the frequency and duty cycle of the circuit.

References

1. Maxim Integrated (2003) Application note 2154: microcontroller clock—crystal, resonator, RC oscillator, or silicon oscillator? © 10 Sept 2003. Maxim Integrated Products, San Jose. http://www.maximintegrated.com/en/app-notes/index.mvp/id/2154. Accessed 29 Sept 2014
2. Fairchild Semiconductor (1974) CMOS oscillators: application note 118. © 1998 Fairchild Semiconductor Corporation, South Portland, 4 pp. https://www.fairchildsemi.com/application-notes/AN/AN-118.pdf. Accessed 31 July 2015
3. Trudgen G (2009) IC crystal oscillator circuits. Copyright © 2009. Rakon, 3 pp. http://docslide.us/documents/ic-crystal-oscillator-circuit-design.html. Accessed 31 July 2015
4. Texas Instruments (2011) AN-400 A study of the crystal oscillator for CMOS-COPS. Literature No.: SNOA676. Copyright © 2011. Texas Instruments, 6 pp. http://www.ti.com/lit/an/snoa676/snoa676.pdf. Accessed 31 July 2015
5. Storr W (2014) Basic electronics tutorials site: multivibrators. Last updated 5 Sept 2014. Copyright © 1999–2014. http://www.electronics-tutorials.ws/sequential/seq_3.html. Accessed 7 Sept 2014
6. Agarwal T (2014) What are the important multivibrator circuits for pulse generation? http://www.elprocus.com/different-types-of-multivibrator-circuits-for-pulse-generation/. Accessed 7 Sept 2014

7. Roon T (2010) 555 timer tutorial. http://www.sentex.ca/~mec1995/gadgets/555/555.html. Accessed 7 Sept 2014
8. Texas Instruments (2014) LM 555 timer. Copyright © 2014. Texas Instruments. http://www.ti.com/lit/ds/symlink/lm555.pdf. Accessed 7 Sept 2014

Chapter 7
Electrostimulation Pulse Generators

Abstract The electrical stimuli stand out as the most widely applied among the physical and chemical aids used in medicine. This has primarily transpired owing to their similarity to natural biological stimuli. Stimulation is a technique in which low-level electrical currents are applied via electrodes for exciting nerve cells or muscle fibers. For treatment by electrostimulation, simple and complex schemes of stimulation must be evolved. To evolve such schemes, manually operated as well as computer-controlled/microcontroller-based programmable stimulators are necessary. These stimulators must have multichannel outputs. In turn, for building stimulators, high-performance, manual/programmable, and low-cost pulse generators are needed. In this chapter, timer IC-based pulse generator is first described for easy understanding. Then microcontroller-based pulse generators are dealt with. Both types of pulse generators permit adjustability of pulse parameters. But in the microcontroller-based pulse generators, the stimulation parameters can be programmed. Important parameters are the frequency of stimulation, width of the pulse, inter-pulse duration, and amplitude of the pulse. These pulse generators provide more convenient, automated adjustments.

Keywords Pulse parameters · Pulse generator · Timer IC · Microcontroller · User interface · Counter · Active resistance

7.1 Introduction

Pulse generators constitute the core facility for implantable electronics. Delivery of pulses of correct parameters at the relevant biological site is the key task performed by many implantable devices. Therefore, the importance of pulse generators is ineffable.

V. K. Khanna, *Implantable Medical Electronics*,
https://doi.org/10.1007/978-3-031-99868-3_7

7.2 Electrical Pulse and Pulse Parameters

An electrical pulse is a burst of voltage or current. It is a change of short duration in voltage or in current intensity. By a short duration is meant an interval of time comparable in extent to the transient processes in electrical circuits. Digital pulses are represented by well-defined regular geometrical shapes. Examples of these shapes are squares and rectangles on voltage versus time or current versus time graphs. This is purposefully done for easy presentation. The square and rectangular shapes denote ideal pulses. The ideal pulses have a 0-s rise time on the leading edge, a constant amplitude, either voltage or current, followed by a 0-s fall time. But actual real-world pulses, as observed on an oscilloscope, widely differ from such ideal shapes. They have neither 0-s rise and fall times nor do they have the absolutely constant fixed, flat amplitude. Nevertheless, their representation by simple geometrical shapes helps in studying several of their characteristics. It is also helpful in understanding the operations of digital circuits. But it must be realized that these are only approximate representations. Further, by proper circuit design, it is possible to shape pulses. So, one can reduce or entirely eliminate the nonideal features of pulses. As another option, one can work with them by tolerating their deviations from ideality. Figures 7.1, 7.2, and 7.3 illustrate the wave shapes and associated terminology.

Pulse levels: For specification of pulse output, two extreme levels are defined. These are the pulse top and pulse base. The pulse top is called the high level. The pulse base is known as the low level. In digital electronics, often there are two voltage levels: a more positive voltage and a less positive or more negative voltage.

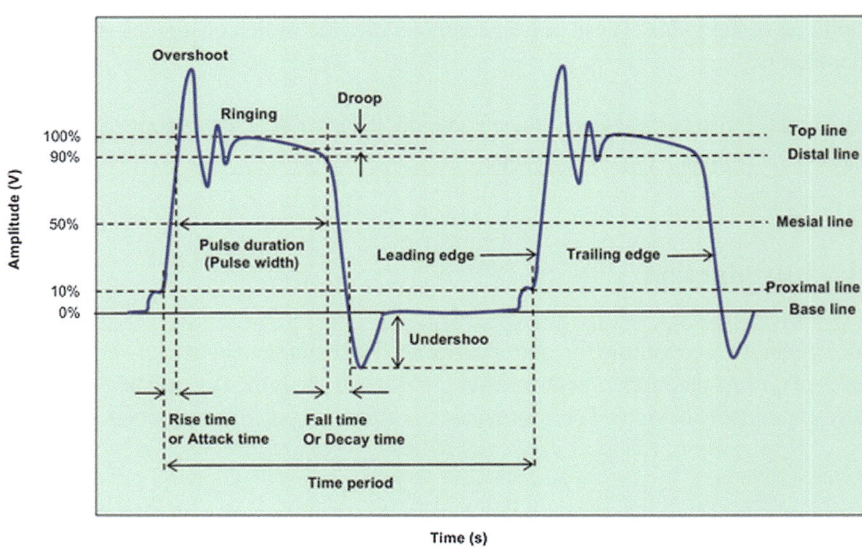

Fig. 7.1 Representative pulse waveform and various parameters used in specifying its characteristics

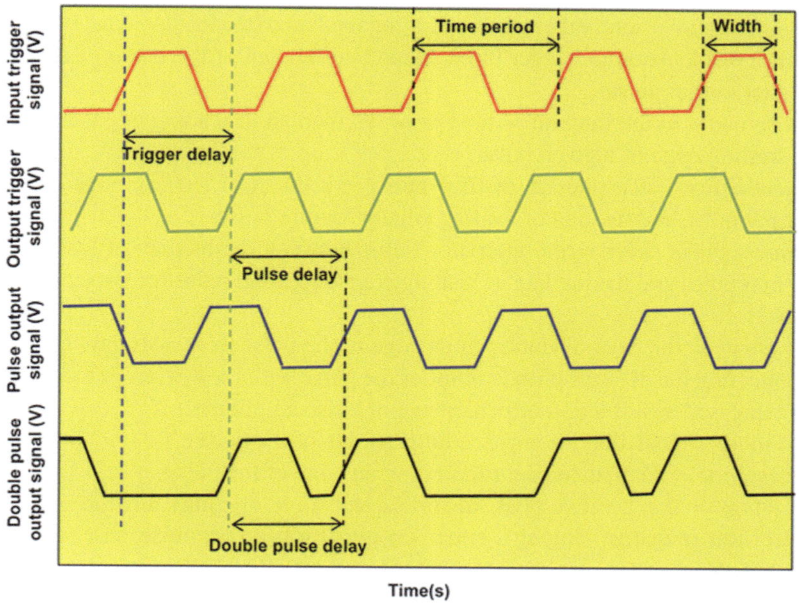

Fig. 7.2 Explaining some parameters of pulses

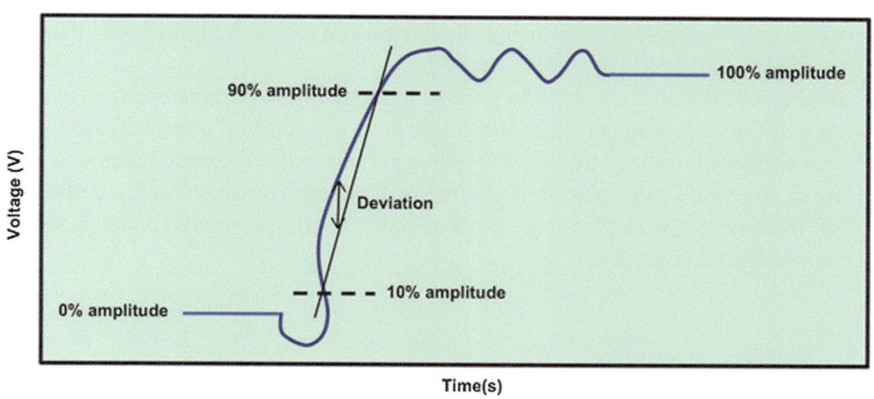

Fig. 7.3 Defining the linearity of a pulse

The former is referred to as the high state, while the later is termed the low state. Pulses are also specified in terms of peak-to-peak amplitude and median offset.

Important parameters for characterizing pulses are defined below:

- *Pulse period* = the interval of time intervening the median of leading edge of one pulse and the median of leading edge of the successive pulse, in a group of pulses.

- *Trigger delay* = interval of time separating median of the leading edge (the trigger point) of the external trigger input signal and median of the leading edge of the trigger output pulse.
- *Pulse width* = the interval of time between median of leading edge and median of trailing edge of a given pulse.
- *Pulse delay* = the interval of time between median of leading edge of trigger output pulse and median of leading edge of output pulse.
- *Double pulse delay* = the interval of time between the median of leading edge of first pulse and the median of leading edge of second pulse for two consecutive pulses.
- *Linearity* = the peak deviation of an edge of the pulse from a straight line drawn connecting the 10% amplitude point of the pulse with its 90% amplitude point. It is expressed as a fraction or percentage of pulse amplitude.
- *Preshoot* = peak distortion preceding an edge of the pulse.
- *Overshoot* = peak distortion succeeding an edge of the pulse.
- *Ringing* = the positive-peak distortion (positive ringing) and negative-peak distortion (negative ringing), either on pulse top or on pulse base, excluding overshoot.
- *Jitter* = the short-term instability of one edge of the pulse with respect to a reference edge. For a period jitter, the reference edge is the foregoing leading edge. Jitter is usually specified as a root mean square (RMS) value.
- *Amplitude droop* = the sagging from the beginning to the end of the pulse. It causes the level detection to disqualify the pulse if it lies at a level lower than the acceptable threshold when it is sampled.
- *Stability* is measured through the average variability of a parameter over a drawn-out lengthy, specified period of time, e.g., hour or year with jitter excluded.
- *Repeatability* is defined as the width of a band inside the accuracy window within which the value of a measured parameter from an instrument is located when the instrument operates with the same adjustments of its controls under the same environmental conditions.

7.3 Pulse Generator

Pulse generator is a fairly familiar contrivance in electronic laboratories, research corporations, etc. It may be a simple electronic circuit with minor adjustability. It is sometimes a piece of electronic equipment. Its primary function is to generate rectangular pulses. It enables control of the pulse repetition rate or frequency. Pulse width too is a variable parameter. Pulse delay can be provided with respect to an internal or external trigger signal. Additional controls are values of the high- and low-voltage levels of the pulses. In some types, the rise and fall times of pulses can be set at required values. Among the diverse features of such circuits and equipments, different applications pose dissimilar requirements. Some individuals want a source of square waves. They may also stress on relatively high frequencies. Others may use the pulse generator for producing short pulses. For some others, short rise/fall times of the pulses are of principal interest. These pulses are injected into the human tissue and used as a stimulus. When evaluating the excellence of pulse generators for implantable device applications, electrical or time characteristics and character, mono- or biphasic, of the pulses are decisive. Also, shape of the signal begotten needs attention.

Pulse generator circuits utilize either digital or analog techniques. They may use a combination of both digital and analog sections as well. As an example, the digital controlling section is given the task of varying pulse repetition rate and duration. The assignment of analog circuitry in the output stage of the pulse generator determines the pulse amplitude and rise and fall times. All pulse generators consist of two main components. These are a power supply unit and a pulse timing control unit. A detailed description of these components will be presented in the following sections.

7.4 Power Supply

It is sometimes necessary to inject current pulses of amplitude ~ 200 mA into the muscle. For this high current injection, high voltage levels around 100 V are needed. These are derived from a rechargeable low voltage DC battery. With this battery, several DC-to-DC converters are connected in series to produce the required higher voltage.

It must be mentioned here that the task assigned to a pulse generator is quite broad in scope. It is not merely limited to producing and sending a pulse to the identified biological site. Through a sensing circuit, it also receives an electrical pulse from the concerned biological site, suppose the heart, via electrodes. The received pulse furnishes information regarding the status of the heart. Before sending this pulse to the pulse generator, the needed signal conditioning is performed by the sensing circuit on the received pulse from the human organ. This signal conditioning transforms the pulse into a suitable form. By such transformation, the pulse becomes acceptable to the pulse generator. Therefore, the sensing circuit acts as an interface between the leads and the pulse generator. After the pulse generator has acquired information about the organ's electrical activity, it responds in the manner it has been preprogrammed to provide the therapy. Thus the pulse generator has a dual role: sensing and pulse delivery. The pulse delivery must be according to the health status of the organ, as determined from the received biological signal.

To reduce problems in recording the small biological signal under study, the stimulus ground must be isolated from the equipment ground [1]. These problems arise when the output of an electronic pulse generator or computer digital-to-analog output is applied directly to a tissue. Its application creates electrical noise in the recording system. It adds a large DC offset too. It may also produce a stimulus artifact. This artifact completely overshadows the comparatively smaller size biological signal under examination. To overcome this concern, the stimulus current is made to follow the path from the positive extremity of the pulse generator through the biological

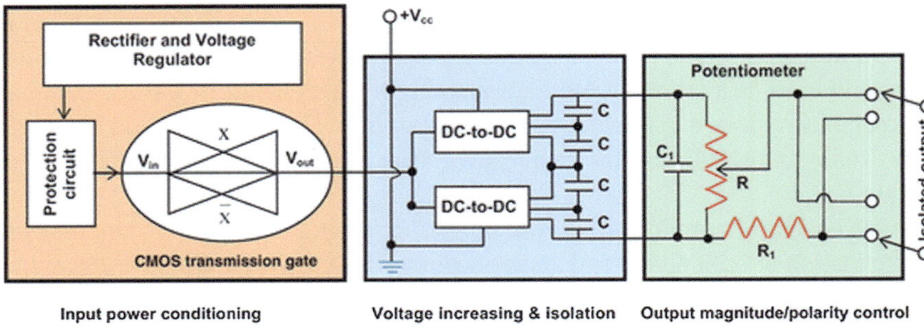

Fig. 7.4 Power supply unit

tissue back toward its negative extremity. Thus the stimulus current bypasses the equipment ground, which is the ground of the recording electrode. Therefore, no disturbance occurs in the biological signal.

The power supply comprises three stages (Fig. 7.4):

1. *Input conditioning stage*: This stage uses an input logic-level pulse. The magnitude of this pulse is ~ 3–12 V. It drives the control inputs of the DC-to-DC converters. Circuit protection from large voltage transients is provided by the input transistor and diodes. These devices restrict, amplify, and invert the input pulse. The output of the transistor is a logic-level pulse. The magnitude of this pulse should be suitable for driving a CMOS quad transmission gate. A high voltage at the pulse input produces a low voltage at the output of the transistor. This low voltage turns off the CMOS transmission gate. It turns on the converters in the succeeding stage.
2. *Voltage augmentation and isolation stage*: This stage uses one or more DC-to-DC converters. Each converter produces a higher voltage output from a smaller battery supply. The capacitors are connected close to each converter. They suppress the switching noise from converters.
3. *Output level and polarity control stage*: This stage is a voltage divider. It sets the stimulus output level. It employs a switch to control stimulus polarity. The potentiometer controls the output level. In addition, it is associated with the capacitor to form a low-pass filter. The filtering of the signal further reduces switching noise from the converters.

7.5 Pulse Timing Control Unit

Two methods to control pulse timing will be described (Tables 7.1 and 7.2). Both these methods are easy to use. The first design is a low-cost, conventional timer IC-based pulse generator (Table 7.1). It is built with 555 timers, common switches, and adjustable voltage dividers or potentiometer knobs. This version is limited to generation of a few waveforms. These waveforms typically consist of one or two

Table 7.1 Features of timer IC-based pulse generator and microcontroller-based pulse generator

S. No.	Property	Timer IC-based pulse generator	Microcontroller-based pulse generator
1.	Cost	Low	High
2.	Supervisory device(s)	Timer ICs	Microcontroller
3.	Scope of pulse parameter variations	Limited	Larger
4.	Programmability	No	Yes
5.	Interactive user interface	No	No

Table 7.2 Controlling pulse parameters in timer IC-based pulse generator and microcontroller-based pulse generator

S. No.	Pulse parameter	Timer IC-based pulse generator	Microcontroller-based pulse generator
1.	Pulse frequency	First timer IC	Counter
2.	Pulse delay	Second timer IC	Microcontroller-operated timing circuits
3.	Pulse duration	Third timer IC	Microcontroller-operated timing circuits
4.	Pulse amplitude	Voltage divider in power supply	DAC and I-to-V converter

shapes of pulses. The duration of the pulses is same. They recur at a fixed rate. The time delay separating the pulses can be varied.

The second more advanced method uses a keyboard with an LCD or a PC (Table 7.2). The LCD or PC helps to produce a convenient user interface. A microcontroller is used to generate accurately timed pulses to trigger the stimulus. It can produce various types of pulses. Single pulses, pairs of pulses, and trains of pulses can be generated. Wide ranges of pulse rates and durations are achieved. Variable delays for single pulses can be produced. Manual triggering of single pulses is also allowed. In both methods to be discussed, it will be assumed that the communication between the controlling part of the implanted device outside the body and the implant inside the body is via wireless link.

7.6 Timer IC-Based Pulse Generator

In this pulse generator (Fig. 7.5), the stimulator is under the supervisory control of a timer device. This timer device produces logic-level pulses of the planned timing [1, 2]. For single pulses, the foremost timing parameters that are necessary are the holdup or delay after commencement of the stimulus command, duration of pulse and pulse repetition time. For double pulses and pulse trains, the pulse interval

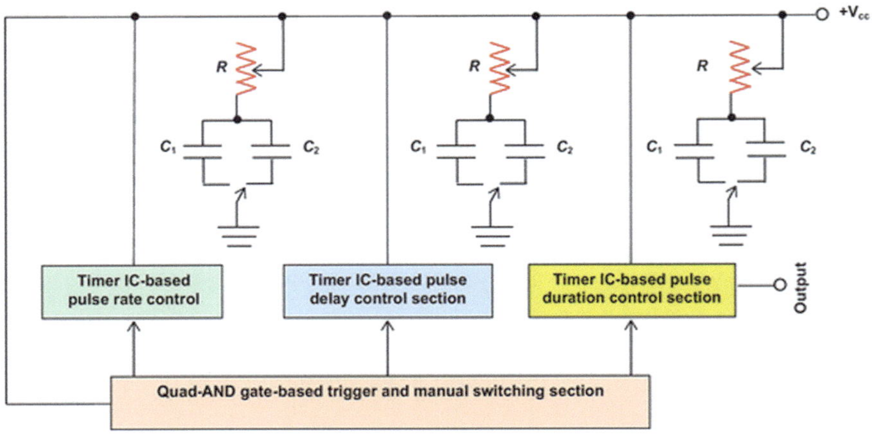

Fig. 7.5 Timer IC-based analog pulse generator

and pulse train duration must be defined. The pulse generator uses a 555 timer chip in each of the three sections. These chips help to organize changes in pulse rate, delay, and duration. In each section of the circuit, there are capacitors and variable resistors. These components are assigned the critical roles of determining the timings of pulses. By altering their values, these timings are varied. The size of the capacitor connected to the threshold pin of the 555 timer is either incremented or decremented. These changes linearly increase/decrease the aforesaid times. The first timer dictates the pulse rate if the stimulator is in pulse train mode. It is disjointed in other circumstances. The second timer plays different roles in single and dual pulse modes. In the single pulse mode, it oversees the pulse delay. In dual pulse mode, it watches over the spacing between two pulses. The third timer administers the duration of the pulse. The pulse amplitude is adjusted by the voltage divider in the power supply section. Thus a combination of timer ICs helps to provide a capability of pulse generation in the desired format.

7.7 Microcontroller-Based Pulse Generator

7.7.1 Why Microcontroller-Based Pulse Generators?

Today, microcontroller-based electronic stimulators are broadly adopted. They can provide moderately sophisticated stimulus timing at lower expenditure [3]. Microcontrollers offer the correct blend of features. These features include programmability options, cost-effectiveness, timing accuracy performance, and efficient power utilization. These are the principal features that are yearned to realize such stimulators. A microcontroller featuring ultralow power consumption must be chosen. There are wide choices for high-priority, all-important medical implants. Microcontroller with perfected and enhanced software is needed. Additionally, it must have the required memory space. A popular microprocessor in state-of-the-art embedded system design is the ARM processor. Advanced RISC machines (ARM) have developed this processor.

ARM fabricates a family of 32-bit and 64-bit processors. These processors are based on a progressive architecture. This architecture is known as the reduced instruction set computer (RISC) architecture. The architecture has the specialty that it uses a small, highly optimized instruction set. It incorporates advanced attributes like extremely low power consumption. It also provides high speed and optimization of code size. Moreover, it is furnished with dynamic voltage scaling (DVS) functionality. This assists in optimizing the energy consumption [4]. The microcontroller is chosen from consideration of the desired features for the application (Table 7.3).

7.7.2 *User Interfaces*

Triggering of the implantable device is controlled by either of the two methods as given (Fig. 7.6): (1) first is by using a standalone microcontroller circuit. Here, the programming is done in assembly language. A matrix keyboard and alphanumeric LCD display provide the user interface. The interface indicates the type of current as well as duration of the pulses. (2) Second is by using a microcontroller operated by a graphic user interface (GUI). For this operation, a Microsoft Windows-based PC is used. The microcontroller guarantees a response in real time, the actual time of occurrence of an event. The ubiquitous PC has implements to provide a more convenient interface with the user. The timers mainly interact with CPU through interrupts. Hence, the waveform of the pulses to be injected into muscle is spawned by a timer 0 interrupt of the microcontroller. The resolution of the pulse times is determined by the time in which the timer 0 overflow interrupt service routine (ISR) executes. If it is 32 μs, the resolution of all pulse times will be 32 μs. The microcontroller in the stimulator has programmable features. Main features of interest are pulse amplitude, pulse-width, inter-pulse duration and stimulation frequency. It is feasible to vary the frequency of the output rectangular pulse waveform. The duty cycle, contingent upon the duration of a pulse and spacing between pulses, is variable too. So is the amplitude of the pulse. It is also possible to select the number of pulses in a cluster of pulses. External synchronization of the generator can also be effectuated.

7.7.3 *Main Tasks of Microcontroller*

The microcontroller performs two main tasks [5]: (1) it controls the keyboard and the display. So, the control over the settings of the generator is in the hands of the user. Frequency of repetition of pulses, trigger delay, etc., can be altered in an interactive

Table 7.3 Microcontroller specifications

S. No.	Feature	Meaning
1.	Data bus	Number of bits of bidirectional conduction paths on which data or instruction codes are transferred. Examples: 4, 8, 16, … 128
2.	Clock speed	Maximum frequency. Examples: 8, 10, 12, 18, 25, 33, 40, etc. MHz
3.	Number of interrupts	Number of asynchronous electrical signals from a peripheral to the processor requiring immediate attention by the microcontroller. Examples: 3, 6, 8, etc.
4.	RAM size	64, 128, 256, etc. bytes
5.	ROM type and size	ROMLess, MaskROM, erasable programmable read-only memory (EPROM), electrically erasable programmable read-only memory (EEPROM), flash, factory advanced service technique read-only memory (FASTROM), one time programmed memory (OTP) Examples: 1 k, 2 k, 4 k, 8 k, etc.
6.	Supply voltage	Examples: 3 V, 5 V, etc.
7.	Number of timers	Examples: 1, 2, 3, etc.
8.	Number of bits in the timer	Examples: 8, 16, 24, 32, etc.
9.	Serial interfaces	Controller area network (CAN), serial peripheral interface (SPI), serial communications interface (SCI), universal asynchronous receiver/transmitter (UART), etc.
10.	Serial port channels	Number of serial ports
11.	I/O ports	Number of input ports, output ports, and input/output ports combined. Examples: 8, 16, 14, 17, 26, etc.
12.	Number of A/D converters	Example: 4×10 bit
13.	Number of bits in A/D converters	Examples: 8, 16, 24, 32, etc.
14.	Package	Examples: bare die, ball-grid array (BGA), flip chip ball-grid array (FCBGA), plastic pin-grid array (PPGA), ceramic dual in-line package (CDIP), plastic dual in-line package (PDIP), plastic leaded chip carrier (PLCC), etc.
15.	Operating temperature	Example: -40 to $+125$ °C
16.	Operating range	Commercial, industrial, military, etc.
17.	Pin count	Number of pins. Examples: 10, 16, 20, 30, 32
18.	Other functions	Watchdog timer, direct memory access (DMA) channels, power saving modes

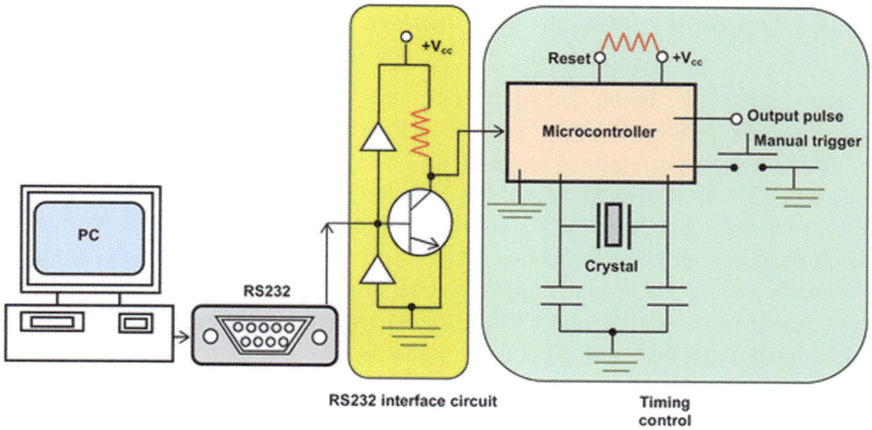

Fig. 7.6 Microcontroller-based pulse generator

manner. (2) It produces a square wave of utilitarian dependability. For obtaining the square wave, the system clock frequency is suitably divided in an internal timer/counter. As a consequence, the output signal frequency can be set over a broad range of values. These values extend from below kilohertz to over megahertz range. The forethought that a crystal oscillator is the reliable source from which the system clock is derived and that the timer operates in a hardware mode pledges the availability of a stable frequency of output signal.

If a PC is used, an additional role undertaken by the microcontroller is sending information to PC and receiving information from it. This is done via the RS-232 interface. Measurements can then be performed automatically. The RS-232 interface is the standard laid down by the Electronic Industries Association (EIA). This standard is meant for the exchange of serial data in binary form between two devices, viz., data terminal and data communications equipment. It was originally developed to stereotype the connection of computers with telephone line modems. RS-232 is chosen because it is a unanimous, well-understood, and supported data interface. In this application, relatively slow data transfer is a nonissue and quite tolerable. In-system programmable microcontrollers provide the opportunity of programming after installation in the system, instead of being programmed first and later installed. Such programming is also called in-circuit serial programming. Building the circuit with in-system-programmable microcontroller and RS-232 interface provides easy accessibility of firmware upgrades of the device. Hence, the pulse generator can be used in self-regulating applications.

7.7.4 Frequency Division by Counters

A frequency divider is referred by various names in the literature. It is sometimes called a clock divider. Other names are scaler and prescaler. It is a circuit which produces an output = the clock input divided by a specified value. Stated in another way, it is a circuit in which a signal having a frequency f_{in} is fed at the input and a signal with a frequency $f_{\text{out}} = f_{\text{in}}/n$ (where n is a numerical integer) is received at the output. Internally, the frequency divider uses an N-bit up-counter. A counter is essentially a dedicated register or pattern generator. This register delivers a particular repetition or succession of binary values known as states in the output when an input pulse signal is applied to it. This input signal is called the clock signal or simply clock [6].

Counters of various capabilities can be implemented by connecting different numbers of flip-flops (Fig. 7.7). The modulus of a counter, written as MOD in brief, is the number of states of output signal across which the counter changes before reaching again to zero. It is defined as the number of output states in one complete cycle. The modulo or MOD number of a counter is simply written as MOD number = 2^n. A counter with$n$ flip-flops has 2^n different states. It is called a modulo-n counter or MOD-n counter. It can count decimal numbers from 0 to $2^n - 1$. A counter with two flip-flops has $2^2 = 4$ different output states. It is called a modulo-4 counter or MOD-4 counter. It can count decimal numbers 0–3, given by $2^n - 1 = 2^2 - 1 = 3$. A counter with three flip-flops has 8 different output states. It is called a modulo-8 counter or MOD-8 counter. It counts from 0 to 7 as $2^3 - 1 = 8 - 1 = 7$. The modulo number of a counter is increased by the addition of a larger number of flip-flops to it. A way of building a counter of higher modulus number is by cascading several flip-flops. Thus a counter with 8 flip-flops has 256 different output states. It is called a modulo-256 counter or MOD-256 counter. It counts from 0 to 255 as $2^8 - 1 = 256 - 1 = 255$ and so on.

7.7.5 Changing Other Parameters of the Pulses

During the production of pulses, the operation of the microcontroller is as follows: immediately after the instrument is switched on, the initialization procedure is accomplished. All through the process of initialization, the activities of various parts of the microcontroller are properly arranged and sequenced. So also are the values of different variable parameters involved in operation. Further, initialization of display unit is performed. Display indicates the data which is required to be fed. Activation of the "amplitude entering function" follows. The voltage waveform to be generated by the stimulator is entered into a program. This program is written in Microsoft Visual C for Windows using a keyboard. If a microcontroller is used, the programming is done in assembly language.

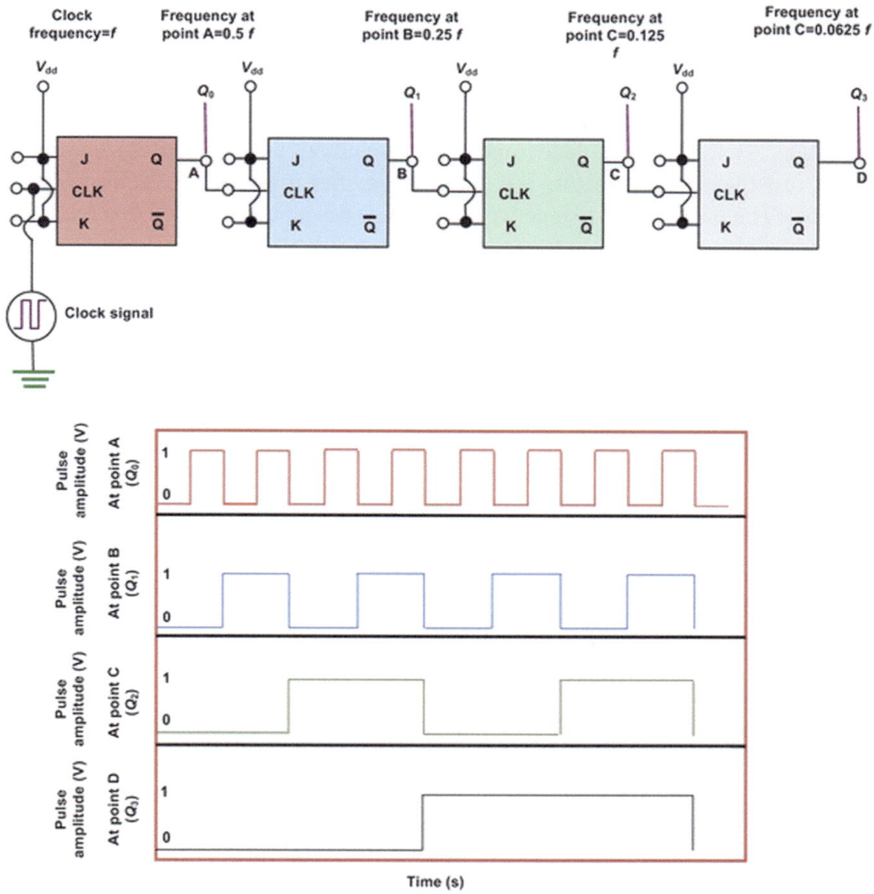

Fig. 7.7 Frequency division

The waveform is entered as a list of times. Together with this list is the corresponding decimal representation of the binary number to be written to each D/A converter. The binary number is referred to as voltage amplitude. Following the entry of the timing parameters, the microcontroller carries out computation of the basic parameters. It adjusts the timing circuits in accordance with the computed values. Suppose generation of a designated group of pulses is to be performed. Then a counter is used to count the selected number of pulses. Depending on the microcontroller used, it is possible to generate up to a certain number, say 256 pulses in the group.

Subsequent to entry of the value of amplitude into the microcontroller unit, the microcontroller computes the applicable 8-bit control number corresponding to the received value. The calculated number is conveyed by the microcontroller to the input of the integrated circuit for 8-bit digital-to-analog converter (DAC). The output current received from the DAC is transformed to the equivalent voltage through the agency of the current-to-voltage converter IC.

Very frequently, the amplitude of the output signal from the pulse generator is changed via multiple stages or steps [7]. The parameter to be adjusted is pulse amplitude. A choice can be made from available methods to vary the pulse amplitude. Use of resistive step attenuators is a common practice in analogue electronics. The attenuator is a passive resistive network to reduce the power supplied to a load by a fixed or variable amount or by a series of swapping steps.

Another way to vary the intensity of current is by using a sliding or rotary potentiometer. The astute reader is acquainted with variable resistors, rheostats, and mechanical potentiometers. Wiper potentiometers of high quality may be comparable to precise attenuators in performance.

Similar to these analog devices, the counterpart device in digital electronics is a digital potentiometer. This potentiometer is also called digital pot or digipot or resistive digital-to-analog converter (RDAC). It is a compact device used to adjust and trim electronic circuits. It is constructed using an R–2R integrated circuit. Other digital-to-analog converters may also be used. The resistor ladder configuration is a very familiar circuit. As indicated by its name, this structure has the appearance of resistor ladder. It comprises a number of steps on the ladder. Every step is provided with its own switch. This switch enables the connection of the particular step to the output pin of the potentiometer. The resistance ratio of the potentiometer is decided by the specifically selected step on the ladder.

Traditional analog potentiometers are not usable. The reason is the obvious difficulty in interfacing them with the microcontroller.Digital potentiometers solve the interfacing problem. They are able to do so by allowing the control of a voltage splitter with digital signals. Another convenient feature of digital potentiometers is their higher adaptability. The enhanced adaptability is due to the absence of a predetermined sweep profile. This is readily evident on realizing that they are not controlled mechanically. The fact that these potentiometers do not have a prefixed profile means that one can run the desired type of profile. The profile running depends on the way the code is written. Then the potentiometer will sweep in either a linear or logarithmic fashion. It may also sweep according to any other profile one has aspired for. Digipots are mostly volatile memory devices. However, in some versions of digipots, nonvolatile memory is provided. This allows the retention of the final position that was programmed after the digit was cycled for power.

Undesirably high power losses taking place during current flow through the resistance constitute the principal deficiency of potentiometric control technique. This happens more than ever in pulse generators when the amplitude of the output signal is large and the matching load resistance is small. From power cutback standpoint, a better way is to change the voltage of the power supply used for feeding the pulse generator circuit. Otherwise active signal shapers or wave shaping circuits may be used. In another option, the voltage of the output amplifier may be varied.

It is not advisable to alter the voltage setting of power supply of pulse generator circuit. This is because it is likely to change the oscillation frequency of the pulse generator. Further, pulse signal generators do not at all times use active signal shapers.

Therefore, variation of the power source voltage of the output amplifier or the last stage of this amplifier seems to be the best approach for controlling the amplitude of the pulse.

A voltage-regulating circuit delivers power to an oscillator. This oscillator is constructed across the primary coil of a transformer. After rectification, the secondary coil of the above transformer supplies the high voltage signal to switched current mirrors in the driving stage. Overall power consumption is minimized. The minimization is possible because the high compliance voltage is proportional to the stimulation current. Control of the pulsed current amplitude is achieved by adjusting the regulated voltage [8, 9].

A typical screen for pulse specification might include the pulse amplitudes and pulse widths. Inter-pulse intervals as well as the number of pulses must be shown. The sequence can then be displayed in a separate window. As new pulse sequences are developed, they can be saved individually or as a list of sequences. Hence, they can be subsequently retrieved.

7.7.6 FET-Based Methods of Amplitude Control

Amplitude control is also achievable through active resistance [10]. The idea of active resistance is based on the use of an active device such as a transistor for resistance control. The active resistance R is realized by taking the help of an active device, viz., the field-effect transistor (FET) (Fig. 7.8). A fixed resistor r is connected across source to drain of the FET. The controlling voltage V_{GS} is applied to the gate electrode of the FET. It alters the drain-to-source resistance of the FET. The active resistance is obtained across the external drain-to-source resistor r. The drain-to-source resistance of the field-effect transistor comes in parallel connection with this external resistor.

For active R realization, the current to be driven into the muscle is first decided. The necessary voltage is converted into equivalent analog voltage by the digital-to-analog converter. This voltage is supplied by the microcontroller as the control voltage to drive the gate of the FET.

In another method of amplitude control, the microcontroller creates a pulse-width modulated (PWM) output. This output is proportional to the value of amplitude required. The pulse-width modulated output is subjected to low-pass filtering. The low-pass filtered signal is converted into a current. This current is used to impel an optocoupler. The output of the optocoupler is taken from a photoFET. On applying this output to a voltage divider circuit, a voltage output is produced. The voltage output is more or less linearly varying with the PWM input [11].

Fig. 7.8 Realization of active *r*

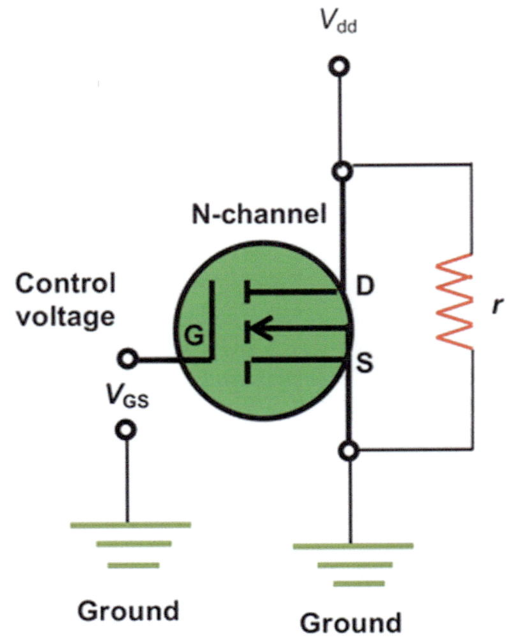

7.8 Discussion and Conclusions

This chapter discussed how to create circuits which will generate precise pulses in rectangular or square waveforms. As electronic designs increase in complexity, it becomes difficult to build the complete circuit. So, prebuilt circuits are used. These circuits are available in different packages. The prebuilt circuits are called ICs. For generating the different waveforms described in this chapter, timer ICs and microcontrollers were utilized. The gimmicks of maneuvering pulse parameters were also treated.

Review Exercises

7.1 Define the following: (1) an electrical pulse and (2) an ideal positive pulse. Is an ideal pulse ever observed?

7.2 Define pulse period, pulse width, and pulse delay. Explain the following terms with respect to a pulse: linearity, preshoot and overshoot, ringing, jitter, and amplitude droop. What is meant by stability and repeatability of a pulse?

7.3 What is a pulse generator? What types of control features are usually built in a pulse generator?

7.4 Name the two main components of a pulse generator. Why DC–DC converters are used in the power supply?

7.5 What are the functions of the three stages of the power supply of a pulse generator? Why is it necessary to isolate the stimulus ground from the equipment ground of the power supply?
7.6 Describe the method of generation of pulses using timer ICs? How are the parameters of the pulses controlled in this method?
7.7 What desirable qualities will you look for in selecting a microcontroller for a microcontroller-based pulse generator? In what respects this pulse generator is better than a timer IC-based pulse generator?
7.8 What does "RS-232 interface" mean? Give reasons for its selection to build the pulse generator.
7.9 What is an electronic counter? What is meant by the modulus of a counter? How is the modulus related to the number of flip-flops that have been employed to build the counter? A counter can count from decimal 0 to 255. How many flip-flops have been used in the counter?
7.10 What happens when the microcontroller-based pulse generator is switched on? How is the amplitude of pulses in this pulse generator controlled?
7.11 What is a digital potentiometer? How does it differ from an analog potentiometer? What is the disadvantage of controlling amplitude with a potentiometer, digital or analog?
7.12 How is amplitude of the pulse adjusted using active resistance?

References

1. Land BR, Johnson BR, Wyttenbach RA et al (2004) Tools for physiology labs: inexpensive equipment for physiological stimulation. J Undergrad Neurosci Educ 3(1):A30–A35
2. Pasca S, Sztojanov I (2006) Microcontroller based pulse generator for neuro-stimulation purposes. In: ELECTRONICS' 2006, Sozopol, Bulgaria, 20–22 Sept 2006, pp 71–76
3. Kocer S, Canal MR, Güler I (2000) Design of low-cost general purpose microcontroller based neuromuscular stimulator. J Med Syst 24(2):91–101
4. Chede S, Kulat K (2008) Design overview of processor based implantable pacemaker. J Comput 3(8):49–57
5. Starecki T, Misiaszek S (2006) Low cost programmable pulse generator with very short rise/fall time. In: Proceedings of SPIE. Photonics applications in astronomy, communications, industry, and high-energy physics experiments, vol 6347, 12 Oct 2006, 63472J, 5 pp
6. Storr W (2014) Basic electronics tutorials. Frequency Division. http://www.electronics-tutorials.ws/counter/count_1.html. Accessed 31 July 2015
7. Hruškovic M, Hribik J (2009) Pulse generator controlled by microcontroller. In: 19th international conference, Bratislava, 22–23 Apr 2009, pp 39–41
8. De Lima JA, Cordeiro AS (2001) A simple constant-current neural stimulator with accurate pulse-amplitude control. In: Proceedings of the 23rd annual EMBS international conference, Istanbul, 25–28 Oct 2001, pp 1328–1331
9. De Lima JA, Cordeiro AS (2002) A low-cost neurostimulator with accurate pulsed-current control. IEEE Trans Biomed Eng 49(5):497–500
10. Cellatoglu A, Karuppanan B, Subramaniam S (2010) Programmable muscle stimulator for activating paralytic patients. In: The international conference on computing, communications and information technology applications (CCITA-2010). CCITA-2010, 6 pp. Ubiquitous Comput Commun J

11. Physiological stimulator. Cornell University. http://people.ece.cornell.edu/land/PROJECTS/Stimulator2/. Accessed 13 Sept 2014

Chapter 8
Biomaterials for Implants

Abstract Essential to the success of an implantable electronic device is the choice of its constructional biomaterial. This biomaterial must neither corrode in the body nor elicit any adverse response from it. Due to immune response from the host body, tissue inflammation occurs along with fibrous encapsulation of the implant. Functionality of the implant is thereby inhibited. Surface modification techniques have been devised for improving the biocompatibility of the implant. Besides the age-old "stainless steel," implants generally use metals like platinum, iridium, titanium, and tantalum. Bioceramic materials like aluminum oxide are also used. Polymeric materials, e.g., polyimide (PI), polyvinylidene difluoride (PVDF), poly(p-xylylene) (parylene), polyetheretherketone (PEEK), polydimethylsiloxane (PDMS), liquid crystal polymer (LCP), etc., have appeared as substitutions for metallic and ceramic biomaterials to serve as the draping covers of medical devices to be lodged inside the human body.

Keywords Biomaterials · Biocompatibility · Foreign body reaction · Fibrosis · Metals · Bioceramics · Polymers

8.1 Aims and Scope of Biomaterials

A biomaterial is either a natural material or a synthetic one prepared by artificial means. It can be a metal, a ceramic or a polymer. It is introduced into the living tissue principally as a part of a medical device, e.g., an implant or prosthesis. The implant is designed to treat/augment/replace an aged, malfunctioning, or appearance-wise unacceptable native bone/tissue/organ/activity inside the human body. The objective of implantation is to bring about improvement in the quality of an individual's life [1, 2]. According to Williams [3, 4],

> A biomaterial is defined as a chemical substance, element, mixture or compound, that has been designed, concocted or contrived to acquire a suitable form. It is used either unaccompanied or as a component of a multipart/multifaced system, to guide the progression of any curative or diagnostic procedure in human or veterinary medicine (concerned with

V. K. Khanna, *Implantable Medical Electronics*,
https://doi.org/10.1007/978-3-031-99868-3_8

animals, especially domestic). It does so by regulation of interactions and exchanges with the mechanisms of living organisms.

The broad scope of biomaterials is vast in extent. It encompasses materials for intraocular lenses. Bone/joint replacements are also included among biomaterials. Materials for making heart valves, pacemakers, etc., too come under its umbrella.

8.2 Defining Biocompatibility

By convention, the notion of biocompatibility is interlinked with implantable devices that are made for lodging inside the human body. These devices may be loged up to the lifetime of the host. In the beginning stages, biocompatible materials were chosen on several negative qualities. These were the qualities of being inactive or inert. They were characterized by properties like nontoxicity, non-immunogenicity, non-thrombogenicity, noncarcinogenicity, non-irritability, and many others [5]. Toxicity signifies the degree to which a substance is harmful to humans. Five properties that must be absent in a material for biocompatibility are cytotoxicity (toxicity to cells), genotoxicity (toxicity to genetic information within the cell, causing mutations), mutagenicity (causing damage to genetic material DNA, thereby increasing the frequency of mutations), carcinogenicity (causing cancer), and immunogenicity (ability to induce an immune response) (Fig. 8.1). Materials possessing these negative properties were therefore classified as biocompatible. It was firmly believed that nonreactive materials fulfilled these negative attributes best. Naturally, some favorite materials were good-quality stainless steel, titanium, and platinum, and polymers such as PTFE, PMMA, polyethylene, and silicones.

As elucidated by Williams [3, 4], biomaterials are being increasingly used in engineering tissues which involves combining scaffolds, cells, and other biomolecules to construct fully functional synthetic or semisynthetic tissues meant to overhaul tissue or organ failure by tissue implantation. Innovatory systems for transporting cells, drugs, and genes constitute another vital application area. In biotechnological applications, interactions between biomaterials and components of tissues are needed. These interactions are of definite and direct nature. Hence, a new model for biocompatibility has come forward. The concept of biocompatibility underwent critical reexamination consequent upon three factors. Firstly, it was realized that biocompatibility was not solely decided by material characteristics. To the contrary, it was also governed by the circumstance in which the material is used. Secondly, it was found that instead of being inert, the material should purposely act in response to the tissues. It should not be disregarded by the tissues. Thirdly, the gradual degradation of material with passage of time in the body (instead of its permanence) is desired in many applications. Then it will not reside permanently inside the body. So, the synonymy of biocompatibility with biological well-being was inadequate to ascribe this quality to any material. Accordingly, biocompatibility was redefined by Williams [3, 4] as "the quality of a material to execute its assigned function in a given

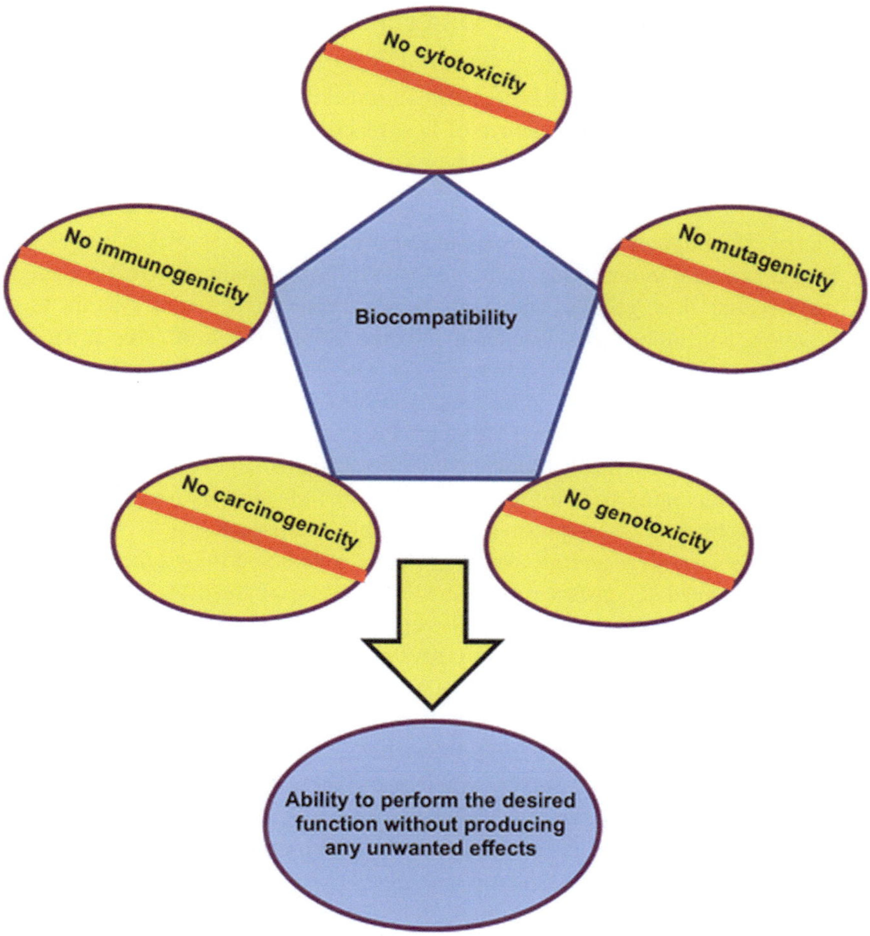

Fig. 8.1 Defining the biocompatibility of a material

therapy for restoration of normal health," under the condition that it does not "elicit any unwanted effects, either local or systemic, in the recipient or beneficiary of that therapy" but is successful in "producing the most suitable advantageous response related to cellular or tissue functions, in that particular state, and making efforts to reach toward perfecting the clinically or scientifically pertinent effectuation of that therapy."

8.3 Responses of Tissues to Materials

Following the implantation of a biomaterial, a prosthetic aid, or a medical device in the living tissue, a complex sequence of host reactions takes place. This sequence involves many biochemical pathways. These constitute *the tissue response continuum* [6, 7]:

1. *Injury*: The implantation injures tissues or organs.
2. *Blood–material interactions*: Damage to vascularized connective tissue changes the vascular flow, caliber, and permeability. From the vascular system, fluid, proteins, and blood cells leach out into the wounded tissue. This procedure is known as exudation. Alterations in the vascular system also embrace those produced in blood and its ingredients. The influence of wound on plasma or cells triggers chemical effects. These effects facilitate many responses from the blood vessels and cells of the swelled region. The onslaught of wound starts the response associated with inflammation. But this response is interceded by the chemicals liberated from plasma, cells, and bruised tissues.
3. *Provisional matrix formation*: Within minutes to hours of injury, the provisional matrix is formed. It consists of fibrin, a fibrous, non-globular protein, produced by the system of blood coagulation. It also contains inflammatory spin-offs released by the reciprocal supportive system. An important product consists of activated platelets. The other products are the cells taking part in the inflammatory response and cells forming the endothelium, the thin layer which covers the interior surface of blood vessels; hence termed endothelial cells. The reorganization and mending processes start. These processes entail recruitment of inflammatory cells and fibroblasts (spindle-shaped cells). They are actuated by the parts, which are inside the temporary matrix, i.e., network of fibrin, or those liberated from it.
4. *Acute inflammation*: Inflammation aids in restraining, neutralizing, diluting, or walling off the injury-causing process. It also triggers a series of events. The triggered events repair and reconstruct the location of implant through substitution of the wounded tissue. The substitution is carried out by regenerating inhabitant cells from parenchyma, the specialized tissue characterizing an organ in contrast to connective or supportive tissues. These cells are called parenchymal cells. Repairing is also done by forming thick dense tissue called scar tissue from cells producing collagen and other fibers, hence, known as fibroblastic tissue. Another method of repairing involves the combined action of the above two routes.

   Acute inflammation extends over a moderately short stretch of time. This duration is typically from a few minutes to a few days. The period depends on the degree of injury. It is characterized by the exudation of fluid and plasma proteins and the ejection of leukocytes (white blood cells). The outflux of leucocytes contains mainly neutrophils (a kind of white blood cells containing enzyme-filled sacs for digesting microorganisms) from the blood vessels to the tissues surrounding them and the injured part.

5. *Chronic inflammation*: Chronic or long-term inflammation is a long-standing response to an on-going medical problem, whereas acute or short-term inflammation described in (4) was an immediate response to an injury. Histologically, chronic inflammation shows less uniformity than acute inflammation. It is characterized by the presence of macrophages, the large specialized, mononuclear, white blood cells, which overwhelm and gulp down foreign invading particles. It also contains monocytes (large phagocytic white blood cells) and lymphocytes (small white blood cells responsible for immune responses), with the multiplication of blood vessels and connective tissues.
6. *Granulation tissue*: Within a period of one day subsequent to implantation of a biomaterial, the response for restoring prior healthy condition begins by the action of monocytes. Other vital contributors to healing are macrophages. This response is trailed by abundance of fibroblasts (cells of connective tissues secreting collagen and other proteins which constitute the extracellular matrix) and vascular endothelial cells at the place of implant, consequential in the development of granulation tissue.
7. *Foreign body reaction*: The foreign body reaction comprises foreign body giant cells (compilation of merged macrophages) and the components of granulation tissue (collagen-rich connective tissue). Thus it consists of macrophages, fibroblasts, and very thin blood vessels or capillaries. The above ingredients are present in varying quantities according to the form and topography of the implant (Fig. 8.2). In comparatively flat and smooth surfaces, the foreign body reaction consists of a 1–2 cells thick layer of macrophages. For somewhat rough surfaces, this reaction comprises macrophages and foreign body giant cells at the surface.
8. *Fibrosis and fibrous encapsulation*: With few exceptions, the final stage of healing response to biomaterials is normally fibrosis (formation of connective fibrous tissues) or fibrous encapsulation (encasing by dense layer of connective fibrotic tissue). The encapsulation safeguards the implant from the immune system. It segregates the implant from the neighboring tissues. Healing of places of implantation involves two distinctive processes. The first process is regeneration of injured tissue through parenchymal cells, viz., those of a gland or organ of the same type. The second process is substitution by connective tissue constituting the fibrous capsule. These processes are controlled by either the multiplicative capacity of the cells in the implant-receiving tissue or persistence of the tissue framework of the injury site.

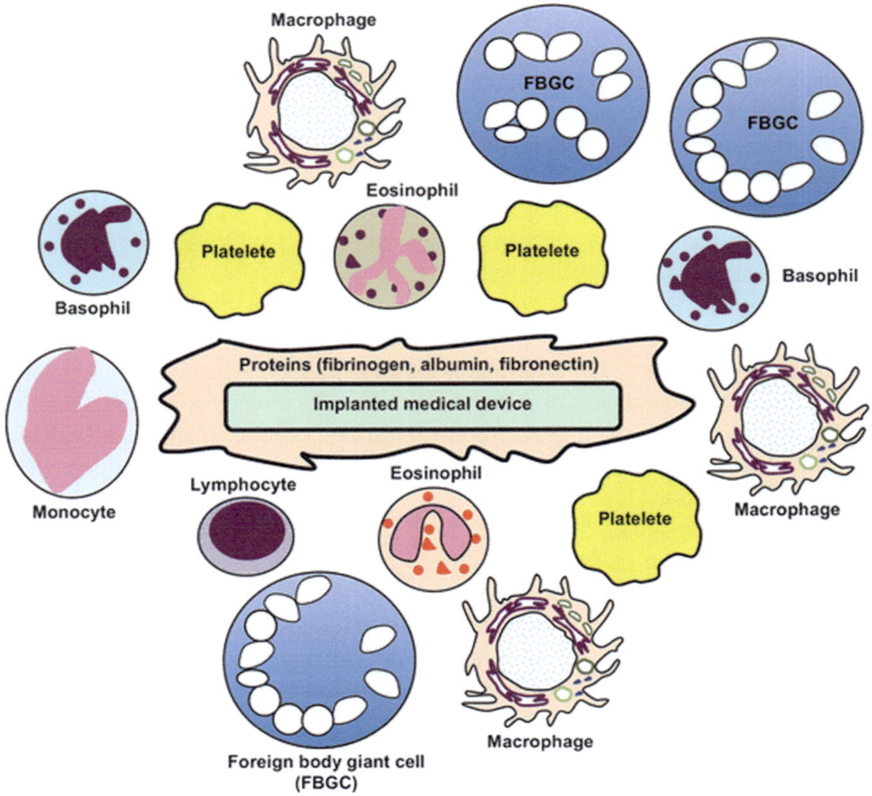

Fig. 8.2 Illustration of the molecular events taking place in the foreign body response to an implanted material or device

8.4 Metallic Biomaterials

8.4.1 Commonly Used Materials

Corrosion and strength-related problems faced by metals in early developmental phase were obviated by the advent of stainless steel. Metallic implants are nonmagnetic. They have a high density. So, they do not interfere with MRI (magnetic resonance imaging) scans and can also be seen with X-rays. Stainless steel, CoCr alloys, and titanium are the three metallic biomaterials that have found widespread use. The type of stainless steel first used in implants contained 18 wt% Cr and 8 wt% Ni. Addition of Mo for corrosion resistance led to 316 steel. Reduction of carbon from 0.08 to 0.03 wt% led to 316L steel. The 316L steel is a popular implant material. It has applications from cardiovascular to otorhinology (study of ear, nose, and throat). The CoCrMo alloy is better for high wear-resistance applications such as in artificial joints [8]. As compared to the density of 7.9 g/cm^3 for 316 steel, that of titanium is only 4.5 g/cm^3. This shows that titanium is a comparatively lightweight metal. Ti and its alloy (Ti6Al4V) display superb tensile strength (maximum stress that a material can withstand on pulling without breaking) and admirable resistance to corrosive effects (deterioration of physical properties by chemical reaction with the environment). Its alloy with nickel (Nitinol) shows shape memory effect, deforming at one temperature and recovering its original shape upon heating. The CoNiCrMo alloy is used for greatly burdened joints, e.g., the ankle joint. Ta has good visibility under X-ray exposure and low magnetic susceptibility.

Contemporary practice for hermetic packaging of implantable pacemakers uses titanium, iridium, platinum, tantalum, and their alloys. Electrical feedthroughs are also hermetically sealed.

8.4.2 Corrosion

Metal implants deteriorate by fretting, pitting, and fatigue types of corrosion. Corrosion of an implant impairs its function. It also harms the surrounding tissues by releasing undesirable chemical substances. Therefore, in order to be biocompatible, the implant must be corrosion-free. Corrosion of a metal/alloy is dependent on its composition and the environment in which it is located. Titanium is covered by a passivating oxide layer. This layer protects it against corrosion. Nonferrous metals and stainless steel have a high corrosion resistance. Chromium present in the steel oxidizes to chromium oxide. The oxide serves as a protecting film.

8.4.3 Processing of Metals

Main products from metals are produced by subjecting the ingot to rolling, milling, forging, or hammering (wrought alloys) and by casting into molds or billets (cast alloys). Powder metallurgy is another important process. Super plastic deformation and isothermal forging come under advanced processes.

8.4.4 Surface Treatment

Surface treatment is of two types. These types are: (1) surface morphological modification such as by electropolishing or (2) surface chemical modification. Ability of cells to adhere to surfaces of implants is controlled by roughness, texture, and porosity. The aim of chemical modification is to provide a particular biological response and stabilize the biomolecules on the surface. It is done on stainless steel by a process called hybrid plasma surface alloying. In this process, mixtures of nitrogen and methane gases are used at temperatures < 450 °C. A double layer structure is formed. In this structure, an exceptionally tough nitrogen-reinforced layer is created over a rigid carbon-enhanced layer [9].

8.4.5 *Surface Coating*

Biocompatibility of metal implants is improved by coating their surfaces by plasma spray or sol–gel technique. Coating of Ti6Al4V alloy by hydroxyapatite (HAP) or carbonated apatite (CAP) provides firm anchoring to the bone. It promotes in-growth of bone into the implant. Titanium nitride coating on Ti by plasma spray or physical vapor deposition helps to prevent it from surface degradation.

8.4.6 *Cleaning and Sterilization*

For cleaning, descaling is done mechanically by sand blasting. Chemical cleaning involves pickling with NaOH and H_2SO_4. Processes for sterilization include autoclaving (high-pressure saturated steam exposure at 121 °C for 15–20 min), treatment with glow discharge Ar plasma (luminous discharge between two parallel electrode plates in argon gas at low pressure), and exposure to γ-rays [10].

8.4.7 *Biodegradable Metals*

These are metals, which are supposed to provide a temporary support for healing. Subsequently, they decay of their own. Mainly alloys of Mg and Fe with other metals have been proposed [11, 12]. MgAl- and MgCa-based alloys and FeMn alloys are prominent; pure Fe is also useful.

8.5 Bioceramics

For communication between the implanted device and the external controlling device without wires or cables, the antenna of the implant has to be located outside its metal package. It thus avoids RF signal blockage by the walls of the package. But this unnecessarily increases the volume of the implant. For placing the antenna inside the package, the package material must be transparent to RF. Ceramics such as alumina serve as a substitute to metals in such applications. They permit easy RF communication of the implant with the outside world. In fact, ceramics find many more applications of different nature in implants.

8.5.1 Types of Ceramics

In terms of their chemical reactivity in physiological ambience, ceramics are generally classified into three groups: (1) bioinert ceramics, e.g., aluminum oxide showing no chemical reactivity with human body; (2) bioactive ceramics, e.g., bioactive glass which react positively with local cells and get attached by chemical bonding forces; and (3) resorbable ceramics in the form of porous or nonporous structures that are gradually substituted by bone, e.g., tricalcium phosphate (TCP) and $Ca_3(PO_4)_2$, and exhibit pronounced reactivity.

Compositionally, ceramic materials are either oxide-based or non-oxide-based. In the former class are included those based on silicon oxide (SiO_2), aluminum oxide (Al_2O_3), and zirconium oxide (ZrO_2). In the latter group are placed the black-colored silicon carbide (SiC). Other examples are silicon nitride (Si_3N_4) and aluminum nitride (AlN).

8.5.2 Dental Ceramics

Ceramics are traditionally recognized as rigid, restorative materials in dentistry. Here, they are shaped by processes such as sintering, casting, pressing, or milling. Showing good chemical durability, they are insoluble or at the most slightly soluble in oral liquid preparations. Degradation takes place either by mechanical wear or by chemical onslaught of weak acids or bases. Among composite dental ceramics may be mentioned Al_2O_3-ceramic skeleton finished with SiO_2 ceramic using lanthanum glass with 39% lithium oxide, as a coupling agent infiltrating Al_2O_3. An additive like lucite restricts the propagation of cracks. Also used as additives are chemical cleaning or fluxing agents and coloring pigments (fine powders added to a vehicle or binder), e.g., fluorescent oxides of cesium and samarium. Notable applications of ceramics are: (1) as pyrolytic carbon (wear-resistant carbon or carbon resistant to wear or normal use and biologically compatible with blood, produced by combined pyrolysis of hydrocarbon with compounds of halogens and metals) coatings for prosthetic heart valves [13], (2) as bone replacement (bioactive or surface reactive glass ceramics, able to bond with soft and/or hard tissues, and calcium phosphate ceramics employed as hard tissue prosthetic aids), (3) for wear applications in joint substitutes (alumina and zirconia ceramics), and (4) in neurosurgical cranioplasty refurbishing of the cranial bone faults to fix the skull deformity. In implantable electronics, dental ceramics are either used as full ceramic implants or as coating materials on titanium-encased implants.

8.5.3 Corrosion of Ceramics

Corrosion of ceramics releases inorganic ions such as silicon, aluminum, and potassium into the surrounding tissues [14]. The ionic leakage takes place in aqueous medium. This leakage depends on the composition of glass and environmental factors. Under severe conditions of high alkali ion concentration, breakage of Si–O–Si bonds takes place. Damaging of glass structure ensues. Selective leakage of alkali ions and dissolution of glass network are the two principal pathways of aqueous corrosion of ceramics.

8.5.4 Toxic Effects

Likelihood of systemic toxicity is extremely remote because of the exceedingly low percentages of lithium and lead. During in vitro studies, cytotoxicity has been observed only in a few ceramics. Ceramics do not cause any neoplasia or abnormal tissue growth. In general perception, ceramics are looked upon as bio-friendly materials.

8.6 Biocompatible Polymeric Materials

8.6.1 Need of Polymeric Materials

After studying metallic and ceramic biomaterials, it is understandable that the entry of polymers in this field must be supported by genuine reasons. Metals and ceramics promise long lifetime implants. But there are limitations in miniaturization, input/output pin density, mechanical flexibility, and cost-effectiveness. These limitations need to be resolved. As we are aware, small-size implantable electronic devices with a vast number of pins for input/output connections pose difficulties in adopting traditional high-temperature techniques of metal molding and welding as well as ceramic sintering. An example of such tiny electronic gadgets is the complex retinal implant. Further, the rigid contours and sharp corners of metal or ceramic implants may cause internal injury and bleeding in the patient's body. It is in these niche applications where polymeric materials have made a dent. These materials have succeeded by providing very light, miniscule, bendable implants with highly densified input/output pins [15].

8.6.2 Special Properties of Polymers

The polymers that have made ingress into the regime of metals and ceramics are polyimide (PI), polyurethane, parylene, polydimethylsiloxane (PDMS), and high-performance liquid crystal polymer (LCP). Besides various epoxies, silicones are also widely used. The above polymers offer manifold advantages. Some of these are: (1) their high resistance to biological reactions, making the implants acceptable guests in the host body; (2) their low densities making them extremely lightweight; (3) their easy molding and processing making possible the fabrication of small-size implants with larger number of input/output pins; (4) their high mechanical flexibility helping in easily joining the human tissues with implants; (5) their easy combination with nanostructured materials rendering possible the enhancement of their characteristics through nanomaterial incorporation; and (6) their low cost enabling the availability of implants at affordable prices.

Polyimide has electrical resistivity > 10^{16} Ω cm, dielectric strength of 2×10^{16} V/cm [16], and dielectric constant 3.5–4 at 1 kHz [17]. Being mechanically flexible, it is good for implantable devices. But it may buckle during implantation due to lack of stiffness.

Polyvinylidene difluoride (PVDF) is chemically nonreactive. It is highly resistant to hydrolysis. It degrades slowly but shows weak adherence to other materials [18]. It cannot form smooth films. It shows low thermal stability (up to 80 °C maximum), has varying dielectric constant (6–10) in the 1 kHz–1 MHz range [19], and is expensive.

Polyetheretherketone (PEEK) provides high mechanical strength, stiffness, fracture toughness, and corrosion resistance. It is amenable to sterilization. It is easily processable by molding or machining. Its dielectric constant is 3.1–3.6 from 50 Hz to 50 kHz. Its density (1.32 g/cm^3) is comparable to that of human bones [20]. It is a good replacement for Ti/Ti alloys [21]. High cost and dimensional instability are its main drawbacks.

PDMS gives high strength, durability, flexibility, and dielectric strength (14 MV/cm) [22, 23]. It is chemically nonreactive and nontoxic. It can be spin-coated on substrates or molded. Its high gas permeability prevents its usage for hermetic packaging. Also, it has a variable dielectric constant (2.3–3.8) from 0 to 50 Hz.

Parylene-C (poly-monochloro-para-xylylene) shows unique barrier properties from both electrical and chemical perspectives. Its dielectric performance is satisfactory, allowing very low leakage current. It has a low dielectric constant (3.15). Its chemical and biological stabilities are also agreeable. It is useful as a chemical coating, which is able to withstand both inorganic and organic media. Its biocompatibility is outstanding. It serves as a conformal protective coating. But it shows a high rate of moisture adsorption, which precludes its use in humid conditions.

LCP exhibits a highly ordered structure in solid and molten states. It is a highly crystalline thermoplastic or thermo-softening plastic. Its dielectric constant ~ 3 remains constant from 0.5 to 110 GHz. It maintains a low loss factor of 0.002–0.0045 up to 100 GHz [24]. Hence, for packaging of implantable devices, it does not affect radio-frequency performances of the devices. Its high heat resistance, chemical resistance, and mechanical flexibility are attractive qualities. It can tolerate weathering and radiation exposure. Apart from these qualities, its small coefficient of thermal expansion, low moisture uptake, stable dielectric constant, and low cost are extra advantages. Its molding temperature is 70–130 °C and melding temperature is 280–330 °C. It has grabbed the place of steel in some medical applications. On the downside, it has weak adhesion with metals and is difficult to pattern.

8.6.3 Polymer Integration

To produce polymer–metal systems, their integration with metal parts is done through surface-activated bonding (SAB). Compared to thermocompression and fusion bonding, application of pressure and heat is not required in this method. Nevertheless, it provides high strength of adherence between bonded pairs. They are stuck together by the atomic forces between surfaces which have been activated, e.g., by fast atomic beam or plasma irradiation, before bonding.

8.7 Discussion and Conclusions

Metals, ceramics, and polymers constitute the major ingredients of the body of an implanted device (Fig. 8.3).

Review Exercises

8.1 What is a biomaterial? Give some examples of biomaterials.
8.2 What negative properties must a material possess in order to be classified as a biomaterial?
8.3 Outline the three main factors due to which the concept of biocompatibility had to be reexamined and its new definition had to be framed? State this new definition.
8.4 What is meant by tissue response continuum? What types of blood–material interactions take place after an electronic device is implanted?
8.5 What is the role of inflammation in restraining the injury-causing process? What is meant by: (a) acute inflammation and (b) chronic inflammation?
8.6 What does foreign body reaction consist of? What is the part played by fibrosis as a host reaction to the medical implantation?
8.7 Name three metallic biomaterials that have been widely used. Discuss their principal features.
8.8 What are the harmful effects of corrosion of an implant? What protective films formed on titanium and chromium help in preventing them against corrosion?
8.9 Name two types of surface treatment done on implantable medical devices. How is chemical modification of steel done?

Fig. 8.3 Examples of main biomaterials for implantable devices

8.10 How does surface coating of an implant improve its biocompatibility? Give an example.
8.11 What surface coating is done on a titanium implant for biocompatibility?
8.12 How are cleaning and sterilization of a medical implant done?
8.13 What is a biodegradable material? Give examples.
8.14 Discuss with examples some important medical applications in which bioceramics are required in place of metals.
8.15 Describe by giving examples the three categories into which bioceramics are subdivided. Give some examples of oxide-based and non-oxide-based bioceramics. What are dental ceramics?
8.16 Why are ceramics said to be bio-friendly materials?
8.17 What are the main reasons which have made the study of polymeric biomaterials important?
8.18 Name five biocompatible polymers, and discuss their main properties.

References

1. Tathe A, Ghodke M, Nikalje AP (2010) A brief review: biomaterials and their application. Int J Pharm Pharm Sci 2(4):19–23
2. Wang X (2013) Overview on biocompatibilities of implantable biomaterials, chap 5. In: Pignatello R (ed) Advances in biomaterials science and biomedical applications. InTech, pp 111–155. https://doi.org/10.5772/53461
3. Williams DF (2008) On the mechanisms of biocompatibility. Biomaterials 29(20):2941–2953
4. Williams DF (2009) On the nature of biomaterials. Biomaterials 30:5897–5909
5. Elshahawy W (2011) Biocompatibility, chap 15. In: Sikalidis C (ed) Advances in ceramics—electric and magnetic ceramics, bioceramics, ceramics and environment. InTech, pp 359–378
6. Anderson JM (2001) Biological responses to materials. Annu Rev Mater Res 31:81–110
7. Anderson JM, Rodriguez A, Chang DT (2008) Foreign body reaction to biomaterials. Semin Immunol 20(2):86–100
8. Hermawan H, Ramdan D, Djuansjah JRP (2011) Metals for biomedical applications. In: Fazel R (ed) Biomedical engineering—from theory to applications. InTech, pp 411–430
9. Sun Y, Haruman E (2008) Influence of processing conditions on structural characteristics of hybrid plasma surface alloyed austenitic stainless steel. Surf Coat Technol 202(17):4069–4075
10. Serro AP, Saramago B (2003) Influence of sterilization on the mineralization of Ti implants induced by incubation in various biological model fluids. Biomaterials 24(26):4749–4760
11. Xin Y, Liu C, Zhang X et al (2007) Corrosion behavior of biomedical AZ91 magnesium alloy in simulated body fluids. J Mater Res 22(7):2004–2011
12. Schinhammer M, Hänzi AC, Löffler JF et al (2010) Design strategy for biodegradable Fe-based alloys for medical applications. Acta Biomater 6(5):1705–1713
13. Sarkar R, Banerjee G (2010) Ceramic based bio-medical implants. Int Ceram Rev 59(2):98–102
14. Elshahawy W, Watanabe I, Koike M (2009) Elemental ion release from four different fixed prosthodontic materials. Dent Mater 25(8):976–981
15. Qin Y, Howlader MMR, Deen MJ et al (2014) Polymer integration for packaging of implantable sensors. Sens Actuators B 202:758–778
16. Stieglitz T, Schuettler M, Meyer J-U (2000) Micromachined, polyimide-based devices for flexible neural interfaces. Biomed Microdevices 2:283–294
17. Jiang J-S, Chiou B-S (2001) The effect of polyimide passivation on the electromigration of Cu multilayer interconnections. J Mater Sci 12:655–659
18. Scheirs J, Burks S, Locaspi A (1995) Development in fluoropolymer coatings. Trends Polym Sci 3:74–82
19. Mark JE (2009) Polymer data handbook. Oxford University Press, New York, 1264 pp
20. Kurtz SM, Devine JN (2007) PEEK biomaterials in trauma, orthopedic, and spinal implants. Biomaterials 28:4845–4869
21. Rivard C-H, Rhalmi S, Coillard C (2002) In vivo biocompatibility testing of peek polymer for a spinal implant system: a study in rabbits. J Biomed Mater Res 62:488–498
22. Lötters JC, Olthuis W, Veltink PH et al (1997) The mechanical properties of the rubber elastic polymer polydimethylsiloxane for sensor applications. J Micromech Microeng 7:145–147
23. Armani D, Liu C, Aluru N (1999) Re-configurable fluid circuits by PDMS elastomer micromachining. In: MEMS'99. Twelfth IEEE international conference on micro electro mechanical systems, Orlando, pp 222–227
24. Thompson DC, Tantot O, Jallageas H et al (2004) Characterization of liquid crystal polymer (LCP) material and transmission lines on LCP substrates from 30 to 110 GHz. IEEE Trans Microwave Theory Tech 52:1343–1352

Chapter 9
Batteries for Implants

Abstract Batteries for implants must possess characteristics such as safety, reliability, high volumetric energy density, low self-discharge, and long duration of service, which represent essential commitments from manufacturers. The state of discharge must be indicated. In the primary batteries, lithium metal anodes are used. The cathode systems include iodine, manganese oxide, carbon monofluoride, silver vanadium oxide, and crossbreed or hybrid cathodes. This choice of batteries caters to the power levels required by implantable devices, which are spread over a broad range of current values from microampere to ampere levels. Limited battery life is a major impediment to the development of advanced medical implant devices, e.g., when a pacemaker battery runs out, it has to be replaced by surgery. With progressive shrinkage of implant size, more emphasis is laid on building smaller, longer-lasting batteries. Applications involving high power usage rates such as neurostimulators working at milliwatt powers employ secondary rechargeable batteries to achieve longer life span with reduced size.

Keywords Battery · Lithium battery · Iodine cathode · Manganese dioxide cathode · Carbon monofluoride cathode · Silver vanadium oxide cathode · Lithium-ion battery

9.1 Introduction

The battery is a system for storing electrochemical energy. Strictly speaking, the battery comprises a multitude of interconnected cells, joined together in a series combination to yield the necessary voltage, and in a parallel connection for achieving the desired current capacity. Each cell comprises two electrodes: a positive electrode and a negative one. These electrodes are separated by a solution of an electrolyte, usually a salt which dissociates into constituent ions, to allow transference of ions from one electrode to the other. Upon connecting these electrodes electrically, the chemical reactions advance in tandem at the two electrodes. These chemical reactions, occurring at the same time, release electrons and enable the valving of current,

V. K. Khanna, *Implantable Medical Electronics*,
https://doi.org/10.1007/978-3-031-99868-3_9

either its controlling or halting, by the device connected to battery. To facilitate understanding, the battery terminology is briefly presented in Table 9.1.

Because of their capability to provide high energy density and afford wide design flexibility, Li-based batteries (Table 9.2) have superseded other battery systems.

Table 9.1 Battery terminology

S. No.	Term	Meaning	Unit/measure (if applicable)
1.	Battery	A device used for transformation of chemical energy into electrical energy in the form of direct current and, conversely, electrical energy into chemical energy	–
2.	Primary battery	A non-rechargeable or non-reenergizeable battery	–
3.	Secondary battery	A rechargeable or reenergizeable battery	–
4.	State of charge (SOC)	Available battery capacity expressed as a fraction of its rated capacity	%
5.	Depth of discharge (DOD)	Complement of SOC, it expresses how deeply a battery is depleted. Specified as a fraction of battery capacity that has been discharged/its maximum capacity	%
6.	Terminal voltage	Output voltage determined across the poles of the battery when load is applied	V
7.	Open-circuit voltage	Output voltage ascertained across the poles of the battery without application of load	V
8.	Nominal voltage	Reported or imprinted voltage of the battery	V
9.	Cutoff voltage	Voltage defining the fully discharged state of the battery	V
10.	Capacity/ nominal capacity/ coulometric capacity	Electric charge stored in the battery = current drawn from a battery × the number of hours the current flows	Ah or mAh
11.	Gravimetric energy density/ specific energy	Energy storage efficiency expressed as nominal battery energy per unit mass; energy-to-weight ratio	W/kg
12.	Volumetric energy density	Energy storage efficiency expressed as nominal battery energy per unit volume; energy-to-volume ratio	W/L
13.	Power density	Maximum available power per unit volume	W/L
14.	Shelf life	Period during which a battery preserves a certain fraction of its capacity when not being used	h

Today, they lead in portable battery sales worldwide [1]. Primary and secondary lithium batteries have already become the premier prototypical sources of power. They offer opportunities for uncontaminated and efficient storage as well as conversion of electrochemical energy. Every day, they are getting more importance for powering modern society. The term "lithium batteries" encompasses the entire gamut of battery systems in which lithium is used as the anode material. However, these batteries differ in features such as cathode material, electrolyte, cell design, and other mechanical aspects. Further, these batteries are wrapped in an assortment of form factors such as coin, cell, and film or prismatic. Each lithium system is distinguished by its inherent characteristics, e.g., electrical properties, energy density, operating temperature, reliability, and safety.

A fundamental reason for the use of lithium is that it is the lightest metal. It has a density about half that of water $= 0.534\ g/cm^3$. Hence, if lithium was not so reactive, a chunk of lithium metal will float on water. However, it reacts vigorously with water. Other notable features of lithium are that it is the third element in the periodic table. Lithium has the smallest atomic number of any metal = 3. So, the number of protons in the lithium atom = 3. Electronic configuration of lithium is $1s^2\ 2s^1$. It belongs to the alkali metal group, the Group 1 of the periodic table of chemical elements. It is soft, silver-white in color, and flammable.

A lithium atom loses one electron in its outermost shell to acquire a stable atomic configuration. In chemistry, this tendency is described as an electropositive behavior. Electropositive elements are those which lose electrons to become positively charged. Lithium is one of the highly electropositive elements. In aqueous solution, the standard potential for the cathode half-reaction Li^+ (aqueous) $+ e^-$ (electron) = Li (solid)

Table 9.2 Some lithium-based batteries

S. No.	Battery	Anode	Cathode	Electrolyte	Open-circuit potential (V)
1.	Li/I_2–PVP	Lithium	I_2/PVP = 30:1 to 50:1	Lithium iodide	2.8
2.	Li–MnO_2	Lithium	MnO_2 + conductive agents	Organic solvent mixture	3.1–3.3
3.	Li/CF_x	Lithium	CF_x	$LiBF_4$ dissolved in γ-butyrolactone or GBL ($C_4H_6O_2$)	3.1
4.	Li/SVO	Lithium	SVO	Propylene carbonate or PC having formula $CH_3C_2H_3O_2CO$ and dimethoxyethane ($C_4H_{10}O_2$) with a lithium salt in the form of a solution	3.2
5.	Li-ion	Graphite (carbon)	Lithiated cobalt oxide	Lithium salts, such as $LiPF_6$, $LiBF_4$, or $LiClO_4$, dissolved in ether ($C_4H_{10}O$)	3.7

versus the hydrogen electrode is − 3.04 V. The negative sign indicates that Li has a tendency to be oxidized rather than reduced in comparison with hydrogen. In other words, lithium is a reducing agent. The standard potential of lithium far exceeds that of magnesium, calcium, and sodium. The extreme lightness of lithium combined with its electropositive character has led to its prolific use in battery technology. Due consideration is taken of its reactivity. Among the other properties that speak in favor of lithium for use in batteries may be mentioned its low volume resistivity (9.5×10^{-6} Ω cm), high ionization energy (5.39 eV), and low melting or liquefaction point (180.5 °C). Its comparative plenitude (0.0017% in earth's crust), as shown by its occurrence in several materials, further strengthens its case. Examples of these materials are continental, geothermal, and oilfield brines. The spectacular electrochemical equivalent of lithium is 0.259 g/Ah. It clearly overrides all other contestants. Notably, Mg, Ca, and Na have electrochemical equivalents of 0.45 g/Ah, 0.75 g/Ah, and 0.86 g/Ah, respectively. Although Al is a close follower with 0.34 g/Ah, its low standard potential of − 1.7 V leads to its exclusion from the race.

Lithium shows almost ideal fundamental properties, amply justifying its use as the anode in both primary and rechargeable batteries. Therefore, scientists have associated the well-accepted and voguish lithium anode with a large number of cathode materials and different electrolytic solutions. This pairing has resulted in the spacious choice of various chemistries. Lithium batteries are subdivided into two classes: primary cells for one-time use and secondary or rechargeable cells for multiple usages. The primary cells contain lithium metal anodes. They represent the preferred battery for cardiac pacemakers. Secondary cells exploit lithium-ion chemistry. They are used with neurostimulators [2].

9.2 Lithium/Iodine–Polyvinylpyridine Battery

Beginning from the work of Gutmann et al. [3] in 1967, patented by Schneider and Moser in 1972 and Moser in 1972, it supplies current in the microampere range. It is the favorite choice for implantable cardiac pacemakers. Since 1972, the battery has undergone drastic improvement. Its energy has reached three times that of the first battery. It has an exemplary track record of reliability and safety [4].

The anode of this battery is made of lithium (Fig. 9.1). The cathode is formed by the thermal reaction of iodine and polyvinylpyridine, $(C_6H_9NO)_n$, a water-soluble polymer made from *N*-vinylpyrrolidone, C_6H_9NO. The (I_2/PVP) ratio varies from 30:1 to 50:1. The reaction between iodine and PVP is accompanied by evolution of heat. It is a highly exothermic reaction. This reaction starts a little below the melting point of iodine (113.5 °C). In the resulting polymeric structure, one iodine molecule is attached with nitrogen atom on the pyridine ring (an aromatic heterocyclic compound characterized by a 6-membered ring structure composed of 5 carbon atoms and 1 nitrogen atom; chemical formula C_5H_5N), and an iodine atom substitutes the alpha hydrogen atom (a hydrogen atom on an alpha carbon, i.e., a carbon atom directly bonded to an atom, group, functional group, or other moiety in an organic molecule) on the polymer chain. This polymeric structure is a conductor showing electronic conductivity. The conductivity varies with the I_2/PVP ratio. The peak value of conductivity occurs at the I_2/PVP weight ratio of 8:1. Since the starting value of this ratio is considerably higher, the electronic conductivity increases until it attains a weight ratio of 8:1. Subsequently, it begins to decrease with the progression of cell reaction.

The electrolyte in this battery is lithium iodide. It is a solid electrolyte or superionic conductor showing a high ionic conductivity due to the fast diffusion of one ionic species in a framework or lattice constituted by ions of opposite sign. The electrolyte also functions as a cell separator. It is a product of the basic cell reaction between lithium and elemental iodine: $Li + ½I_2 \rightarrow LiI$. The Gibb's free energy of this reaction, i.e., the energy that can be used to do work, is − 64.45 kCal/mol. The resultant open-circuit voltage from the cell is 2.8 V. The construction of the battery involves formation of a LiI layer at the anode by the addition of molten cathode material (I_2–PVP) to the Li anode. In one time-consuming method, a solution of the organic

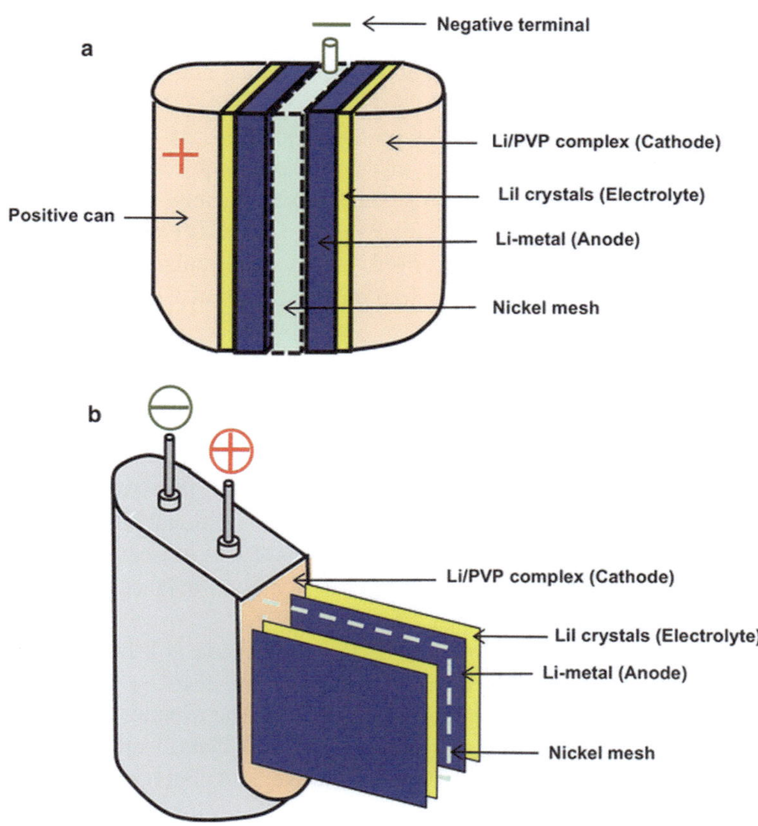

Fig. 9.1 Layers in the Li/I_2–PVP battery consisting of two cells: **a** front view of the cell and **b** cutaway view of cell opened from side to show the layers

material in a solvent is carefully smeared on the anode by painting or brushing. In another method, a sheet of the organic material is performed by hot pressing. The sheet is applied to the anode surface using an adhesive. This requires a greater amount of organic material. Also, it is difficult to retain flexibility of the sheet under dry conditions. In a further improved method [5], the substrate material is flexible. It is preferably a synthetic open mesh fabric. The applied film is also flexible. It can be applied over a smooth or irregular surface. The method provides greater manufacturing efficiency, better uniformity, and reduced cell-to-cell variability.

Discharging of the battery is accompanied by growth and thickening of the LiI layer. Exhaustion of Li and I_2 takes place. The cell impedance increases. The battery voltage decreases. The same is easily detected by the electronic circuit in the pacemaker. Thus the discharge profile of the battery is an indicator of its condition. It allows the doctor to know the end-of-service point of the battery. The doctor can therefore plan timely preventive scheduling of the surgery for battery replacement.

Historically, the first commercially triumphant Li/I_2–PVP battery was large in size (14 mm × 45 mm × 52 mm). It had a capacity of 3.5 Ah. The second smaller size design had a capacity of 1.4 Ah. As mentioned in Table 9.1, the ampere-hour (Ah) representing a unit of electric charge, is the rating of the battery capacity. A battery which is able to supply a current of 1 A to a load nonstop for 1 h is said to have a capacity of 1 Ah.

The first case-grounded battery was produced in 1975. Also, noting that stainless steel was not attacked in absence of water, the unrequired inert material was eliminated from the design. The energy density was thereby tripled. Subsequent improvements included the increase of I_2–PVP ratio from 10:1 to 30:1 for achieving higher energy density. The use of a grooved, wrinky anode enlarged the surface area and hence the capability of the battery for delivering current.

9.3 Lithium–Manganese Dioxide Battery

Launched into commercial market by Ikeda et al. [6, 7], this battery is used for medium rate applications. Some of the applications are additional functionality pacemakers, neurostimulators, and drug delivery systems. These devices/systems require power at milliwatts level. Li–MnO_2 system offers a good balance of performance and safety for consumer applications.

The anode is made of pure lithium metal. The cathode is a high conductivity mixture of heat-treated electrolytic manganese dioxide (MnO_2) with conductive agents. The electrolyte is an organic solvent mixture. The mixture contains mainly ethers, with an alkali metal salt dissolved in it (Fig. 9.2). This electrolyte is liquid at room temperature and normal pressure. It is nontoxic and noncorrosive. Electrolyte filling is done by creating a vacuum inside the cell. The cell is hermetically sealed. There is no internal cell pressure. The safety vent opens if the internal temperature or pressure of the cell exceeds predefined limits. A separator is placed between the

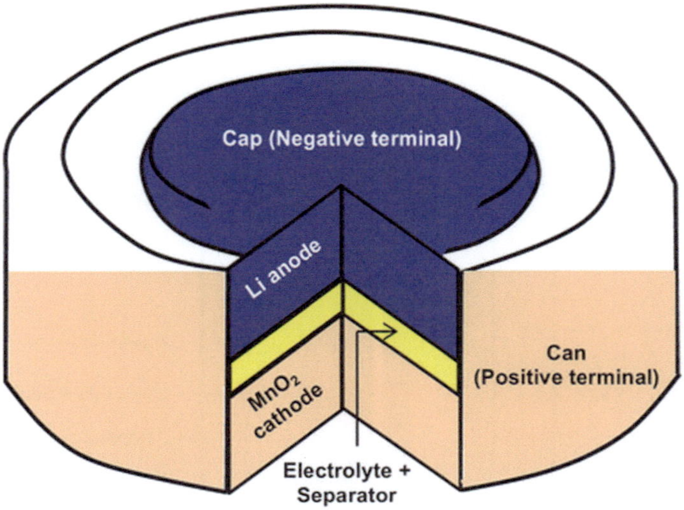

Fig. 9.2 Li–MnO_2 button cell

anode and the cathode. In one version of separator, a polyethylene layer is sandwiched between two polypropylene layers.

A larger battery voltage in the initial stage as well as during discharge is possible due to the high conductivity of the cathode. Also, reliable performance after long storage periods is assured by the thermodynamically stable MnO_2. The cell has a long shelf life up to 10 years. High ionic conductivity together with low viscosity of the electrolyte permits efficient utilization of cathode over varying temperatures and during rapid discharge periods.

At the anode, lithium is oxidized to Li^+ ions: $Li \rightarrow Li^+ + e^-$ (electron). Inside the cell, the Li^+ ion diffuses through the electrolyte and separator toward the cathode. Outside the cell, the electron moves from anode to cathode. At the cathode, Li^+ ion, tetravalent MnO_2, and electron combine. Tetravalent MnO_2 is reduced to trivalent MnO_2. The solid product formed during the discharge reaction is left behind in the cathode. The discharge reaction is not accompanied by the liberation of any gaseous product. Hence, there is no buildup of pressure inside the cell. So, a pressurized situation does not develop, increasing the safety. The cell reaction is $Li + (MnO_2)^{4+} \rightarrow Li^+ + e^- + (MnO_2)^{4+} \rightarrow Li^+(MnO_2)^{3+}$. Open-circuit voltage of the cell ranges from 3.1 to 3.3 V. Therefore, a single Li–MnO_2 cell replaces two conventional cells.

The voltage profile during discharge comprises three distinctive regions. The first region is characterized by a gradually decreasing voltage. The second relatively long-duration region is marked by a flat voltage–time graph showing constancy of voltage. In the third region, the voltage slowly decreases toward the end of cell life [8–10]. These three regions correspond to the three stages of insertion of lithium into the manganese dioxide lattice. The first and the third regions represent the insertion of

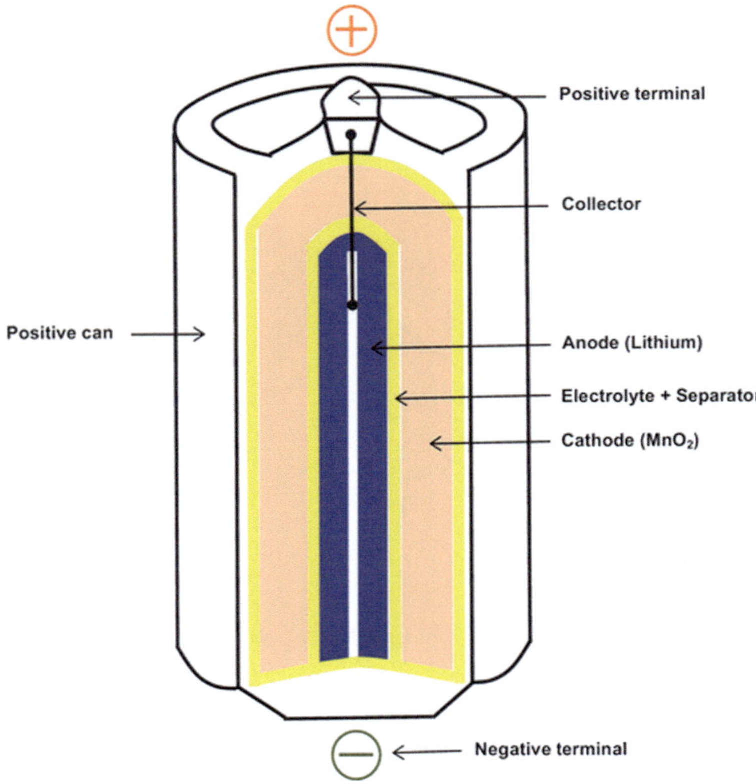

Fig. 9.3 The basic components of a lithium solid-core battery

lithium into the MnO_2 lattice via a homogeneous reaction, i.e., one that occurs in a single phase. The second region pertains to a two-phase reaction.

Lithium batteries are made in either of the two versions: solid core (Fig. 9.3) or wound (Fig. 9.4).

9.4 Lithium/Carbon Monofluoride Battery

This battery is an alternative choice for milliwatt power range because of its low internal impedance. Its high energy density guarantees long life operation for implantable devices working at medium rate currents. Predictability of its discharge profile allows electronic determination of end point. Low self discharge and reliable cell chemistry are additional benefits.

The cell is made of a lithium anode, a carbon monofluoride (CF_x) cathode, and lithium tetrafluoroborate ($LiBF_4$) electrolyte dissolved in γ-butyrolactone. Highly stable carbon monofluoride is obtained by fluorination of carbon with fluorine at

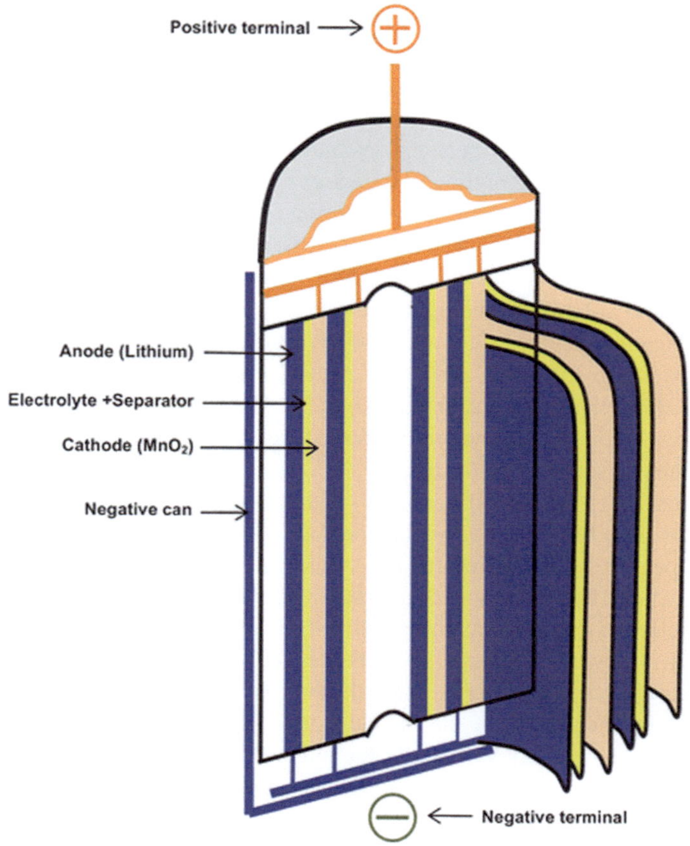

Fig. 9.4 A wound lithium cylindrical battery

400–600 °C. Usually, a carbon substrate such as graphite powder is exposed to fluorine gas at high temperature. This exposure creates a material where fluorine is intermixed with carbon at a molar ratio near 1:1, but generally not precisely 1:1 [11]. These materials are collectively referred to as carbon monofluoride. They often range from $CF_{0.68}$ to $CF_{1.12}$. The structure of CF_x contains sp^3 hybridized layers of carbon in which the fluorine atoms are covalently bonded to carbon. A higher fluorine percentage decreases the electrical conductivity of CF_x. Therefore, a conductive additive is included along with a binder to allow pressing of the cathode pellet. Li/CF_x batteries permit customization or tuning of the cathode according to specific applications. Variation of the incorporation of fluorine inside the carbon structure at the atomic scale at the time of bulk production leads to these changes conforming to individual or personal specifications. Thus properties of the battery are modified to obtain higher energy densities. By such customization process, characteristics of the battery can also be altered in other ways to achieve the performance desired by the user.

The cell operates through the chemical reaction: $x\text{Li} + \text{CF}_x \rightarrow x\text{LiF} + \text{C}$. It gives an open-circuit potential of 3.1 V [12]. The discharge occurs initially through inclusion of solvated Li^+ ions within the fluoride stratum. Consequently, a transitional phase is formed. This phase consists of carbon, fluoride, and lithium ions. It is a diffusion layer which slowly decomposes into LiF, C, and solvent molecules.

The discharge potential of the cell increases by heat treatment of CF_x at a temperature lower than that causing its decomposition [13]. Carbothermal treatment using a combination of heat with carbon black increases the discharge performance [14].

Comparing Li–MnO_2 battery with Li/CF_x type, Li–MnO_2 batteries have lower gravimetric and volumetric energy densities, typically 200 W/kg and 550 W/L than Li/CF_x contenders whose values are ~ 700 W/kg and 850 W/L. Ranges of working temperature for the Li–MnO_2 and Li/CF_x batteries are – 20 to – 60 °C and – 60 to – 160 °C, respectively. Corresponding shelf life ranges are 5–10 and 15 years. The only disadvantage of Li/CF_x battery is its moderately higher price. This higher price far outweighs its superior performance.

Rangasamy et al. [15] found that a new solid, bifunctional electrolyte Li_3PS_4 (LPS) can generate 26% higher capacity than the theoretical maximum if each component acted independently. It also extends the life span of the device. The increase in capacity is caused by the supportive interactions between the electrolyte and cathode. The solid electrolyte serves as the chemically inactive electrolyte at the anode. It also acts as an active ingredient at the cathode. With the discharging of the battery, a lithium fluoride salt is produced. This salt promotes the electrochemical activity of the electrolyte. It converts the electrolyte from an inactive to an active component of the battery. This finding supersedes the hitherto prevalent belief that in a battery the anode, cathode, and the electrolyte can play only single roles. This is because of the dual role shown for the electrolyte. The revolutionary idea will help in designing batteries of unparalleled energy densities.

9.5 Lithium/Carbon Monofluoride–Silver Vanadium Oxide Hybrid Battery

This battery supplies superior capability for pulsatile current flow than a Li/CF_x battery can do on its own. Besides supplying a higher power density than Li/CF_x cells, the provision of a broad high-level region or tableland at a lower potential toward the completion of discharge affords easier end point detection. It has mixed-breed cathodes conglomerated from CF_x and silver vanadium oxide (SVO), $Ag_2V_4O_{11}$. In one design, the two are physically mixed together. In the other design for high current pulses, a laminated cathode is used. In the physically mixed hybrid cell, at low current rates, first the partial reduction of SVO takes place. In the next stage, CF_x is completely discharged. Finally, the residual SVO is reduced [16]. At higher rates of current, SVO delivers a major portion of energy. The reason is its high current-carrying capability. This decreases its potential with respect to CF_x. So, it is recharged by CF_x. By that means, the potentials of SVO and CF_x are in equilibrium. Depending on the application, the composition of the mixture is varied to fulfill the desired capacity or power density need.

9.6 High-Rate Lithium/Silver Vanadium Oxide Battery

This battery has the capability to supply intermittent current pulses of magnitudes 2–3 A for charging the capacitors of implantable cardioverter defibrillators as well as supplying a continuous current of small value for pacing the cardiac system. Therefore, it is commonly used in defibrillators. The cell has a Li anode. The silver vanadium oxide cathode is prepared by a solid-state synthesis route. Its layered structure embodies edge and corner sharing, distorted, octahedral VO_6 structures. The electrolyte consists of propylene carbonate ($C_4H_6O_3$) and dimethoxyethane ($C_4H_{10}O_2$) with a lithium salt in solution. There are two cell designs: (1) either multi-plate cathode type in which several cathodes are parallelly connected to increase the output or (2) a flattened coil in which the anode and cathode strips are wound together. The discharge reaction is: $7Li + Ag_2V_4O_{11} \rightarrow 2Ag^0 + Li_7V_4O_{11}$. It provides an open-circuit potential of 3.2 V [17, 18]. The discharge product is $Li_xAg_2V_4O_{11}$, where x denotes the number of moles of lithium intercalated into SVO. Intercalation is the reversible insertion of a material between layers in a lattice. The Ag^+ ion is reduced to Ag^0 from $0 < x < 2.4$. Crystallinity is lost during this process. Silver reduction improves the capacity. So, high-rate applications are catered to. Earlier in the discharge process, vanadium reduction takes place. But from $2.4 < x < 3.8$, the reduction of V^{5+} to V^{4+} dominates. At $x > 3.8$, during further reduction, V^{4+} changes to V^{3+}. Thus, mixed valence materials containing V^{3+}, V^{4+}, V^{5+} ions are formed. Due to this multistage reduction, the potential profile is a stepped one. This profile is beneficial for fibrillator use. It allows physicians to envisage the terminal stage of battery.

The superior electrochemical performance of nanostructured SVO as compared to bulk SVO can be utilized to enhance the battery function [19]. If nanoscale SVO particles are used, the movement of Ag^+ and Li^+ ions in the SVO skeleton becomes faster. The transport becomes faster due to their shorter diffusion lengths. In addition, the exceptional electronic properties of nanoparticles result in higher capacities and cell voltages.

9.7 High-Rate Lithium–Manganese Dioxide Battery

This battery has the capability to deliver current pulses of large magnitudes. It is a useful optional choice for implantable cardioverter defibrillators. Here, the cathode is a mixture of MnO_2 with chromium oxide and lead oxides. This mixture produces a profile of battery discharge, which clearly signposts the point at which cell replacement is warranted [20]. Augmentation of cell performance is achieved by employing a double cell design [21]. In this design, two cathode-limited series-connected cells are employed in a single battery enclosure for reducing the volume. Still further performance amelioration is accomplished by integrating cathode-limited double cells into a mixed structure. Consequently, a greater value of low rate capacity is obtained at a reduced consumption of volume. There is a trivial deterioration in the capability to provide pulsed power. Pulsed power is provided by storing energy for a long time and releasing it suddenly in the form of a high-energy pulse so that instantaneous power delivered has a very high value [22]. The discharge curve of Li–MnO_2 cell for use in defibrillator devices has been mathematically modeled [23]. The model enables prediction of cell behavior.

9.8 High-Rate Lithium/Carbon Monofluoride–Silver Vanadium Oxide Hybrid Battery

It can deliver high current pulses like Li–MnO_2 and Li/CF_x–SVO batteries and serves identical function. A novel electrode design-based Li/CF_x–SVO hybrid cell gives high current pulse output for defibrillation. It employs a cathode consisting of CF_x juxtaposed between two SVO layers [24]. The cells exhibit a stepped discharge profile with plateaus at 3.2, 2.8, and 2.6 V. These plateaus pertain to the discharging of SVO, CF_x, and SVO layers, respectively. At the beginning, SVO is reduced. Then CF_x supplies a major portion of energy for low rate currents. But in the event of heavy load pulsing of the cell, preferential discharging of SVO takes place. Then the resultant potential of silver vanadium oxide is less than that of carbon monofluoride. The profile of cell discharge has the shape of a series of steps; hence, it is called a stepped profile. This kind of profile readily signifies the condition when the battery is down.

9.9 Secondary Lithium-Ion Battery

It is a battery for milliwatt power level. It can be recharged in the implanted state to the original pre-charge condition. It is used in neurostimulators working in the milliwatt power range. As for primary batteries, essential qualities expected for these batteries are also safety and reliability. Over and above, the battery must have high energy density. Low self-discharge is another virtue. These qualities cannot be least compromised with for implantable devices.

The working of secondary cells is based on lithium-ion cell technology [25]. Early efforts to produce lithium batteries that could be recharged for reuse were frustrated by their substandard cycling characteristics. Problems concerning safe handling of metallic lithium, a reactive caustic metal, further aggravated the issue. These attempts were not successful until a carbon anode was used. This anode permitted the intercalation (insertion) of ionic lithium and its pairing with a high-voltage cathode. This was done using lithiated cobalt oxide. A substitute cathode material is lithium iron phosphate ($LiFePO_4$). It has a marginally lower capacity as rivaled against lithium cobalt oxide ($LiCoO_2$). But its conductivity is appreciably lower. It is inexpensive and also displays lower chemically reactivity.

The Li-ion cells have a carbonaceous anode. The cathode is made of a metal oxide (Fig. 9.5). The electrolyte is a mixture of lithium salts, e.g., $LiPF_6$, $LiBF_4$, or $LiClO_4$. These salts are dissolved in an organic solvent like ether. In these cells, both the electrode materials can intercalate lithium ions reversibly. Materials used in cathode manufacture are nickel-, manganese-, and cobalt-based oxides. $LiCoO_2$ is very frequently used. Its structure is composed of interspersed layers of $LiCoO_2$, interchanging repetitively with one another [26]. The layered structure facilitates the intercalation of lithium ions. The cell construction involves the use of a discharged cathode as the Li^+ ion source. Initial charging digs out lithium from the cathode and infuses it into anode (Fig. 9.6). Upon discharging (Fig. 9.7), a reversal of this process occurs. Unnecessary heating and rupturing of the cell may take place on over charging the cell. The same happens in case of a short-circuiting incident. Such an incident must be always avoided. A high energy density is obtained from cells using $LiCoO_2$. The cycle life is ~ 500–700 discharge cycles. These cycles are of deep discharge nature.

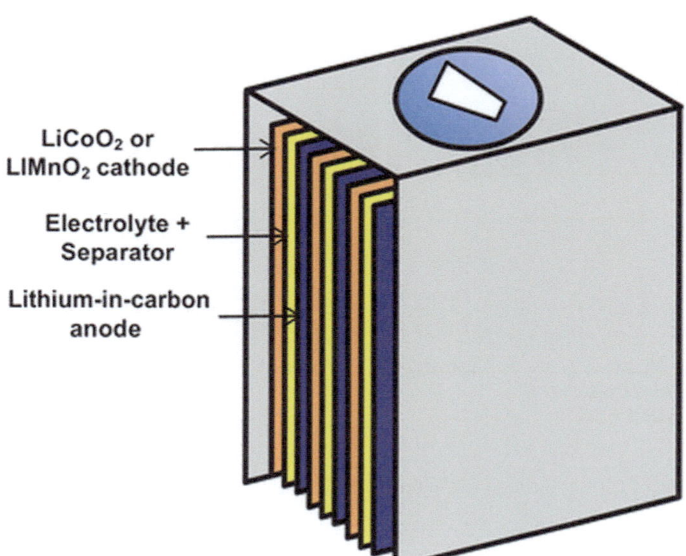

Fig. 9.5 Lithium-ion battery

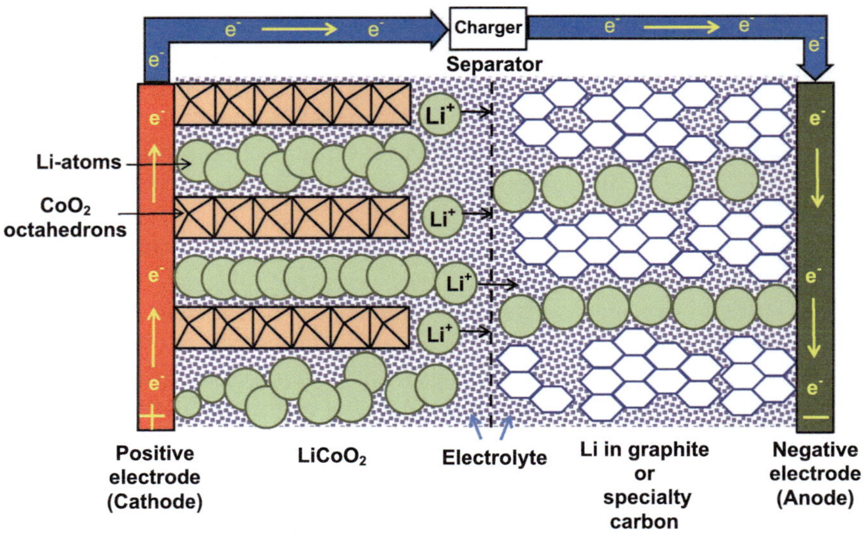

Fig. 9.6 Flow of current through a lithium-ion battery during charging. In $LiCoO_2$, layers of lithium are placed between broad flat pieces of octahedrons formed by cobalt and oxygen atoms

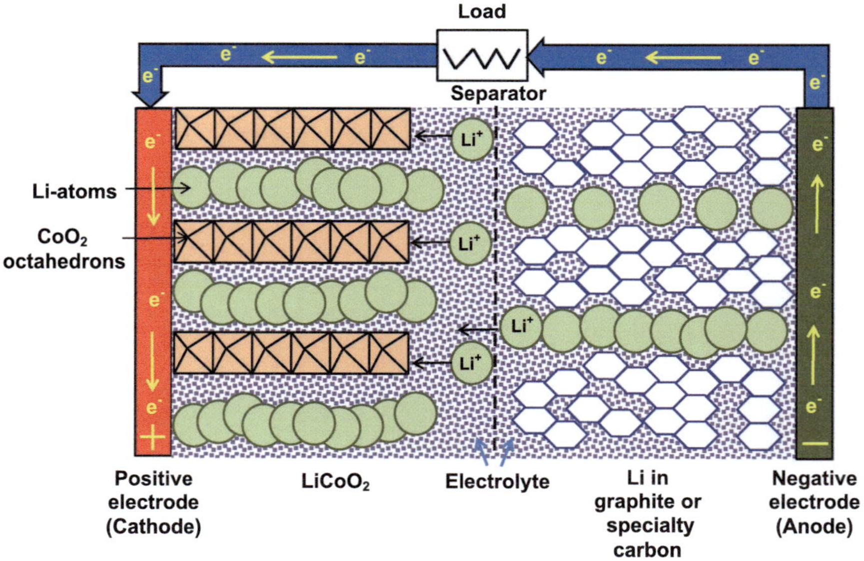

Fig. 9.7 Flow of current through a lithium-ion battery during discharging

9.10 Discussion and Conclusions

A broad panorama of primary cells has been developed for implantable devices. These primary cells use lithium anodes in combination with various cathode materials. They are distinguished by their capabilities to deliver currents ranging from microamperes to a few amperes in magnitude. Single-ingredient cathodes include those made of I_2, MnO_2, CF_x, and $Ag_2V_4O_{11}$. Among cathodes of hybridized variety, mention may be made of cathodes made by interleaving CF_x and $Ag_2V_4O_{11}$ layers. Apart from the aforesaid primary cells, which are not chargeable, there exist secondary rechargeable cells based on Li^+-ion chemistry. These can be recharged from outside while remaining implanted in the human body.

Irrespective of the technology used, performance parameters of the battery such as cell potential, capacity, or energy density depend upon the deep-rooted, innate properties of the formative materials of the anodes and cathodes. The lifetime is a function of the properties of interfacial regions separating the electrodes from the electrolyte. Safety is determined by the stability of the constructional materials used for making electrodes and also depends on the nature of interfaces within the cell. Improvements can be made by changes in chemical design of the battery and engineering of the cell structure. The optimum performing electrode/electrolyte/electrode combination is found through the selective use of prevailing and up-and-coming materials to fabricate the two electrodes of the cell. Suitable electrolyte arrangement is one that minimizes adverse reactions occurring at the confluence and interaction planes between the electrode and the electrolyte.

Review Exercises

9.1 What is a battery? How does it differ from a cell? Define the following terms for a battery: (1) nominal voltage, (2) open-circuit voltage, and (3) cutoff voltage.

9.2 What are primary and secondary batteries? Are secondary batteries indefinitely rechargeable?

9.3 What is meant by capacity of a battery? Explain the meanings of the terms, "gravimetric energy density" and "volumetric energy density".

9.4 Name the lightest metal. To which group of the periodic table does it belong?

9.5 What are the reasons for the widespread use of lithium in batteries? Is lithium an electropositive element? What is its standard electrode potential? Explain the significance of the negative sign in the electrode potential.

9.6 Is "lithium battery" same as "lithium-ion battery"? If not, what is the difference? Which one is discarded after one-time use?

9.7 Name the anode, cathode, and electrolyte materials in the following batteries: (1) Li/I_2–PVP, (2) Li–MnO_2, and (3) Li/CF_x.

9.8 How is the I_2/PVP cathode made in the lithium battery? How does the I_2/PVP ratio influence the conductivity of the electrolyte?

9.9 What is the basic cell reaction in a Li/I_2–PVP battery? What value of open-circuit potential is obtained?

9.10 What charge carrier processes occur in a Li–MnO_2 cell at different locations: (1) at the anode, (2) inside the cell, (3) at the cathode, and (4) outside the cell.
9.11 What is the chemical reaction through which a Li/CF_x cell operates? Prepare a comparative chart highlighting the relative advantages and disadvantages of Li–MnO_2 and Li/CF_x cells.
9.12 How is it possible to cross the theoretical barrier of capacity of a Li/CF_x cell using a bifunctional electrolyte? What are the two roles played by the electrolyte?
9.13 What hybrid batteries have been developed with capability of supplying higher values of pulsed currents than CF_x alone? What other notable advantage is offered by these batteries?
9.14 Describe the discharge behavior and characteristics of Li/CF_x–SVO hybrid cells using the physically blended cathode model.
9.15 Point out the constructional difference of a (Li/CF_x–SVO) hybrid battery from a simple Li/CF_x battery? What significant advantage is derived by this difference?
9.16 What is the discharge reaction taking place in a Li/SVO battery? In what ways does the use of SVO nanoparticles help in improving the battery performance?
9.17 Which battery is able to comply with the high power demands of ICDs for supplying high current pulses of magnitudes 2–3 A to rapidly charge the capacitors of the device? How is the SVO cathode material prepared? Why is the discharge profile of this battery a stepped one? How is it beneficial?
9.18 How does a high-rate Li–MnO_2 battery differ from a simple Li–MnO_2 battery?
9.19 Explain the origin of plateaus in the stepped discharge profile of a Li/CF_x–SVO hybrid battery.
9.20 What is the typical application of a lithium-ion battery? Of what materials are the anode and cathode made? What is the source of lithium ions?

References

1. Tarascon J-M, Armand M (2001) Issues and challenges facing rechargeable lithium batteries. Nature 414:359–367
2. Takeuchi ES, Leising RA, Spillman DM et al (2003) Lithium batteries for medical applications. In: Nazri G-A, Pistoia G (eds) Lithium batteries—science and technology. Springer, New York, pp 686–700
3. Gutmann F, Hermann AM, Rembaum A (1967) Solid-state electrochemical cells based on charge transfer complexes. J Electrochem Soc 114:323–329
4. Holmes CF (2007) The lithium/iodine-polyvinylpyridine battery—35 years of successful clinical use. ECS Trans 6(5):1–7
5. Rudolph FW (1990) Anode coating for lithium cell. US patent 4,934,306
6. Ikeda H, Sato T, Tamura H (1977) Lithium-manganese dioxide cell. Denki Kagaku oyobi Kogyo Butsuri Kagaku, 45:314–318
7. Ikeda H, Hara M, Narakuwa S (1978) Lithium-manganese dioxide cell, Proceedings of the 28th Power Sources Symposium, 12–15 June, Atlantic City, NJ, ECS, Inc., pp 210–213

8. Pistoia G (1982) Some restatements on the nature and behavior of MnO_2 for Li batteries. J Electrochem Soc 129(9):1861–1865
9. Ohzuku T, Kitagawa M, Hirai T (1989) Electrochemistry of manganese dioxide in lithium nonaqueous cell I. X-ray diffractional study on the reduction of electrolytic manganese dioxide. J Electrochem Soc 136(11):3169–3174
10. Nardi JC (1985) Characterization of the Li/MnO_2 multistep discharge. J Electrochem Soc 132:1787–1791
11. Usrey ML, Chen X, Pena Hueso JA et al (2010) Lithium/carbon monofluoride batteries with organosilicon electrolytes. United States patent US 8,486,569 B2
12. Greatbatch W, Holmes CF, Takeuchi ES et al (1996) Lithium/carbon monofluoride (Li/CF_x): a new pacemaker battery. Pacing Clin Electrophysiol 19(11):1836–1840
13. Zhang SS, Foster D, Read J (2009) Enhancement of discharge performance of Li/CF_x cell by thermal treatment of CF_x cathode material. J Power Sources 188:601–605
14. Zhang SS, Foster D, Read J (2009) Carbothermal treatment for the improved discharge performance of primary Li/CF_x battery. J Power Sources 191:648–652
15. Rangasamy E, Li J, Sahu G et al (2014) Pushing the theoretical limit of Li-CF_x batteries: a tale of bifunctional electrolyte. J Am Chem Soc 136(19):6874–6877
16. Chen K, Merritt DR, Howard WG et al (2006) Hybrid cathode lithium batteries for implantable medical applications. J Power Sources 162:837–840
17. Takeuchi ES, Thiebolt WC III (1988) The reduction of silver vanadium oxide in lithium/silver vanadium oxide cells. J Electrochem Soc 135(11):2691–2694
18. Leising RA, Thiebolt WC, Takeuchi ES (1994) Solid-state characterization of reduced silver vanadium oxide from the Li/SVO discharge reaction. Inorg Chem 33:5733–5740
19. Cheng F, Chen J (2011) Transition metal vanadium oxides and vanadate materials for lithium batteries. J Mater Chem 21:9841–9848
20. Fehrmann G, Frömmel R, Wolf R (1996) Galvanic cell having improved cathode. US patent 5,587,258
21. Drews J, Wolf R, Fehrmann G et al (1997) High-rate lithium manganese-dioxide batteries—the double cell concept. J Power Sources 65(1–2):129–132
22. Drews J, Wolf R, Fehrmann G et al (1999) Development of a hybrid battery system for an implantable biomedical device, especially a defibrillator cardioverter (ICD). J Power Sources 80(1–2):107–111
23. Root MJ (2010) Lithium–manganese dioxide cells for implantable defibrillator devices—discharge voltage models. J Power Sources 195(15):5089–5093
24. Gan H, Rubino RS, Takeuchi ES (2005) Dual-chemistry cathode system for high-rate pulse applications. J Power Sources 146:101–106
25. Brodd RJ, Bullock KR, Leising RA et al (2004) Batteries, 1977 to 2002. J Electrochem Soc 151:K1–K11
26. Orman HJ, Wiseman PJ (1984) Cobalt(III) lithium oxide, $CoLiO_2$: structure refinement by powder neutron diffraction. Acta Crystallogr C C40:12–14

Chapter 10
Wireless Communications and Powering of Implants

Abstract Communication and powering facilities augment the capabilities of the implants by providing remote monitoring of therapy and charging of implant batteries to avoid replacement by surgery. At short distances in the range of a few centimeters, inductive links are used. The transference of data and power pose conflicting requirements. These requirements are sometimes fulfilled by using separate coils. The cost is a larger footprint and increased electromagnetic interference. Load-shift keying (LSK) technique is applied for uplink data transmission. Downlink data transmission is implemented by one of the three techniques: binary amplitude-shift keying (BASK), binary frequency-shift keying (BFSK), or binary phase-shift keying (BPSK), with BASK representing the plainest approach. Long-distance telemetry > 2 m is restricted to the agreed 402–405 MHz band for therapeutic implants or the industrial, technical, and medicinal/curative radio bands: 902–928 MHz, 2.4–2.4835 GHz, and 5.725–5.875 GHz frequency bands with transmission range up to 10 m.

Keywords Inductive charging · Resonance charging · Radio charging · Biotelemetry · ASK · FSK · PSK · LSK · AC-LSK · Adaptive LSK · PPSK · PHM

10.1 Introduction

After implantation of the electronic device in the human body, two basic concerns are: (1) to feed power to the implant for its uninterrupted operation, and (2) to maintain unidirectional/bidirectional exchange of data amidst the implant and the outside world. Power- and information-carrying signals often pose oppositely directed requirements. A rigorous study of power and information transmission is therefore useful to appreciate their roles. It also helps to optimize their respective operations, thus improving the performance of the medical implant as a whole.

V. K. Khanna, *Implantable Medical Electronics*,
https://doi.org/10.1007/978-3-031-99868-3_10

10.2 Powering the Implant

Powering of the implant is the process of supplying energy to the implant. By receiving energy, it can function properly in the desired manner. As implants cannot derive energy from body heat or nutrients in the blood, they must be supplied energy from a source either outside the body or placed internally (Table 10.1).

10.2.1 Through Percutaneous Leads

The easiest conceivable, though most inconvenient, way to power the implant is by connecting a wire passing from the external device percutaneously (across the skin of the patient) to the implant. It impairs the mobility and freedom of the patient. The tethered patient feels uncomfortable and tied up due to restrictions on movements. Moreover, there is risk of infection at the site where the wire is inserted. So, breaching the skin is unsafe for the patient as well as the implant [1]. On these grounds, it is long given up except for temporary testing.

10.2.2 Wireless Charging

Many applications require more ampere-hours than a single battery can provide. Using a larger number of batteries in the implant is not possible. There are two underlying reasons. Firstly, only a limited space is available in the implant. Secondly, extra weight unduly burdens the patient. Hence, it is an unaffordable solution and outright rejected. In place of percutaneous charging, the concept of "wireless charging" of the

Table 10.1 Methods of powering implants

S. No.	Through percutaneous leads	Wireless charging
1.	Disagreeable and embarrassing to the patient	More acceptable and convenient to the patient
2.	Totally restricts patient's movement	Restricts movement within limited area
3.	Chances of infection	No such chances
4..	More energy efficient	Less efficient
5.	Less resistive heating	Increased resistive heating
6.	Faster charging	Slower charging due to lower efficiency
7.	Less complex and less costly	More complex and expensive as it requires drive electronics and coils in both external part and implant

implant battery has gained acceptance. This charging is done by a wireless process. The wireless process is based on electromagnetic induction across the skin instead of a wire penetrating through the skin [2]. Without harming the skin, the battery can draw power from the external source. Obviously, wireless charging is less efficient. It is also slower than charging through wires because of the energy lost in the intervening medium. But wireless power transfer can provide power for indefinite periods. Also, there is no risk of infection associated with a percutaneous lead.

Wireless charging is not a new idea. In 1831, Michael Faraday discovered electromagnetic induction. He demonstrated that electromagnetic energy can propagate through space. Afterwards, in the late eighteenth century and early nineteenth century, Nikola Tesla conducted experiments on wireless transmission of power. Wireless charging is subdivided into three classes (Table 10.2): (1) inductive charging, (2) resonance charging, and (3) radio charging. Of these, only the first two are of wide interest in implantable electronics. The third class uses higher frequencies. The higher frequencies may be associated with injurious biological effects. Notwithstanding, depending on the application chosen and communication distance, several different frequencies are used in medical electronics and implantable electronics. They range from kHz to MHz range.

10.3 Inductive Charging

As opposed to conductive charging using a conductor for transferring power, inductive charging represents a technique of repositioning power across a short distance wirelessly [3]. It involves the use of a pair of induction coils. One of the coils is placed in the implanted device in the human body (receiver coil). The other coil is placed in the external device located outside the body (sender coil). Commonly, solenoidal coils are used, but for systems resting over the brain, planar spiral "pancake" coils are more suitable [4].

The two induction coils, not connected to each other, serve as the loosely coupled coils of a transformer. The coil in the implanted device is the secondary or pickup coil, while that in the external device is the primary coil. When the externally located primary coil is energized by passing an alternating current through it at the appropriate frequency, an electromagnetic field is set up surrounding this coil. This field also embraces the close-lying secondary coil inside the body. Hence, it can capture a shared segment of the embracing field, which causes a current to flow through it. This current is amplified and rectified to charge the battery.

10.3.1 Frequencies Used

Inductive charging is done at frequencies below radio frequencies or in the low radio-frequency range. Radio-frequency band stretches from 3 kHz to 300 GHz. AM radio

Table 10.2 Three types of wireless charging

S. No.	Inductive charging	Resonance charging	Radio charging
1.	Based on magnetic induction	Works on resonance-enhanced magnetic induction	Radio waves received through an antenna are used for the charging process
2.	Uses an induction coil in the external part to produce an electromagnetic field, which is alternating in polarity. Another induction coil is placed inside the implanted medical device. This coil draws power from the alternating electromagnetic field of the external part and transforms it into an electrical current. The current thus created charges the battery of the implant. The implant works with this battery as a source of energy	Uses an inductance-capacitance tuned circuit for resonance coupling. By doing so, the coils in the transmitter coil (located externally) and receiver coil (placed inside the implant) undergo oscillations at the resonant frequency. Through these coupled oscillations, power is transferred from the external to the internal part, wherein it is used for battery charging. This battery is the source of energy for the implant	Uses an inductance-capacitance tuned circuit for resonance coupling (like resonance charging) whereby the transmitter and receiver coils oscillate at the same frequency. The power transferred from the transmitter coil of the external part to the receiver coil of the internal part inside the implanted medical device is used for charging the battery of the implant. The charged battery feeds the implant
3.	Uses sub-radio frequency or the lower frequencies in radio-frequency band of the spectrum of electromagnetic waves	Uses sub-radio frequency or the lower frequencies in radio-frequency band of the electromagnetic wave spectrum	Uses radio-frequency waves
4.	An electrical transformer is formed by the two induction coils in proximity	Resonant-coupled transformer	Radio transmitter/receiver
5.	Less efficiency	Highest efficiency	Least efficiency
6.	Near-field effect	Near-field effect	Far-field effect
7.	Typical charging distance ~ 5–7 mm	Typical charging distance ~ 7–40 mm	Typical charging distance ~ 2–10 m
8.	Less spatial freedom. The patient cannot move around easily while charging	More spatial freedom	Largest spatial freedom
9.	Useful for shallow implants	Useful for shallow/deeper implants	Useful for shallow/deeper implants
10.	No electromagnetic pollution	No electromagnetic pollution	Creates electromagnetic pollution
11.	Less harmful to the body	Less harmful to the body	Harmful to the body

broadcasts within 535–1700 kHz and FM radio between 88 and 108 MHz. As the frequencies used for inductive charging lie in the low radio-frequency range, therefore it does not produce any electromagnetic interference in electronic devices. It is a near-field magnetic field-based system. Further, inductive charging involves magnetic field phenomena. Since magnetic fields interact weakly with living organisms, it is biologically safe.

At 125 kHz, the skin depth in titanium, the widely used housing material for implants, is 0.9 mm. Therefore, at an operating frequency of 125 kHz, the coupling is effective up to 1 mm thick titanium [5]. Typically, the distance separating the participating coils is < 3 cm. The coefficient of coupling is ~ 0.1–0.3 [6]. Around 175 kHz, the bandwidth is 50 kbps and the range is 8 cm [7].

10.3.2 Coupling and Loading Variations

Depending on coil displacements, the coupling is not constant. Also, the loading power within the implant varies with stimulation data changes [8]. Coupling and loading variations may lead to excessive implant heating. The heating effect may injure the adjoining tissues. Inadequate power transmission due to coupling/loading variations may cause diminution in implant voltage or current supply. This decrease is followed by implant malfunctioning and shutdown. To take care of these variations, more power may be needed, which is a power loss. Also, power fluctuations expose the tissues unnecessarily to strong electrical and magnetic fields.

10.3.3 Design Considerations

The inductive link is basically an air-core transformer. After making a beginner's diagram for the equivalent circuit of the inductive linkage, the coil geometry, dimensions, number of turns, wire type, etc., are decided. Quality factors (*Q*-factors) of the coils and the coefficient of mutual induction between them are vital bounds in the design of inductive link. To enhance the coupling between the coils, planar circular coils are more favored than solenoids. As compared to class A, B, or C amplifiers, the amplifiers of class E category display comparatively higher efficiency. These are highly efficient switching power amplifiers. In these amplifiers, the transistor works as an on/off switch. The waveforms are shaped to avoid simultaneous existence of high voltages and currents in the transistor.

Realization of the ideal, theoretical 100% efficiency is possible because of the mutual displacement between the voltage and current transitions on the time scale. At an instant at which voltage is high, the current is low. Further, at a low voltage, current becomes high. Thus the voltage and current are never high or low at the same instant. This non-overlap of voltage and current in time or their nonsimultaneous acquisition of maximum or minimum values together helps to decrease the power dissipation. Power dissipation is minimized specially during the switching changes.

10.3.4 Applications

This method is widely practiced in medical implants because interaction between the two coils can take place through biological tissues like fat, muscle, bone, or blood. Besides implantable devices, many commonly used electrical appliances such as electrical toothbrushes and mobile phones avail of inductive charging method.

10.4 Resonance Charging

This method involves resonance coupling of coils. The two coils, sending coil connected to the power source and the receiving coil in the implant, are tuned to the same resonance frequency. This tuning enables energy transfer from the former to the latter at small distances of several inches or more. The power is not picked up by neighboring objects unless they are tuned to the same frequency. In comparison to conventional magnetic charging, resonance charging allows higher charge transfer rates. Additionally, resonance charging allows more flexibility in the charger location or configuration outside the body. To give an example, the charger for an eye implant can be mounted on the spectacles. The charger for the ear implant can be placed below the pillow. Also, the charging can be made more effective on an implant placed deeper inside the body.

10.5 Radio Charging

10.5.1 Similarity to Radio Transmission and Reception

Similar to a radio transmitter, in this method a radio source sends radio waves at 915 MHz. The waves are captured by the receiver in the implanted device. Thereafter, they are demodulated to recover the transmitted signal and charge the battery after suitable processing. Due to the weaker capturing rate, low power devices within a range of 10 m from the transmitter can use this facility for battery charging. In the 402–405 MHz band, the bandwidth is 250 kbps and read range is 2–5 m [7]. But the method spreads electromagnetic pollution.

10.5.2 Safety Limits

According to human safety standards, the power transmitted to the body must be less than 10 mW/cm^2. Hence, there are restrictions on the carrier frequency to transfer power safely through the living tissue. If the frequency of electromagnetic waves is very high and in the range of radio frequencies from 3 kHz to 300 GHz, rapid heating of the tissues occurs. This heating is very much like the cooking of food in a microwave oven. For RF waves at high RF power densities ~ 100 mW/cm^2, considerable heating of tissues takes place. The body is not able to dissipate the large quantity of heat away. The rate of absorption by the human body has the maximum value at 80–100 MHz. Delicate regions like the eyes and testis need special care.

With reference to IEEE standards [9], below 100 kHz, the RF waves produce electrostimulation. Above 100 kHz, RF waves cause a sensation of heating. The 100 kHz frequency represents the frequency of thermal crossover. At frequencies < 100 kHz, electrostimulation effects dominate. On the opposite side, at frequencies > 100 kHz, temperature effects are predominant in the event of uninterrupted exposure to waves. In case of waveforms shaped in the form of pulses and having a small duty cycle, the frequency of thermal crossover can lie in the megahertz region.

Many global standards for electromagnetic protection have laid down the confinement ranges for power that can be dissipated safely inside the human body without producing unpleasant effects at frequencies > 100 kHz. The limits have been prescribed with reference to the parameter known as specific absorption rate (SAR). The SAR is defined as the amount of power which is dissipated per unit mass of tissue. It is expressed in W/kg. Another defining parameter is the upper limit of allowable exposure (MPE) for current and fields [10]. Generally, at frequencies > 100 kHz and < 6 GHz, SAR limits are articulated with regard to two SAR parameters: (1) SAR, which has been determined as a mean or midline value considering any 1 g of tissue shaped as a cube, and also (2) the SAR averaged over the full body. For the general people, the IEEE standard decrees a tolerable ceiling value of 1 g average SAR = 1.6 W/kg. For SAR determination in the human body, numerical analysis and approximations or investigational methods based on experiments and phantoms (models of the human body) are applied. An utmost diffused numerical method to perform the above evaluation is the finite-difference time-domain (FDTD) technique. Over the past 15 years, this method has found widespread usage for several dosimetric assessments. Mathematical models or replicas of the body, as obtained by MRI scanning of patients offering their voluntary services, are employed for calculations along with FDTD method.

High frequencies are despised because they are harmful to the human body. Another adverse effect of high frequency is that a large portion of the transmitted power is lost by virtue of thermal losses in heating the tissues. Therefore, an unjustifiable drain of power takes place. As an outcome, the efficiency of the system decreases. Hence, besides the electromagnetic fields set up in the human body, the induced current in the implanted coil can dissipate substantial power within the coil itself. The finite resistivity of the coil is the underlying cause of this dissipation. This power wasted in the coil in turn raises the temperature in the surrounding tissues. Application of numerical or experimental methods helps in answering this query.

Apart from the power losses through thermal effects, at high frequencies, power-conditioning circuits in the implant as well as in the external transmitter consume more power. Necessarily, the power transmission efficiency decreases.

10.6 Biotelemetry

Telemetry is the transmission of measured parameters of the patient to the external device and vice versa.

10.6.1 Active Telemetry

Active telemetry systems make available comparatively effective data transfer in both directions over long distances using on-board battery [11]. Naturally, these systems offer increased size and decreased life. They are actually radio transmitters employing analog or digital modulation over a carrier signal. The carrier signal is amplified and transmitted from the transmitter to the receiver antenna [12]. Aboard signal processing circuits for amplification, mixing, information superimposition on the carrier signal by modulation and its extraction, and various other operations on signals are assimilated into the implant. For power supply, these systems depend on batteries, especially for long-term use. Hence, their regular maintenance becomes unavoidable. Overall system cost is increased by the added cost of the battery and signal processing circuitry. A further expenditure is incurred in the assembly of the different components. Moreover, prevention of heating of the tissues proximate to implants becomes troublesome. This happens owing to the lossy characteristics of the active systems.

10.6.2 Passive Telemetry

Passive telemetry decreases the distance up to which communication is effective. But it permits battery-free operation providing unrestricted time of operation. No on-board batteries are needed. Battery-less operation increases the autonomous life expectancy of the system. It also decreases the occupied space significantly. This is because the battery is a major contributor to the volume of the system. Absence of the hulking batteries makes passive systems ideal candidates for biomedical implants that have uncompromising size constraints. Passive wireless systems usually include less on-chip signal processing. So, the device size is eventually smaller. By discarding the batteries and a larger fraction of on-chip signal processing, the size and the final

Table 10.3 Active and passive telemetry [13, 14]

S. No.	Active telemetry	Passive telemetry
1.	The implant has a local oscillator producing radio-frequency waves that carry data	The implant does not produce radio-frequency waves itself. Instead, it receives energy by engaging in mutual interaction with the waves produced by the outside part. The same energy is used for data transference
2.	Power attenuation is less and smaller antennas suffice	Power is significantly reduced leading to high attenuation, and therefore large antennas are required
3.	Useful for smaller implants located deeper inside the body	Usually used in larger implants and placed underneath the skin
4.	Higher power consumption	Lower power consumption
5.	Used when a huge amount of data is to be transmitted at a fast speed. Disfavored whenever lower data rates are acceptable due to power consumption and space required by antenna and circuitry	Favored for low data rate transmission, e.g., when a small quantity of data is to be conveyed at a slow pace

cost of the device decrease. A comparative study between the active and passive systems is made in Table 10.3.

10.7 Data Telemetry Uplink: From the Implanted Medical Device to Its External Part

Telemetry uplink involves the contactless transmission of patient's vital data to the external device.

10.7.1 Digital Modulation Techniques: A Quick Relook

In binary amplitude-shift keying (BASK) (Fig. 10.1), the amplitude A of the carrier signal is altered in concurrence with the information-containing modulating signal. However, its frequency and phase are kept constant. As an example, transmission of 0 is done using one particular value of amplitude A_1 of the carrier. Transmission of 1 is done by using another amplitude value A_2. If one amplitude value A_1 or A_2 is taken as zero, the transmission takes place through the presence or absence of the carrier. This is a particular form of BASK called on/off keying (OOK).

In binary frequency-shift keying (BFSK) (Fig. 10.2), the frequency f of the carrier signal is altered in response to the modulating signal. Its amplitude and phase are kept constant. One particular frequency f_1 represents 0. Another frequency f_2 denotes 0.

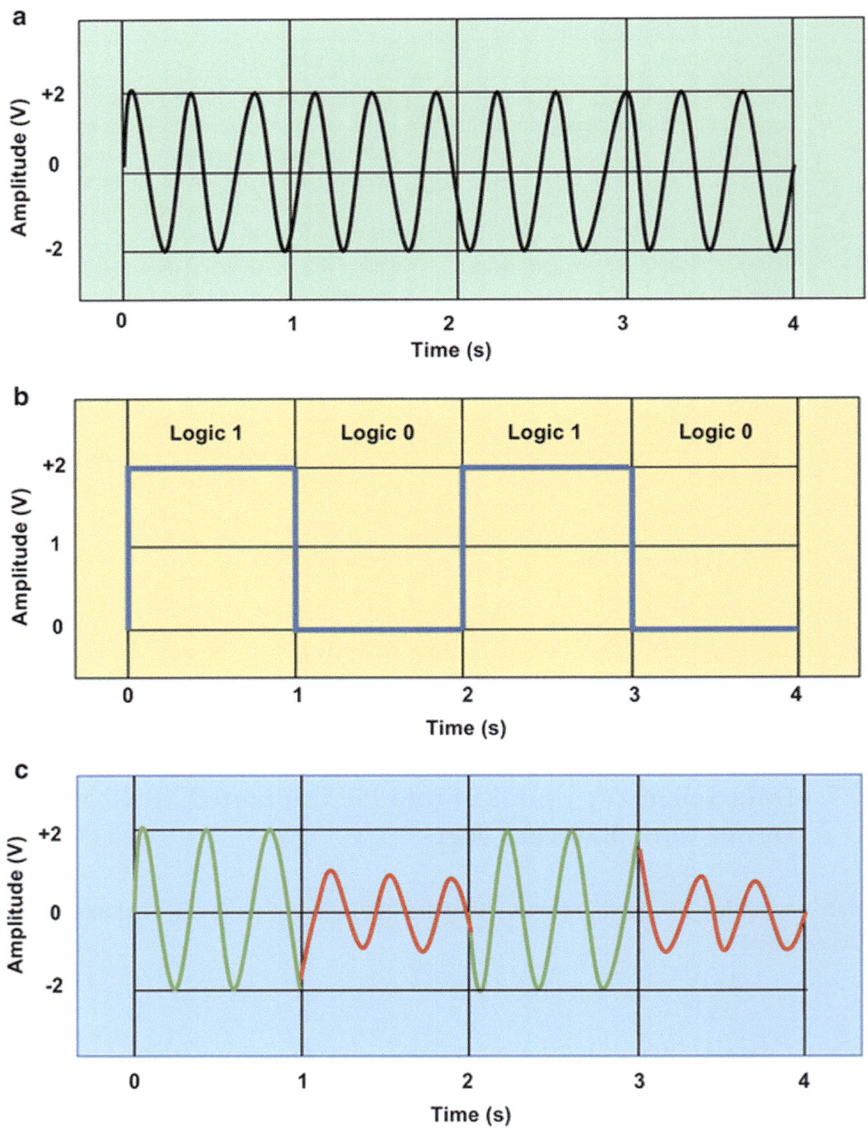

Fig. 10.1 Amplitude-shift keying (ASK): **a** sine wave carrier, **b** modulating digital signal, and **c** modulated wave

In binary phase-shift keying (BPSK) (Fig. 10.3), the phase ϕ of the carrier signal is shifted to indicate information. Its amplitude and frequency are kept constant. Phase is the starting angle of the sinusoidal waveform. Thus one phase value may correspond to $\phi_1 = 0$. The other phase value may correspond to $\phi_2 = 180°$.

Table 10.4 displays the characteristics of BASK, BFSK, and BPSK.

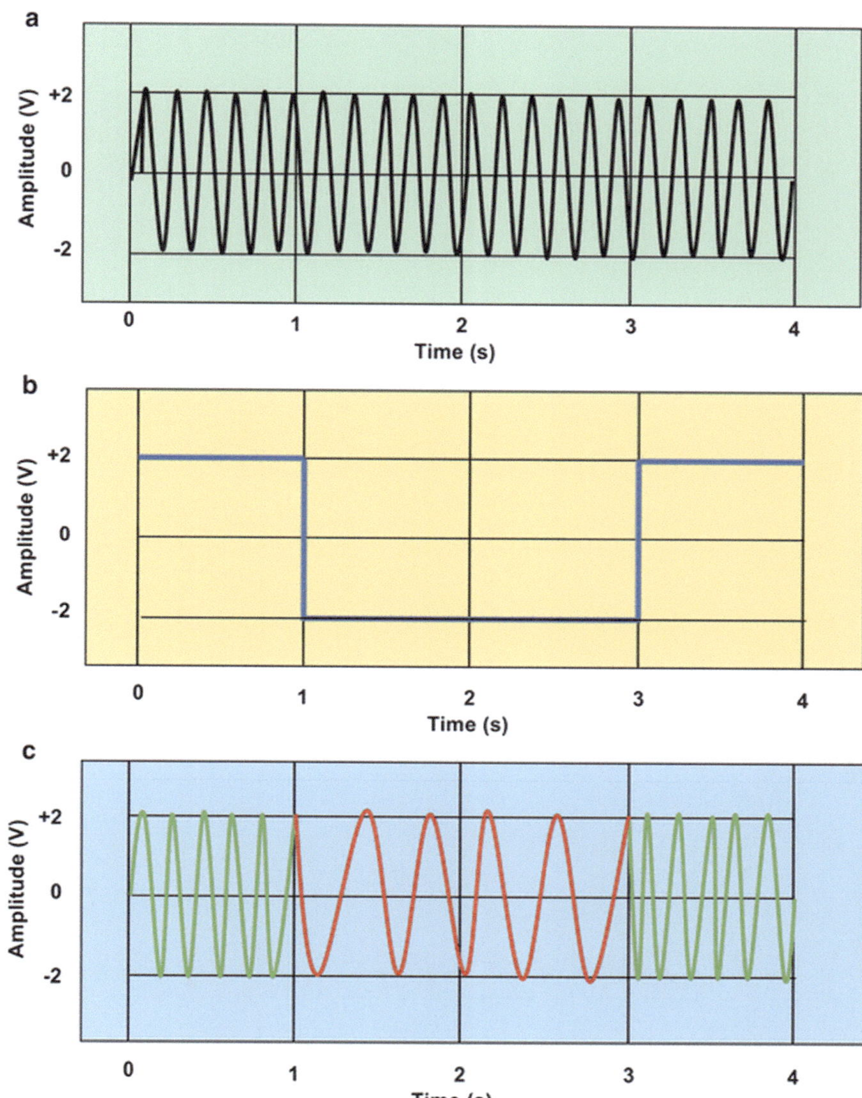

Fig. 10.2 Frequency-shift keying (ASK): **a** sine wave carrier, **b** modulating digital signal, and **c** modulated wave

Reader's attention is drawn to the prefix "binary" before ASK, FSK, and PSK. This prefix signifies two amplitudes, two frequencies, and two phases, respectively. Note that in BASK, BFSK, and BPSK, each symbol represents one bit. To increase the bandwidth over that achievable with BASK, BFSK, and BPSK, multilevel or *M*-level ($M = 2^N = 2^2, 2^3, 2^4, 2^5, \ldots$) modulation schemes are used. In these schemes, each symbol represents *N* bits. These are called *M*-ary ASK, *M*-ary-FSK, and *M*-ary

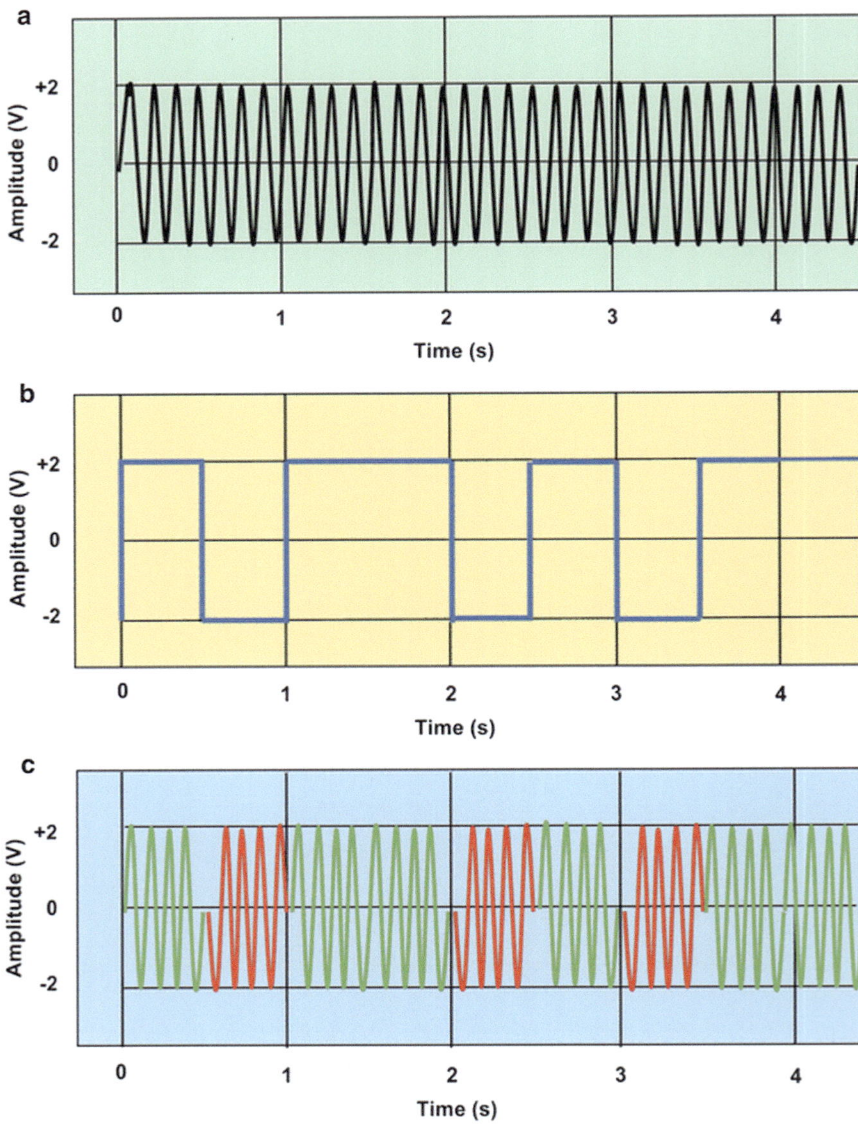

Fig. 10.3 Phase-shift keying (ASK): **a** sine wave carrier, **b** modulating digital signal, and **c** modulated wave

PSK modulation schemes, respectively. As an example, an M-ary ASK scheme uses M amplitudes in place of 2 amplitudes in BASK. M-ary FSK and M-ary PSK have similar meanings.

Table 10.4 Digital modulation schemes at a glance

S. No.	BASK	BFSK	BPSK
1.	Uses two different amplitudes to represent 0 and 1	Uses two different frequencies near the carrier frequency for representing 0 and 1	Uses two different phases separated by 180° to represent binary digits 0 and 1
2.	Simple	More complex than ASK	More complicated process of detecting and recovering signal than either in ASK or FSK
3.	Performs poorly. It is profoundly influenced by noise and interference effects	Better than ASK but less than PSK	Superior in performance in comparison to both ASK and FSK
4.	Low bandwidth	FSK spectrum is 2× ASK spectrum	Needs or extends over similar bandwidth as ASK. Possible to achieve more efficient bandwidth utilization and obtain higher rate of data transfer, as compared with FSK
5.	Sensitive to interference	Less susceptible to errors than ASK	Less susceptible to errors than ASK
6.	Used to transmit up to 1200 bps on voice grade lines and very high-speed digital data over optical fiber	Used to transmit up to 1200 bps on voice grade lines, for 3–30 MHz RF range and at even higher frequencies on local area networks with coaxial cable	Used in satellite communication systems, being a robust mode

10.7.2 Load-Shift Keying and Multilevel Load-Shift Keying

For data transmission from the medical implant to the exterior device, load-shift keying (LSK) is used. LSK is a special form of amplitude-shift keying (ASK). Also called "reflectance modulation," it is a communication arrangement. It allows simultaneous transference of power and data on the same inductive linkage (Fig. 10.4). LSK is based on variation of loading of the secondary coil in accordance with the serialized stream of data. The secondary coil is heavily loaded when logic "high" is to be sent and kept unloaded for sending a logic "low" state. By mutual inductance of the coils, any variation in the load of the secondary coil is reflected as a change in impedance of the primary coil. If this impedance change is resistive, amplitude-shift keying of the backscattered signal results. If it is capacitive, phase-shift keying takes place.

In a typical implementation, the binary data stream short-circuits the implant-side coil. The alteration in impedance is reflected in the transmitter. This is because the load due to the implant is much larger than the resistance offered to current flow

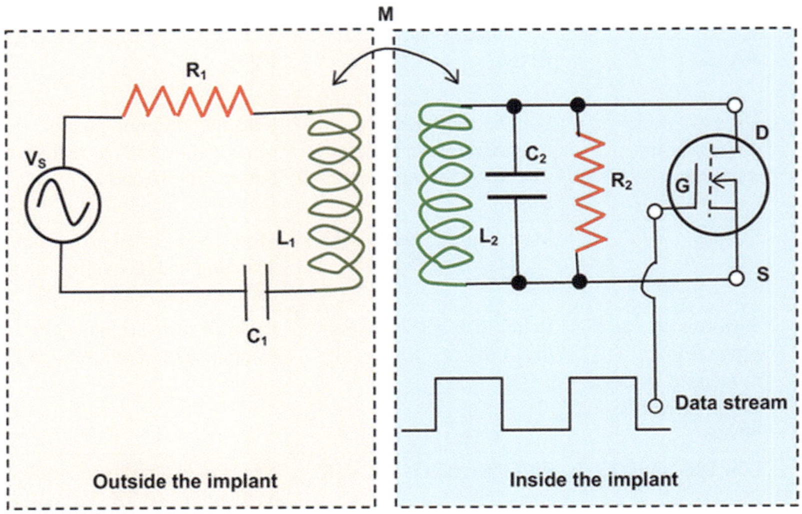

Fig. 10.4 LSK modulation principle

by the switching transistor during its on-state. For a system using only two coils for both power and data, there is a risk of discontinuation in power supply if the short-circuiting of the implant-side coil is done for too long a period. An optimization of the characteristics of the link ought to be done to obtain maximum transference of power under the condition when communication is inoperative.

The main shortcoming of LSK is the sizeable deviation created by it in the voltage induced in the secondary coil with respect to its normal value. Due to this disturbance, difficulty is experienced in voltage regulation to provide a dependable source voltage. The bandwidth of LSK is held back by the coefficient of coupling, the coil parameters, and the transitory behavior of the inductive link. The bandwidth limit of the inductive link due to the quality factors of the coils fixes the maximum data rate. A large value of Q assists efficiency of transferring power but restrains the rate of data transport [15].

For data rate enhancement, multilevel LSK is preferred, as demonstrated in [16] with discrete components. Herein, 2× improvement in data rate was indicated for a back telemetry system based on passive LSK and comprising four levels in comparison with usual back telemetry of bilevel type (Table 10.5).

10.7.3 Auxiliary-Carrier Load-Shift Keying

To overcome the deficiencies of LSK for reverse data telemetry from the medical implant to the exterior environment, auxiliary-carrier load-shift keying employs two carriers. These carriers are called the main carrier and the auxiliary carrier. They are

Table 10.5 Bilevel and multilevel LSK [16]

S. No.	Bilevel LSK	Multilevel LSK
1.	In essence, opens or short-circuits the secondary coil located on the side of the implant	Creates > 2 different backscattering patterns whose properties are varied by multibit/symbol of data transmitted
2.	Provides back telemetry at a limited rate of data transfer	Provides back telemetry at a high rate of data transmission

given the responsibilities of power transfer from the exterior device to the medical implant and data telemetry from the medical implant to the exterior device, respectively [17]. The need for two carriers arises because power transference is more efficient at low frequencies at which a smaller bandwidth suffices. Contrariwise, data telemetering efficiency increases with increasing bandwidth. The large bandwidth is available at the higher frequencies. So, a single frequency cannot meet both the requirements. If a single-carrier frequency is used, efficiency of one function is maximized at the sacrifice of efficiency of the other function due to their opposing frequency dependences. The two carrier waveforms are superimposed on each other. A dual-frequency resonant inductive link has therefore to be set up between the external device and the implant. This link has two resonant frequencies, one resonant frequency to carry out the function of transmission of power (f_M) and another resonant frequency for the role of transmission of data (f_A). The principle of auxiliary-carrier-LSK is illustrated in Fig. 10.5.

Auxiliary-carrier LSK (AC-LSK) is very beneficial for power retrieval on the implant side as compared to simple LSK. The simple LSK shows degradation in quality, as pointed out above. Thus continuous transfer of power can take place without any disturbance from reverse data telemetry. In the system developed by Karimi et al. [17], carrier frequencies of 1 MHz and 10 MHz were selected for telemetry of power and data, respectively. Similarly, depending on particular applications and keeping the human safety considerations in view, one lower and one upper frequency can be selected for building the auxiliary-carrier LSK system. This gives more freedom to the implant system designer. The auxiliary-carrier LSK technique introduces no difficulties with the coil designs for the inductive link. In the opposite way, some complexities in power amplification as well as matching and reverse data telemetry circuits are unavoidable. Table 10.6 compares AC-LSK with the conventional LSK.

10.7.4 Adaptive Control Load-Shift Keying

In an ordinary LSK-based system, large power fluctuations are induced by displacements in coil positions, loading changes, or both. To take care of all possible situations, excess power must be supplied to the implant. This unduly lowers the power

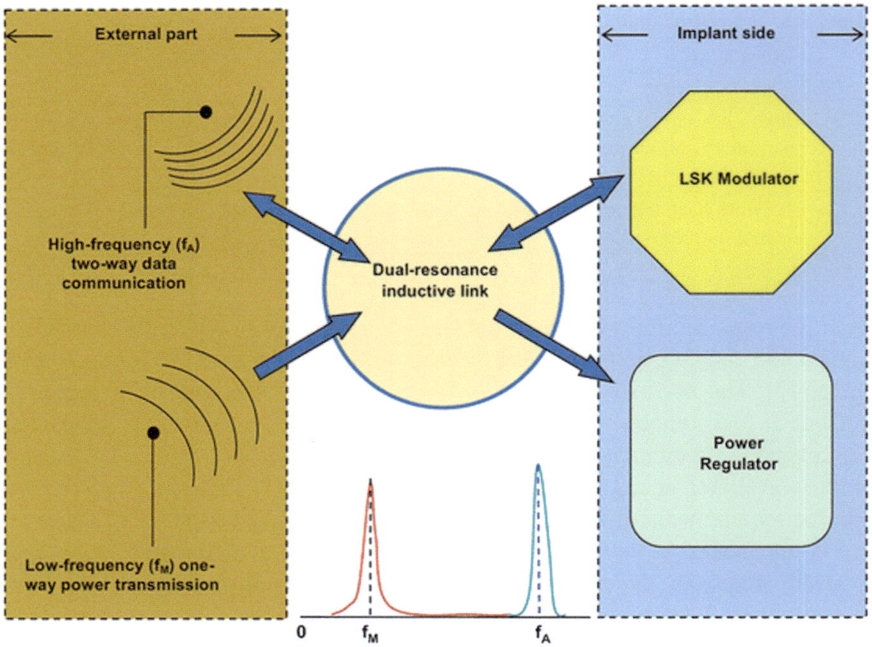

Fig. 10.5 Auxiliary-carrier load-shift keying approach

Table 10.6 Conventional and auxiliary-carrier LSK [17]

S. No.	Conventional LSK	Auxiliary-carrier LSK
1.	Uses single-carrier link	Uses dual-carrier link
2.	Uses one single carrier for telemetry of both power and data	Uses two carriers: main carrier for power and an auxiliary carrier for data telemetry
3.	Restriction on the choice of frequency for data	No such restriction because power and data are handled by different carriers
4.	Suffers from low data transfer rates	Since frequency for carrier telemetry can be chosen to be much higher than one used for power telemetry, high data rates are achievable
5.	Degradation in power transferred to the medical implant	No conspicuous deprivation in power

efficiency. Wang et al. [8] designed and implemented an adaptive control LSK system for a retinal prosthetic device. The proposed system removes power fluctuations induced by coupling coefficient or loading variabilities (Table 10.7). The system is a dual band telemetry link using a lower frequency for transmission of power and a higher frequency for transfer of data. Information regarding the power reaching the

Table 10.7 Conventional and adaptive LSK

S. No.	Conventional LSK	Adaptive LSK
1.	Large fluctuations occur in the power transferred to implant with coil displacements, load variations, etc.	Mitigates such fluctuations
2.	Gives low power efficiency because excessive power is supplied to overcome worst-case situations	Provides improved power efficiency
3.	Involves undesirable exposure of tissues to high power	Unwanted exposure of tissues to harmful power levels is avoided

implant side is sent back to the external device in a closed-loop network. This information flow enables the feeding of the exact amount of power needed by the medical implant from the exterior device. Up to 250 mW of power could be supplied over a coil distance from 0.7 to 1.5 cm. The data transfer rate from the medical implant to the external humankind was 3.3 kbps.

10.7.5 Passive Phase-Shift Keying

The passive phase-shift keying [18], with acronym PPSK, utilizes the transitory response of the inductive linkage caused by the switching of the load and maneuvering of power. By opening or closing a switch at regular intervals, data from the implant are transmitted to the exterior device. Using a carrier frequency of 4 MHz, data transmission rates up to 222 kbps could be achieved. A monolithic PPSK modulator [15] for telemetry by inductive coupling was fabricated using CMOS technology with gate length of 0.6 μm. It used a solitary set of coils to execute both the activities of transference of data and supply of power. For a carrier frequency of 13.56 MHz, it was found to transmit up to 847.5 kbps (1/16 of the carrier) along with power transfer. It recorded the speediest rate of data transmission reached by a single inductively coupled wireless link applied concurrently for delivery of power and conveyance of data in medical implants. Contrasting with LSK, in PPSK, the switch across the secondary coil (Fig. 10.6) shuts up in synchrony with the carrier for half the carrier cycle. The momentary response in the primary coil current is perceived as a logic "1" signal.

10.7.6 Pulse Harmonic Modulation

Hitherto described modulation techniques for inductive telemetry links were carrier-based. These carrier-based methods were used in the early implants. So, the carrier waveform of low frequency employed for data transfer could also be utilized for implant powering. In high-performance, wider bandwidth implants, on the other

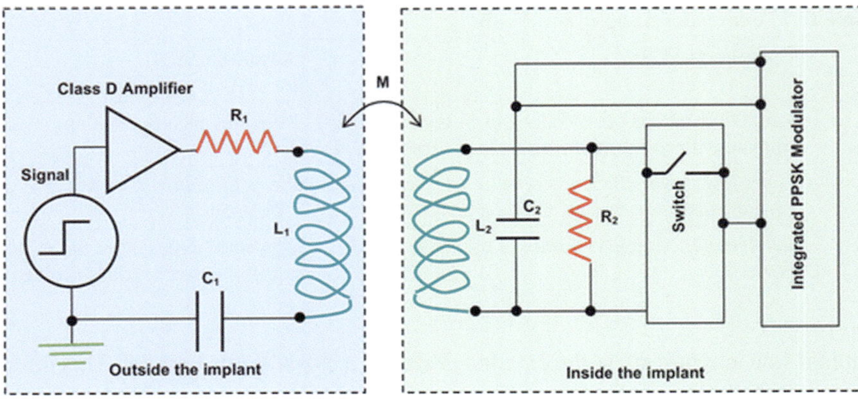

Fig. 10.6 PPSK modulation system

hand, the carrier wave for power was estranged from that for data. This separation was done because the frequency of the power carrier could not be increased due to the losses in the biological tissue at high frequencies. The deployment of data carrier waves of high frequency for broadband communication in the implants requires the undesirably complicated RF circuits for stabilizing the frequency. An example is provided by the phase-locked loops (PLLs). The RF circuits are more power hungry.

For obtaining lower power consumption and higher data rates in near-field telemetry links, a pulse-based data transmission technique has been devised [19]. It is called pulse harmonic modulation (PHM). The transceiver for PHM employs two constricted pulses of particular amplitudes and timings at the transmitter section. This is required to smother intersymbol interference at the *LC* tank circuit of high-quality factor at the receiver segment. The interference is repressed for achieving high rates of data transfer without decreasing the *Q*-factor of the inductive link to enlarge the bandwidth. Consequently, better range and selectivity are achievable with the link. The PHM receiver architecture utilizes a pulse-based automatic gain control circuit. This circuit appreciably lessens the power consumption and inter-symbol interference at short coupling distances. By using the PHM technique and carefully designing the transceiver, high data rates of 20 Mbps could be attained across a 1 cm inductive link with a carrier frequency of 66.7 MHz.

Pulse harmonic modulation represents a high-speed data transmission technique for inductive links. But the drawback is the requirement of a separate link for its implementation because the method is carrier less.

10.8 Data Telemetry Downlink: From the External Part to the Implanted Medical Device

Telemetry downlink is the transmission of information from the external device to the patient.

10.8.1 *Amplitude-Shift Keying*

Due to the inherent simplicity of designing the modulator at the external device and the demodulator at the implant, amplitude-shift keying was the first preferred technique used for data transmission. But the data transmission rates were exceedingly low. The bit rates are generally limited to 1/10th of the carrier frequency. An ASK demodulator for implantable electronic devices [20] energized through an inductive link was tested with carrier frequencies in the 1–15 MHz range. Data rates up to several 100 kbit/s could be supported. Several ASK-based systems have been reported [21, 22].

The main limitations of ASK-based methods are:

1. In applications requiring high-bandwidth data transmission, it is not possible to integrate the large-value capacitors of high-order filters. These filters are necessary to obtain spiky cutoff frequencies in the low radio-frequency part of the spectrum.
2. The voltage induced between the terminals of receiver coil varies inversely as the third power of the distance separating the coils, presupposing that all other parameters are constant. The implication is that the amplitude of the induced signal, which is the entity entrusted with the responsibility of carrying information in ASK, is exceptionally susceptible to the relative distance changes in implants. Such amplitude fluctuations are very infuriating.
3. The signal is impaired even when the relative distance is fixed. This impairment takes place because a constant power is received. Any alterations in the current consumed by wireless chip, as caused by current impulses of the digital circuits induce changes in the envelope voltage of the carrier signal received. Hence, the quality of the ASK signal is degraded.

10.8.2 *Frequency-Shift Keying*

The superior immunity of FSK over ASK against noise sources and interference is well known. This ruggedness of FSK becomes further apparent in inductively coupled devices. These devices obtain both data and power from the same carrier wave. Unlike amplitude in ASK, the relative distance or changes in current do not influence the frequency of the induced signal in FSK schemes. The modus operandi of FSK modulation and demodulation is utilized to propel data to inductively power wireless medical implants at rates of data transfer > 1 Mbps. The carrier frequencies used are on a par with those used in ASK [23]. This FSK demodulator was checked by experimentation up to 2.5 Mbps. The bit error rate was 10^{-5}. This testing was done at the same time as receiving an FSK carrier signal of 5 or 10 MHz frequency.

BFSK enables a higher ratio of data rate to carrier frequency than BASK. Notwithstanding, the need of a broad passband in the inductive linkage to permit the copious frequencies imposes restriction on transference of power.

10.8.3 Phase-Shift Keying

PSK is a more appropriate communication procedure for data exchange and power transfer than either ASK or FSK. This suitability is because the amplitude and frequency are constant. Moreover, a carrier signal of constant amplitude ensures steady transfer of power with high efficiency. Antennas of a fixed size can be used because the carrier frequency is constant. These antennas are easily designed. The objective of design is to optimize the transference of power and data. Thus BPSK provides stable and efficient power transfer. The shortcoming of BPSK is the complexity of demodulator using commonly a Costas loop [24].

In the prototype of a demodulator architecture employing binary phase-shift keying (BPSK) [25], the highest data rate of the demodulator, found by derivation to be 1/8th of the frequency of carrier wave used, viz., 13.56 MHz, was 1695 kbs.

10.9 Discussion and Conclusions

While laying down specification of implantable devices, many parameters have to be dealt with in design and for satisfactory implementation. A few such parameters are power requirement, communication range (long-range or short-range), bandwidth, the location and environment of the medical device inside the human body, data transfer rate, frequency, authoritarian standards, and operating life of implants. These parameters guide toward the strategy to be planned for choosing uplink and downlink power and data telemetry schemes and for building the hardware/software for realization (Fig. 10.7).

Review Exercises

10.1 After a medical device has been implanted in the human body, what are the two main needs that must be fulfilled?

10.2 Prepare a table bringing out the main advantages and disadvantages of supplying power to an implanted device through percutaneous leads and by the wireless method.

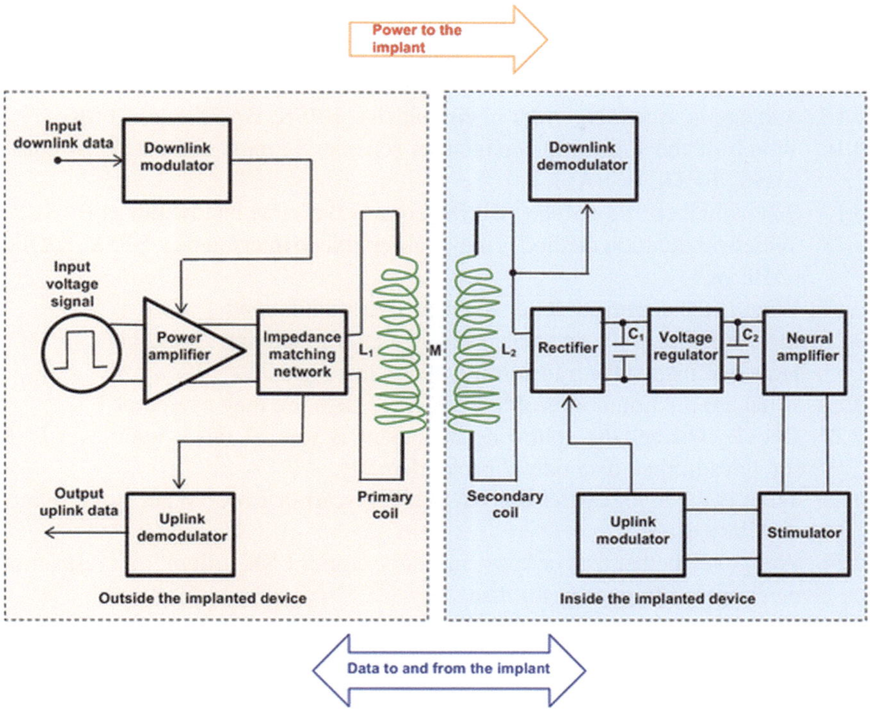

Fig. 10.7 Organization of a wireless biotelemetry system for transmission of power to an implant and data to and from the implant via an inductive link

10.3 Name the three types of wireless charging techniques. Which of these techniques are widely used in implantable electronics?

10.4 During inductive charging of an implanted device, what shape of coils is generally used? What shape is preferred for systems placed over the head?

10.5 How does magnetic induction used in inductive charging differ from resonance-enhanced magnetic induction used in resonance charging?

10.6 Arrange in ascending order of efficiency: resonance charging, inductive charging, and radio charging.

10.7 Arrange in descending order of charging distance: inductive charging, radio charging, resonance charging.

10.8 Arrange in ascending order of frequencies used: resonance charging, radio charging, and inductive charging.

10.9 Comment on the statement, "Theoretically, 100% efficiency is achievable with inductive link."

10.10 Why is the use of high frequencies disfavored for implantable devices?

10.11 Define specific absorption rate (SAR) of tissue. How are SAR limits expressed?

10.12 Which telemetry system, active or passive, produces radio waves?

10.13 Which telemetry system, active or passive, requires larger antennas?
10.14 In a medical implant, a large capacity of data is to be conveyed at a high transference rate. Which telemetry system, active or passive, will be suitable?
10.15 Arrange in ascending order of complexity: BPSK, BASK, and BFSK.
10.16 Which of the following modulation schemes is most vulnerable to noise: BPSK, BFSK, or BASK?
10.17 State whether true or false: BPSK requires the same bandwidth as BASK?
10.18 Which modulation method is most susceptible to interference: BPSK, BASK, or BFSK?
10.19 What is the common name of reflectance modulation?
10.20 In LSK, how is the secondary coil loaded for sending "logic high signal"? How is it loaded for transmitting "logic low signal"?
10.21 What are the limitations of bilevel LSK? How are they overcome?
10.22 Decide whether the following statement is true or false: multilevel LSK provides limited data rate of back telemetry.
10.23 What is the main carrier used in auxiliary-carrier LSK? What is the role of auxiliary carrier?
10.24 Point out whether true or false: auxiliary-carrier LSK suffers from restriction on choice of frequency for data.
10.25 In which of the two schemes, LSK or auxiliary-carrier LSK, high data transfer rates are not attained?
10.26 Which of the two schemes, conventional or adaptive LSK, suffers from the problem of large fluctuations in power transferred to the implant? Which of the two gives low power efficiency? Which one produces overexposure of tissues?
10.27 What is the principle of PPSK? How does it differ from LSK?
10.28 Name a modulation scheme which is not carrier-based. What is its advantage? What is the shortcoming?
10.29 For data telemetry downlink, what was the first preferred technique? What were the problems faced with this technique?
10.30 Discuss the relative merits and demerits of the following for data telemetry downlink: ASK, FSK, and PSK.

References

1. Galbraith DC, Soma M, White RL (1987) A wide-band efficient inductive transdennal power and data link with coupling insensitive gain. IEEE Trans Biomed Eng 34(4):265–275
2. Buchmann I (2012) Charge medical devices without wires. MED Medical Electronics Design. http://www.medicalelectronicsdesign.com/article/charge-your-medical-devices-without-wires. Accessed 26 July 2014
3. Chawla N, Tosunoglu S (2012) State of the art in inductive charging for electronic appliances and its future in transportation. In: 2012 Florida conference on recent advances in robotics, Boca Raton, FL, 10–11 May 2012, pp 1–7

4. Harrison RR (2007) Designing efficient inductive power links for implantable devices. In: IEEE international symposium on circuits and systems, ISCAS 2007, New Orleans, 27–30 May, pp 2080–2083
5. Bradley PD (2011) Wireless medical implant technology—recent advances and future developments. In: Proceedings of the ESSCIRC, 12–16 Sept 2011, pp 37–41
6. Si P, Hu AP, Malpas S et al (2008) A frequency control method for regulating wireless power to implantable devices. IEEE Trans Biomed Circ Syst 2(1):22–29
7. Daniluk K, Niewiadomska-Szynkiewicz E (2012) Energy efficient security in implantable medical devices. In: Proceedings of the federated conference on computer science and information systems, pp 773–778
8. Wang G, Liu W, Sivaprakasam M et al (2005) Design and analysis of an adaptive transcutaneous power telemetry for biomedical implants. IEEE Trans Circ Syst I Regul Pap 52(10):2109–2117
9. IEEE standard for safety levels with respect to human exposure to radio frequency electromagnetic fields, 3 kHz to 300 GHz: IEEE standard C95.1™-2005. Copyright © 2006 by the Institute of Electrical and Electronics Engineers. http://ieeexplore.ieee.org/stamp/stamp.jsp?arnumber=1626482. Accessed 26 July 2014
10. Lazzi G (2005) Thermal effects of bioimplants: power dissipation characteristics and computational methods. IEEE Eng Med Biol Mag 24(5):75–81
11. DeHennis AD, Wise KD (2005) A wireless microsystem for the remote sensing of pressure, temperature, and relative humidity. J Microelectromech Syst 14(1):12–22
12. Sezen AS (2006) Passive wireless sensing strategies for high frequency biomedical sensing applications. Ph.D. thesis, University of Minnesota, pp 4–5
13. Cimirakis D, Demosthenous A (2014) Design considerations of biomedical telemetry devices, chap 2. In: Nikita KS (ed) Handbook of biomedical telemetry (sec. 2.3). IEEE Press series on biomedical engineering. Wiley, Hoboken
14. Ghovanloo M, Atluri S (2008) An integrated full-wave CMOS rectifier with built-in back telemetry for RFID and implantable biomedical applications. IEEE Trans Circ Syst 55(10):3328–3334
15. Cirmirakis D, Jiang D, Demosthenous A et al (2012) A fast passive phase shift keying modulator for inductively coupled implanted medical devices. In: Proceedings of the ESSCIRC, 17–21 Sept 2012, Bordeaux, pp 301–304
16. Xu W, Luo Z, Sonkusale S (2009) Fully digital BPSK demodulator and multilevel LSK back telemetry for biomedical implant transceivers. IEEE Trans Circ Syst II Express Briefs 56(9):714–718
17. Karimi M, Sodagar AM, Mofrad ME et al (2012) Auxiliary-carrier load-shift keying for reverse data telemetry from biomedical implants. In: 2012 IEEE biomedical circuits and systems conference (BioCAS), Hsinchu, 28–30 Nov 2012, pp 220–223
18. Zhou L, Donaldson N (2003) A fast passive data transmission method for ENG telemetry. Neuromodulation 6:116–121
19. Kiani M, Ghovanloo M (2013) A 20-Mb/s pulse harmonic modulation transceiver for wideband near-field data transmission. IEEE Trans Circ Syst II Express Briefs 60(7):382–386
20. Gudnason G (2000) A low-power ASK demodulator for inductively coupled implantable electronics. In: Proceedings of the 26rd European solid-state circuits conference, 19–21 Sept 2000, Stockholm, Sweden, pp 385–388
21. Yu H, Najafi K (2003) Low-power interface circuits for bio-implantable microsystems. In: IEEE international solid-state circuits conference, San Francisco, CA, 13 Feb 2003, 10 pp
22. Naghmouchi F, Ghorbel M, Hamida AB et al (2004) CMOS ASK system modulation dedicated to cochlear prosthesis. In: IEEE 1st international symposium on control, communication and signal processing, Hammamet, Tunisia, 21–24 Mar 2004, pp 267–270
23. Ghovanloo M, Najafi K (2004) A wideband frequency-shift keying wireless link for inductively powered biomedical implants. IEEE Trans Circ Syst I Regul Pap 51(12):2374–2383
24. Sawan M, Hu Y, Coulombe J (2005) Wireless smart implants dedicated to multichannel monitoring and microstimulation. IEEE Circ Syst Mag 5(1):21–39

25. Luo Z, Sonkusale S (2008) A novel BPSK demodulator for biological implants. IEEE Trans Circ Syst I Regul Pap 55(6):1478–1484

Chapter 11
Cyber Security and Confidentiality Concerns with Implants

Abstract Many lifesaving implantable devices are equipped with wireless technology. This technology enables remote device checks and relieves patients from recurrent consultant visits. But this convenience is associated with unforeseen hazards. These hazards are the security and privacy of data. The labor needed to defend patients from exploits of stealing or nastiness gains more significance. This is especially so with increasing use of wireless telecommunication facilities and the services of global computer network or Internet by implanted devices. The susceptibilities of medical devices are of two types, viz., control or privacy susceptibilities. In control susceptibilities, an unauthorized person acquires control of device operation. The unlicensed person reprograms the device without the patients' knowledge to disable its therapeutic services. In privacy susceptibilities, confidential patient data are disclosed to an unsanctioned party. Both vulnerabilities are detrimental to patient's health outcome. Both are avoidable by incorporating well-thought-out measures in device design.

Keywords Security · Confidentiality · Privacy · Encryption · Cryptography · Jamming · Hijacking · Insulin pump · ICD · Biosensor · Shield

11.1 Introduction

Security is freedom from risk or danger from adversaries. Security should be clearly differentiated from safety, a somewhat similar term that is often confused with security. Safety is concerned with design errors or system failures. Security of data means that its storage and transference are protected. Cyber security or information technology security is the organization of preventive know-hows, procedures, and rehearses. This body is devised for protecting computers, nettings, software packages, and information from unapproved access, change, or destruction. Cyber security of a system along with its reliability constitutes its trustworthiness.

V. K. Khanna, *Implantable Medical Electronics*,
https://doi.org/10.1007/978-3-031-99868-3_11

Confidentiality or privacy is freedom from observation, disturbance, and interference by others. Confidentiality of data implies its accessibility by and availability to authorized personnel only. They may access it either for viewing or using the same.

11.2 Apprehensions of Patients Receiving Implants

Apprehensions of defiance of cyber security and confidentiality are not unreasonable. These anxieties arise because many implantable electronic devices contain computing and communication modules. Wireless and Internet connectivity is an intrinsic feature of these devices (Fig. 11.1). As we are aware, the malicious incidents of invasion of computers by viruses are frequently heard. Hacking of accounts or theft of laptops is common too. Therefore, it is plausible that any unauthorized person can take liberty to gain control of an implant. The evil person can go the extent of crippling its functioning. Exposure of a patient's vital data may also tempt a person with mala fide intensions to deliberately make changes in data that are harmful to the patient. By hacking, a person may operate the drug delivery pump of a patient to administer lethal doses of drugs. A wicked person may gain control over a defibrillator and impart an unnecessary high-voltage electric shock to the patient's heart. Therefore, such trepidations in the minds of patients and doctors need to be dispelled. They can only be allayed by building defensive mechanisms to thwart threats. Additional impregnable features have to be introduced into the systems to make them secure and private.

11.3 Security Requirements

Medical implants should ensure continued, reliable service through secure communication and functionality. Following are the main requirements that need to be fulfilled [1]:

1. Device-Existence Privacy: No one except the authorized personnel should be able to detect that a patient has an implanted device.
2. Device-Type Privacy: If someone knows that a patient has an implanted device, the type of device should not be disclosed.
3. Specific Device ID Privacy: No unauthorized person should be able to track any individual device.
4. Data Integrity: Access to the private details of a patient such as name, diagnostic/therapeutic parameters, and other stored data should be invincible to unauthorized people.

Security requirements vary from device to device. In devices like cochlear implants, malfunctioning poses less risks to human life. So, they are deemed to be less life-threatening. Hence, user identification and data validation may be adequate.

Fig. 11.1 Wireless body area network

But for devices such as pacemakers and insulin pumps, any security contravention may endanger the life of the patient. So, elaborate security features must be built into the device. Rigorous testing and verification of these features must be done prior to marketing to avoid any remote chances of security violation.

11.4 Causes of Security Breaches

Probable reasons are either deliberate or unintentional [2].

11.4.1 Deliberate Breaches

Causes could be jealousy against competitors, e.g., for damaging the reputation of a firm; seeking financial advantages by accessing private data; sabotage by a dissatisfied employee or customer; and a terrorist attack.

11.4.2 Unintentional Breaches

Causes include inadvertent collateral damage spawned by a virus, worm, or other malicious software. The software might have been designed to disrupt other computers but invaded the medical network also. This may occur during updating software of medical devices through the Internet. Although updating is aimed at improving the device functionality, it also provides a portal for software contamination.

11.5 Types of Adversaries

An adversary is an opponent, enemy, competitor, or combatant. Adversaries are characterized in accordance with their objectives, capabilities, and resources at their hands (Table 11.1). Security designers assess the different threats in terms of definite criteria. The criteria laid out include their values and the efforts applied by adversaries to gain access to the implanted device.

Secret, purposeful real-time interception of a phone call, videoconference, or other private messages, and listening to their conversation without consent fall under eavesdropping. A passive eavesdropper can capture data but does not disturb or modify it. An active adversary is one endowed with augmented capabilities to compromise with the data. This adversary indulges in erroneous controlling actions. An oscilloscope, software radio, directional antenna, etc., may be used in these actions [3]. Such an

Table 11.1 Passive and active adversaries

Sl. No.	Passive adversary	Active adversary
1	One with limited facilities who violates privacy of data without any interference. This could be a person listening to radio communication from an implanted device	One with advanced resources. This person is not only capable of receiving the radio signals from an implanted device but also able to modify the operation of the device in an unfriendly manner.
2	Less dangerous	More dangerous because this adversary can acquire control of the device. He/she can guide it to erroneous life-threatening therapies

adversary may produce RF traffic for blocking signals. By such blocking, the signals are prevented from performing their assigned tasks. Blocking of wireless communication called jamming may be continuous. Continuous jamming is characterized by a nonstop signal of a fixed power. Jamming may be periodic. Periodic jamming is marked by pulsating action to damage a packet if hit. Jamming can also be reactive. In reactive jamming, the decisions are based on the present and past channel states. Another damaging capability involves binary analysis. This analysis is done for disassembling the software of the device. The intention is to know its operation. Then the same knowledge is used with malevolence. An adversary possessing an external device constructed for use with the implant could undesirably use the device for wicked intents, e.g., for disabling therapies [1].

11.6 Design Principles for Implant Security

Designers should keep security in mind from the very beginning [4]. The ideas of security in computers and computer networks do not represent a new field. The already proven concepts have to be applied to implants according to the particular application. If security is imbibed after the system has been built, unexpected failure forms may be encountered. Thus security is a preplan rather than an addendum. Security is an essential or integral component of both the product and the manufacturing company. The organization and the management should enforce procedures to prevent leakage of vital information that may lead to flaws. One tactic that assists in tackling the confrontation of the implant designers with dreadful intentions goes by the name of "the principle of defense in depth." This principle recommends multiple tiers or layers to toughen security. The chosen measures embody the broad field covering from corporal security and admission controllers to security of the network.

If the system design is simple, it can be readily understood. Therefore, different likely routes to be adopted by the adversaries can be preconceived. Such a preconception enables provision of suitable mechanisms to ward off the dangers. A difficult design may make it too complex for the designer to foresee the weak links.

A source code is the initial program written in a programming language. It is readable by humans. It is later translated into a machine language. This translation gives it a form called the machine code with binary 1s and 0s for execution by the computer. Therefore, designers of implants should use standard source codes. These standard source codes have proved their worth over time. New untested codes need not be followed. Further, designers should not rest assured that the source code is too difficult to understand. They should always remember that it can be deciphered by someone to create nuisance.

Any information that is to be kept under the veil of secrecy is classified as sensitive data. It should be disclosed only to commissioned parties. Cryptography deals with techniques of information hiding and verification. It aims at protecting communication from adversaries. One form of cryptography is called encryption. It converts the given data known as *plaintext* into a form termed *ciphertext*. It does so, for example, through substitution of numbers by letters using an algorithm. In cryptography, a cipher is an algorithm for encryption or decryption. Only legitimate persons are able to read the encrypted data. For reading the data, a secret key is given. This key is actually an algorithm that unwraps or undoes the encryption. For safeguarding against threats, cryptographic building blocks are combined to construct a cryptosystem.

Encryption is the transformation of data from its original native format called the plain text format into a format known as the cipher text format. This format is not easily comprehensible to people that do not have official clearance or approval to do so. Decryption is the reverse process. It involves reverting encrypted data back into its original native format. This enables it to be unreservedly understood by common people.

Designers should use regular cryptographic building blocks in place of impromptu designs. Customary ciphers and security protocols must be used. But cryptographic keys must be carefully protected. The vast expertise of the cryptographic community is gainfully utilized by taking recourse to cryptosystems that have undergone scrutiny by professionals. Cryptographic specialists with years of experience must be employed. Building homemade encryption or key management systems is disastrous. Moreover, use of algorithms that have been broken long ago is precarious. Encryption technology leveraging new algorithms should be utilized.

A perilous practice of database encryption relates to the storage of the key used for encrypting the data or the authentication credential. In this practice, the key is stored in the same database along with the encrypted data. Such key storage should never be done. Indeed, the management of encryption key must be kept unconnected with the database that was used for storing the data encrypted with that key. Encryption keys are better protected by the hardware. Then the encryption key does not at any time quit the device. Hence, unlawful personnel or data thieves are neither able to retrieve the key nor the cryptographic functions and operations in which the keys are used. To reiterate, encryption places a high burden on a network and its users. However, data encryption is not difficult. But allowance of access to protected files for ratified users while keeping unwanted people away is complicated.

Similar to the security practices followed for one's own devices, any devices from third parties must also prove their capability of safeguarding before acceptance. Any encryption claims must be validated prior to use.

Threat modeling is concerned with studies of the different kinds of possible threats to security and their behavioral pattern to build suitable threat models. These models can be used to devise countermeasures for the expected menace.

Prima facie policy planning appears to fall outside the domain of device designers. But the long series of changes in the life of a device developing into a product and preventive regulatory surroundings have made policy a design time issue. Implanted devices are pushed into market after undergoing validation tests. Supposing that new security threats are discovered after launch of a product, it becomes the responsibility of manufacturers to plan for any changes such as software updates. These changes are normally done in a clinical environment. Permitting updating in an unrestricted or poorly verified setting may cause security problems. During policy formulation, the regulatory environment under which the devices are placed in their market life-times must also be considered. Fresh clearance is required for significant updates to devices that are already in the market. Thus, it is essential that the designers should tackle with foresight the likely future threats. They should solve such problems at the design time itself. As threat modeling helps to plan for future complications, designers can advocate for security statements and policies at the company level. These considerations imply that policy planning must be kept in view at the design stage of a product. The above ideas are presented in a concise form in Table 11.2

Table 11.2 Extension of fundamental security ideas to implantable devices

Sl. No.	Idea	Explanation
1	Integration of security from creation	Security should be built in the system during its construction. It should not be stuffed after the system is completed
2	Simplicity of security designs	Simple security systems are easy to understand. Then the possibilities of attacks can be argued. Necessary retaliatory methods are suggested and readily put into effect
3	Adoption of industry-standard source code techniques	Designers must refrain from using nonstandard source codes. This self-enforced restraint is necessary because their ruggedness is not guaranteed
4	Non-reliance on obscurity	Obscurity is not an assurance for security. It should always be presumed that someone may break open the source code and cause havoc
5	Encryption of sensitive data	Encryption is an unsurpassed method to avert the appalling inconveniencies with stolen data. If the data is encrypted, nothing useful is taken even if experienced hackers penetrate a system. To bestow the uppermost ranks of security, encryption should be invariably supplemented with proper management of the key
6	Use of standard cryptographic building blocks	Carefully deliberated and established building blocks must only be used. New technology-based algorithms must be applied, wherever applicable. Old algorithms that are already broken must be avoided
7	Authentication of third-party devices	Third-party device vendors must prove their cryptographic claims
8	Modeling of threats	Threats should be prioritized. The most appealing targets for attackers must be identified and defended against. This must be followed by less frail targets

11.7 Expository Examples of Security Breach Possibilities

Several researchers have conducted mock drills of penetrating through the security of commercial implantable devices. Some interesting examples of such studies are presented here.

11.7.1 Hijacking an Open-Loop Procedure: The Insulin Infusion Pump

The insulin pump arrangement is an open-loop procedure. In this system, a patient varies the pump settings, as per requirement. The system has both implanted and external components. A subcutaneous glucose sensor is used for measuring the instantaneous glucose concentration in the blood. The measured concentration is wirelessly transmitted to the external control. Based on this measurement, a wirelessly coupled insulin infusion pump subcutaneously delivers insulin to the patient. This pump works under the supervisory control of the patient. The patient reads the display on a wireless remote control for necessary information. Then he/she alters the pump settings through this remote control. It is this remote control interface carrying the patient glucose level and control information that is open to security threats.

Demonstration of security infringement was done by Li et al. [5, 6] on the wireless communication link of a commercially available blood glucose measurement and delivery system. To launch the attack, the frequency of the radio link was found online in the public domain from Federal Communications Commission (FCC) as 915 MHz. By intercepting the radio signal, the modulation scheme was decided as on/off keying. For data reception by the insulin pump, a code of 6 digits in hexadecimal code must be entered by the user. These digits constituting the personal identification number of the device were stamped on the backside of the glucose meter or its remote control accessory. The data packets between the remote control and the glucose meter were intercepted. After synchronizing the sequence of binary zeroes and ones, it was ascertained that the communication packet contained 80 information bits whose roles are explained below: the first 4 bits, device type; next 36 bits, device PIN; next 12 bits, information bits; next 12 bits, counter bits, repeating after 256 reckons; succeeding 12 bits, random cyclic redundancy check (CRC) bits; and final 4 bits, 0101. The counter was found to be an 8-bit counter. The device PIN was found to be transmitted without encryption. After conducting several trials, the CRC parameters were determined. Regarding replay attacks, the defense methodology was that the system did not accept any packet if the counter had the same value as the preceding packet. Hence, two packets are intercepted and transmitted in an alternating fashion for acceptance by the system. After discovering the packet format and CRC parameters, it was possible to design a valid packet. This designed packet will be acceptable to the insulin pump. It will enable full control of the operation of the pump to fall in the illegitimate hands of the attacker.

Both passive and active mode attacks were brought into the realms of possibility. One could imitate and pose as an empowered user to fool around with the device controls. For carrying out these manipulations, one could use easily available tools such as Universal Software Radio Peripheral (USRP). The USRP provides an inexpensive hardware platform for software radio. It enables the users worldwide to undertake wide-ranging applications in research, academics, and industry. Without knowing the device PIN, it was possible to successfully launch three types of attacks:

1. System Privacy Attacks: Eavesdropping on the communication channel could be done. The attacker can decode information about the device type, its PIN, and medical status of the patient.
2. System Integrity Attacks: In a replay attack, the attacker acquires the control of the pump by alternative transmission of two succeeding packets. The attacker can report an erroneous glucose reading to it. This reading would cause operation of the system in a malfunctioned manner.
3. System Availability Attacks: The attacker jams the communication channel. Then the attacker makes certain that either remote control ceases to work or no data is transmitted.

By knowing the device PIN, the following kinds of attacks could be launched: (1) stoppage of insulin infusion inside the body, in which the blood glucose level of the patient is elevated; (2) restart of a stopped pump, in which the pump would push insulin into the patient's body; or (3) release of an indiscriminate dose of insulin into the body. Thus misconfiguring of the pump could be done leading to erratic behavior. Such incorrect response through wireless forgery can cause heavy insulin pumping during hyperglycemia. It may cause insulin stoppage during hypoglycemia. These disturbances could result in a serious risk to patient's life.

In this open-loop system , the user is a part of the loop. The threats target the user. Such a system comes under the subfield of usable security. It aims at building a secure user interface design. Users must be given unambiguous indications regarding the status of the device. They must be provided with tools to take informed security decisions. They must also be educated about the potential security risks to the devices which they are using.

Two solutions were also proposed for countering the attacks: (1) Rolling code technique: In this technique, the same PIN is not sent always. In lieu of the same PIN being sent, a rolling code encoder is firmly rooted in the remote control. Another encoder is embedded in the insulin pump. Such an arrangement is more difficult to break. An encryption key is shared between the rolling code and the remote control. The data sent are encrypted. Also, the rolling code keeps changing every time. Hence, PIN extraction is a hard nut to crack. (2) Body-coupled communication: Here, there is no communication through the air medium. Instead, communication takes place through the patient's body. Thus, the communication is effected in the restricted region nearby the patient's body. Therefore, there is less likelihood of eavesdropping. Persons in close vicinity of the patient can, however, do this, but they may get caught. A bonus advantage is the lower power consumption by the communication system. The power consumption is low because the system works over a shorter range.

11.7.2 Security Analysis of a Closed-Loop System: The Implantable Cardioverter Defibrillator

The ICD extends the capabilities of a pacemaker by delivering a large shock to the patient to arrest an unsustainable heart rhythm. Unlike the insulin pump, the ICD is a closed-loop system. In this system, the sensing of an abnormal rhythm prompts the actuation. Using simple software and radio tools, researchers could record the communicative messages between the ICD and a programming console used in clinics [3, 7]. Patient data were revealed easily. It was found that the signals were not coded in any way against undesired access. There was no evidence of encryption. Replaying the translated commands, it was possible to control or immobilize the ICD action. A sequence of radio transmissions was found to keep the ICD in highly active mode. The packets were transmitted regularly for an indefinite period. Enormous power was thus drained out. By employing this type of gimmick, it is possible to prematurely exhaust the battery of an implanted device.

In contrast to open-loop systems where the user interface is the weakest link in the chain, in the closed-loop system, the biggest challenge is the automated decision-making. To avoid security violations, some manufacturers allocate secret keys to their devices before deploying them. If any such pre-distributed key becomes compromised during deployment, every device with that key must be updated. A viable option is to let customers generate their own keys before using the devices. The issue of key compromising is done away with. But a way to install the key on a device without user interface must be found. Halperin et al. [7] put forward a credible method employing transcutaneous acoustic coupling to exchange the key material. Audio alerts or beep warnings are used for communication with a system without a user interface.

RF interface is not the only unsafe zone. Analog sensors responsive to electromagnetic interference (EMI) serve as unrestrained admission points into an otherwise guarded system. These sensors allow an attacker to influence sensor readings. The amended sensed data appear directly at the application layer of the device. It thus dodges usual security apparatuses and gives the assailant some chances of governing the system. Foo Kune et al. [8] showed that in open air, preconceived, conscious electromagnetic interference under 10 W could hold back pacing. It could provoke shocks for defibrillation at interspaces up to 1–2 m on ICDs. Then the sensing leads along with the medical devices were submerged in a brackish bath for closer approximation to conditions inside the human body. In this case, the distance for the similar trial lessened to <5 cm. Sometimes the attacker cannot match the wavelength of the EMI signal with the length of the sensing leads. Nevertheless, an increase in power of EMI transmission can induce signals in millivolt range at the sensing leads of the implantable device. Therefore, sensed time-dependent voltages are prone to contamination by analog signal injection through EMI. Moreover, sensing in ICDs is done in the subkilohertz band. Due to this reason, filters in this frequency range cannot be used. In addition, coaxial design is prevented by mechanical constraints on the sensing leads. These issues make EMI more embarrassing to the implanted device.

To combat the EMI, a suggested defense strategy is to detect the spurious sensor input. This detection is done by checking its consistency with the refractory period of the heart tissue.

11.7.3 Security and Privacy of Implantable Biosensors Used for Data Acquisition

Biosensors span a broad category of signals and techniques for processing signals. They cover a range of data rates, from the low-data-rate glucose sensors to the high-data-rate optical imaging devices [3]. These biosensors can detect biomarkers for various diseases. They can also measure pH, temperature, and other parameters. They communicate wirelessly through the tissue to supply the information to the external monitoring systems. Many of these sensors are also powered wirelessly. The security concerns associated with these sensors are different from those discussed above. The data acquired by these sensors are used for actuating therapeutic devices or drug delivery systems. Therefore, confidentiality of these data must be strictly guarded to avoid their unethical use.

The main concern with implanted biosensors is that their wireless transmissions are small in information content and take place at less frequent intervals as opposed to the continuous transmissions of other devices. These small, infrequent transmissions are more difficult to protect. A sensor might require a few minutes to accomplish its assignment. After completion of task, it delivers only few bytes of data. Cautious usage of cipher is essential in case the plaintext data from the sensor acquire only a small number of dissimilar denominations. There is only a little intrinsic redundancy in the small quantity of data. So, error correction must be done.

Sometimes, small biosensors are injected into a patient (Fig. 11.2). They are powered inductively through a bandage-like external patch on the skin to relay the sensed data outside. When a biosensor is paired with a patch, different risks arise. In view of the short transmission range around a few millimeters for a subcutaneous biosensor, eavesdropping may be difficult. However, a fraud through caricature of either the clinical reader or patch is a likely possibility. If a patient is unconscious, the patch can sometimes be easily detached and swapped with a fraud one. Likewise, a deceitful sensor can feed wrong information to a dependable patch. Proper cryptographic mechanisms should make sure that all the components involved in the treatment are well authenticated.

Implanted biosensors with stringent space and energy restrictions pose special difficulties in protecting against security and privacy attacks. The imposed restrictions make routine encryptions unusable. New cryptosystems for energy-constrained implanted devices offer a ray of hope. Notwithstanding, the available algorithms for these devices are very few as opposed to general-purpose solutions, which are of little help.

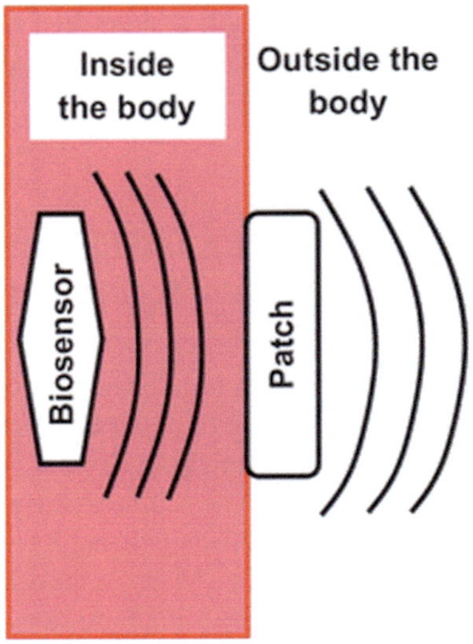

Fig. 11.2 Biosensor injected into a patient and powered via inductive coupling by a patch, which sends the received data to a physician

Table 11.3 Security challenges posed by different systems

Sl. No.	Feature	Type of system		
		Open-loop systems	Closed-loop systems	Implanted biosensor systems
1	Example	Insulin pump	Implanted cardioverter defibrillator	Subcutaneous glucose sensor
2	Risky situations	User interface	RF interface violation and EMI contamination	Impersonation and fraudulent practices
3	Suggested remedies	Rolling codes, body-coupled communication	Secret keys, detection of fake sensor input	Cryptography

Table 11.3 gives a comparative depiction of the security precariousness of open- and closed-loop and implanted biosensor systems and the obliteration of the Achilles' heels.

11.8 Conflict of Security with Safety, Efficiency, and Usability

There is an obvious antagonism between patient security and safety [Li]. Tighter and stricter security norms may sometimes hinder the provision of required medical attendance to the patient. This may be especially so in emergency situations when the patient is incapacitated and needs urgent attention. Then if the doctor is unable to adjust the device parameters, valuable time is lost. Clearly, authentication mechanisms should be context-aware and flexible. The reason is that during an emergency when a patient is unconscious, he/she cannot cooperate with the doctor to provide password for the device.

Cryptographic solutions must be as undemanding and frivolous as conceivable, in terms of both calculation time and storage requirements. Otherwise, the power and storage space will be drained quickly. Thus, there is a rivalry between security and efficiency.

Usability becomes difficult as security increases. Operation by omitting some manual steps simplifies usage but weakens the security. Long-distance usage over wireless enhances usability by monitoring the patients at home, instead of the clinic. At the same time, there is increased risk of meddling by eavesdroppers.

11.9 Negative Aspects of Security Scheme

The need of authentication of the user to start the operation of a device strengthens its security. But let us imagine an emergency situation when the patient is unconscious. The attending doctor may not be able to reprogram the device according to patient's condition. In such cases, the built-in authentication or encryption algorithms act as a hindrance to the doctor in providing immediate medical care. Hence, the patient may be denied the valuable service required in a critical situation.

Elaborate encryption schemes unduly load the power source. A lot of power is drained. The implant should primarily prove its worth in medical treatment. Therefore, heavy-duty encryption schemes may be a boon to security. But they may act as a burden on the battery causing it to run out fast. This calls for a balancing of cyber security and confidentiality with safety and utility of the device [7]. Safety means that the implant should be more beneficial than harmful. Utility implies its usefulness to patients and doctors. Please see Table11.4 for a side-by-side delineation of security and usability.

11.10 Protection Without Device Modification

Several patients have already received implants. If any security measures are to be adopted, the question arises how they will be incorporated in devices which are already implanted in patients. It is not easy to surgically remove and recall such devices. Therefore, an alternative approach is logical. In this scheme, the responsibility of protection is assigned to a personal base station. This base station will serve as a jammer of implant messages for others, thus preventing their decoding efforts [9]. But for the genuine users, this jamming action is ineffective. Thus it is a jammer-cum-receiver. The base station is a shield against illegitimate users (Fig. 11.3).

Table 11.4 Incompatible scenario

Sl. No.	Situation	Security	Usability
1.	Medical emergency	Requires the proper certification of the user for altering its functionality to provide the required emergency treatment	Authentication process may render it impossible to provide treatment if the patient is unconscious. Treatment cannot be given in the absence of a programming device with the shared secret
2.	Energy drainage	Heavy-weight encryption unduly loads the power supply. It exhausts the battery prematurely and necessitates battery replacement	Energy consumption by the encryption must not overburden the normal availability of the device for the intended use
3.	Cost	Costly encryption may make the large-scale deployment of subcutaneous biosensors prohibitively impractical	Encryption costs must be a proportionately reasonable fraction of the price of implantable device

11.11 Discussion and Conclusions

Ubiquitous health monitoring (UHM) through wireless body area networks (WBANs) provides e-health care for in-home monitoring and diagnosis. It frees the patients from frequently visiting hospitals [10]. It is more important in countries with shortage of medical infrastructure. Its relevance increases especially during natural disasters and calamities, when several precious human lives can be saved.

Many available devices have shown weakness and defenselessness against attacks. Security and privacy protection of patient information mandates the deployment of various techniques. They are essential both during transmission and storage in the network. They are applied carefully in accordance with the degree of risk resulting from the data tampering to the patient for the particular medical implant. In addition, the associated power constraints must be considered.

Cryptography is not a cure for all security risks. Many questions still remain unsolved. Some of these unresolved questions are: If an implantable device uses encryption of data, how is the necessary key distributed? Who should distribute the key? How does an implantable device distinguish between external agents that are allowed to communicate with it benignly and those that may interact wickedly? Since security of some implantable devices needs to be disabled for emergency health care, how to protect these devices under nonemergency circumstances? Should a device raise hue and cry as an audible alarm during an attack?

Review Exercises

11.1 What is the difference between security and safety? Define cyber security. What does privacy of data mean?

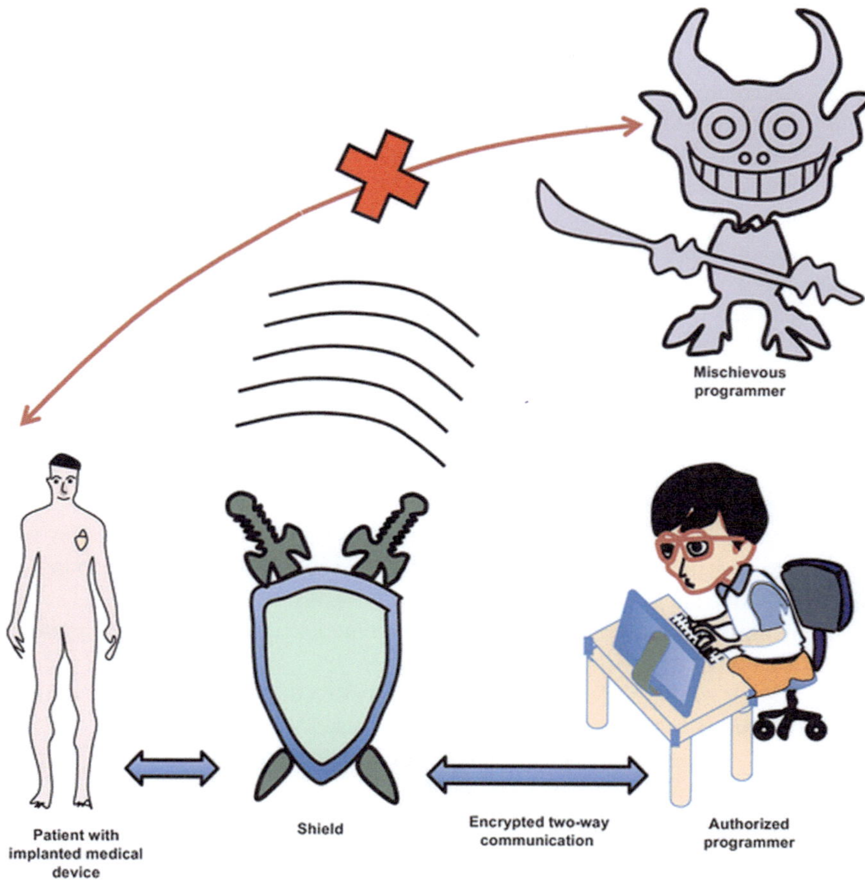

Fig. 11.3 Protection of a medical implant without modification by inserting a shield, which paralyzes any straight interfacing or interaction with the implant, whereas a ratified person is able to establish a secure channel with the implant through the shield

11.2 "A logical fear irks the minds of patients that cyber security and confidentiality of implantable medical devices can be contravened." Do you agree with this statement? If yes, please explain why?

11.3 State the four principal security requirements that must be fulfilled for implantable medical devices. Do cochlear implants and cardiac pacemakers present the same security risk?

11.4 How do you classify the causes of security breaches? Give examples of each class.

11.5 What is an adversary? Into how many groups will you place the different adversaries for implantable devices? Which type of adversary is most dangerous?

11.6 It is said that the planning for security of implantable device should commence from the very beginning. Give your arguments in support of this assertion.
11.7 Why should the security system be simple? How does it help you?
11.8 What is a source code? Why may happen if the designers of implantable devices use nonstandard source codes?
11.9 What is cryptography? What do you understand by encryption and decryption of data? Explain the terms: "plain text" and "cipher text."
11.10 Why is it said that building homemade encryption can cause havoc? Give reasons.
11.11 Why should the encryption key be stored in a different database than the one used for storing the encrypted data?
11.12 What is threat modeling? How does it help in providing security to implantable devices?
11.13 Why is policy planning considered to be a job of the implantable device designer? Give arguments justifying your answer.
11.14 Explain the working of an insulin pump system for diabetes relief. How was a passive attack on this system demonstrated? How was it possible to launch an active attack on this system?
11.15 What is the most attractive target for the attacker in an open-loop system? What is usable security?
11.16 What is rolling code technique for security of a device? What makes the PIN difficult to crack when the rolling code is used?
11.17 What is meant by body-coupled communication? How does it help in protecting against eavesdropping?
11.18 How was the possibility of security violation of an ICD demonstrated? How can an attacker deplete the battery of an ICD very fast, inflicting suffering on the ICD patient?
11.19 What is the weakest link in the closed-loop system? How does distribution of secret keys prevent security breaches in a closed-loop system? Highlight some problems associated with the keys.
11.20 What kind of attacks can be done through intentional EMI? How can EMI disturb the operation of an ICD? Suggest a suitable remedy to counteract the EMI menace.
11.21 Bring out the main worries of implantable device designers regarding the security of implanted biosensors. Emphasize the differences between these security concerns with those normally faced with implantable devices. Suggest the necessary remedial steps.
11.22 Explain by giving examples on how the severity of security measures falls in the way of providing efficient treatment to the patients and hence conclude that a balance needs to be struck between the security and utilization of implantable devices.

11.23 A large number of patients today are getting benefitted by different types of devices implanted in their bodies. If therefore a new security scheme is launched, it may not be applicable to these patients. How can such patients be protected against security threats? Suggest a scheme applicable to such patients.

References

1. Daniluk K, Niewiadomska-Szynkiewicz E (2012) Energy efficient security in implantable medical devices. In: Proceedings of the federated conference on computer science and information systems, 9–12 Sept, Wroclaw, pp 773–778
2. Maisel WH, Kohno T (2010) Improving the security and privacy of implantable medical devices. N Engl J Med 362(13):1164–1166
3. Burleson W, Clark SS, Ransford B (2012) Design challenges for secure implantable medical devices. In: 49th ACM/EDAC/IEEE design automation conference (DAC), 3–7 June, San Francisco, CA, pp 12–17
4. Ransford R, Clark SS, Foo Kune D et al (2014) Chapter 7: Design challenges for secure implantable medical devices. In: Burleson W, Carrara S (eds) Security and privacy for implantable medical devices. Springer, New York, pp 157–173
5. Li C, Raghunathan A, Jha NK (2011) Hijacking an insulin pump: security attacks and defenses for a diabetes therapy system. In: 13th international conference on e-health networking, applications and services. Columbia, MO 13–15 June, pp 150–156
6. Li C, Zhang M, Raghunathan A et al (2014) Chapter 8: Attacking and defending a diabetes therapy system. In: Burleson W, Carrara S (eds) Security and privacy for implantable medical devices. Springer, New York, pp 175–193
7. Halperin D, Kohno T, Heydt-Benjamin TS et al (2008) Security and privacy for implantable medical devices. Pervas Comput 7(1):30–39
8. Foo Kune D, Backes J, Clarke SS et al (2013) Ghost talk: mitigating EMI signal rejection attacks against analog sensors. In: Proceedings of the 34th IEEE annual symposium on security and privacy, Berkley, CA, 19–22 May, pp 145–159. https://doi.org/10.1109/SP.2013.20
9. Gollakota S, Hassanieh H, Ransford B et al (2011) They can hear your heartbeats: non-invasive security for implantable medical devices. In: SIGCOMM'11, 15–19 Aug, Toronto, 12 p
10. Li M, Lou W, Ren K (2010) Wireless data security and privacy in wireless body area networks. IEEE Wireless Commun Mag 17(1):51–58

Part II
Applications

Chapter 12
Neural Amplifier Circuits in Implants

Abstract The neural signals are low-frequency (mHz–kHz) and low-amplitude signals (μV–mV). Therefore, the amplifiers for these signals must be low-noise circuits. Additionally, the front-end amplifiers must reject interference due to the common-mode signals as well as electrode effects. Amplification techniques based on clocking and continuous-time approaches are described. The clock-based techniques include switched-biasing, chopper, and auto-zeroing methods. The traditional continuous-time circuit is the AC-coupled-operational transconductance amplifier-based neural amplifier endowed with capacitive feedback. The unavoidable trade-off between input capacitance and area of the chip against the gain of the amplifier can be relaxed. This is achieved when a clamped T-capacitor network replaces the feedback capacitor.

Keywords Neural amplifier · Biopotential amplifier · Switched-bias amplifier · Chopper-stabilized amplifier · Auto-zeroing amplifier · Continuous-time amplifier · Zero-drift amplifier · IMD · Operational transconductance amplifier

12.1 Introduction

A low-noise biopotential amplifier is the main component of a bio-signal recording system designed for implantable applications. It significantly influences the power and noise performance of the arrangement. To make amplifiers for implantable devices, neuroscientists and clinicians yearn that power dissipation should be as small as possible. Then the neighboring tissues will not suffer any damages by the associated heating effects. Low power consumption together with high precision and low $1/f$ noise is mandatory features of CMOS amplifiers. Many techniques are applied to mitigate $1/f$ noise. The most straightforward way is to use large input device geometries. This approach is successful in processes, which give low surface state densities in the beginning. For processes giving high surface densities, the approach becomes unjustified because of the very large geometries that are needed. Another way of noise reduction is the use of buried channel devices. In these devices,

V. K. Khanna, *Implantable Medical Electronics*,
https://doi.org/10.1007/978-3-031-99868-3_12

the channel is located deeper inside and far removed from the surface. At this depth, the surface states cannot exert their influence. This calls for process steps, which are not usually practiced in a standard fabrication sequence. One more option is a circuit-based technique to translate the noise energy from the base band to some higher frequency. By this translation, noise will be unable to contaminate the signal. In the treatment below, the reader will have the opportunity to learn about amplifiers employing different approaches of noise elimination.

12.2 Clock-Based Amplifiers

12.2.1 Switched-Biasing Amplifier

Switched-biasing amplification is based on switching or cycling of the transistor between two defined states [1]. It may be recollected that 1/*f* noise, also referred to as low-frequency noise, flicker noise, or pink noise, is an electronic noise whose power spectral density (PSD), describing the distribution of power as a function of frequency, varies inversely as the frequency *f* (− 10 dB/decade) [2, 3]. It may be further recalled that this noise is produced from imperfections of the interface between the gate oxide and the silicon substrate in the MOSFET. Two theories of 1/*f* noise have been proposed [4], viz., the carrier number fluctuation theory and the carrier mobility fluctuation theory. The carrier number fluctuation theory ascribes the 1/*f* noise to the unsystematic capturing and release of carriers in the conducting channel. Its prediction of noise density referred to input is as follows: The noise density does not depend on the gate biasing voltage and is proportional to the second power of thickness of gate oxide. The carrier mobility fluctuation theory attributes the 1/*f* noise to fluctuations in the carrier mobility. Its forecast on noise voltage referred to input is as under: the noise voltage increases with gate biasing voltage and varies as thickness of gate oxide. Each theory has been partially successful in interpretation of experimental observations, and neither does so completely.

In accordance with the carrier number fluctuation theory of charge trapping and de-trapping, a distribution of trapping times originates through the transference of electrons from the surface of the semiconductor to the traps in the oxide. The power spectral density results from the superposition of relaxation activities having Lorentzian spectra and a dispersal of time constants. Bloom and Nemirovsky [5] found that it was possible to reduce the 1/*f* noise by switching between two states. The first state is a noisy state produced by biasing the transistor into inversion for on-state operation . It is characterized by appreciable generation of 1/*f* noise. The second state is a noise-free state generated by biasing the transistor into accumulation for off-state operation. In this state, the 1/*f* noise is negligible. Insertion of the off-states between the noisy on-states reduces the 1/*f* noise. This reduction of noise is caused by interference with the long time constants of electron transitions in trapping and de-trapping processes. It is likely that when the transistor is switched off, the traps at the Si–SiO_2 interface are vacated. When switched on, the traps still remain empty for some time. During this period, the noise is reduced.

The idea of switched-biasing technique is illustrated in Fig. 12.1. In this technique, a constant gate bias is not applied. Instead, the transistor is continuously switched between an operational or active state to a nonoperational, resting, or inactive state. This switching condition is created by the application of a voltage source having a square wave output across the gate-source terminals of a MOS transistor. The high level of the square wave is greater than the threshold voltage of the transistor. Hence, the transistor is biased into inversion regime. The low level is lesser than threshold voltage to enable biasing of the transistor into accumulation. Thus by cycling the transistor between strongly inverted and strongly accumulated states, the $1/f$ noise is decreased. The resting period between the operational states not only diminishes the $1/f$ noise but also the overall power consumption. For a square wave signal having 50% duty cycle, the baseband noise decreases to 0.5 times at a high switching frequency. Also, any modulation effects lie outside the bandwidth of interest. They are separated and removed by filtering.

To distinguish from techniques like chopping (described in the next section), which reduce the noise effect on a circuit level, switched biasing does so on the MOSFET device level itself. The former only mitigates the effect of noise leaving the noise source alone.

Switching transistors between two states may not be always possible. But this kind of freedom is available in some circuits, which are continuously processing the signals. It is also available in circuits in which bias current is required for certain intervals of time. Thus switching off the MOSFETs during periods of time in which they are not contributing toward circuit operation helps in weakening $1/f$ noise.

12.2.2 Chopper-Stabilized Amplifier

12.2.2.1 Chopper Amplification Concept

Extremely low-magnitude DC signals need very high-gain DC amplifiers. Such DC amplifiers are difficult to build with low DC offset and noise. Desired bandwidth and stability are tough to achieve. However, AC amplifiers are easily realized. The small-value DC signals can therefore be amplified by first converting them into AC signals. For this conversion, a chopper circuit is used to break up the signal by multiplying by a signal $M_1(t)$. Then it resembles an AC signal. In this process, the signals V_{os} and V_n denoting the DC offset and noise of the amplifier, respectively, are inserted into the signal as additives. The obtained AC signal is augmented by a gain factor G by feeding it to an AC-coupled amplifier. After amplification, the signal is integrated to restore the DC signal by multiplying by signal $M_2(t)$. Thus by chopping the input signal, it is amplified as an AC signal. From the amplified AC signal, the equivalent DC signal is recovered in amplified form after passing through low-pass filter (LPF). Precision electronic instrumentation utilizes this concept for stable amplification.

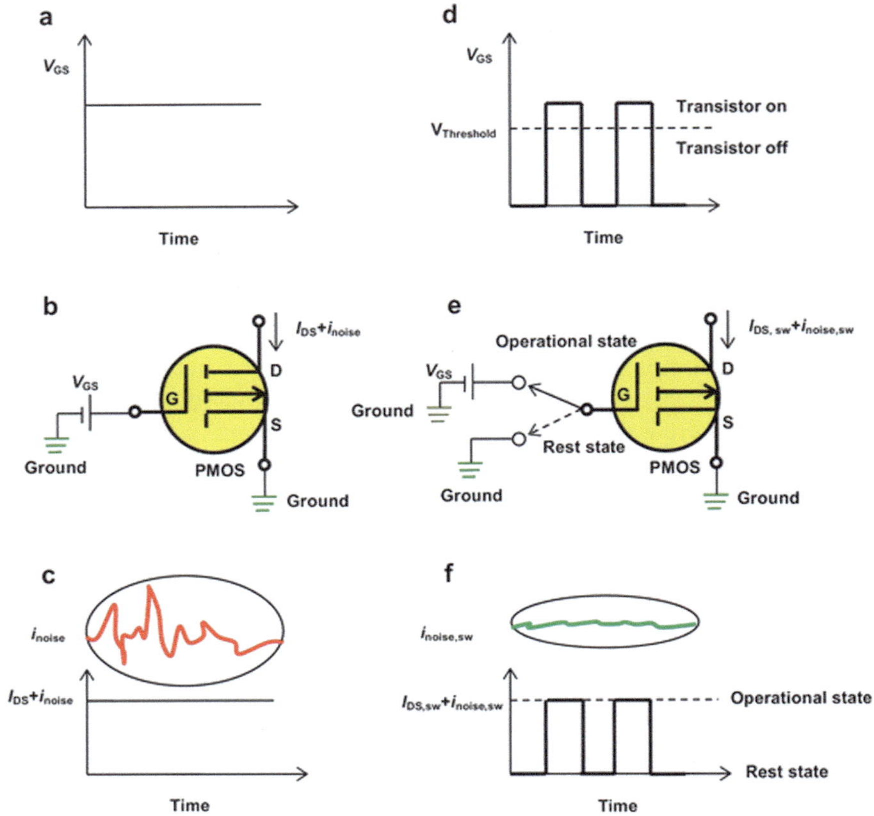

Fig. 12.1 Switched-biasing concept: **a–c** constant bias; **d–f** switched-bias

12.2.2.2 Conventional Chopper-Stabilized Amplifier

Chopper-stabilized (CHS) amplification is a technique based on modulation of the signal. The basic principle of this technique (Fig. 12.2) is that byamplitude modulation, the given signal is swapped to a higher-frequency region where $1/f$ noise is low. In this modulation, the chopping frequency is the carrier frequency and the input voltage is the modulating signal. After modulation, the signal is amplified at this higher frequency. Subsequently by demodulation, the signal is translated back to thebaseband frequency. As the signal is modulated twice, it shifts to even harmonics of chopping frequency. On the other hand, the noise introduced during amplification along with the offset has undergone modulation barely on one single occasion. Hence, the (noise + offset) is reordered to odd harmonics of the frequency of chopping waveform. The low-frequency noise is thus reallocated outside the band of interest from low- to high-frequency band. Finally by low-pass filtering, the amplified signal obtained is noise-free. The modulated noise, which had been shifted to high frequencies, is sieved out. It must be remembered that the baseband is the maximum usable bandwidth of the input signal. Baseband noise is made up of input offset voltage, 1/*f* noise, and white noise in the low-frequency range. Low-frequency white noise persists in the output signal. This happens because amplitude modulation does not affect the spectral density of white noise.

Flow diagram of a chopper-stabilized amplifier is shown in Fig. 12.2. Before entry into the amplifier, the signal is transposed using a CMOSFET switch modulator to the chopper frequency by the chopping signal. The chopping signal is a rectangular periodic signal $M_1(t)$ having a much higher-frequency f_{CH} than the signal frequency and the noise corner frequency. During the above translation of the signal by the chopping signal, it is shifted to the odd harmonics of the chopping frequency. It is well known that modulation of a square wave causes appearance of subbands on both sides of the chopper frequency. Hence, presence of 1/*f* noise leads to the introduction of subbands on both sides of odd harmonics of the frequency f_{CH}. In the next step, the signal is amplified by feeding to an amplifier of gain *G*. The amplified signal spectrum lies at higher frequencies than the 1/*f* noise corner frequency. After amplification, the

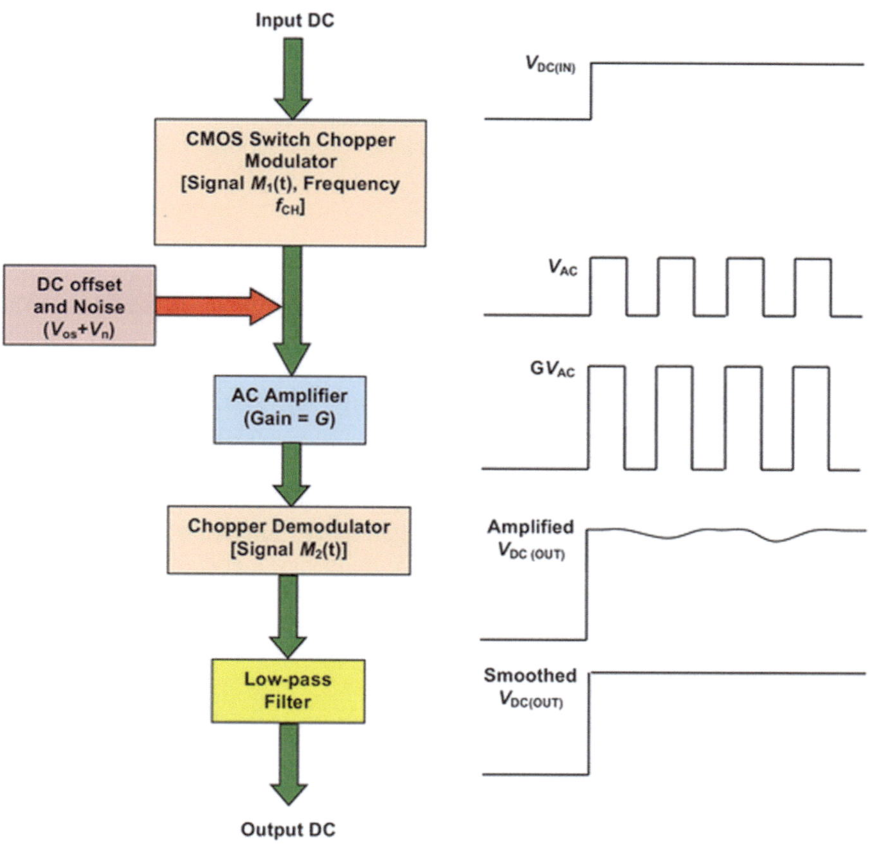

Fig. 12.2 Flow diagram of a chopper-stabilized amplifier

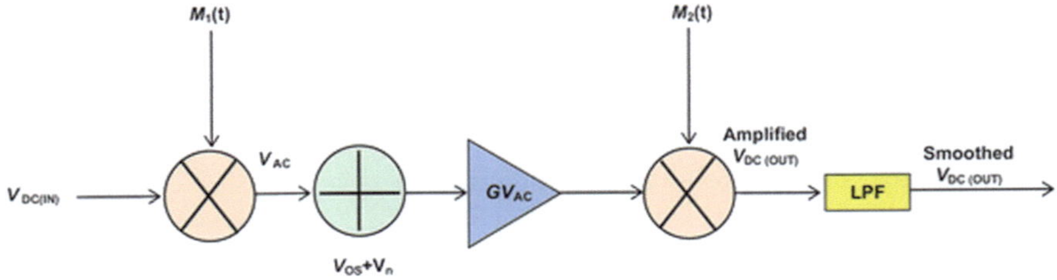

Fig. 12.3 Symbolic representation of the principle of chopper amplification

amplified signal plus noise produced by the amplifier is reconverted back to baseband form by the demodulator. This reconversion is done by multiplying by switching waveform $M_2(t)$ similar to $M_1(t)$ that was used previously for upconversion. As a consequence of this reconversion, the signal is demodulated to even harmonics of the chopping frequency and sidebands appear around the even harmonics of f_{CH}.

It must be noted that the DC offset and noise ($V_{\mathrm{os}} + V_{\mathrm{n}}$) before the amplifier (preamplifier) are modulated on one occasion only and moved to the odd harmonics of the chopping frequency. Finally, low-pass filtering restores the signal at the output. The modulated noise and artifacts are removed. In this process, up-modulated noise is subdued. On the whole, there is minimal aliasing of signal or noise. For achieving maximum DC gain, phase shift between input and output modulators must be = that due to the amplifier.

Figure 12.3 illustrates the principle of chopper amplifier symbolically. The cross symbol denotes a modulator/demodulator.

It must be mentioned here that the modulation/demodulation functions are carried out by simple pairs of MOSFET switches controlled by the clock signal [6]. The parasitic effects due to the switching transients in the modulation process are able to pass through the low-pass filter without cancelation. Nonidealities arising from MOSFET switches are clock feedback and charge injection. These nonidealities lead to residual offset generation. This residual offset can be decreased appreciably without much loss of gain. This is possible if the amplifier bandwidth is restricted to two times the chopper frequency [7]. The chopper frequency itself cannot be chosen to be very high. It is fixed by AC amplifier gain-phase limitations. Errors induced by modulator response time also influence the frequency. Generally, frequencies in the low kHz range are used for chopping. This obviously implies the imposition of a limitation on bandwidth of the amplifier. This is an issue that must be dealt with for all-embracing adoption of this kind of amplifier.

12.2.2.3 Wide-Bandwidth Chopper-Stabilized Amplifier

The bandwidth issue is tackled by building an amplifier that takes care of both low-frequency and high-frequency signals. This essentially means that the amplifier must consist of two sub-amplifying segments, one segment catering to low-frequency signals and the other dealing with high-frequency signals [8]. Then the outputs from these sub-amplifying segments are added together. This summation yields the amplified wide-bandwidth signal. The two sub-amplifying segments must be matching in their gain characteristics. By this matching is meant that at the frequency at which the gain of the high-frequency amplifier begins to decrease, that of the low-frequency amplifier should increase sufficiently so that the total gain is constant. In other words, for smooth gain–frequency characteristics, the lower cutoff frequency of the wideband amplifier must be equal to the upper cutoff frequency of the chopper amplifier. Following up on this basic idea of a two-segment amplifier, the circuit (Fig. 12.4), drawn after [8, 9], consists of two parallely connected amplification paths: (a) a wideband amplifier A_1 providing constant amplification in a broad spectrum of frequencies and therefore suitable for the high-frequency components of the signal and (b) a conventional chopper-stabilized path A_2 characterized by high-gain, high accuracy for low-frequency signal components. The chopper-stabilized amplifier is called the stabilizing amplifier. It biases the positive terminal of the wideband amplifier so that the potential at the summing point is zero. Hence, any offset and noise are eliminated. Thus this circuit of dual amplification paths succeeds in enlarging the amplifier bandwidth. Typically, bandwidth of several megahertz is achieved with the low drift chopper amplifier characteristics. On the downside, this circuit is capable of inverting operation only since the control of positive terminal of the wideband amplifier rests solely with the chopper-stabilized amplifier. Moreover, any residual noise of the chopper-stabilized amplifier is also amplified by the wideband amplifier. This amplification of noise fouls the output signal.

A comparison between switched-biasing and chopper-stabilized amplifiers is given in Table 12.1.

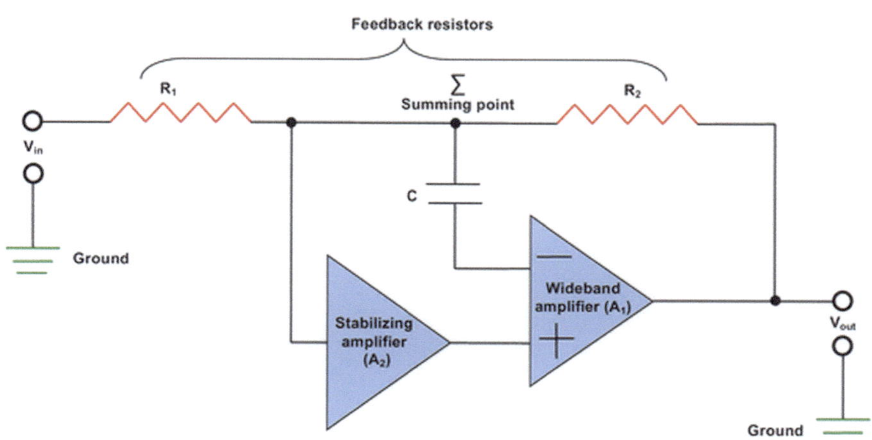

Fig. 12.4 A wideband chopper-stabilized amplifier

Table 12.1 Switched-biasing and chopper-stabilized amplifiers

Sl. No.	Switched-biased amplifier	Chopper-stabilized amplifier
1	It is a cycling technique	It is a modulation technique
2	It reduces the 1/*f* noise*itself*	It reduces the effects produced by 1/*f* noise in electronic circuits
3	It reduces power consumption	It leads to more power consumption
4	Due to reduction of 1/*f* noise at its corporeal genesis, i.e., the physical roots of noise, the circuits operating at high frequencies, wherein 1/*f* noise undergoes upconversion, are also benefitted	Its performance is solely effective in the low-frequency range. This kind of amplifier is not helpful to downgrade issues faced with noise that has gone through upconversion

12.2.3 *Auto-zeroing Amplifier*

Auto-zeroing amplification is a signal sampling-based technique. Sampling is the study of a process from a few selected items called samples. Here, the sample of interest is the input offset voltage and noise. Like the chopper-stabilized amplifier, the auto-zeroing amplifier (AZA) (Fig. 12.5) comprises two amplifiers. One amplifier is a wideband amplifier called the main amplifier A_M. The other amplifier is an offset and noise correcting amplifier, known as the nulling amplifier A_N. The remaining components are switches S_1, S_2 and storage capacitors C_1, C_2.

Each cycle of the auto-zero clock consists of two phases: the sampling phase A and the amplification phase B. During the sampling phase A, a sample of the noise

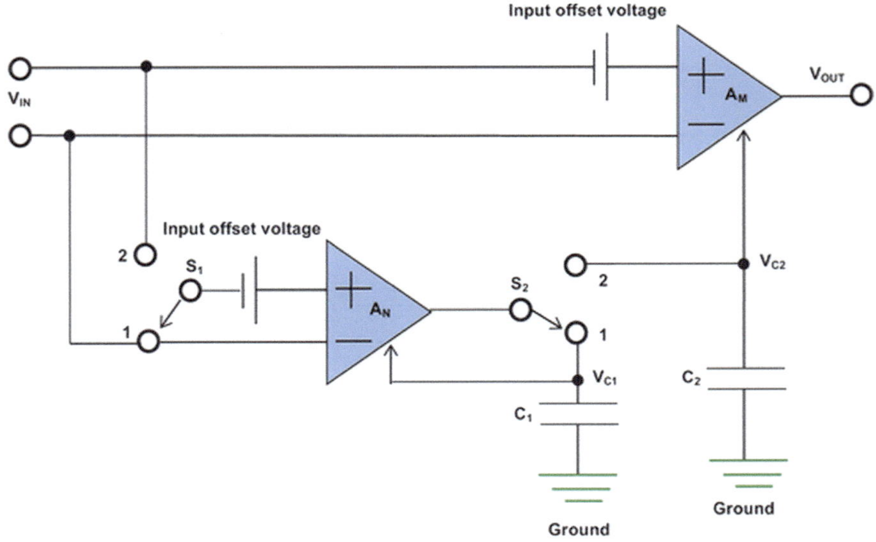

Fig. 12.5 Auto-zero amplifier

is collected. The oscillator starts this phase by placing switches S_1, S_2 in position 1 whereby the inputs of the nulling amplifier are short-circuited together and connected to the inverting input terminal, while the output of the amplifier is connected to the capacitor C_1. As a result, the input noise is sampled in the capacitor C_1. Thus this phase involves storage of noise sample in the capacitor.

In phase B, the amplification phase, the oscillator places both the switches in position 2. The nulling amplifier nulls its own offset voltage. It amplifies the voltage stored on C_1, denoted by V_{C1}, and subtracts the same from the amplified input signal. Also, capacitor C_2 is charged to voltage V_{C2} by the output from the nulling amplifier. This voltage serves as an offset correcting voltage for the main amplifier.

Thus the sampled noise is subtracted from the instantaneous noise of the amplified signal. Because the sampling frequency is chosen to have a value $\gg$ the noise frequency, the sampled noise and the low-frequency continuous noise are intensely correlated throughout a period of sampling. The $1/f$ noise is removed through cancelation in this subtraction process.

Total amplifier gain ~ gain of the nulling amplifier × gain of the wideband amplifier and offset voltage ~ (offset voltage of the nulling amplifier × offset voltage of the wideband amplifier) / gain of the wideband amplifier. The gain of the wideband amplifier is made very large to reduce the offset voltage.

12.3 Zero-Drift Amplifiers

The two low $1/f$ noise amplifier categories discussed in preceding subsections, namely, chopper-stabilized and auto-zero amplifiers, are able to achieve voltage offsets at nanovolt level [10]. Further, they undergo extremely low offset drifts with time. Also, they experience very small variations with temperature. The input offset voltage is a basic parameter for the amplifiers. The exact meaning of this term is that it is the differential DC voltage applied between the inputs of an amplifier to make the output voltage = 0 V with respect to ground or voltage between differential outputs = 0 V, according to the circuit type.

As the name suggests, a zero-drift amplifier is one yielding a near zero offset voltage drift. The zero-drift amplifier employs the self-correcting circuit architecture. It delivers very high accuracy over both time and temperature. The amplifier constantly corrects itself for any DC errors. The correction makes it as accurate and truthful as possible. It works by dynamically modifying the offset voltage to the correct value and reshaping the noise density. As already seen, the $1/f$ noise of the amplifier is also a DC error. It is therefore simultaneously eliminated.

Realizing that the annoyances such as thermal drift and $1/f$ noise are very awkward to remove, engineers glean several advantages by using zero-drift amplifiers. Some of these are low offset voltage < 10 μV maximum, super-low offset voltage drift < 40 nV/°C on time and temperature scales, negligible $1/f$ noise, very high open-loop gain, very high power supply rejection ratio (PSRR), very high common-mode rejection ratio (CMRR), high input impedance, capability of high temperature operation to as much as 200 °C, and absence of outside trimming requirement [11]. Over and above, the overall output error of a zero-drift amplifier is much lower than that provided by an accustomed top-grade precision amplifier of similar configuration.

Each technique used for zero-drift amplifier, auto-zeroing or chopping, has its benefits and drawbacks and is used in different applications (Table 12.2). As already described, auto-zeroing uses the sample-and-hold technique. Inasmuch as the sampling operation results in noise foldback to the baseband, auto-zero amplifiers are characterized by more in-band voltage noise. Also, more current is used to suppress the noise resulting in higher power dissipation. Alternatively, chopping based on signal modulation and demodulation has lower baseband noise. But it produces noise spectra at the chopping frequency and multiples or harmonics of this frequency. Therefore, the low-frequency noise of chopper-stabilized amplifiers shows consistency with their flat-band noise. However, they generate a sizeable quantity of energy at the chopping frequency and multiples of the same. It is often necessary to filter the output. This feature promotes the suitability of these amplifiers toward DC or low-frequency applications in distinction to auto-zero amplifiers. In wider band applications, auto-zero amplifiers are preferred.

In applications requiring low noise and wide bandwidth along with switching glitch-free operation, the optimal amplifier is one combining both auto-zero and chopping techniques. A combined zero-drift amplifier is one that employs auto-zeroing together with chopping technique to bring down the energy at the frequency

Table 12.2 Chopper-stabilized and auto-zeroing amplifiers

Sl. No.	Property	Chopper-stabilized amplifier	Auto-zeroing amplifier
1	Principle	It uses modulation and demodulation	It uses sample-and-hold technique
2	Offset	Its offset is very low	Its offset is also very low
3	Noise	It gives a lower noise level at low frequencies, which is in agreement with its flat-band noise	It gives more low-frequency noise because of aliasing effect
4	Power consumption	Its power consumption is low	Its power consumption is high because more current is used for suppressing noise
5	Bandwidth	Its bandwidth is narrow	Its bandwidth is wide
6	Ripple	It gives high ripple	It gives low ripple
7	Energy	It produces a huge quantity of energy at the chopping frequency and multiples thereof	It produces diminutive energy at auto-zero frequency
8	Applications	It is preferred for applications dealing with low power levels and working in low-frequency ranges, i.e., at sub-100 Hz frequencies	It is favored for applications in which wideband operation is demanded

at which chopping is done. At the same time, it restricts the noise to extremely small magnitudes in the range of low frequencies. This combined amplifier gives a larger bandwidth in divergence to that realized with traditional zero-drift amplifiers.

Zero-drift amplifiers find applications in circuits having an anticipated lifetime of design > 10 years. They are also used in signal chains where closed-loop gains are > 100 at low frequencies < 100 Hz and with low-amplitude levels [11]. Debatable issues raised in the circuit applications of zero-drift amplifiers are concerned with charge injection and clock feedthrough. Also, worthy of attention are the intermodulation distortion and overload recovery time.

Charge injection takes place by way of the switching activities of choppers and auto-zero amplifiers. It appears at the inputs of the amplifier. Due to input switching action, small ripples appear at the output terminals of the amplifier when it is constructed in a noninverting configuration. The amount of charge injected is not temperature dependent. It increases with rising values of gain of the circuit, source resistance, or gain setting resistor. Clock feedthrough may occur in an improperly designed amplifier. It can also occur in an amplifier using a pure chopping technique. The clock feedthrough results in the appearance of inaccuracies in the form of artifactual offshoots of the internal clock over the frequency domain.

Intermodulation distortion (IMD) or intermodulation (IM) is a standard metric of the linearity of amplifiers. It is the amplitude modulation of electrical signals containing two or more different frequencies. It is obtained from the ratio of the power of fundamental tones and third-order distortion products. It is expressed in decibels (dB). As the frequency of input signal moves nearer to the chopping or auto-zero frequency, IMD is introduced into the signal. Further, when the input frequency reaches nearer to the clock frequency, the errors become appreciably large. The intermodulation distortion between the input signal of high frequency and the chopping frequency produces tones at frequencies = sum and difference of chopping and input frequencies. Application of clever methodology for design with a suitable concoction of chopping and auto-zeroing strategies helps to reduce IMD.

Typically, the overload recovery time of zero-drift amplifiers is greater than that of the regular amplifiers made using CMOS designs and fabrication techniques. In case of a wide gap between the inputs of an auto-zero amplifier, the output is forced into saturation. The nulling amplifier looks at this difference as a type of offset. It attempts to make the inaccuracy zero. This attempt thrusts the main amplifier deeper into the state of saturation. The time of recovery is lengthened.

Amplifiers having integral intellect and judgment, which identifies overloading, can expedite recuperation from being overburdened in a time span as short as 30 μs. However, the amplifiers deprived of intelligence will need longer time to recover.

12.4 Continuous-Time Amplifier Circuits

The amplifier circuits so far presented fall under the general category of clock-based circuits. The reason is that they use a clock generation circuit. The use of circuit subsystems generating high frequencies makes them susceptible to interference by high-frequency effects and clock feedthrough. This susceptibility is the result of coupling between control signals. It also adds to design complexity. This paves the way toward continuous-time amplification techniques to overcome the aforesaid frequency-related limitations. Continuous-time and clock-based amplifiers are compared in Table 12.3.

12.4.1 Operational Transconductance Amplifier

In the class of continuous-time amplifiers, an important circuit is the AC-coupled operational transconductance amplifier circuit. This circuit is constructed across a single CMOS OTA platform, employing capacitive feedback, as designed by Harrison and Charles [12].

The operational transconductance amplifier (OTA), also called a transconductor, a diamond transistor or a macrotransistor, is a versatile building block. It is a high-impedance differential input amplifier. In this amplifier, a differential input voltage (the difference of voltages between two points with reference to a third point, viz., the ground or common point) produces a current at the output. Hence, it acts as a current source controlled by voltage, i.e., a VCCS [13]. A supplementary input for current enables controlling of transconductance of the amplifier. Main features of OP-AMP and OTA are given in Table 12.4.

12.4.2 Comparison of OTA with Bipolar Transistor

In similarity to a bipolar transistor, the three terminals of OTA serve the functions of a high input impedance base terminal *B*, a low impedance input/output emitter terminal *E*, and a current output collector terminal *C*. In dissimilarity to a bipolar transistor,

Table 12.3 Clock-based and continuous-time amplifiers

Sl. No.	Property	Clock-based amplifier	Continuous-time amplifier
1	Clock generator	Yes	No
2	Design	Complicated	Simple
3	High-frequency interference	Yes	No
4	Clock feedthrough effect	Yes	No

Table 12.4 OP-AMP and OTA

Sl. No.	Property	OP-AMP	OTA
1	Voltage/current source	It is an amplifier	It is also an amplifier
		Its output is a voltage signal	Its output is a current signal
		The voltage obtained at its output terminals is directly proportional to a voltage difference = that between the voltages applied to its two input pins	The current obtained at its output pins bears a direct proportionality relationship with a voltage difference = differential input voltage
		It serves as a voltage source that is itself controlled by a voltage, i.e., as a VCVS. It is known as voltage-differencing amplifier (VDA)	It acts as a current source that is controlled by voltage, i.e., as a VCCS
2	Gain	It is a fixed-gain voltage-amplifying device	It is a variable-gain voltage-to-current amplifier
3	Output impedance	Low	High
4	Circuit design	Its circuits are designed using negative feedback	It is possible to design circuits that do not employ negative feedback
5	Frequency	It works satisfactorily for applications restricted to operate at low frequencies. These include audio and video circuits and systems. In the higher-frequency range, OP-AMP-based designs are arduous owing to their frequency restrictions	At high operating frequencies, OTAs have the potential to supersede OP-AMPs as the basic structural elements for various analog signal processing circuits and systems
6	Load	It is used to drive loads which are resistive or capacitive	It is used to drive small, purely capacitive loads

the operational transconductance amplifier is self-biased, getting its biasing voltage from the circuit itself, and provides bipolar output, i.e., the output terminal either sources or sinks the current.

12.4.3 OTA Versus Operational Amplifier

Like a conventional operational amplifier, it may be used as a negative feedback amplifier. But unlike this amplifier, it delivers a current signal at the output, not a voltage signal. For linear applications, it is generally used in "open-loop" configuration without negative feedback.

12.4.4 OTA Operation

If $V_{\text{in}+}$ stands for the voltage at the noninverting input of OTA and $V_{\text{in}-}$ denotes the voltage at its inverting input, the output current I_{out} of the OTA being proportional to the differential input voltage $(V_{\text{in}+} - V_{in-})$ is given by

$$I_{\text{out}} = (V_{\text{in}+} - V_{in-})g_{\text{m}} \tag{12.1}$$

where the constant of proportionality g_{m} is the mutual conductance or transconductance of the amplifier.

The output voltage V_{out} of the amplifier is = output current (I_{out}) × load resistance (R_{L}). Hence, we may write

$$V_{\text{out}} = (I_{\text{out}} \times R_{\text{L}}) = (V_{\text{in}+} - V_{in-})g_{\text{m}} \times R_{\text{L}} \tag{12.2}$$

The voltage gain is then the ratio = output voltage/differential input voltage. So, it may be expressed as

$$A_{\text{v}} = \frac{(V_{\text{in}+} - V_{in-})g_{\text{m}} \times R_{\text{L}}}{(V_{\text{in}+} - V_{in-})} = g_{\text{m}} \times R_{\text{L}} \tag{12.3}$$

The transconductance g_{m} of the amplifier is directly proportional to the input bias current. This property makes it useful for electronic control of amplifier gain. CMOS technologies are very convenient for OTA implementation. This is because MOSFETs are intrinsically and symptomatically current source devices controlled by voltages applied at the gate terminal. Naturally, therefore, a diversity of CMOS technology-based OTAs having various geometrical and configurational designs is available for a multiplicity of applications.

The operation of differential I/O topology of OTA can be understood with reference to Fig. 12.6. This topology uses two current mirrors to improve balance between differential paths [14]. These current mirrors have size ratio η to increase output current by a factor of η. As$v_{\text{i}+}$ increases, the value of $i_{\text{d}+}$ increases. The enhanced value is transferred to the output side with η-times multiplication due to current mirror on the left side. Similarly, as $v_{\text{i}-}$ increases, the value of $i_{\text{d}-}$ increases. The same is transferred to the output side with η-times multiplication due tocurrent mirror on the right.

12.4.5 Single OTA-Stage CMOS Amplifier

This amplifier [12] is an OTA-based amplifier with capacitive feedback (Fig.12.7). It is useful for amplifying signals in the milli-to-kilohertz frequency range. It secures the elimination of large DC offsets. Any DC offset is removed by the capacitive

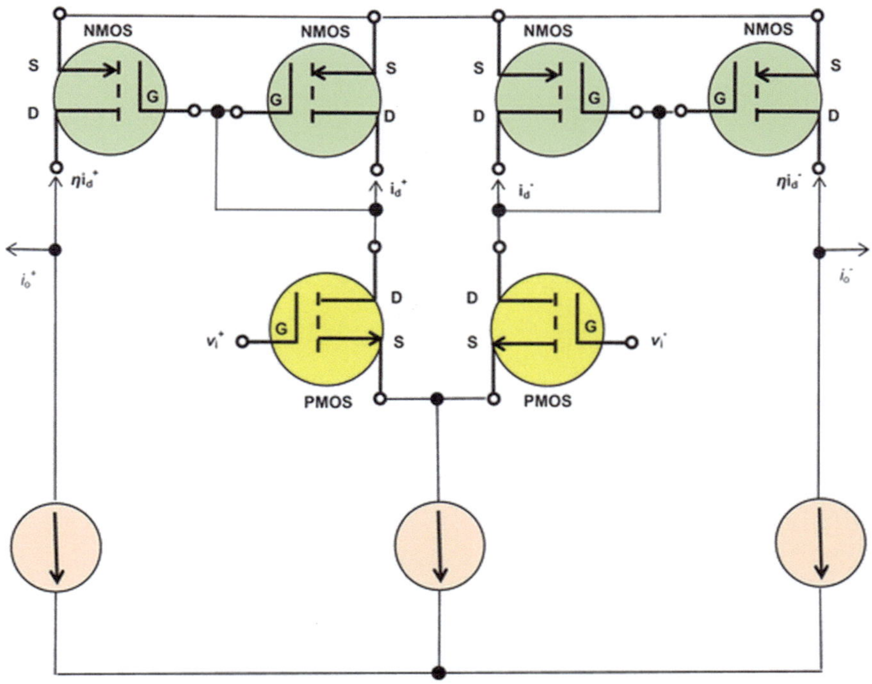

Fig. 12.6 Differential input/output circuit topology of the operational transconductance amplifier

coupling of the input through C_1. The mid-band gain A_M of the amplifier is fixed by the capacitance factor C_1/C_2. IfC_1, $C_L \gg C_2$ (where C_L is the load capacitance), for a transconductance g_m of the OTA, the amplifier bandwidth ~ $g_m/(A_M C_L)$.

The circuit topology is built around an OTA stage. Transistors M_a–M_d are MOS-bipolar devices. They act as pseudoresistors having large incremental resistances. By fabricating P-channel, large area devices, $1/f$ noise is reduced. When gate-source voltage V_{GS} is > 0, each device acts as a diode-connected bipolar junction transistor (BJT) consequent to the activation of the parasitic source–well–drain PNP BJT. But for $V_{GS} < 0$, each device acts as a diode-connected P-channel MOS transistor. At low voltages across this MOS-bipolar device, the incremental resistance r_{inc} has a very high value > 10^{12} Ω.

The cutoff frequency of the input high-pass filters for AC coupling is adjustable in the millihertz region by virtue of the very large incremental resistances of the pseudoresistors. The upper cutoff frequency is decided by C_L, g_m and the ratio C_1/C_2, whereas the lower cutoff frequency is set by capacitance C_2 multiplied by the pseudoresistor created by the transistors M_a and M_b. The reported amplifier was fabricated using 1.5 μm CMOS process in an area of 0.16 mm^2. Signals in the range 0.025 Hz to 7.2 kHz could be passed. The gain was 39.5 dB, input referred noise was 2.2 μV$_{RMS}$ (root mean square), and the bandwidth was 7.5 kHz. The powerdissipation was 80 μW.

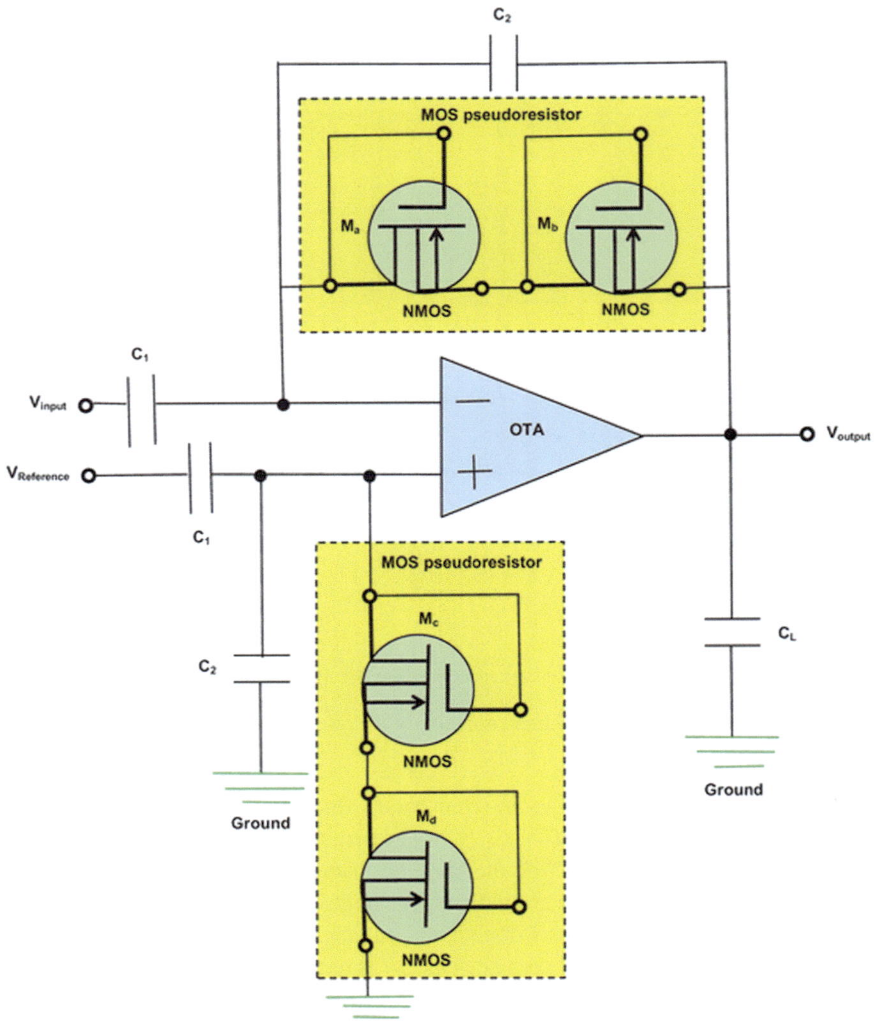

Fig. 12.7 OTA-based CMOS neural amplifier with capacitive feedback

12.4.6 Low Input Capacitance Amplifier

Noise is reduced by increasing the gain. The mid-band gain is the ratio of input capacitance-to-feedback capacitance. Hence, any effort to obtain high gain involves increasing the input capacitance or decreasing the feedback capacitance. If the input capacitance is increased, the input impedance decreases and the area of silicon consumed increases. The feedback capacitance is restricted by factors such as parasitic capacitance and fabrication technology. A decrease in input capacitance is beneficial from input impedance and area viewpoints. But if the feedback capacitance change is restricted, the above has to be done only by sacrificing the gain.

It was shown by Ng and Xu [15] that by replacing the feedback capacitor by a clamped T-capacitor network, the same value of mid-band gain is achievable. This is possible by using a smaller input capacitance with associated higher input impedance and smaller area. The reverse-biased diodes D_1, D_2, connected to the floating node, provide leakage paths to remove any induced charges. Hence, they keep the bias voltage within safe limits.

This amplifier was fabricated by 0.35 μm CMOS process in an area of 0.056 mm^2. The mid-band gain was 38.1 dB, the input capacitance was 1.6 pF, the input referred noise was 13.3 μV_{RMS}, the bandwidth was 8 kHz, and the power rating was 6 μW [15].

12.5 Discussion and Conclusions

This chapter familiarized the reader with the main circuit topologies used in designing neural amplifier circuits. One of the primary noise sources is 1/*f* noise. Along with input offset voltage [16], it dictates the accuracy of the amplifier. Discussions were centered on clocked-based and continuous-time circuits.

Review Exercises

12.1 How does 1/*f* noise originate conforming to the carrier number fluctuation theory? How is this noise debilitated by switched-biasing? Is the noise removal by switched-biasing done at circuit level or device level?

12.2 Explain the concept of chopper amplification. Why is chopper-stabilized amplification said to be a modulation-based technique? Describe the steps involved in this kind of amplification.

12.3 Explain giving diagram the working of a chopper-stabilized amplifier. Why does this amplifier have a limited bandwidth? How is the bandwidth enlarged?

12.4 Name an amplifier based on signal sampling. What are the roles of the wideband and nulling amplifiers in this circuit? What are the two phases in the operational cycle of this amplifier called? How is the offset voltage related to the gain of wideband amplifier?

12.5 What is meant by input offset voltage drift of an amplifier? What is a zero-drift amplifier? Name the two dependent variables with respect to which the drift is measured. Mention some benefits derived from zero-drift amplifiers.

12.6 What are the two types of amplifying circuits used for zero offset voltage drift? Prepare a comparative chart listing the performance merits of each type.

12.7 Compare the performance of clock-based with continuous-time amplifiers.

12.8 What is an operational transconductance amplifier? How does it differ from: (a) a bipolar transistor and (b) an operational amplifier? Write the equation for its voltage gain.

12.9 Describe with circuit diagram the working of a single OTA-stage CMOS amplifier with capacitive feedback.

12.10 How does a low input capacitance amplifier help to reduce noise? How are the changes in input and feedback capacitances restricted beyond a limit? What is the remedy?

References

1. Klumperink EAM, Gierkink SLJ, van der Wel AP (2000) Reducing MOSFET 1/f noise and power consumption by switched biasing. IEEE J Solid State Circuits 35(7):994–1001
2. Tian H, Gamal AE (2000) Analysis of 1/f noise in CMOS APS. In: Blouke MM, Sampat N, Williams GM (eds) Proc SPIE 3965. Sensors and camera systems for scientific, industrial, and digital photography applications. SPIE, San Jose, CA, pp 421–430. https://doi.org/10.1117/12.385433
3. Miguez M, Arnaud A (2008) A study of flicker noise in MOS transistor under switched bias condition. J Integr Circuits Syst 3(2):63–68
4. Hung KK, Ko PK, HU C et al (1990) A unified model for the flicker noise in metal-oxide-semiconductor field-effect transistors. IEEE Trans Electron Devices 37(3):654–665
5. Bloom I, Nemirovsky Y (1991) 1/f noise reduction in metal-oxide-semiconductor transistors by cycling from inversion to accumulation. Appl Phys Lett 58(15):1664–1666
6. Bilotti A, Monreal G (1999) Chopper-stabilized amplifiers with a track-and-hold signal demodulator. IEEE Trans Circuits Syst I Fundam Theory Appl 46(4):490–495
7. Said AH (2010) Design of a chopper amplifier for use in biomedical signal acquisition. M.S. Thesis, Southern Illinois University, Edwardsville, pp 10–12
8. Kugelstadt T (2005) Auto-zero amplifiers ease the design of high-precision circuits. Analog Applications Journal © Texas Instruments Incorporated 2Q 2005:19–28.http://www.ti.com/lit/an/slyt204/slyt204.pdf. Accessed 31 July 2015
9. Tuite D (2009) Chopper-Stabilized Op Amps. Electron Des.http://electronicdesign.com/analog/chopper-stabilized-op-amps. Accessed 10 Jul 2014
10. Wong V (2011) Lowest noise zero drift amplifier has 5.6 nV/$\sqrt{}$Hz voltage noise density. Analog Devices: Application Note AN-1114, ©2011–2014 Analog Devices, Inc, 8 p. http://www.analog.com/media/en/technical-documentation/application-notes/AN-1114.pdf. Accessed 30 Jul 2015
11. Moghimi R (2010) Zero-drift amplifiers: a keystone to the elusive ideal amp? Electron Des.http://electronicdesign.com/power/zero-drift-amplifiers-keystone-elusive-ideal-amp. Accessed 6 Oct 2014
12. Harrison RR, Charles C (2003) A low-power low-noise CMOS amplifier for neural recording applications. IEEE J Solid State Circuits 38(6):958–965
13. Ramus X (2013) Demystifying the operational transconductance amplifier. Texas Instruments, Application Report SBOA117A, 13 p. http://www.ti.com/lit/an/sboa117a/sboa117a.pdf. Accessed 30 Jul 2015
14. Sheikh ST, Dahigoankar DJ, Lohana H (2012) Comparative analysis of CMOS OTA IOSR. J VLSI Signal Proc 1(3):01–05
15. Ng KA, Xu YP (2013) A compact, low input capacitance neural recording amplifier. IEEE Trans Biomed Circuits Syst 7(5):610–620

16. Palmer R (2001) Texas Instruments Application Report SLOA059: DC Parameters: Input Offset Voltage (VIO). Copyright 2001, Texas Instruments Incorporated, 24 p.http://www.ti.com/lit/an/sloa059/sloa059.pdf. Accessed 12 Jul 2014

Chapter 13
Implantable Sensors

Abstract Sensors are the primary components of a monitoring system. Micro- and nanofabrication technologies have now advanced to the stage at which wireless sensor systems can be included in the implants with minor modification. These systems provide unique, personalized data for each patient to be used for optimizing outcomes. An acceleration sensor mounted on an artery is used for blood pressure measurement. Coupling a pressure transducer to the right ventricle (RV) lead of a pacemaker or defibrillator helps in continuous intracardiac pressure monitoring. Implantable chemical sensors are employed for real-time monitoring of clinically important species, e.g., blood gas measurements (pH, pO_2, and pCO_2). Subcutaneously implanted enzymatic glucose sensors enable continuous glucose monitoring. Single-walled carbon nanotubes (SWCNTs) encased in alginate work as inflammation sensors, which can be implanted for detection of nitric oxide.

Keywords Blood pressure sensor · Accelerometer · SAW sensor · Foreign body response · Oxygen shortfall · Enzyme · Blood gas sensor · Diabetes · Artificial pancreas · Glucose biosensor · Inflammation sensor

13.1 Introduction

Micro- and nanoelectronics have created gigantic opportunities for data processing. However, the sensors continue to be the frailest link in the sequence of data collection and handling components [1]. Implantable sensors should be drift-free, besides having high sensitivity and selectivity. Usually, an analog sensor interface circuit amplifies the feeble signals from the sensors to acceptable magnitudes. These analog signals are transformed to the digital domain to avail the accompanying mathematical and decisive capabilities. To cope up with the inescapable long-term drift, the only way out is to provide a bidirectional communication link. The far-flung controlling station can thereby adjust the offset voltage of the amplification stage to compensate for the undesirable drift. The ability of reprogramming the installed sensor is

V. K. Khanna, *Implantable Medical Electronics*,
https://doi.org/10.1007/978-3-031-99868-3_13

very beneficial. Telemetry is of crucial importance in fine-tweaking the otherwise inaccessible, fully sealed systems.

Additional ICs are required to correct for nonlinearities of the sensor and temperature-induced deviations of readings. Piezoresistive pressure sensors commonly connected in a Wheatstone bridge circuit configuration require temperature compensation. Also, to decrease the power consumption, sampling techniques with short duty cycles must be applied to these sensors.

Capacitive sensors used with switched-capacitor techniques have fulfilled power consumption criterion. Switched-capacitor circuits are discrete-time systems using capacitors and switches to imitate the behavior of resistors. They work by moving charges into and out of capacitors through opening and closing the switches. To cite an example, a switched-capacitor resistor comprising a capacitor and two switches works by connecting the capacitor at a given frequency alternately to the input and output. Hence, its value can be adjusted by changing the switching frequency, thus yielding a programmable resistance. Such circuits allow easy implementation on integrated circuits.

13.2 Implantable Blood Pressure Sensor

A person is said to suffer from hypertension or high blood pressure if the maximum and minimum values of pressure exerted by blood fall outside defined limits as follows: the maximum pressure exerted by the blood on the walls of blood vessels when the left ventricle contracts (systolic blood pressure) should be measured as more than 140 mmHg, and/or the minimum pressure exerted by the blood on the walls of blood vessels when the ventricles undergo relaxation and dilation (diastolic blood pressure) should be found to exceed 90 mmHg. Hypertension can cause cerebrovascular disease, heart attack, and congestive heart failure. Kidney catastrophe and retinal damage transpire. When it cannot be controlled by changes in the lifestyle of the patient such as diet regulation, body weight reduction, exercise, etc., a personalized drug treatment is practiced under frequent surveillance. Constant tracking of the blood pressure of patients facilitates effective disease management and reduces hospitalization costs by diagnosing troubles related to physical fitness at an initial stage and by therapy optimization.

The common methods for blood pressure measurement involve the use of an external cuff (noninvasive) or a catheter-based system (invasive). Invasive method is more accurate than the noninvasive one. Both these cumbersome methods fail to provide long-term, continuous blood pressure measurement. They not only forbid free movement of the patient but also lead to complications. Chief complications are occlusion of blood circulation, trauma to the arterial vessel, and infection risks. For blood pressure measurements at a specific site over extended periods, implantable sensors stand as the only possible devices [2].

13.2.1 Capacitive Pressure Sensors

A capacitive sensor fabricated by a fusion bonding process [3] is based on a SiGeB diaphragm, utilizing a silicon membrane heavily doped with boron. This membrane is epitaxially grown on a silicon substrate, as shown in Fig. 13.1 drawn after [3]. Germanium is also incorporated during epitaxial layer growth. This incorporation is done to compensate for boron-induced strain. The interfacing circuit for the sensor is an IC containing a capacitance-to-frequency conversion section and a voltage regulator to curtail supply voltage fluctuations. As the sensor is based on capacitive principle, its power consumption is very small ~ 4 mW in a simulated arterial environment. Its in vivo performance was not investigated.

A pilot study of sensor implantation in the femoral artery of 12 sheep via standard catheterization was conducted on a blood pressure sensor. This sensor comprised two functional components: the sensor tip, a stainless-steel capsule containing the capacitive pressure sensor (accuracy = ± 1.0 mm of mercury in the range of 30–300 mm of mercury), and the telemetric transducer, communicating with the external readout unit [4]. The two components were connected via a data cable of 1 mm diameter, and the full length of the system was 20 cm. High-quality blood pressure recordings could be obtained over a 6-month period.

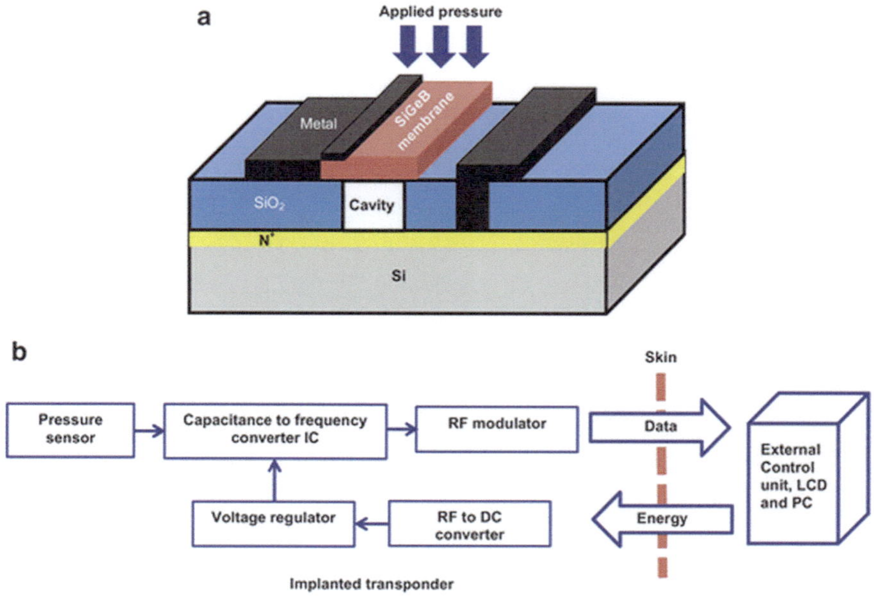

Fig. 13.1 Blood pressure measurement system: **a** capacitive pressure sensor and **b** interface circuit and passive telemetry link

13.2.2 Accelerometers

Blood pressure is related to pulse wave velocity. To measure pulse wave transit time, two sensors appear to be necessary. But pulse waves undergo reflections at bifurcations in the blood circulation system. Therefore, a single sensor can be used to detect both waves: the incident and rebounded waves. The interval of time separating the incident and reflected waves can thus be measured. The reflected wave transit time (RWTT) is obtained from this time delay. The RWTT value is linked to blood pressure. Hence, measurement can be carried out using only a single accelerometer.

For avoiding motion artifacts, two accelerometers are mounted, as shown in Fig. 13.2. The accelerometers (2 mm × 2 mm) are mounted on an artery at two different locations by minimally invasive techniques [5–7]. The implant shown in Fig. 13.3, drawn after [7], is based on a flexible, stretchable polyimide substrate. Photolithographically defined copper tracks are epoxy-glued to both sides of the substrate. The electronics for data telemetry is connected by reflow soldering. It is strengthened with a biocompatible epoxy and stiffened by a glass epoxy. Finally, it is passivated with biocompatible parylene-C for protection against body fluids.

For evaluation, the sensor was implanted in a rabbit for over 2 weeks. The RWTT was strongly correlated to the systolic blood pressure with a mean deviation of 4.3 % (5.0 mmHg) for 1800 pulses; the correlation coefficient was 0.96. The power consumption of the implantable unit was only 312 μW.

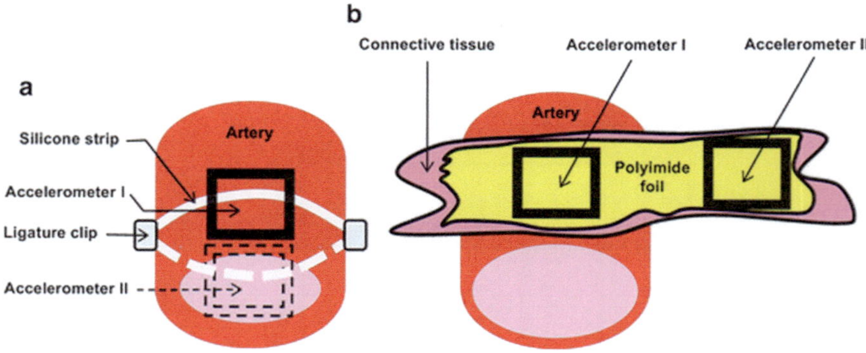

Fig. 13.2 Placement of accelerometers for blood pressure measurement for avoidance of motion artifacts: **a** diametrically opposite locations on an artery using elastic silicone strip tied by ligature clips and **b** one on connective tissue above the artery and another on connective tissue nearby where there is no arterial influence. Difference between signals from the two accelerometers gives the acceleration caused by expansion of the artery

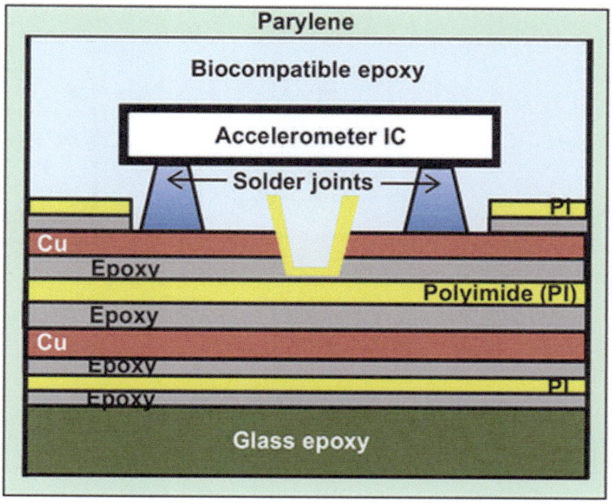

Fig. 13.3 Layered structure of the polyimide substrate with Cu tracks, mounted 2 mm × 2 mm × 1 mm accelerometer IC, and encapsulation

13.2.3 SAW Sensors

The SAW sensor system (Fig. 13.4) is a wireless, battery-less microsystem made of three principal components:

(a) The SAW Resonator: It is fabricated on a piezoelectric ST-cut quartz substrate mounted on an elastic diaphragm [8, 9]. This diaphragm covers a pressure chamber into which blood flows through a small orifice at the bottom. Blood flow distends the diaphragm. The resonator has 171.5 finger electrode couples as an interdigitated array. It has 140 electrodes for one reflection grating and the same number for the other grating. The gratings are photoetched on the substrate.
(b) An Oscillating Circuit: This circuit is connected to the SAW resonator constituting a SAW oscillator.
(c) RF Power Converter: It contains a rectifier and regulator to convert RF power into DC power. The RF power is collected through a pickup coil. This coil is spiraled around the shell of the sensor and powered by inductive coupling to DC power. The same is fed to the SAW oscillator.

The blood pressure changes exert a bulging action on the elastic diaphragm. This action distorts the substrate underlying the SAW resonator. It therefore resonates at a different frequency. Thus the changes in resonance frequency of the SAW resonator closely follow the blood pressure variations. The RF signals from the oscillating circuit of SAW resonator carry the information on blood pressure of the patient. These signals are transmitted by frequency modulation from a tiny antenna for further processing. The received signals undergo processing operations by a computer, and the waveform of blood pressure is flaunted on its monitor. A sensitivity of 1.75 kHz/mmHg has been reported. For demonstrating its biomedical application, in vivo studies have been performed on rats.

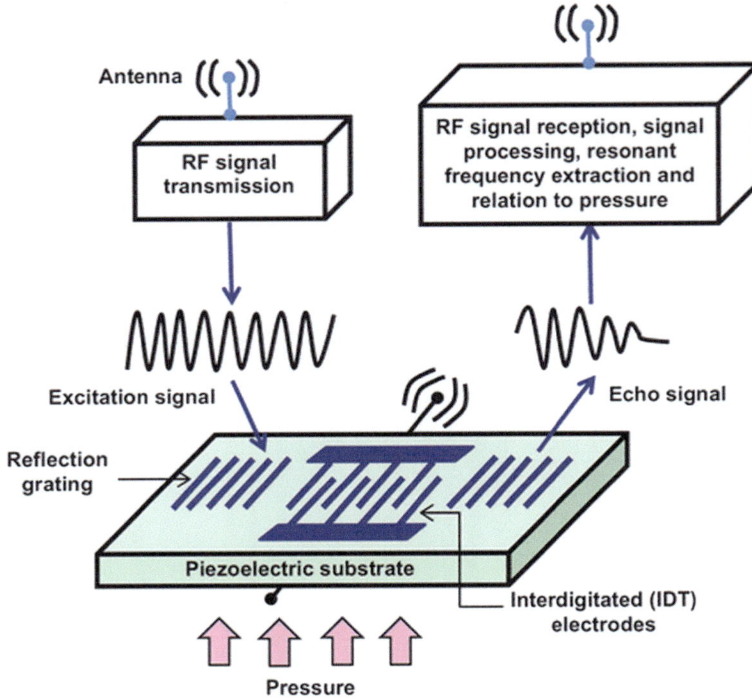

Fig. 13.4 Passive remote blood pressure measurement by SAW resonator

In another investigation [10], the sensor was modeled as a suspended 60-μm thick and (5.3 × 4)-mm-size top plate in the form of a slim membrane of quartz. It has a SAW resonator at its middle and is suspended in air above a cavity. By bonding process, it is joined securely everywhere around the periphery to form a closed cavity with a bottom substrate of greater thickness. In absence of pressure, the quartz membrane is not deflected. The resonant frequency is f_0 (say). When the pressure is increased, the membrane bends into the cavity. The resultant strain produced in it changes the resonant frequency of the SAW resonator to $f_0 + \Delta f_0$ (suppose). The sensor was used as a component of an implanted device for acquisition of blood pressure information in animal by wireless communication. This experiment showed its capability of providing endless monitoring of blood pressure in real time. Further, it is capable of wireless operation. Therefore, it is useful for a patient who can walk around and is not admitted to hospital; hence, called an ambulatory patient in contrast with an inpatient who requires hospitalization.

13.3 Predicaments of Implantable Biosensors

While the physical sensors are encapsulated inside safe packages, chemical sensors have to interact directly with body fluids. Besides foreign body response, issues that need attention are oxygen deficit and enzyme stability.

13.3.1 Foreign Body Response

By way of foreign body response already described in the chapter on biomaterials, within a short period of time after implantation, the surface of the sensor is fouled. It is covered with a 10–100-μm thick layer of proteins and cells [11]. This encasing by cells builds up a barrier to mass transfer. The barrier opposes analyte diffusion to the sensing element. Consequently, the in vivo performance of the sensor and its stability over a period time are severely impaired. When active cells are deposited on the surface of an implantable sensor, there is a meaningful change in the oxygen concentration nearby. The altered oxygen level introduces pronounced inaccuracies in the readings taken from the implanted biosensor. For a biosensor, the problem of foreign body response is more perplexing than concerns related to device, e.g., an electrical or component fiasco and enzyme degradation.

An initial approach to improve the in vivo sensor operation was to use specially designed biocompatible or antifouling coatings, e.g., Nafion [12], polyurethanes with phospholipid polar groups [13], and polyvinyl alcohol hydrogels [14]. These coatings avert biofouling. They partially mitigate sensor degradation by maintaining a desired, continuous flow of analyte molecules over an extended time span. This mitigation is achieved by reducing protein adsorption and by fostering unification of the sensor with the enclosing tissues. The coatings used should be sufficiently thin and porous to permit the sensor to respond speedily to the variations in analyte concentration. However, even when nontoxic biocompatible materials are used, a variety of immune responses are induced by implantation.

In rivalry to a smooth surface, a textured sensor surface improves in vivo sensor performance. It does so by increasing vascularity (blood vessel content) in the region around the implant. Nanostructured membranes prevent the adsorption of proteins on the implant surface. The repulsion takes place by virtue of their having a different contact angle and hence wettability as compared to a smooth, unwrinkled membrane. Therefore, their degree of hydrophobicity or hydrophilicity is dissimilar to the smooth surface. Coatings made from polymer composites and having functional materials entrenched within a fine-grained matrix are also used. Another way of reducing inflammation and quelling fibrous enclosing consists in using tissue response modifiers (TRMs). Eco-friendly, decomposable microspheres containing TRMs are rooted within a polymeric material, which is both biocompatible and antibiofouling [15]. The degradation of microspheres is activated by a change in either hydrogen ion concentration or ionic strength or by electrical incitement. A stable liberation of TRMs to the tissue contiguous to the implant is thereby initiated.

13.3.2 Oxygen Shortfall

Oxidase enzymes are used in many biosensors as biorecognition elements. The enzymes catalyze the oxidation of analytic species, forming hydrogen peroxide. The released hydrogen peroxide is detected by amperometry involving measurement of current. For most oxidase enzymes, oxygen is the cofactor whose presence is essential to complete the reaction. Therefore, the behavior of the sensor depends on the changes taking place from variations in the concentration of oxygen in the vicinity of the sensor. An oxygen imbalance ensues because the quantity of dissolved oxygen in the body is appreciably less than the concentration of analyte such as glucose. Thus, there is a stoichiometric limitation of the enzymatic reaction. The deficiency of oxygen is further intensified by implantation. This intensification stems from the use of anesthesia and the occurrence of inflammatory response resulting in local consumption of oxygen. So, the response of the sensor saturates at higher concentrations of the analyte. This saturation effect disturbs the linearity of its characteristic.

Efforts to decrease the oxygen dependence have led to three generations of biosensors. In the first generation, outer membranes are used. These membranes lower the analyte flux while affecting the oxygen influx negligibly. Such membranes include Nafion, polyurethane, cellulose acetate, etc. A drawback of these membranes is the decrease in sensitivity of the device. Another disadvantage is an increase in its response time. Moreover, these membranes degrade by calcification, delamination, etc. Consequently, two alternative routes have been followed leading to the second and third generations of biosensors. In the second generation, oxygen dependence has been reduced by using redox mediators. These mediators compete with oxygen in the catalysis of the enzymatic reaction. In the third generation, different nanomaterials, e.g., gold nanoparticles, carbon nanotubes, etc., have been used. The nanomaterials serve to connect the redox centers of enzymes to the surface of the electrode. However, none of these approaches have been subjected to in vivo testing.

13.3.3 Enzyme Stability

Two commonly used methods for immobilization of enzyme entail: (a) cross-linking of enzyme/bovine serum albumin with glutaraldehyde and (b) electrochemical growth of a conductive polymer matrix. Presence of low molecular weight serum components and ions of transition metals, e.g., zinc and ferrous ions, can inhibit enzyme catalysis. In glucose biosensors, excess loading of glucose oxidase has been reported to extend the lifetime of the sensor [16]. But any unrestricted, glucose oxidase left in uncross-linked condition, within the enzyme stratum cross-linked with glutaraldehyde gently leaches out. By this means, the sensitivity decreases over time [17]. Thus, to extend the functional in vivo lifetime of the sensor, the relative quantities of the enzyme and cross-linking agents must be optimized with respect to sensitivity.

A germane procedure for enzyme immobilization in implants is biomolecular entrapment. This may be done with or without covalent binding. A hydrogel may be used for the purpose. The technique is not only innocuous, but it is also carried out easily. It needs little-to-no poisonous chemicals. Over and above, it preserves properties of the enzyme in a limited area adjoining an electrode without chemically upsetting their innate structure [18]. Further, to mitigate the problem of enzyme stability, synthetic or artificial enzymes have emerged [19] as also nonenzymatic catalysts [20].

13.4 Implantable Blood Gas Sensors

It is essential to monitor the health status of cardiopulmonary function in hospitalized critically ill patients. For this monitoring, it is necessary to carry out quantification of blood gases such as pO_2, pH, and pCO_2 in blood flowing in an artery. These measurements are performed by a catheter-type sensor positioned in the radial artery of the patient [21].

An indwelling electrochemical oxygen sensor works on the classical Clarke-style amperometric principle. In this sensor, oxygen undergoes reduction at a micro-dimensional working electrode of platinum, silver, or gold. This electrode is confined within a narrow diameter gas-permeable catheter tube made of a polymer. An Ag/AgCl reference electrode is also tucked inside the catheter [22]. Partial pressure of oxygen in the medium surrounding the catheter is proportional to the current. Hence, it is obtained from the current value.

Another potentiometric catheter design entails measurement of the corrosion potential (EMF) of a cobalt working electrode inserted within an analogous constricted diameter tubing [23].

Implantable electrochemical pH and pCO_2 sensors are typically potentiometric devices. These are based on either polymer membrane pH electrode technology or the use of solid state metal-oxide-based pH microsensors [24]. Incorporation of lipophilic proton ionophores, e.g., tridodecylamine, within the walls of plastic tubing offers a convenient means to evolve a dual pH/pCO_2 sensing design. This design has provided truthful in vivo results in animal experiments [25].

One commercially available product is the Paratrend probe, developed by Pfizer/Biomedical Sensors. This indwelling sensor is based on a hybrid design using electrochemical and optical techniques. Oxygen is sensed electrochemically. But pH and pCO_2 are determined by fiber-based optical fluorescence sensors [26]. The performance of the Paratrend probe has been reported to be acceptable [26, 27]. Nonetheless, in situ recalibrations have been suggested to be carried out frequently [27]. Notwithstanding, such an implanted device provides valuable trend or tendency information for monitoring the status of the patient. It may be noted that the analytical results may not always agree with values obtained by in vitro tests on discrete blood samples.

The above deviations may be significant. Therefore, they have prevented implantable devices from finding widespread clinical use. Hence, it is essential to understand the causes of deviations associated with the implanted devices when they are employed for constant measurements in the body. The main cause is the biological response of systems in living condition to the weird external devices inserted in the flowing blood stream. Protein molecules, e.g., collagen, fibrinogen, and von Willebrand factor (VWF—a blood clotting glycoprotein), are found to adsorb on the polymer exteriors of in vivo sensors in no time [28]. The protein coating thus formed by adsorption acts as an intermediary in the adherence and triggering of metabolically agile cells such as platelets [29]. The platelets display a central role in coagulation of blood. Activated platelets recruit circulating cells to the surfaces of the polymer. Eventually blood clots are formed. Active cells present on the covering of these sensors produce local changes in the hydrogen, oxygen, and carbon dioxide concentrations due to cellular respiration, altering their concentrations in comparison to the main blood flow. These local changes vitiate the measurements.

Nitric oxide is an effective vasodilator (a medicine which dilates or opens blood vessels) and preventer of platelet adherence and triggering. It is generated from L-arginine by a category of enzymes called nitric oxide synthases (NOS). Therefore, release or generation of NO locally at the interface between the blood and the sensor may help to control the hemocompatibility problem. For impeding platelet adherence and triggering, the nitric oxide must be released at or over the discharge from regular endothelial cells. Thus bloated formation of thrombus is prevented. On these grounds, a plausible solution that has been suggested to overcome the biocompatibility issues in intravascular estimation of blood gases in real time is through the inhibiting action of locally released nitric oxide at the interface of the sensor with the blood, as elucidated above [30].

For continuous production of NO at the interface between the sensor and the blood, in situ NO generation was planned. In this plan, NO was produced by catalytically decomposing endogenous *S*-nitrosothiols (RSNOs) inside the blood. This was achieved by entrenching metallic Cu particles intended to serve as catalysts. The Cu particles were of two different sizes (3 μm and 80 nm). The Cu-particle entrenchment was done in thin polymer coatings. The coatings were made on the surface of oxygen-sensing catheters working on electrochemical principle. The catheters were intravascular, i.e., placed within the blood vessel. More accurate values were obtained from the sensor thus fabricated as compared to control sensors, which did not produce nitric oxide. This was corroborated by implantation of the two types of sensors in arteries of the pig for 19–20 h. The NO-releasing sensor showed reduced tendency toward surface thrombus formation than the control sensors. The NO-generation approach offers a possible answer to this problem. This approach may be applicable to other sensors, which are always in contact with blood during their functioning.

13.5 Artificial Pancreas Concept

13.5.1 Metabolite Sensors

Certain organic compounds are produced during the life-sustaining chemical processes, e.g., digestion or other bodily chemical transformations, taking place within a living organism. These compounds are produced either as an intermediate or as a product. A well-known example is glucose (simple sugar). Such compounds are known as metabolites.

13.5.2 Treatment Options for Diabetes

Type 1 diabetes (T1D) is a disease of serious nature, which has caused worries to many people and therefore caught the attention of several scientists worldwide. It is brought about by a diminution of performance of pancreatic β-cells. The decreased functionality leads to insufficient insulin production [31]. Transplantation of the whole pancreas or islets of Langerhans is a possible treatment. But such transplantation is riddled with several issues. Main issues are accessibility to transplants and acceptability of long-term immunosuppression. Protracted function of grafted organ or cells is not fail-safe. Traditional therapy employs multiple daily insulin injections. Automatic/semiautomatic systems are required to replace this therapy. A lifelong implanted fully automated artificial or bio-artificial pancreas (BAP) represents a boon for type 1 diabetics. It will substitute the functionality of the endocrine pancreas. The artificial pancreas (Fig. 13.5) will use an accurate in vivo glucose sensor for constant real-time measurement of blood glucose levels. These glucose values will be used to modulate insulin dispensation in a closed-loop diabetes management. An implantable insulin pump will be the source of insulin. Thus the pump will supply the optimum, timely insulin dose in accordance with blood glucose level variations. This arrangement will provide relief to the vast diabetic population around the world. It will replace the present discrete methods of glucose estimation. At present, glucose determination is done from test strips through intermittent blood sampling by painful finger pricking. The artificial pancreas will be a painless and continuous method. It will eliminate patient intervention. It will also give glucose level information during different physiological states of the patient such as sleep, walking, and running.

First suggested in the 1960s [32], there have been persistent attempts for realization of implantable glucose sensors. In several efforts toward this direction in animal and human studies, it was observed that the sensor signal drifted appreciably after implantation. This postimplantation drift makes it impractical to quantitatively correlate the signal with the dynamic blood glucose concentration from the preimplantation calibration characteristic. The drift occurs either due to inherent properties of the sensor or local physiological phenomena in the surrounding tissues.

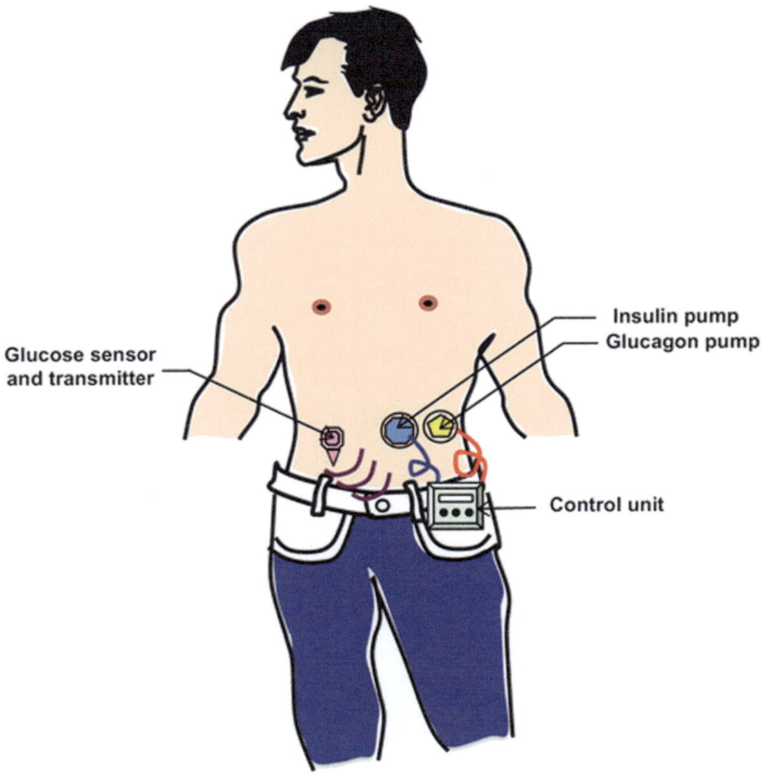

Fig. 13.5 Artificial pancreas containing glucose sensor and transmitter for checking glucose level in the blood every few minutes and sending the glucose information to the controller unit fitted with two infusion pumps, one for insulin for lowering glucose concentration and another for glucagon for raising glucose concentration

13.5.3 Subcutaneously Implanted Glucose Biosensor

Updike et al. [33] applied a polyurethane membrane containing entrapped glucose oxidase on a polarographic H_2O_2 sensor with an O-ring. This membrane polymer material could avert oxygen deficit. For trouble-free sensor operation in the biological environment, they added two coatings over the active sensing membrane layer:

1. *Bioprotective Membrane Layer*: It functions to prevent macrophages from reaching in close vicinity of the enzyme active membrane. Macrophages choreograph the formation of the foreign body capsule (FBC). They undergo degranulation. During this process, a potent albeit momentary oxidative species is released from the myeloperoxidase enzyme. The enzyme produces hypochlorite ClO^- (bleach). The univalent anionic ClO^- group destroys the polyurethane membrane containing the enzyme in active form, thereby quickly terminating the functioning of the sensor.

2. *Angiogenic Layer*: The term "angiogenesis" refers to the process of creation of fresh blood vessels from those prevailing at an earlier time. The angiogenesis layer is based on expanded polytetrafluorethylene (PTFE). It is placed over the bioprotective layer. It allows glucose diffusion by fostering the growth of new fine branching blood vessels serving as passageways ahead of the sensor window. At the same time, it reduces the influence of foreign body reactions caused by macrophages on sensor function. After about 7 days, the abovementioned capillary formation was found to stabilize the environment around the sensor.

The sensors were sterilized with 0.05 % thimerosal ($C_9H_9HgNaO_2S$) or 2.4% glutaraldehyde $CH_2(CH_2CHO)_2$ for 16 h at the comfortable ambient temperature. The sterilized sensors were implanted in dogs. These dogs were rendered diabetic for a limited period by obstructing the secretion of insulin by injecting a synthetic somatostatin, a hormone inhibiting somatotropin secretion. A superfluous payload of enzyme inside the membrane enabled a lifetime of 3–5 months for this sensor.

Updike et al. [33] reported a linear response of the sensor in the clinically relevant blood glucose range 2.2–38.9 mmol/L, which is equivalent to 40–700 mg/dL. The response time was 4–7 min. In the best situation, the interval of recalibration was 20 days, and utmost lifetime was > 160 days. In contradistinction, implantable glucose sensors without the angiogenic layer showed relatively less linearity. They had a dynamic range up to 22 mmol/L glucose. In the best circumstance, the interval of recalibration was found to be 18 days. The maximum lifetime observed was 94 days. These data for sensors without the angiogenic layer showed that their performance was much inferior in comparison to those of sensors with angiogenic layer.

Later pilot trials conducted in diabetic human beings suggested comparable stability in glucose sensing [34]. Thus this subcutaneously implanted glucose sensor has good prospects as a trouble-free durable device for continuously monitoring glucose concentration in the blood.

13.6 NO Detection-Based Implantable Inflammation Sensor

An optical sensor for in vivo NO detection has been reported [35]. It uses semiconducting single-walled carbon nanotubes (SWCNTs). In order to detect NO, the sensor has to perform two jobs: to provide selective molecular recognition and to carry out optical transduction of recognition. To accomplish these tasks, proper functionalization of the SWCNTs is necessary. Besides biocompatibility, this functionalization should also exhibit long-term stability after implantation inside the body. These requirements greatly restricted the choice of available interfaces. It was found that an oligonucleotide $ds(AAAT)_7$ of DNA enabled selectivity of SWCNTs for NO detection. This selectivity was sustained after ligation to a polyethylene glycol (PEG), $C_{2n}H_{4n+2}O_{n+1}$, segment having a molecular weight of 5 kDa. The copolymer ligated with polyethylene glycol could stabilize near-infrared fluorescent sensors based on single-walled carbon nanotubes (SWCNTs) in solution. Hence, these sensors could be introduced by intravenous injection into mice. The sensors emptied from the lungs within 2 h after injection. Thus they evaded in vivo nanotoxicology. The half-life for their withholding in the liver was 4 h. After they were localized within the liver, transient inflammation could be tracked employing nitric oxide as an indicator and gesturing molecule.

The intravenously injected PEG-$(AAAT)_7$-SWCNTs worked on a shorter time scale. In contrast to these shorter time scale sensors, the researchers investigated an implantable alginate-encapsulated sensor platform for localization of $(AAAT)_7$-SWCNTs specific to tissues. They studied its performance on a longer time scale covering several days and months. The $(AAAT)_7$-SWCNTs were found to perform well as inflammation sensors for nitric oxide detection, which were capable of operating in vivo. They gave a limit of detection ~ 1 μM. Due to the nonexistence of photobleaching, they were practically stable for > 400 days. Only minor changes in activity were observed. There was neither intrinsic immune reactivity nor any unfavorable response during this long period.

Two operational modes for functionalized SWCNT sensors were explained: injection trailed by confining inside the liver and straight implantation into a definite tissue.

13.7 Discussion and Conclusions

Implantable sensors provide measurements of target analytes inside the human body in real time. For accurate measurements, highly selective, drift-free, low-cost, energy-efficient sensors are needed. According to reports [35], a nitric oxide sensor can be inserted subcutaneously for a period longer than a year and used for monitoring cancer or inflammation-causing diseases. Time is not far off when implantable sensors will become a part and parcel of healthcare instrumentation and the medical care paradigm will shift to a personalized model in which the remotely located doctor would attend to patient's problems and solve them through wireless networks.

Review Exercises

13.1 Implantable sensors must be drift-free. If an implanted sensor has a small drift, how can the necessary correction for drift be applied externally to a fully sealed implanted sensor?

13.2 How is the low power consumption requirement of implantable sensors achieved using capacitors and switched-capacitor techniques?

13.3 What is meant by the term, "high blood pressure"? What are the potential health risks to a hypertension patient? How is blood pressure commonly measured? Can these methods provide blood pressure values at an internal site of the body? What solution do you propose?

13.4 Explain with the help of a diagram the working of a SiGeB diaphragm-based sensor for blood pressure measurement. Why this sensor has low power consumption?
13.5 Describe the reflected wave transit time (RWTT) method of blood pressure measurement using a single accelerometer.
13.6 What are the three main components of a SAW sensor microsystem? How is this microsystem applied for blood pressure measurements?
13.7 How does foreign body response degrade the performance of an implantable biosensor? What countermeasures are followed to avoid this deterioration?
13.8 Discuss the problem of oxygen shortfall in reference to an implantable biosensor. Describe how efforts to solve this problem have led to development of three generations of biosensors.
13.9 Describe one technique used for enzyme immobilization in implantable biosensors. Mention any particular advantage offered by this technique.
13.10 What measurements are performed using implantable blood gas sensors? What is the significance of measuring blood gases?
13.11 For what applications is a Paratrend probe used? What are the principles utilized in operation of this probe?
13.12 How does the local release of nitric oxide at the sensor/blood interface help in avoiding thrombus formation? In what way is this local release done?
13.13 What is type 1 diabetes? How can it be treated? Explain the concept of an artificial pancreas.
13.14 Describe the roles of the following layers in an implantable glucose biosensor: (a) bioprotective membrane layer and (b) angiogenic layer.
13.15 How does an inflammation sensor work? What gas molecules are used as marker signals of inflammation?

References

1. Puers R (2005) Implantable sensor systems. DISens symposium-book 2005, 14 p. http://www.disens.tudelft.nl/symposium2005/book/6_puers.pdf. Accessed 31 Jul 2015
2. Potkay JA (2008) Long term, implantable blood pressure monitoring systems. Biomed Microdevices 10:379–392
3. Chatzandroulis S, Tsoukalas D, Neukomm PA (2000) A miniature pressure system with a capacitive sensor and a passive telemetry link for use in implantable applications. J Microelectromech Syst 9(1):18–23
4. Cleven NJ, Isfort P, Penzkofer T et al (2014) Wireless blood pressure monitoring with a novel implantable device: long-term in vivo results. Cardiovasc Int Radiol 37:1580–1588. https://doi.org/10.1007/s00270-014-0842-0
5. Theodor M, Fiala J, Ruh D et al (2013) W1C.002: implantable accelerometer for determination of blood pressure. In: Transducers, Barcelona, Spain, 16–20 June 2013, pp 1659–1662
6. Theodor M, Ruh D, Förster K et al (2013) Implantable acceleration plethysmography for blood pressure determination. In: 35th annual international conference of the IEEE EMBS, Osaka, Japan, 3–7 July 2013, pp 4038–4041

7. Theodor M, Fiala J, Ruh D et al (2014) Implantable accelerometer system for the determination of blood pressure using reflected wave transit time. Sens Actuators A Phys 206:151–158
8. Liang B, Fang L, Tu CL et al (2011) A novel implantable saw sensor for blood pressure monitoring. In: Transducers' 11, Beijing, China, 5–9 June 2011, pp 2184–2187
9. Ye X, Fang L, Liang B et al (2011) Studies of a high-sensitive surface acoustic wave sensor for passive wireless blood pressure measurement. Sens Actuators A Phys 169:74–82
10. Murphy OH, Bahmanyar MR, Borghi A et al (2013) Continuous *in vivo* blood pressure measurements using a fully implantable wireless SAW sensor. Biomed Microdevices 15:737–749
11. Vaddiraju S, Tomazos I, Burgess DJ et al (2010) Emerging synergy between nanotechnology and implantable biosensors: a review. Biosens Bioelectron 25:1553–1565
12. Galeska I, Chattopadhyay D, Papadimitrakopoulos F (2002) Application of polyanion/Fe^{3+} multilayered membranes in prevention of biosensor mineralization. J Macromol Sci Pure Appl Chem A39(10):1207–1222
13. Ishihara K, Tanaka S, Furukawa N et al (1996) Improved blood compatibility of segmented polyurethanes by polymeric additives having phospholipid polar groups. 1. Molecular design of polymeric additives and their functions. J Biomed Mater Res 32(3):391–399
14. Onuki Y, Bhardwaj U, Papadimitrakopoulos F et al (2008) A review of the biocompatibility of implantable devices: current challenges to overcome foreign body response. J Diabetes Sci Technol 2(6):1003–1015
15. Vaddiraju S, Singh H, Burgess DJ et al (2009) Enhanced glucose sensor linearity using poly(vinyl alcohol) hydrogels. J Diabetes Sci Technol 1(3):863–874
16. Yu B, Moussy Y, Moussy F (2005) Coil-type implantable glucose biosensor with excess enzyme loading. Front Biosci 10:512–520
17. House JL, Anderson EM, Ward WK (2007) Immobilization techniques to avoid enzyme loss from oxidase-based biosensors: a one-year study. J Diabetes Sci Technol 1(1):18–27
18. Lu D, Cardiel J, Cao G et al (2010) Nanoporous scaffold with immobilized enzymes during flow-Induced gelation for sensitive H_2O_2 biosensing. Adv Mater 22(25):2809–2813
19. Schlosser K, Li Y (2009) Biologically inspired synthetic enzymes made from DNA. Chem Biol 16(3):311–322
20. Dai Z, Liu S, Bao J et al (2009) Nanostructured FeS as a mimic peroxidase for biocatalysis and biosensing. Chem Eur J 15(17):4321–4326
21. Frost MC, Meyerhoff ME (2002) Implantable chemical sensors for real-time clinical monitoring: progress and challenges. Curr Opin Chem Biol 6:633–641
22. Rolfe P (1990) In vivo chemical sensors for intensive-care monitoring. Med Biol Eng Comput 28:B34–B47
23. Meruva RK, Meyerhoff ME (1998) Catheter-type sensor for potentiometric monitoring of oxygen, pH and carbon dioxide. Biosens Bioelectron 13:201–212
24. Espadas-Torre C, Oklejas V, Mowery K et al (1997) Thromboresistant chemical sensors using combined nitric oxide release/ion sensing polymeric films. J Am Chem Soc 119:2321–2322
25. Telting-Diaz M, Collison ME, Meyerhoff ME (1994) Simplified dual lumen catheter design for simultaneous potentiometric monitoring of carbon dioxide and pH. Anal Chem 66:576–583
26. Venkatesh B, Clutton-Brock TH, Hendry SP (1995) Continuous measurement of blood gases using a combined electrochemical and spectrophotometric sensor. J Med Eng Technol 18:165–168
27. Coule LW, Truemper EJ, Steinhart CM (2001) Accuracy and utility of a continuous intra-arterial blood gas monitoring system in pediatric patients. Crit Care Med 29:420–426
28. Andrade JD, Hlady V (1986) Protein adsorption and materials biocompatibility—a tutorial review and suggested hypotheses. Adv Polym Sci 79:1–63
29. Lisman T, Weeterings C, de Groot PG (2005) Platelet aggregation: involvement of thrombin and fibrin(ogen). Front Biosci 10:2504–2517
30. Wu Y, Rojas AP, Griffith GW et al (2007) Improving blood compatibility of intravascular oxygen sensors via catalytic decomposition of s-nitrosothiols to generate nitric oxide in situ. Sens Actuators B Chem 121(1):36–46

31. Ricotti L, Assaf T, Dario P et al (2013) Wearable and implantable pancreas substitutes. J Artif Organs 16:9–22
32. Gough DA, Armour JC (1995) Development of the implantable glucose sensor: what are the prospects and why is it taking so long? Diabetes 44:1005–1009
33. Updike SJ, Shults MC, Gilligan BJ et al (2000) A subcutaneous glucose sensor with improved longevity, dynamic range, and stability of calibration. Diabetic Care 23(2):208–214
34. Renard E (2002) Implantable closed-loop glucose-sensing and insulin delivery: the future for insulin pump therapy. Curr Opin Pharmacol 2:708–716
35. Iverson NM, Barone PW, Shandell M et al (2013) In vivo biosensing via tissue-localizable near-infrared-fluorescent single-walled carbon nanotubes. Nat Nanotechnol 8:873–880

Chapter 14
Cardiac Pacemakers

Abstract Ever since the introduction of the first artificial pacemaker in 1932, pacemaker technology has advanced rapidly. The early pacemakers could not sense the electrogram. They were brainless devices which only paced the ventricles asynchronously. Subsequent advanced devices called demand mode pacemakers contained a sense amplifier. This amplifier measured the cardiac activity of the patient to evade competition of the actual rhythms of the heart with paced rhythms. Furthermore, single-, dual-, and biventricular pacemakers were launched. Single-chamber pacemakers (one lead) were used to set the pace of only chamber of the heart; this single chamber was usually the left ventricle. Dual-chamber pacemakers (two leads) could set the pace of two chambers of the heart. Biventricular pacemakers used three leads. One lead was placed in the right atrium. The other two leads lay inside the ventricles, one lead per ventricle. Another noteworthy feature is that the early devices were an assembly of discrete resistors, transistors, and capacitors wired together on printed circuit boards, whereas the new devices are highly complex and integrated microprocessor-based systems. They are essentially extremely small computers equipped with RAM and ROM facilities. The topical topologies of pacemakers are tremendously complicated. They include two parts: the analog part and the digital part. The analog portion comprises the sense amplifier and an output stage which performs the pacing. The digital portion consists of sections containing the microcontroller with associated circuitry and the storage memory with accessories. The pacemakers are capable of implementing diagnostic scrutiny of the received electrograms. They provide device programmability. Also, they offer adaptive rate pacing, i.e., they are able to change the paced rate in proportion to metabolic workloads using an accelerometer.

Keywords Pacemaker · Sinoatrial node · Unipolar lead · Bipolar lead · ECG · Arrhythmia · Single-chamber/dual-chamber/biventricular pacemaker · Asynchronous/synchronous pacemaker · NBG code · Epicardial/transvenous implantation · Rate-responsive pacemaker · Pacing lead

V. K. Khanna, *Implantable Medical Electronics*,
https://doi.org/10.1007/978-3-031-99868-3_14

14.1 Introduction

A pacemaker of the 1960s typically weighed 250 g, contained a circuit based on two transistors and powered by 4–6 batteries, had a fixed pulse rate (asynchronous), producing 1 ms wide pulses at 70 pulses per minute, was nonprogrammable, and had a service life less than 3 years with uncertain reliability. Remarkably contrastingly, a present-day pacemaker weighs 10 times less; uses avant-garde microprocessor-/microcontroller-based integrated electronic circuitry and memory along with the analog segment containing the sense amplifier and the output pulse for pacing; is fed by high energy density lithium/iodine batteries with life > 10 years; employs novel materials for electrode tips and new structures providing exceptionally low thresholds of stimulation; uses advanced hermetic packaging; can sense intrinsic heart activity and adjust pacing rate; is multiprogrammable, dual-chamber catering to both the chambers of the heart; and is capable of performing diagnostic functions, data collection, telemetry, and a multitude of other tasks [1]. Figure 14.1 illustrates the chambers of the heart. This diagram will help readers in following the contents of this chapter easily.

14.2 Natural and Artificial Pacemakers of the Heart

The sinoatrial (SA) node, also called the sinus node, is a cluster of cells. It is located in the right atrium of the heart in a region near the top of the atrium. It lies close to the junction with the superior vena cava (Fig. 14.2). Electrical discharges released from this node regulate the normal rhythm of the heart. This small area of the heart is therefore known as its natural pacemaker. It performs the task of starting and controlling each cycle of the heart, inclusive of its electrical and mechanical activities. The electrical impulses produced by the SA node trigger the quivering of heart muscles, their contraction, as well as their dilation. They look after the proper synchronization of heart muscles. The synchronization is imperative for forcing the blood into the ventricles. Looking at the spreading paths of the electrical signals from the SA node, these signals travel rapidly throughout the atria. They reach the ventricular muscle through conducting pathways. This is essential to ensure the simultaneity of contraction of all the muscle fibers. The pathways comprise the internodal tracks and atrial fibers. The atrioventricular node (AV node) itself is included in the pathways. Other main components are the bundle of His, the right and left bundle branches, along with the Purkinje fibers.

The normal rhythm of the heart consists of 60–100 contractions per minute. It varies with physical stress or emotional strain, increasing under both these conditions. It decreases when a person is taking rest. The heart beating rate differs from person to person. Abnormal conditions can influence this rate drastically. Among such conditions can be mentioned heart injuries. Generalized infections too can affect the heart beat. In diseased conditions, the electrical system of the heart does

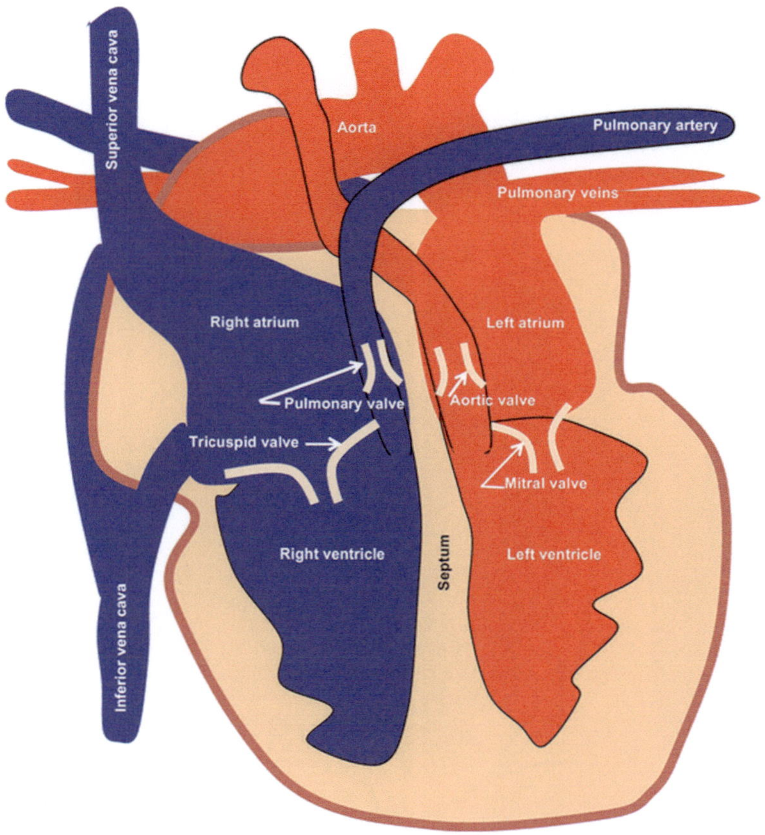

Fig. 14.1 Chambers of the heart

not work properly. This happens either if the SA node is in a diseased condition or the conduction system has become choked or clogged up. Then impulses from the node do not propagate along their normal trajectories or courses. So the heart starts to beat uncharacteristically. If the ventricles (lower chambers) of the heart beat too slowly, the patient needs an artificial heart pacemaker. The pacemaker helps to make it beat regularly again. Thus, enough blood flows around the body.

The artificial pacemaker is an electrical stimulator (Fig. 14.3). This stimulator is implanted immediately beneath the skin of the patient's chest below the collar bone. It produces a periodical train of electrical pulses that are transmitted to the electrodes. These electrodes are situated on the outer layer of the walls of the heart called epicardium, the middle layer of heart walls known as the myocardium, or the inner layer of the heart walls termed the endocardium. The controlled, rhythmic stimuli are delivered to the heart muscle. They cause it to maintain an efficient heart rate over extended periods of time. Thus, the heart functions with its normal pumping capacity. This effect can be used as a prosthetic aid in those diseases in which the

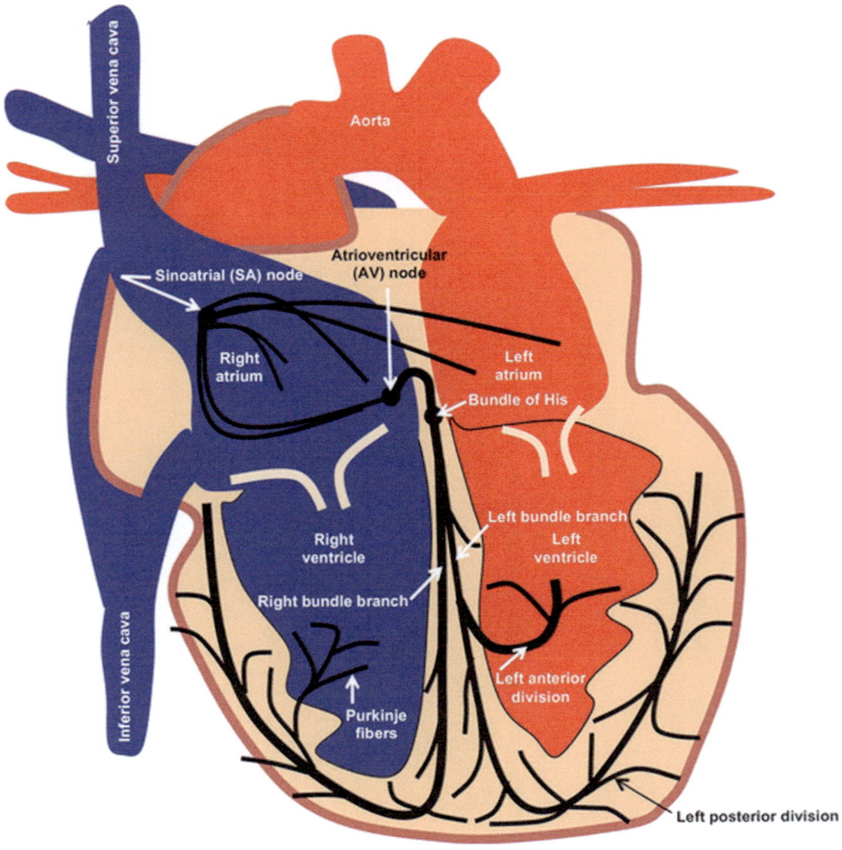

Fig. 14.2 Electrical conduction system of the heart

heart is not intrinsically stimulated at a proper rate [2]. Thus, pacemakers are cardiac rhythm managers capable of correcting a myriad of complex heart abnormalities.

The main function of a typical pacemaker is to detect and investigate the heartbeat of a person to find out if it is normal or irregular. Whenever any such irregularities are detected, it paces the heart via electrical stimulation. Electronically, the artificial pacemaker is an embedded system operating in real time. It has both software and hardware capabilities. It is placed inside hermetically sealed encapsulation. Functionally, it comprises at least three parts: (a) an electrical pulse generator, (b) a source of electrical power (battery), and (c) an electrode (lead) system. The electrode system establishes the electrical connection of the pulse generator with the heart (Fig. 14.4).

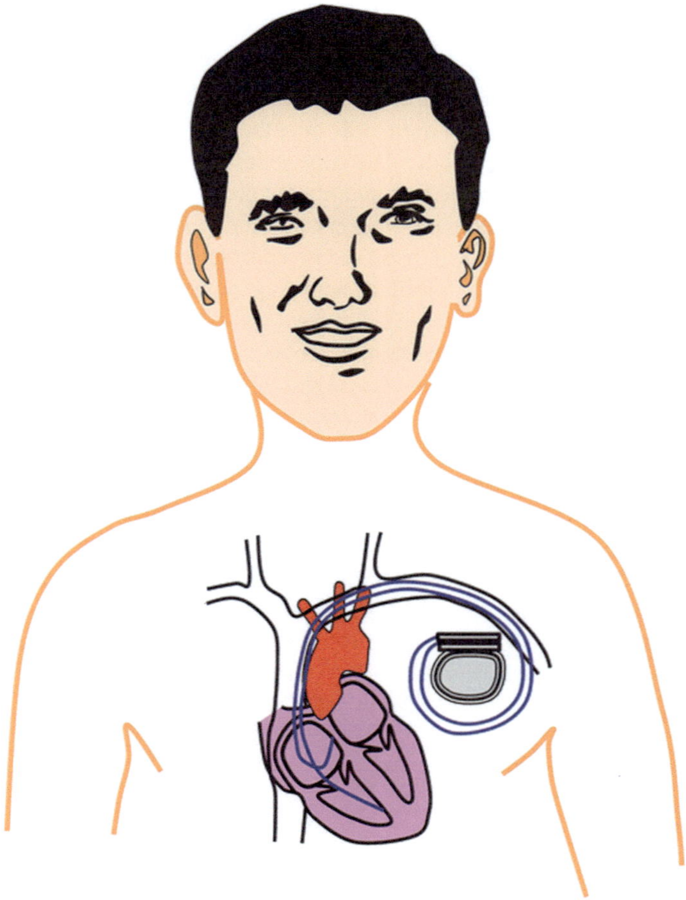

Fig. 14.3 Patient with implanted pacemaker

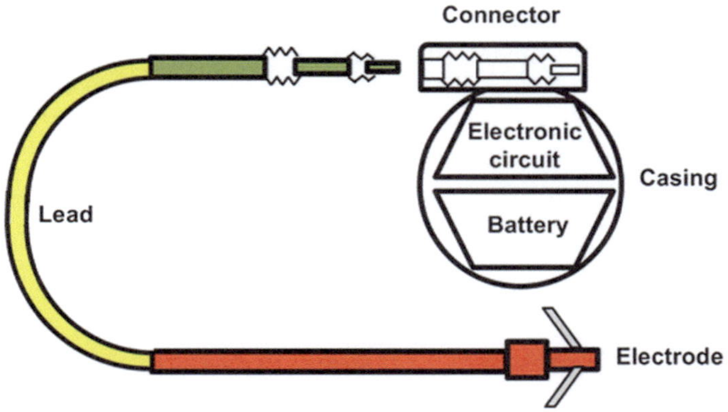

Fig. 14.4 External appearance of the cardiac pacemaker

14.3 Unipolar and Bipolar Stimulation

In a unipolar stimulating device, the electrode tip stimulates the heart. The pacemaker unit serves as the reference. In a bipolar device, the lead has both a stimulating tip, the cathode, and a ring, the anode. The ring generally has a much larger surface area. The separation between the ring and tip is typically 2–3 cm, depending on the pacemaker model. In all modern pacemakers, stimulation occurs at the cathode. The stimulating cathode is situated in or on the myocardium of the heart. The anode is located distally near the heart (bipolar pacing) or remotely as a part of the pulse generator (unipolar pacing).

The current threshold of stimulation is the same for both unipolar and bipolar leads. The voltage threshold, however, is slightly higher for a bipolar lead because of the increased lead resistance. This increased resistance during bipolar pacing is due to the ring area being much smaller than the pacemaker case area (which is the anode in unipolar pacing). Since resistance is a function of conduction path area, the anode in bipolar pacing has more resistance than in unipolar pacing.

A benefit of a bipolar lead configuration is that the signal-to-noise ratio of the sensed heart signals is better than that found with unipolar leads. The bipolar sensing configuration eliminates much of the noise resulting from nearby muscle movement. Most modern devices can be changed from unipolar to bipolar configuration and vice-versa. This changeover is implemented through telemetry.

Table 14.1 presents the salient features of the two types of pacemaker leads.

Table 14.1 Unipolar and bipolar leads

Sl. No.	Unipolar leads	Bipolar leads
1	It consists of a single isolated conductor with an electrode typically protruding at the tip of the lead. This electrode is the cathode. The pulse generator housing is used as the anode. It is usually stationed in the pectoral region so that the anode–cathode distance exceeds 10 cm	It comprises an arrangement of two isolated conductors, which are connected to the two electrodes. The distance between anode and cathode is usually not > 1.5 cm
2	Sensing behavior is inferior to bipolar leads	It outperforms unipolar lead. It has an improved signal-to-interference ratio. When atrial activation is to be sensed, its lower sensitivity to far-field potentials produced by the ventricles is a distinct advantage
3	It is less affected by electromagnetic interference (EMI). Skeletal muscle potentials too cannot influence it harmfully	EMI and skeletal muscle potentials cause significant disturbances
4	It is more flexible mechanically	It is inelastic and more intricate from a mechanical construction perspective

14.4 The Electrocardiogram Waveform

By fixing electrodes on the surface of the human body, the electrical activity of the heart can be recorded accurately. This waveform is called the electrocardiogram (Fig. 14.5).

The voltage range of electrocardiogram (ECG) signals is 2 mV when peak-to-peak values are considered. Its bandwidth is 0.05–150 Hz. The ECG signal of a normal person (Fig. 14.6) contains five distinct waves: (a) the P-wave indicates atrial depolarization, (b) the Q-wave points at septal depolarization, (c) the R-wave signifies early ventricular depolarization, (d) the S-wave hints at late ventricular depolarization, and (e) the T-wave signals repolarization of the ventricles. In some persons, a minor peak occurs either at the end or after the T-wave. It is called the U wave. Its origin is believed to be a repolarization potential.

The P waves should all look similar and not exceed 0.3 mV. The P–Q interval arises from the propagation delay time of the specialized cells. These cells are the AV node and the His–Purkinje system. The P–Q interval usually lasts for 0.2 s. The QRS complex represents the ventricle depolarization waveform. The S–T interval signifies the duration of the action potential. Normally, it persists for about 0.25–0.35 s.

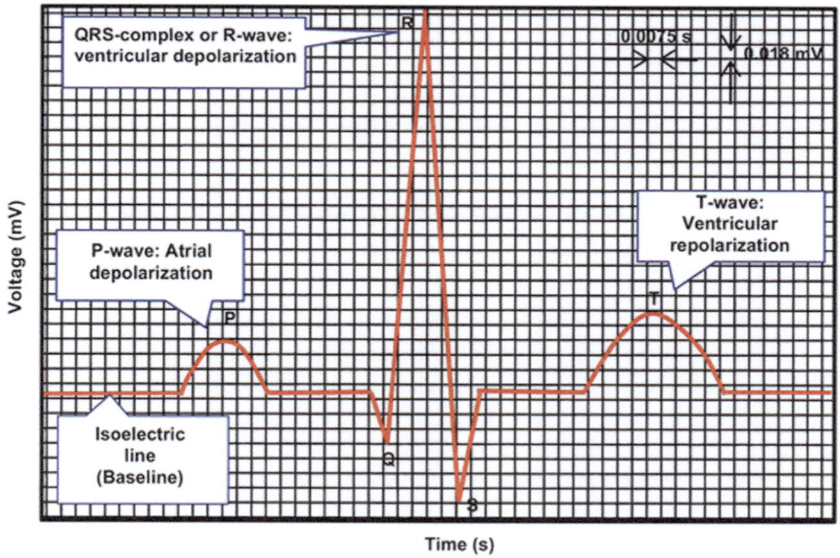

Fig. 14.5 Typical tracing of ECG waveform and related activities taking place in the heart

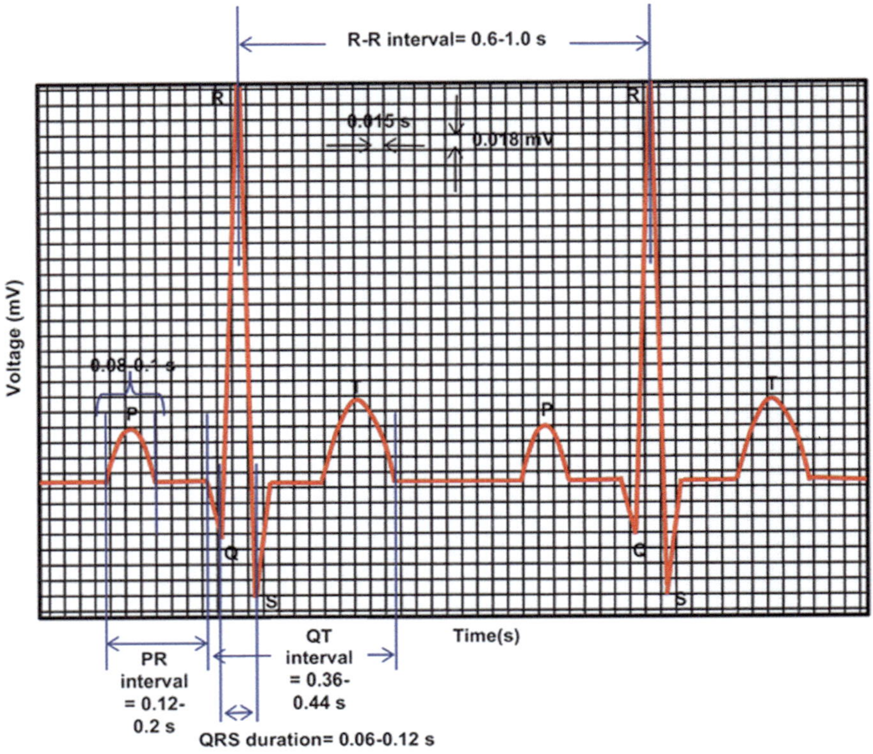

Fig. 14.6 ECG waveform showing the ranges of durations of different intervals for a normal person

14.5 Arrhythmias and Pacemaker Indications

Arrhythmias are cardiac problems in which rhythms of heart beats become abnormal. They increase, decrease, or become irregular. Main types of arrhythmias are: (a) increased heart beat or tachycardia > 100 beats per minute, (b) decreased heart beat or bradycardia < 60 bpm, or (c) irregular heart rhythm. There are two common causes of bradycardia. One cause is the sickness of sinus syndrome. It is a disease affecting the sinoatrial (SA) node, the natural pacemaker of the heart. The other cause is heart block. A block of this kind occurs when the upper chambers of the heart, namely, the atria, and its lower chambers, viz., the ventricles, are not harmonized in action. As a result, atrioventricular (AV) block takes place. The impact of these diseases is that heart beats too slowly. It beats either sporadically or beats at a laggard pace all the time. In all cases, the heart might not be able to propel plenty of blood. The blood pumped might not be ample enough to cope up with the needs of the body. At this declining heart rate, blood supply to the brain decreases profoundly. This may lead to light headedness and sometimes fainting.

Pacemakers are used in the following conditions [3]: (a) sick sinus syndrome, (b) symptomatic sinus bradycardia, (c) tachycardia–bradycardia syndrome, (d) atrial fibrillation with sinus node dysfunction, (e) full atrioventricular block (third-degree AV block), (f) chronotropic inadequacy (incompetence to increase the heart rate to concord with exercise level), (g) lingering QT syndrome, and (h) cardiac resynchronization therapy (CRT), biventricular pacing, or multisite ventricular pacing.

14.6 Types of Artificial Pacemakers

According to the number of pacing leads used, artificial pacemakers are of three types: (a) single chamber, (b) dual chamber, and (c) biventricular (Table 14.2).

Figures 14.7, 14.8, and 14.9 help in understanding the placement locations of the pacing lead in single-chamber, dual-chamber, and biventricular pacemakers, respectively.

Based on the pacing demands, pacemakers are subdivided into: (a) asynchronous and (b) synchronous pacemakers (Table 14.3). An asynchronous pacemaker delivers signals at a fixed speed. But a synchronous pacemaker first senses the endogenous cardiac electrical activity (spontaneous depolarization). It withholds or inhibits its output of electrical stimuli in case of detection of a signal derived from the inbuilt electrical activation of the heart. It activates only on receipt of sensations, which show inadequacy of the spontaneous rhythm by the heart. In this way, the competition of the implanted pacemaker with the patient's own natural pacemaker is avoided. Thus, it minimizes the risk of pacemaker-induced fibrillation associated with rapid, unsynchronized contraction of cardiac muscles. It performs its task by measuring the interval between the native beats of the heart. It delivers a stimulating pulse whenever that interval exceeds a set value.

The normal heart rhythm varies all throughout the 24 h of the day. It beats in a relaxed fashion or quickens up several times during this span. The deceleration or acceleration of heart rate is dependent on a person's level of activity, exercise, and other factors. In the resting or sleeping state of an individual, the heart slows down. In response to exercise, excitement, and elation, it speeds up. Rate-responsive pacing is desired for a patient whose heart fails in adjusting its beating rate to meet the

Table 14.2 Pacing lead-wise classification of pacemakers

Single chamber	Dual chamber	Biventricular
It uses one pacing lead only, placed either in the atria or ventricles. Generally, the lead placement is done in the right ventricle	It uses two pacing leads, one lead in the atria and other lead in the ventricles. Commonly, one lead is stationed in the right atrium and the other lead in the right ventricle	It uses three pacing leads. Position-wise, the first lead is inserted in the right atrium. The second lead lies in the right ventricle and the third one in the left ventricle

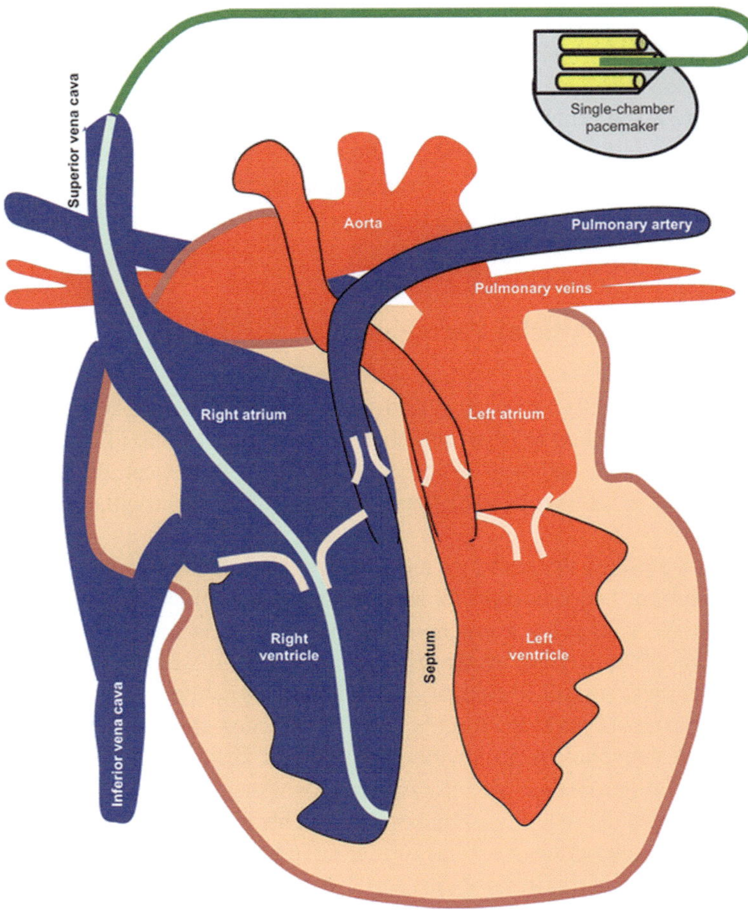

Fig. 14.7 Single-chamber pacemaker

demands of activities carried out by the human body. A rate-responsive or frequency-responsive pacemaker uses an activity sensor to perceive the intensiveness or severity of the physical activities of the body to regulate the pacing rate. This sensor is an electronic device to detect changes in metabolic demand. It does so by measuring some relevant parameter from the body. Such a decisive parameter could be body motion of the patient. Parameters such as respiration rate, pH, and blood pressure too can be used. The sensor forms a component of either the pacemaker device itself and/or its lead. In accordance with the measured value of the parameter, an algorithm in the pacemaker automatically adjusts its output. Modern rate-responsive pacemakers are endowed with the capability of acclimatizing to a broad range of sensor inputs for meeting the physiological needs and/or catering to the activities of the user. Looking at the level of daily activities of a specific patient, the attending physician fine tunes

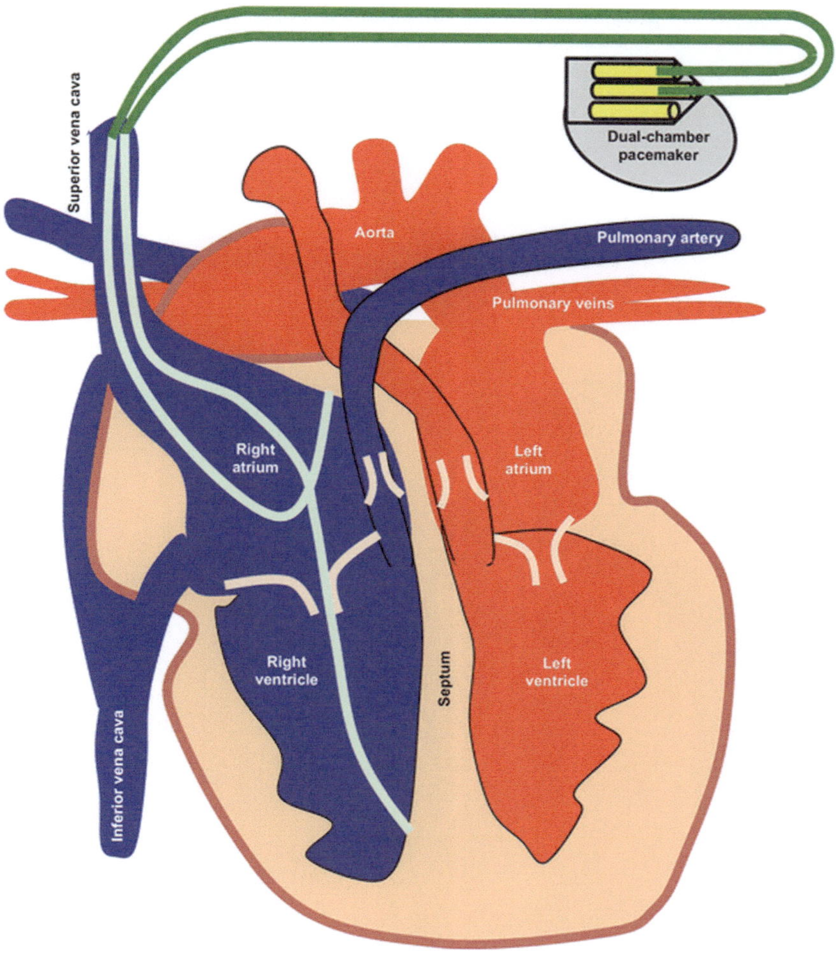

Fig. 14.8 Dual-chamber pacemaker

the sensor(s) to meet individual needs. Rate-responsive pacing meticulously mimics the normal heartbeat maintaining harmony with patient's bodily exertion.

14.7 Pacemaker Codes

A standardized classification code has been developed for pacemakers. Its objective is to facilitate the use and understanding of pacemakers. It is used as a shorthand to explain the reasons for the implantation of a particular pacemaker in a given patient. It does so by providing quick information about the principal features of the pacing system used by the patient. These features are the heart chamber paced and the

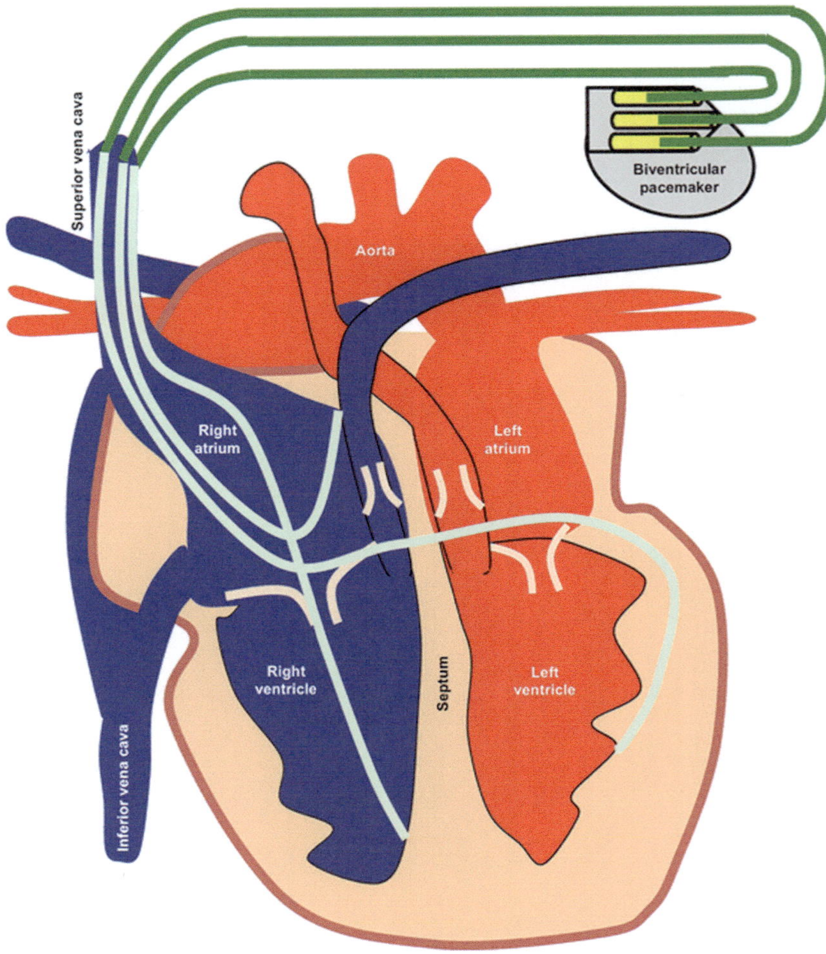

Fig. 14.9 Biventricular pacemaker

chamber sensed. It also tells about the type of response to intrinsic beats and other characteristics.

The NBG code has evolved from a joint project of two esteemed groups. One of these groups is the North American Society of Pacing and Electrophysiology, with acronym NASPE. The other group is the British Pacing and Electrophysiology Group, represented by acronym BPEG [4]. The full form of the NBG code is quite lengthy, and the letters N, B, and G stand for the bold letters (North, British, and Generic) in its expanded form, viz., "*North* American Society of Pacing and Electrophysiology (*N*) and*British* Pacing and Electrophysiology Group (*B*) Generic (*G*) Pacemaker Code." In this code, a sequence of letters is used to describe the mode in which the pacemaker is operating [5].

Table 14.3 Asynchronous and synchronous pacemakers

Sl. No.	Asynchronous or fixed-rate or nondemand pacemaker	Synchronous or inhibited or demand pacemaker
1	It delivers electrical pulses at a fixed rate to the ventricle. The pulse repetition frequency is factory preset and unalterable. No attention is paid to the spontaneous activity taking place inside the patient's heart, i.e., its function is independent of any atrial or ventricular activity	It provides an electrical impulse for stimulating the heart only under one condition. This condition is that the natural heartbeat must be absent. Otherwise, it remains inoperative
2	In this mode, competition between the delivered pulse and the natural heart activity can occasionally provoke arrhythmias. Ventricular fibrillation may also be triggered	It includes a sensing amplifier section along with the asynchronous pacemaker. This amplifier detects intrinsic heart activity. The unwanted antagonism between the two pacing rates is thus avoided
3	It unnecessarily wastes battery power by delivering pulses even when they are not required, thereby shortening its life	It prolongs the battery life of the system. This is because of its need-based activation occurring only when pacing stimuli are actually called for
4	It is rarely used today except to initiate or terminate some tachycardias	It is the commonly used pacemaker

The first two positions of this code represent the chamber(s) of heart paced and chamber sensed. These positions seem relatively straightforward.

The first letter (position I) of the code designates the chamber (or chambers) of the heart being paced. This could be A standing for atria, V denoting ventricles, or D indicating dual or both.

The second letter (position II) of the code stipulates the chamber that is being used for executing the second function of a pacemaker, namely, sensing of native signals from the heart. It indicates from: which chambers the pacemaker is receiving the sensing signal? Again, this could be A for atria, V for ventricles, D for dual, or O for none, i.e., no sensing takes place.

The third letter (position III) tells us the pacemaker's mode of response to a sensing event. In plain words, it explains the action taken by the pacemaker upon detection of a particular type of activity in a "sensed" chamber. The response of the pacemaker could be one of the three types: T standing for "triggers the pacemaker," I for "inhibits the pacemaker," and D for "dual—inhibited and triggered." Rarely encountered, the letter T (triggered) in the third position is an admirable method to watch the locality of sensing of inherent events. It is generally used for device testing. Upon identification of the third letter as I (inhibited), the manner of retorting is to hold back or refuse giving an output from the pacemaker when an event is sensed. The letter D (dual) in the third position is an indicator of the fact that the device will answer back to the sensed signal in one of the two ways: (a) either it will inhibit the pacemaker response and trail the sensed event or (b) it will restrain the output to be provided on the sensed channel. Instead, it will prompt an output to preserve the AV synchronization.

The fourth letter (position IV) conveys information about what parameters of the device are programmable? Let us consider the letter R (denoting "rate response") in position IV. It only gives the information that the pacemaker has the capability of being programmed (and programmed to) a function that is rate responsive. The letter C, standing for "communicating," articulates that the pacemaker is adept at transmission and/or reception of data for the purpose of providing information or for programming. M denoting "multi-programmable" suggests that the device is capable of being programmed in more than three parameters. Programmable parameters of interest are rate, sensing, and output. Refractory periods, mode, and hysteresis are also important. P (meaning "simple programmable") usually indicates that the pacemaker has three or fewer parameters that can be programmed. Seldom does one come across the symbol O (none). This symbol indicates that there are no programmable parameters for this pacemaker.

The fifth letter (position V) hints at the provision of any antitachycardia features in the pacemaker, as a supplementary facility. This letter place is earmarked fully for utilization by the implantable cardioverter defibrillators (ICDs). This position expresses their capability to pull patients out of lethal tachycardias by delivering them pacing or shock pulses. Most present-day ICDs are represented by the letter D in the fifth position. The letter D stands for dual—shocks and paces.

14.8 Fitting the Pacemaker

14.8.1 Surgery for Pacemaker

Pacemakers are mostly fitted bytransvenous implantation; in a few cases, they are fitted byepicardial implantation. Before fitting the pacemaker, the patient is givenantibiotics to avoid any infection even though the procedure is done under sterile conditions.

In transvenous implantation, the surgery takes about an hour as a day case with local anesthesia and under sedation. Making an incision on the left side under thecollarbone, the pacemaker is inserted between the skin and chest muscle. The wires are slipped into a vein and guided under X-ray screening to the relevant chamber of the heart. After connecting the electrode leads to the pacemaker box, the box is fitted into the small pocket between the skin and the chest muscle. The incision is stitched and dressed up. The pacemaker is hardly visible from outside unless, of course, the person is very lean and thin. A small bump in the skin can be seen over the place where the pacemaker has been implanted.

Epicardial implantation is done if it is difficult to get to theveins in the more practicedtransvenous method. Here, the electrode lead is affixed to the outer surface of the heart, the epicardium, through a cut in the abdomen or chest wall. After pacemaker insertion, the patient is checked for blood pressure or any bleeding from incision site. Waking up from the effect of sedation, the patient is allowed to eat and drink and get up on feet with the advice on how to sit up or how far to move the arm to avoid any lead displacement. Some stitches dissolve on their own while others need to be cut after 7 days. Antibiotics/painkillers are continued for a few days. The lead position is checked by a chest X-ray examination. A cardiac physiologist adjusts the pacemaker parameters suitable for the patient.

The patient has to follow the restrictions on heavy weight lifting for several weeks. Extreme motion of the arm on the side of the pacemaker is not permitted. Returning to normal activities is possible within a few days. A pacemaker identification card is given to the patient. This card gives details about the make and model of the implanted device. On this card is written the specific information on the type of leads and pacemaker implanted. It is recommended that patients should carry this card whenever they move out on tour. This card is helpful to healthcare professionals during subsequent patient monitoring and medical evaluations. It can also be shown to the security personnel at airports [6].

14.8.2 Post-operation Follow-Ups

Follow-up appointments with thecardiologist are required throughout life after every 3–12 months. During each consultation, the rate of discharge of the pacemaker is checked. The intensity of the electrical pulse is determined to find any faults and correct them. Modern pacemakers store information regarding the status of the battery and performance of the pulse generator. By accessing this information, the cardiologist reprograms the pacemaker settings for providing best treatment to the patient.

The pacemakers do not impact the lifestyle of the receiving patient adversely. But they evoke anxiety about the permissible activities that can be freely performed. Patients ask questions regarding the motion and effort (whether they can swim? how heavy load can they lift?) and influences of the environment (e.g., can they use a mobile or cellular phone? orcanthey use an electrical razor for shaving?) [7].

14.9 First Pacemaker Implantation

The October 8, 1958, is a red-letter day in the annals of cardiac history. On this auspicious day, the first pacemaker implantation [8] was performed in Sweden. This surgery was undertaken by two specialists: Ake Senning and Rune Elmqvist. Ake Senning was the heart surgeon, in-charge of the Department of Thoracic Surgery, Karolinska Hospital, Stockholm. Rune Elmqvist was the physician inventor. He was a graduate in medicine. However, he had not pursued a medical practice. Instead, he became an engineer [9, 10]. This implantation was executed on a 43-year-old patient, Arne Larsson, an engineer by profession. The patient had been hospitalized with complete heart block and had suffered 20–30 Stokes–Adams attacks daily for 6 months. A Stokes–Adams attack is a sudden collapse of a patient into unconsciousness with or without convulsions. This happens due to an abnormality in heart rhythm, viz., a slow orabsent pulse.

The pacemaker implantation was done at the pleading of Ms. Else Marie. Ms. Marie was the patient's wife who persuaded the doctors to help her despairingly sick husband. Taking cue from journal and media releases about continuing experiments with stimulation of the heart byelectrical pulses, the lady requested and pestered the two scientists for a solution considered plausible but yet nonexistent, namely, the implantable pacemaker. For avoidance of public exposure, the implantation was done in the loneliness of the evening. During this period, the operating rooms were vacant. To carry out the implantation, a left-sided thoracotomy was done. In thisthoracotomy, two electrodes were inserted in the muscular tissue of the heart and tunneled to the box containing the pacemaker. The box was placed in the wall of the abdomen. The first implanted pacemaker worked only for a short span of a few hours.

The second pacemaker functioned in a satisfactory manner for around a week's duration. Then it suddenly showed a plummeting of the magnitude of pacingstimulus. This suggested that probably the lead of the pacemaker had fractured. The failure was not attributed to malfunctioning of thepulse generator. The unit was entirely handmade. It was encapsulated in a new epoxy resin (Araldite). The biocompatibility of this resin was superb. The approximate diameter and thickness of the pacemaker were 55 mm and 16 mm, respectively.

A blocking oscillator was used for building the pulse generator circuit. This circuit provided good operational stability and efficiency. The electrical parameters of the pulse generator were as follows: amplitude of the pulses delivered = 2 V, pulse width = 1.5 ms, and constant pulse rate = 70–80 beats per minute. The pulse generator circuit had small power consumption. As an energy source, two Ni–Cd rechargeable cells were used. These cells had a rating of 60 mAh each. The cells were sealed, packaged, and joined in series. They could be recharged by a 150 kHz radio-frequency current. The charging current for the cells was produced by an externally located vacuum tube working on mains supply. This vacuum tube was connected to the external coil. Via a silicon diode, an internal coil antenna of diameter ~ 50 mm was connected to the cells. This internal antenna was inductively coupled percutaneously to the large external flexible coil. This coil measured 25 cm in diameter and was attached to the patient's abdomen using adhesive ribbon. Charging of the pacemaker was done once a week for 12 h.

Rune Elmqvist soon ceased his engrossment in pacemakers. However, he continued active participation in other fields of medical technology. He breathed his last in 1997 at the age of 90. Ake Senning continued to be very dynamic and agile in the field of heart surgery until he died in 2000, aged 84. Arne Larsson, the patient who received the pacemaker implants, survived and outlived both the engineer and the surgeon whose glorious and commendable efforts had saved his life. During his lifetime, he was fitted with five lead systems. He required 22 pulse generators. These pulse generators were of 11 different models. He died on 28 December 2001 at the age of 86. Providentially, his death was destined from a malignancy. This was totally unrelated to his disease of conduction tissue or his implanted pacemaker.

14.10 Evolution of Pacemaker Electronics

14.10.1 Pulse Generators

The pulse generator is partitioned into several sections: (a) the sensing circuit for detection of the built-in cardiac depolarization signal from theheart chamber (s) being paced; (b) the output circuit to supply the electrical signal of desired energy measured by pulse amplitude and duration to the leads for onward delivery to the heart muscle; (c) thetiming circuit to regulate length of the pacing cycle, refractory as well as blanking period (blanking period is the period in which the sensing circuit is turned off; refractory period is the period during which the circuit can perceive the signal but does not kick off any timing interval), pulse duration, andtiming intervals intervening atrial and ventricular events; (d) the telemetry circuit for conveying data and reports from an RF antenna and receiving messages from an RF decoder; and (e) a power source for input power. All the components reside inside a titanium case that has been hermetically sealed and has a connector block for accepting the leads [11].

Sensing Amplifier: It is the front end of a pacemaker. It has a vital role of detecting the occurrence of natural heart activity [12]. Accurate sensing helps the pacemaker to decide whether or not the heart has produced a beat on its own. This beat is the intrinsic heart beat. A sensing amplifier uses filters to allow sensing of P waves and R waves. The filters reject any inappropriate signals. The electrogram entering the amplifier is examined by a bandpass filter. This filter has a center frequency of 30 Hz. The frequency of the R-wave of ECG lies in the range 10–30 Hz. Amplitude of the R-wave, in general, 5–25 mV (and also P-wave, 2–6 mV for dual-chamber pacemaker), is found by comparing with an adjustable reference voltage. This seeks to ensure that sufficient signal is available. Only signals higher than the reference voltage are sensed. The slew rate of the signal is also determined. It is the time rate of change of voltage and should lie between 0.75 and 2.5 V/s.

Output Circuit: Pulse generators work in the constant voltage mode. The pulse amplitude lies in the wide range from 0.8 to 5 V. Sometimes, they may be as high as 10 V. The current varies according to source impedance. Pulse duration extent is from 0.05 to 1.5 ms. From the archetypal battery voltage of 2.8 V, elevated voltages are obtained by voltage multiplier circuits. As an example, the voltage is doubled by charging two parallely connected capacitors of smaller sizes. This combination of capacitors is discharged into the output series-connectedcapacitor.

Timing Circuit: Internally generated clocks are run by the signal received from the crystal oscillator by the digital timing and control circuit. This operation is carried out at divisions of oscillator frequency. An important circuit is the rate-limiting circuit called runaway protection circuit. It prevents the pacing rate from increasing beyond an upper limit, should any component breakdown turn out by misfortune.

Telemetry Circuit: The pulse generator provides information regarding the pulse and electrical parameters such as amplitude and duration of the pulse, impedance of the lead, and battery impedance to the programmer. The programmer reciprocates by dispatching messages in coded form to the pulse generator. This exchange of information serves to alter parameters and features, and salvage diagnostic data.

Power Source: Lithium iodide batteries have a life as long as 15 years enhancing the pulse generator life to that point. On an average, the battery capacity is in the ambit from 0.8 to 3.0 A h. Battery life depends on several factors such as pulse amplitude and duration, pacing rate, single-/dual-chamber pacing, and lead design. The semisolid lithium iodide gradually thickens and inspissates. The internal resistance of the battery is thereby raised. The output voltage varies linearly with internal resistance. Hence, it decreases slowly from 2.8 to 2.4 V, which is an indicator of 90 % usable life of the battery. Thereafter, the voltage decreases exponentially to 1.8 V. During this span, the internal resistance of the battery rises from 10 to 40 kΩ. As soon as the battery voltage decreases to the critical value of 1.8 V, the operation of pulse generator becomes unpredictable and random. It goes haywire and is said to reachitsend of life.

14.10.2 Pacemaker Miniaturization

Invented by Jack Kilby in 1958 and first used in a pacemaker in 1971, the integrated circuits soon gained widespread acceptance. They earned recognition as the regular-building blocks for pacemaker design [13, 14]. Hybrid microcircuits also played a central role in the miniaturization of pacemakers. Hybrid microcircuits are assemblies of interconnected chips containing devices and integrated circuits, small-size chips ofpassive components, e.g., resistors and capacitors, and films of resistors/capacitors printed by inks and fired. These components are either mounted or fabricated on an insulating alumina substrate.

In the primary phase of pacemaker development, the VLSIs were only mounted on one surface of an alumina substrate. Subsequently their juxtapositioning on both the sides was achieved. Now the chips are fixed in a multilayer organization on the top of one another. Indeed, high-tech double-sided hybrid circuits are common. Power losses originate from the capacitances in wire connections at high frequencies. Hence, the interconnections among ICs carrying high-frequency signals demanded more current. Naturally, there was an inclination toward denser circuit integration. The aims were to decrease current drain and also to save and utilize the area needed to accommodate circuits. In the 1980–1990 decade, ICs had device geometries around severalmicrons, and present-day ICs have feature sizes in the submicron or nanometer dimensions, forging ahead to 14 nm gate length and even 10 nm or 7 nm. The trend toward low-dimensional structures has moved unabated. Complementary metal-oxide (CMOS) ICs featuring low power demand and high trustworthiness have pervaded electronics everywhere. In mixed signal design, the analog and digital components are coexistent on the same chip. Techniques such as switched-capacitor filters have been used. They eliminate off-chip hybrid circuit components and help to stay away from trimming operations that are needed with thick-film resistors for achieving precise values.

Incorporation of CMOS technology-enabled microprocessors into pacemakers nearly transformed pacemakers into implantable microcomputers. A high degree of flexibility is afforded by microprocessors. Other notable design features include crystal oscillator timing. Often a backup oscillator is provided. Additionally, programming and telemetry circuits use the power generated by the programming equipment that is externally located. They do not work on the pacemaker battery and so do not overburden it. Improved filtering techniques were developed to comply with stringent electrical standards. They helped in avoiding the effects ofelectromagnetic interference (EMI). The EMI pollution spreads from proliferating sources of electromagnetic waves in the form of telecommunication equipment. Electronic article surveillance gadgetry and cellular phones have further aggravated the situation. Early microprocessors used much higher current than was required for pacemaker application. Therefore, specialized, very small, energy-efficient microprocessors were designed for pacemaker manufacturers. In today's pacemakers, a single integrated circuit chip may contain a plurality of functions including the microprocessor and memory. Output circuitry and telemetry, plus other customized features, may also be incorporated.

14.11 Software-Based Pacemaker Architecture

As the pacing functions are becoming complicated and multifaceted, circuitry is built in a more generic architecture, which is essentially computer-based. This novel approach consists in specifying the functions in software to provide a greater degree of freedom. The specification thus framed allows noninvasive modification of software by externalwireless control after implantation. By this approach, a family of devices is developed starting from a few basic microelectronic designs. Essentially, the moot idea is that product differences can be made by software changes instead ofhardware variations.

A software-based pacemaker consists of a telemetry system together with decoder and timing circuit. Other functions included are the analog sensing and output and analog rate-limiting circuitry. The microprocessor acts as the controlling element. It contains two kinds of memory. These are the read-only memory (ROM) and random access memory (RAM). The RAM stores the software instructions. Also stocked in the RAM are the data such as serial number and patient identification. Besides, vital diagnostic information too is stored in the RAM. More complex features may also be built in some models. The ROM circuitry is designed to authenticate flow of information in anerror-free style. During each pacing operation, it allows the execution of internalself-testing routines. Upon detection of errors, the pacemaker can toggle to a backup pacing system. Any likelihood of anomalous pacing behavior caused by software errors is thus diminished.

A RAM-based pacemaker seems to be beneficial on first sight. Nonetheless, it must not be forgotten that such a pacemaker will lose its complete program in case of power interruption. Moreover, its operational life is shortened by the increased current consumption.

14.12 Programmability and Telemetry

Programmability is the capability to accept a new instruction set to change parameters. It permits pacing rate and output to be adjusted postimplantation without any surgery. All-inclusive programming techniques involve binary coding. Also provided are error detection features and real-time telemetry capabilities. One of the first diagnostic functions provided was the competence of a pacemaker to transmit the values of diagnostic parameters in real time. Significant parameters are the lead impedance andpulse amplitude. To begin with, measurements of parameters were done for a few parameters only. Transmission of endocardial electrograms soon followed in real time. Annotation of the state of pacemaker was done.Counters were supplied to reckon the pacing and sensing fractions as percentages. Provision of histograms of heart rate history was another feather in the cap. Now, in advanced systems, almost every feature of therapy has correlated data related with diagnosis to demonstrate its functionality. In order to obtain an expressive sequence of events showing types of incidents versus time and stored waveforms, rate profiles or trends can be included and heart rate versus time plots can be generated. Implantable systems store the information received from a host of various types of physiological sensors. This information helps the doctors in diagnosing diseases and prescribing corrective medical therapies. Evolution of pacing has been observed as starting from a system in which data collection was zero to systems allowing the user to put together several types of information. It stands at the pinnacle of information technology. A vast amount of information is available on drug efficacy. Besides, disease progression can be followed.Tachycardia events are recorded too.

14.13 Rate-Responsiveness

During the late 1970s and 1980s, several sensor-based rate-responsive systems were launched. These systems used blood pH, respiratory rate, body vibration or motion, and QT interval in the ECG as the basic parameters for monitoring therapy. With their help, they could quicken or slow down the pacing rate in step with metabolic demands. Today the majority of pacemakers work in a rate-responsive manner. They contain one or more sensors. Among the sensors mentioned above, only those based on measurement of QT interval and body activity sensing are exclusive to pacing.

Others have been used elsewhere. Of these, the method using body activity sensor has become the main technique of sensing for physical exercise. Two robust devices for implementation of body activity sensor are common. These devices utilize piezoelectric crystals for detection of vibration or deflection. In one case, the sensor is mounted on the inner surface of the pacemaker container to detect vibrations in the body. In the other case, a piezoelectric cantilever mounted on the circuit board measures recurring acceleration of the torso or trunk of the human body. A voltage is generated by the mechanical deformation on bending of the cantilever beam. Thus the acceleration is transduced to a voltage signal in either case.

14.14 Automatic Safety/Backup Features

The pacemaker is a true life savior for many people. These people owe their lives to the pacemaker and cannot survive without its service. With the enormous responsibility placed on this tiny gadget, any intermission or haltage of working of the pacemaker for even a moment can endanger the life of the patient. So, backup support has to be provided in the device to attend to such unforeseen calamities. With this objective, a few pacemakers have a bipolar verification function. This function incessantly checks whether the bipolar lead is integral and satisfactorily working during every cycle of pacing. No sooner than a soaring anodic resistance is noticed, the pacemaker switches over to unipolar pacing without human intervention. Then it uses its can as the anode. This change over from bipolar to unipolar mode prevents a severe calamity.

Another type of automatic support feature guards against cataclysmic stoppage of pacemaker due to the failure of logic circuits. In this circumstance, a voltage-dependent timing circuit or RC oscillator comes to rescue. The oscillator provides basic pacing support on the occurrence of microprocessor breakdown. Software errors, crystal debacle, or other disruptions are also taken care of. At any instant when the backup circuit cannot detect any heart-related event for a time span of 2.8 s, it takes over control. Immediately, it makes essential pacing support available to savethe precious human life.

14.15 Pacing Leads and Connectors

14.15.1 Lead Construction and Design

Success of a pacemaker implantation unquestionably depends on the pacemaker device. But the part played by the leads is no less critical and cannot be overlooked. Pacing leads have to tolerate the hostile environment inside the body for the long periods that pacemaker device resides inside the patient's chest. They cannot afford to be fragile because any handling by the implanting surgeon must not upset their function. Thus these leads must be rugged enough to last long and allow easy handling.

Construction-wise, every lead has four maincomponents. These are: (a) the electrode, (b) the conductor, (c) the insulation, and (d) the connector pins. In so far as lead design is concerned, the leads may differ vastly in design. Pacing electrode design is a major issue. Evidently, an adequate quantity of electrical energy must be brought together at the site of excitable tissue in order to stimulate thecardiac muscles. Therefore, a key matter of concern is the density of the current at the interface of the tip of this electrode with the tissue. This current density is influenced by several factors. To cite a few among these, the surface area of the electrode, pulse width, and pulse amplitude are important. The fibrotic encapsulation of the electrode must be considered.

Manufacturers have made genuine efforts to control some of the aforementioned factors. A small-diameter electrode provides increased current density. The result is a lower stimulation threshold. However, it has inferior sensing performance. Satisfactory pacing and sensing performance are achieved by making the tip of theelectrode porous. The tip contains thousands of pores of sizes 20–100 μm. Growth of tissue inwards into the pores increases the effective sensing area while preserving a small area for pacing function. An important improvement has been the steroid-eluting electrode. In this kind of electrode, 1 mg of an anti-inflammatory medicine such as corticosteroid (dexamethasone sodium phosphate) is included in the silicone core encircled by the material of the electrode. The leakage ofcorticosteroid into the heart muscle takes place gradually over several years. The steroid reduces the swelling and soreness caused by lead placement. Consequently, the alarming rise in pacing thresholds with nonsteroid electrodes observed over 8–16 weeks after implantation is avoided. Hence, there is a sizable decrease in energy necessities. A consistent depolarization of the heart is thus enabled. A smaller size battery can therefore suffice to give the same operational lifetime.

Endocardial leads are those tunneled through the venous system into the right atrium of the heart. These leads perform better than epicardial leads. The latter are fixed on the external surface of the heart.

On the opposite end of the lead to the electrode, there has been a drastic miniaturization of connectors. Some earlier connectors had large diameters of 5–6 mm. The pins were ~25 mm long. Connector diameters have been reduced to 3.2 mm. Pin lengths are less than 5.0 mm. Bipolar leads use two electrodes positioned in the heart. A coaxial connector is used for these leads. These leads require only a single receptacle, bringing about a remarkable reduction in the dimensions of bipolar connectors for pacemakers.

14.15.2 Lead Fixation Mechanisms

Postimplantation, the lead mustremain fixed to the myocardium. The fixation mechanisms are either passive or active. These mechanisms differ in the way the lead grasps and seizes the heart muscle. Leads using passive fixation mechanism are knotted with tines. These prongs or sharp points become intertwined with the weblike coating layer of the heart. Leads based on active fixation mechanisms employ corkscrew mechanisms for locking. The barbs or hooks secure firm attachment to the myocardium.

14.15.3 Lead Materials

The conductor must be sturdy but supple and bendable enough to forebear the stress exerted by the pounding heart. It is usually made of a nickel alloy called MP35N. This material is hardenable and strengthening with age, nonmagnetic, and alloy of nickel–cobalt–chromium–molybdenum. The alloy has an inimitable combination of properties. It has ultrahigh strength. Its toughness and ductility are remarkable too. Its corrosion resistance is outstanding. It is suitable for applications where a blend of large strength, high elasticity modulus values, and high-quality resistance against corrosive action are compulsory requirements.

Two prominent insulation materials for the leads are silicone and polyurethane. Silicone leads are thicker, mainly due to low tear strength. Also, the coefficient of friction of silicone is high. Due to this reason, two leads can pass through the same vein only with difficulty. The problem is avoided by applying a special coating on silicone during manufacture.

Electrode evolution has taken place from platinum–iridium electrodes having large-surface-area ~ 30–40 mm^2 to electrodes made of novel materials with moderately small-surface-area ~ 4–12 mm^2. The innovative materials are iridium oxide-coated titanium, titanium nitride, platinum black, etc. Vitreous or pyrolytic carbon coating a titanium, graphite core, etc.,has also been used.

14.16 Pacemaker Myths and Misconceptions

The incidence ofelectromagnetic interference is the lowest of any position tested when cellular telephones are used in the normal position, i.e., at the ear. Such low EMI does not produce any clinically significant meddling [15]. The patient may clasp the phone, keeping it close to the ear, but at a distance from the pacemaker [6]. The telephone should not be placed in a position over the pacemaker. In the on-state, the telephone should not be placed in a pocket, either overlying or in vicinity of the pacemaker.

People with pacemakers may work freely with different household appliances. Microwave ovens and power tools can be operated. Pacemaker patients are allowed to participate in many activities that are overwhelmingly strenuous and exhausting. They can play games like golf, tennis, or basketball. After seeking permission from their cardiologist, they may also take part in sports like marathons or scuba (self-contained underwater breathing apparatus) diving. They can safely move past airport security checks in the normal fashion without nervousness or hesitation. During traveling, they must always remember to carry the identification card given to them at the time of the pacemaker implantation. The pacemaker receiving patient must be clearly told that pacemaker is not a substitute for anyheart medications. It is not a treatment forhigh blood pressure. It is neither a remedy for angina, heart rhythm problems, etc., nor does it offer assurance of protection against cholesterol-laden plaque causing blockages in blood vessels that set off deadly heart attacks.

14.17 Discussion and Conclusions

Pacemaker industry has received a tremendous boost from the advances in many technologies. It has been catapulted by groundbreaking research in fields such as biomaterials, microelectronics, sensors, batteries, digital signal processing, and software developments. Modern pacemakers are smaller (23–30 g) than earlier devices and are fashioned in a less obtrusive, more physiological shape [16]. Duly supported by the knowledge explosion in cardiology, the pacemakers of today represent true marvels of science! [17, 18].

Review Exercises

14.1 Compare a present-day pacemaker with an early pacemaker, bringing out their important features and capabilities.

14.2 At what rate does the human heart normally beat? Does this rate change during exercise? Which area of the heart is called its natural pacemaker?

14.3 What are the main functions of an artificial pacemaker? Name its three principal parts.

14.4 What are unipolar and bipolar stimulating devices? In what ways do the current and voltage thresholds of a bipolar stimulating device differ from those of a unipolar one? Why?

14.5 Which lead configuration, unipolar or bipolar, provides a high signal-to-noise ratio? How does it do so? Which one is more prone to EMI? Which one is more flexible?

14.6 What is the typical voltage range of ECG signal? What is the bandwidth? What cardiac phenomena are represented by the QRS complex? What are the time intervals P–Q and S–T segments?

14.7 What is an arrhythmia? What are the two common causes of bradycardia? Mention five disease conditions in which pacemakers are used?

14.8 How are the pacemakers classified on the basis of the number of pacing leads? At what places are the leads placed in the different pacemaker classes?
14.9 A pacemaker delivers a pulse to the heart only when its natural beat is absent? What is this pacemaker called? How is it able to do so? How does the battery life of this pacemaker compare with that of a pacemaker always delivering pulses at a fixed rate?
14.10 What is the need of a pacemaker code? What kind of information does it provide?
14.11 What do the letters N, B, and G in the acronym NBG represent? What do the first three letters in this code indicate?
14.12 What information is furnished by the fourth and fifth letters in the NBG code?
14.13 Describe the transvenous and epicardial modes of pacemaker implantation? How is the lead position checked after implantation? What does the doctor do during post-operation follow-up of the patient?
14.14 What is a Stokes–Adams attack? When was the first artificial pacemaker implanted? On whom was it implanted? Where was it implanted?
14.15 What circuit in a pacemaker is used to detect a missing heart beat? How does it find out whether a beat is missing? Explain the roles of timing and telemetry circuits of a pacemaker?
14.16 How are higher voltages obtained from a small battery voltage? What is the circuit used for this purpose called?
14.17 What are various factors determining the life of the battery? How does the battery voltage decrease with its utilization?
14.18 What are the different approaches that have contributed toward miniaturization of modern era pacemakers?
14.19 What is meant by software-based pacemaker architecture? What is a programmable pacemaker?
14.20 A pacemaker is rate responsive. What does this mean? How is rate-responsiveness built into a pacemaker? Name the commonly used sensors used for this functionality?
14.21 Mention one automatic backup feature used in a pacemaker? How is it implemented?
14.22 What are the four main components of a pacing lead? What is the advantage of making the tip of a pacing lead porous?
14.23 What is a steroid-eluting electrode? How does it reduce battery consumption?
14.24 What are the two types of fixation mechanisms of leads to the myocardium? How do they differ?
14.25 Name two materials that are used for insulation of the pacing leads.

References

1. Bolz A (2012) Cardiac pacemaker systems. Handbook of medical technology. Springer, Berlin, pp 767–783

2. Martínez GA (2007) External pacemaker of diagnose and research, experience in military hospital center Mexico. In: Proceedings of the world congress on engineering and computer science (WCECS 2007), 24–26 Oct 2007, San Francisco, pp 12–15
3. Beyerbach DM. Pacemakers and implantable cardioverter-defibrillators. http://emedicine.medscape.com/article/162245-overview#aw2aab6b8. Accessed 13 Dec 2014
4. Bernstein AD, Daubert J-C, Fletcher RD et al (2002) The revised NASPE/BPEG generic code for antibradycardia, adaptive-rate, and multisite pacing. J Pacing Clin Electrophysiol 25(2):260–264
5. Love CJ (2006) Cardiac pacemakers and defibrillators, 2nd edn. Landes Bioscience, p 174
6. Wood MA, Ellenbogen KA (2002) Cardiac pacemakers from the patient's perspective. Circulation 105:2136–2138
7. Rassin M, Zilcha L, Gross DA (2009) Pacemaker in my heart—classification of questions asked by pacemaker patients as a basis for intervention. J Clin Nurs 18(1):56–62
8. Aquilina O (2006) A brief history of cardiac pacing. Images Paediatr Cardiol 8(2):17–81
9. Woollons DJ (1995) To beat or not to beat: the history and development of heart pacemakers. Eng Sci Educ J 4(6):259–268
10. Furman S (2003) The early history of cardiac pacing. Pacing Clin Electrophysiol 26(10):2023–2032
11. Ridgely P, Forde M (2006) Ch 54: Implantable cardiac pacemakers. In: Bronzino JD (ed) Medical devices and systems. CRC, Boca Raton, pp 54-1–54-12. https://doi.org/10.1201/9781420003864.ch54
12. Haddad SAP, Houben R, Serdijn WA (2003) An ultra low-power dynamic translinear cardiac sense amplifier for pacemakers. In: Proceedings of the IEEE international symposium on circuits and systems, ISCAS'03, Bangkok, 25–28 May, pp V-37–V-40
13. Sanders RS, Lee MT (1996) Implantable pacemakers. Proc IEEE 84(3):480–486
14. Haddad SAP, Serdijn, WA (2009) Ch2: The evolution of pacemakers: an electronics perspective. In: Ultra low-power biomedical signal processing: an analog wavelet filter approach for pacemakers. Springer, pp 13–31
15. Hayes DL, Wang PJ, Reynolds DW et al (1997) Interference with cardiac pacemakers by cellular telephones. N Engl J Med 336:1473–1479
16. Woodruff J, Prudente LA (2005) Update on implantable pacemakers. J Cardiovasc Nurs 20(4):261–268
17. Soontornpipit P (2012) An implantable biotelemetry review: technology for a cardiac pacemaker device. Asia J Publ Health 3(3):111–120
18. Mittal T (2005) Pacemakers—a journey through the years. Ind J Thor Cardiovasc Surg 21:236–249

Chapter 15
Implantable Cardioverter Defibrillators

Abstract Implantable defibrillators represent the most significant innovation to thwart sudden cardiac death caused by ventricular arrhythmias. Devices have also been designed for the treatment of atrial fibrillation. ICDs are generally implanted in subjects who have withstood more than one incident of ventricular tachycardia or fibrillation. They are also applicable in subjects whose clinical synopsis indicates highly probable, up-and-coming persistent ventricular tachycardia or fibrillation. Chosen highly suggestive subjects with atrial fibrillation too can benefit from ICDs. In contrast to pacemakers, ICDs are endowed with capability of dispensing electrical shocks of high energy to the heart. These shocks are imparted to set right serious, life-threatening, rapid, and sustained arrhythmias. Ventricular fibrillation and ventricular tachycardia are such arrhythmias. Atrial fibrillation too has devastating effects. Such abnormalities are unmanageable by pacing with electrical pulses of low energies and are often irredeemable if not bothered about. The battery chemistry of ICDs uses silver vanadium pentoxide. Large and bulky capacitors are necessary to change the 3–6 V battery output into the 750 V shock required to defibrillate a heart.

Keywords ICD · Defibrillation · Sudden cardiac death · VT · VF · Cardioversion · Epicardial/endocardial/subcutaneous lead · Arrhythmia · Single-chamber/dual-chamber ICD

15.1 Introduction

The predicament of sudden cardiac death (SCD) is multifaceted having many aspects and requiring multipronged attack. Many questions have been posed concerning its physiopathological mechanisms. These inquiries at the convergence of physiology with pathology are still evading answer [1]. Very unexpectedly and unpredictably, patients ailing with congestive heart failure pass away owing to arrhythmia leaving their beloved ones shocked and bewildered. This may happen even to patients undergoing confirmed beta-blockade medical therapies. Approaches developed specifically to prevent sudden death include treatment with amiodarone hydrochloride or

V. K. Khanna, *Implantable Medical Electronics*,
https://doi.org/10.1007/978-3-031-99868-3_15

shock delivery by an implantable cardioverter defibrillator (ICD). The competence of amiodarone to trim down death risk is uncertain and unresolved. Plain, shock-only ICD therapy provides better survival chances than afforded by modern drug therapy [2, 3].

A superfluity of medical researches has reinforced the benefits of ICD in clinical practice [4]. Implantable cardioverter defibrillator has metamorphosed the treatment of patients who stand at the risk of sudden cardiac death [5–7]. It is a small battery-powered, pager or a mini cassette-sized electronic device. Typically, it weighs approximately 85 g. It occupies 40 cm^3 volume. The location for its placement is underneath the skin in the chest of the patient. Surgery for ICD implantation is done under the influence of local anesthesia and sedation only.

The patients who are recommended ICD implant are primarily those facing the danger of sudden cardiac death (SCD). These are the patients suffering from cardiac ailments known as ventricular tachycardia or ventricular fibrillation. The term "ventricular tachycardia or VT" means rapid regular beating of the ventricles, the lower chambers in the heart. "Ventricular fibrillation or VF" implies rapid uneven pounding of the ventricles. It may be noted that the adjective "rapid" is common to both disease states, but in one case the heart beat is regular and faster than normal; in the other case, it is irregular and quicker than usual.

15.1.1 Explanation of VT and VF

VT is defined as three or greater than three ventricular extrasystoles in succession at a rate > 120 bpm (beats per minute). Often, it may exceed 150 bpm. An extrasystole is a heartbeat outside the normal rhythm of the heart that occurs in response to an impulse from its natural pacemaker, the sinoatrial node. It is a premature contraction of the heart arising from some part of the heart other than this node. The extrasystole is followed by a stoppage or pause period. During this period, the electrical system of the heart resets itself. After the pause, the heart contracts with greater force. This forceful contraction of the heart is termed palpitation.

In VF, the ventricles contract in a much disorganized, ineffective fashion. Essentially, the heart ceases to pump. If untreated, survival is unlikely. VT is often diagnosed and treated. But, unfortunately, VF almost always results in cardiac arrest and death if not stopped quickly. This kind of death is known as "sudden cardiac death." VT frequently degenerates into VF if prompt remedial action is not initiated. Thus, VT and VF are two ghastly and frightening heart arrhythmias that cause the heart to beat very fast, either regularly or irregularly.

15.1.2 Cardioversion

ICD continuously monitors the rhythm of the heart. It inspects and invigilates the heart beat with great caution. Any abnormality is unable to escape its scrupulous attention. It quickly identifies any abnormal heart rhythm. No sooner than it detects a very fast, abnormal heart rhythm, it supplies an electrical shock immediately to restore a normal rhythm of the heart. However, when the heart is beating normally, the device remains calm and quiescent. It wakes up from its state of dormancy only when called upon to do so. Thus, the overall function of ICD is described as "cardioversion." The term "cardioversion" implies the conversion of one cardiac rhythm to another. It is the transformation of an abnormal heartbeat into a normal heartbeat. In general, cardioversion implies the discharge of comparatively high-energy electrical shocks into or across cardiac tissue to apprehend a tachyarrhythmia of a heart chamber, i.e., a heart rate > 100 bpm [8]. The term "cardioversion" is also applied to the cessation of high rate tachycardias using electrical pulses or bursts of lesser energy. Defibrillation is the halting of atrial or ventricular defibrillation by higher energy shocks. It has been considered in the past as a form of cardioversion. ICD systems provide synchronous cardioversion shocks and/or asynchronous defibrillation shocks. Another technique known as antitachycardia pacing (ATP) involves the application of petite spurts of pacing impulses at rates greater by a 0.1–0.2 factor than the tachycardia. It terminates the arrhythmia in 0.6–0.9 fractions of episodes. Then shocks are unnecessary. The quality of life of the patient is improved. Moreover, the battery life of ICD is lengthened.

15.2 Difference Between ICD and Pacemaker

It must be clarified that both ICDs and pacemakers considered here are implantable medical devices. Further, it must be recapped that they are both implanted under the skin of chest or abdomen of heart patients (Table 15.1). But ICD is capable of imparting a greatly effective bump to the heart if it starts to beat in a dangerous way. This may be looked upon as a grueling, punishing jolt to bring back the departure of heart beat to normalcy. The pacemaker uses only low-energy pulses to prompt the slow beating heart. These pulses can be viewed as attempts to cajole and coax the heart to function properly, not as severe impacts.

In the newer-generation ICDs, the capabilities of supplying high-energy shock as well as low-energy pacing pulses are provided in the same device. These ICDs operate in two modes. They have modes for defibrillation, wherein the device imparts a high-intensity shock to reinstate regular rhythm of heart. They also have facilities for pacemaker function, wherein the device conveys a minute electrical incitement to the heart at predetermined rates.

With progress in underlying technologies, ICD sizes have continuously decreased. These defibrillators are made in considerably shrunken versions. Consequently, the

Table 15.1 Comparison between pacemaker and ICD

Sl. No.	Pacemaker	ICD
1	It is recommended in patients, whose heartbeat slows down to an unwholesome low rate, beating too leisurely	It is counseled in unambiguous patients who are at jeopardy for possibly grave ventricular arrhythmias
2	Its appearance is similar to ICD, but it is smaller in size than ICD	It looks much alike a pacemaker, except that it is a little bigger in size. Colloquially or informally, it is the pacemaker's big brother
3	It is implanted for treatment of less dangerous heart rhythms. These arrhythmias usually occur in the atria, the upper two chambers of the heart. It boosts up the slow heartbeats to standard level by supplying low-energy electrical pulses to restore normal heartbeat, i.e., to maintain the heart thumping at the appropriate rate	It treats fast, dangerous heartbeats because of a frightening rhythm disorder from the ventricles, the two bottom chambers of the heart. The ventricles start to quiver rather than contract strongly. Initially, it sends pulses of low energy to reestablish heart rhythm, but switches to pulses of higher energy when the low-energy shocks are unsuccessful in preventing death from a cardiac arrest
4	The low-energy pulses delivered by it do not produce any pain. They are faint and painless	Although lasting for only a fraction of a second, the high-energy pulses supplied by it are painful to the patient
5	It is not shocking the heart all the time. It only imparts feeble prompting signals to the heart muscle	It shocks the heart whenever it needs to be shocked
6	It is a device producing a mild action	It is a device causing an aggressive action

miniaturized defibrillators are almost always implanted in the pectoral area rather than in the abdomen, as they used to be done earlier. This is highly convenient and opportune for the patient. It has also lowered implant morbidity. Long-term lead-related complications have also faded away and disappeared.

15.3 Necessity of ICD

Questions that come to the mind of the reader are: What is the problem if the defibrillation shocks are delivered from outside? Why is it necessary to implant the defibrillator inside the body? Of course, the defibrillation shocks can be given externally, but it is a matter of how soon they are provided in order to be effective. A patient may suffer from ventricular fibrillation, a rapid and disorganized activation of the ventricles in the heart, while in home, in office, on the road, or anywhere, and help to deal with medical emergency may take time to reach the patient. But this disease does not forgive any deferments. It must be stated that despite the progress in exigency medical systems and expertise of resuscitation techniques, sudden cardiac death almost always follows ventricular fibrillation [9]. Survival rates are very low for persons who have cardiac arrest outside the hospital [10]. Even the patients who recover after resuscitation may writhe with severe, lasting weakening of cognitive and motor functions. This happens due to holdups that are often incurred before restoration of a steady rhythm. Therefore, an implantable device is essential to keep an eye on and interpret cardiac rhythm like a robot and to provide necessary shocks for defibrillation within no time after detection of ventricular fibrillation. Consequently, the implantable cardioverter defibrillator has matured from a therapy of last opportunity for patients suffering from frequent cardiac arrest to a standard for management of arrhythmias. This standard is useful in crucial prevention of a life-threatening event occurring for the first time. It is also used in secondary prevention cases. It is a tool preventing the repetition of a likely lethal arrhythmia or cardiac arrest in patients who have a history of coronary heart disease.

Patients who can potentially be benefitted from an ICD are separated into two broad categories [11]:

(a) Secondary Prevention: This category includes patients who have endured a dangerous ventricular arrhythmia. Also reckoned are the patients who have sustained ventricular tachycardia (VT). For these patients, the ICD is implanted for the secondary avoidance of sudden cardiac arrest. Sustained VT is usually defined as VT instigating hemodynamic symptoms, chiefly fainting, pre-fainting, or chest pain prolonging for more than 30 s.
(b) Primary Prevention: This category consists of patients who have hitherto not suffered sudden cardiac arrest, albeit stand at high risk for the same. In this case, an ICD is implanted for prevention of this mishap as a primary facility.

15.4 Historical Background

Drs. Michel Mirowski and Morton Mower and their coworkers were distraught and gloomed by the death of a close associate and advisor. The mentoring colleague had been admitted to the hospital with complaints of persistent ventricular tachyarrhythmias. Stricken by this grief, they spearheaded the development of a life-prevention device in the late 1960s [12]. A novel idea evoked from their disenchantment with the capabilities of the treatments at hand during those times for individuals poised at high death risk. They conceptualized an implantable device. This device would incessantly watch the heart rhythm and provide corrective action in the event of any deviation of the rhythm from its course. For the correction procedure, the device would be embellished with the ability to transport shocks for defibrillation whenever ventricular tachyarrhythmias occurred. In the decade 1970–1980, many tentative, investigational models were fabricated and improved by these workers. After years of vigorous testing and evaluation, the first implantation was accomplished in 1980 in a young woman agonized with recurrent ventricular fibrillation. This woman had experienced two preceding cardiac arrests. Several years elapsed and the ICD treatment was offered in only a few centers. It was mainly aimed at persons with recognized cardiac arrest due to ventricular fibrillation. Commercial defibrillation devices were approved by the US Food and Drug Administration in 1985. Expertise of these devices grew by leaps and bounds and so did validation of the usefulness of defibrillators to terminate malignant ventricular arrhythmias. Since that time, ICDs have become the preferred treatment for patients at high danger from terrifying arrhythmias. This is attributed to rapid refinements and sophistications in device technology. Stockpiling of support from randomized medical assessments also helped the enterprise. Loss of confidence in the universal efficacy of drug therapy further promoted the adoption of ICDs. From the treatment of last resort, ICDs have advanced as an embodiment of supreme standard therapy for patients at high threat for ventricular tachyarrhythmias. High-risk patients are those who have endured life-threatening arrhythmias. They include persons with heart diseases who are in fear of such arrhythmias, but have not shown any symptoms.

15.5 ICD Construction

Broadly, the implantable defibrillator consists of two main components: (a) the pulse generator and (b) one or more leads for pacemaker function and electrodes for defibrillation. The pulse generator is housed in a sealed titanium can. This pulse generator can is the abode of electronic circuitry and power supply. It contains microprocessors and associated integrated circuitry for signal filtering/analysis of the cardiac rhythm and the delivery of pacing pulse or shock. Memory chips inside the can are used to store electrograms and patient data. A module is included for telemetry through bidirectional transmission of information. The can encloses a Li–Ag–V_2O_5 battery with DC–DC voltage converters for power conditioning. It also contains aluminum or aluminum chloride electrolytic capacitors to store charges. On the top cover of the pulse generator can lies a header made of epoxy resin. Leads for pacing and defibrillation leads are connected here. The leads must be capable of transferring shocks of high energy to the heart without any damage to the myocardium.

15.6 Epicardial Versus Endocardial (Transvenous) Lead Systems

Early defibrillators used epicardial patches, but transvenous leads are now more commonplace (Table 15.2).

Figure 15.1 shows epicardial ICD, and Fig. 15.2 depicts endocardial ICD.

Another ICD category is subcutaneous ICD (Fig. 15.3). In subcutaneous ICD implantation [13], the lead is stationed in the tissue of the chest underneath the skin. Its orientation is in the vertical direction. It is kept parallel to and at a distance 1–2 cm to the left sternal midline. Horizontally, its position is at the level of the sixth rib, until reaching the left anterior axillary line. The lead carries an 8-cm shock coil.

Table 15.2 Epicardial and endocardial leads

Sl. No.	Epicardial leads	Endocardial leads
1	They offer superior distribution of the defibrillating electric field	Their field distribution is inferior
2	They need less energy for defibrillation	They need greater energy for defibrillation. For adequate defibrillation, separate shocking components were often demanded. Examples are superior vena cava and coronary sinus leads. Besides, subcutaneous patches or arrays are also used. These weaknesses have been largely overcome by advancement in two areas. First area is lead design. Another area pertains to waveforms used for defibrillation
3	They require a thoracotomy for inserting the device. Thoracotomy is a riskysurgical procedure	They have avoided the need for thoracotomy. They provide demoted implant morbidity, along with improved long-term lead reliability

The coil is surrounded by two sensing electrodes. The pulse generator is twice the size of the traditional type. It is placed in the tissue of the chest beneath the layers of the skin. Its location is above the sixth rib flanked by the left midaxillary and the left anterior axillary lines.

15.7 Arrhythmia Detection

The obvious question at this stage is: What is the sure sign of the occurrence of a ventricular arrhythmia for which a heavy shock is urgently required. Otherwise, there are chances of false alarms and unnecessary wasteful shocks leading to great discomfort of the patient. To this intent, two rhythm characteristics are widely used for detection of a ventricular arrhythmia. One characteristic is heart rate; the other is duration of arrhythmia, typically 1–3 s. These characteristics can be programmed according to the requirements of the individual patient. The rate indicator makes a distinction between a tachyarrhythmia and a normal cardiac rhythm. The duration criterion enables the revealing of continuous incidents only.

During ventricular fibrillation, the bipolar electrogram has small amplitude or its amplitude is unstable. For detection of such low-amplitude signals, a specially designed method called the dynamic sensing technique is employed. In this technique, the ICD allows sensitivity-gain adjustment. It transforms the amplitude of the intracardiac signal in the range which the device can detect during the full

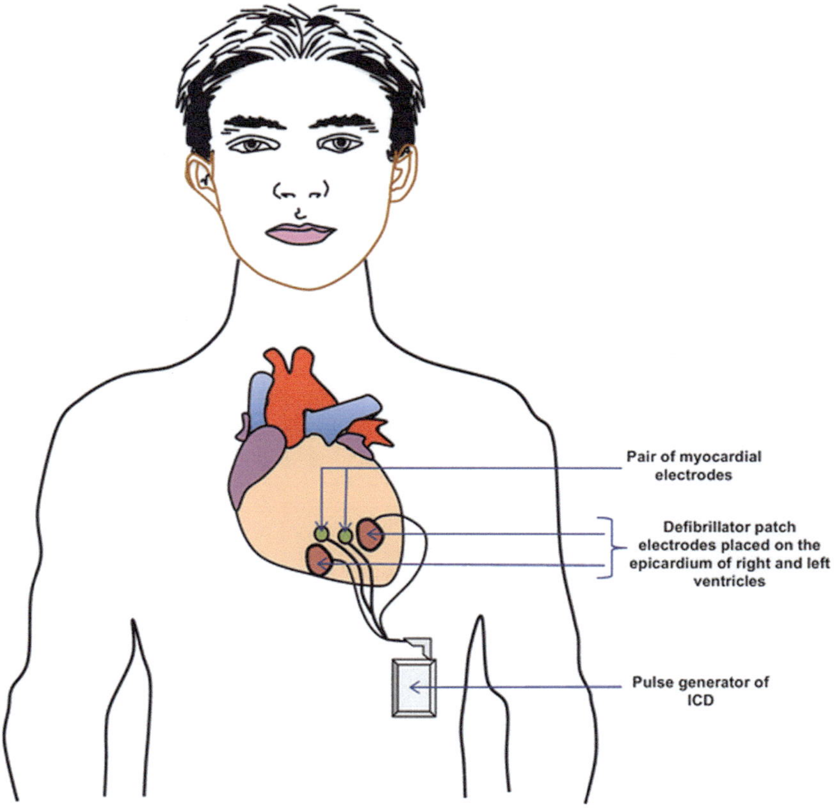

Fig. 15.1 Epicardial ICD implantation

cardiac cycle. By employing this dynamic gain or sensitivity threshold, QRS detection becomes permissible. After QRS detection, the sensitivity is decreased. Immediately, it begins to decrease to the value that was programmed. Hence, T-wave counting is eschewed while consenting to extremely responsive detection during a major part of the cardiac cycle. Thus the recognition of a ventricular fibrillation event is facilitated.

15.8 Detection Zones

ICDs offer several detection zones that are classified on the basis of the seriousness of arrhythmias. Independently programmable treatments are ascribed to each zone. The swift tachycardias, e.g., ventricular fibrillation, are accorded the highest priority of attention. They are treated more aggressively. An immediate shock is the only retort. The slower ventricular tachycardias are retaliated with overdrive pacing which is not

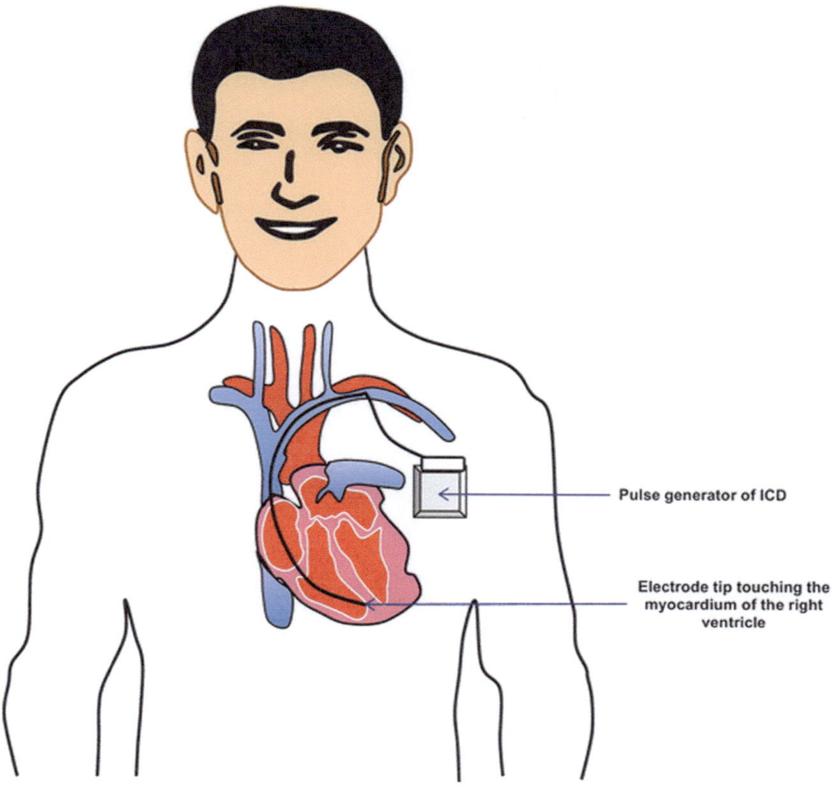

Fig. 15.2 Endocardial (transvenous) ICD implantation

painful. Frequently, the requirement for a shock is eliminated. By further division of the ventricular tachycardia region into low speed and high speed, dissimilar treatments are applied to ventricular tachycardias of dissimilar speeds. Thus, ICDs deal with various kinds of malfunctions in different ways. Obviously, a high-intensity shock is not necessarily the cure for all disorders. It may be superfluous on some occasions.

15.9 Algorithms for Detection of Arrhythmias

Distinguishing between different types of arrhythmias is possible by step-by-step enunciation of the procedure to be followed in reading the patterns of ECG signals. For solving this problem, ICDs use various algorithms which continuously work on the ECG signals to classify whether a certain characteristic of ECG waveform represents a benevolent or a dangerous situation. These algorithms depend on the type of ICD, whether single or dual chamber. The former uses information from its

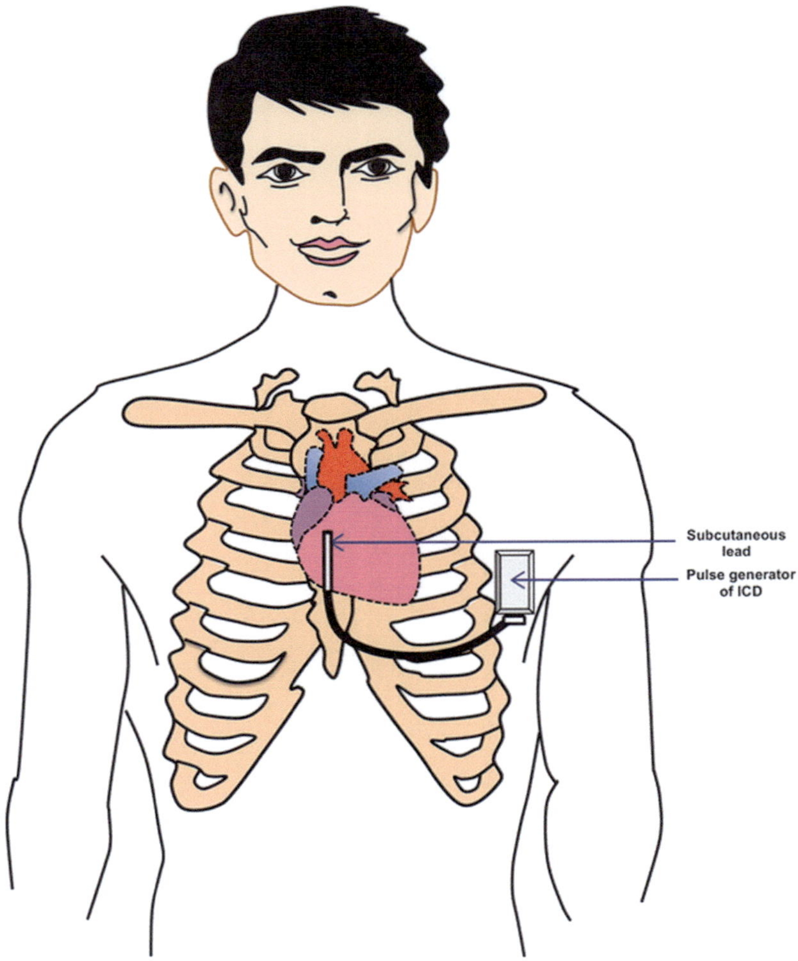

Fig. 15.3 Subcutaneous ICD implantation

single lead, whereas the later operates on information supplied by both the leads. The algorithms and computer programs aid in arriving at the correct decision regarding the degree of threat to the patient.

Algorithms also vary among devices, depending on the manufacturer. However, they share some general trends, which will be discussed here.

15.9.1 Algorithms for Single-Chamber ICDs

To discriminate whether the arrhythmia is ventricular or supraventricular, a single-chamber ICD uses enhancements. These are based on rhythms of cardiac patterns or QRS waveform, its shape, and its structure. Features unique to an arrhythmia are carefully scrutinized in identifying its outbreak. It is beneficial to find whether the onslaught of arrhythmia was gradual or sudden, whether the arrhythmia cycle was stable or unstable, or whether the intracardiac QRS width was normal or unduly long.

Unanticipated and astonishing outburst of arrhythmia aids to tell apart the episode of a ventricular tachycardia from that of sinus tachycardia. A ventricular tachycardia has a fast start, while sinus tachycardia is marked by a slow beginning.

Constancy of the length of arrhythmia cycle is a vital characteristic which assists in separating the regular ventricular tachycardia from the usually irregular atrial fibrillation. For demarcating between different QRS wave morphologies, a comparison is performed between the intracardiac electrogram of the QRS complex in tachyarrhythmia with the form and structure of QRS for the patient during normal sinus rhythm. Resemblance of the two electrograms supports the development of a rapid supraventricular tachycardia. Divergence between electrograms is an alert for ventricular tachycardia.

The intracardiac QRS width is a parameter which is used to extricate ventricular from supraventricular tachycardias. In case of a broad QRS complex tachycardia, there is a definite probability of ventricular tachycardia.

15.9.2 Algorithms for Dual-Chamber ICDs

As already said, detection algorithms for tachyarrhythmia ICDs exploit the information obtained concomitantly from the two leads: the atrial lead and the ventricular lead. During atrioventricular dissociation, the atria and ventricles are not activated synchronously. This condition forewarns of ventricular tachycardia if the ventricular rate is >the atrial rate. Atrioventricular association indicates the existence of a supraventricular rhythm. If fibrillation has been recognized in the atria, a jagged, uneven ventricular rhythm can only be assigned to atrial fibrillation.

15.10 Therapies Administered

There are three principal ways to terminate *ventricular arrhythmias*. These are by delivering synchronized shock (cardioversion), by imparting nonsynchronized shocks (defibrillation), or by overdriving (antitachycardia pacing).

Tachyarrhythmias within the scope of ventricular fibrillation are dealt with by furnishing instant defibrillation.*Ventricular tachycardias* are treated by supplying concatenation of overdrive pacing. This is especially true for the slower types. As a repercussion to failure of overdrive pacing, cardioversion is done at low energies. However, circumstances may arise wherein the rhythm worsens or proves to be unmanageable by employing less forceful steps. In the aftermath, defibrillation is imperative.

Ventricular fibrillation or fast speed ventricular tachycardia requires defibrillation. In ventricular fibrillation, the efficacy of defibrillation is > 98%.

For stopping*ventricular tachycardias* that are not controlled by overdrive pacing, low-energy cardioversion is frequently a solution. Many monomorphic tachycardias (in which all the QRS waves look the same) are brought to a halt with shocks of 1 J or less. Because arrhythmias are sometimes hastened by low-energy shocks, low-energy cardioversion should at all times be succeeded by high-energy shocks.

In existing devices, facility for administration of 4–8 consecutive shocks is available. Maximum energies of the shocks imparted lie in the range 25–42 J. Using endocardial leads with biphasic waveforms, this energy is adequate to defibrillate the mainstream of patients. Actually, the average energy required for victorious defibrillation is ~ 10 J. To ensure accomplishment, a methodical testing of the system is done during implantation. Induced fibrillation must be quickly replicated by an energy that is a minimum 10 J less than the utmost output of the device. Defibrillation should be reproduced with the standard 10 J safety allowance. If it is not obtained, suitable changes are indispensable and have to be carried out. The lead configuration may be altered. The electrodes may be repositioned. More defibrillating elements may be added. However, this is hardly ever the case with biphasic waveforms.

15.11 Postimplantation Patient Follow-Up and Monitoring

For postimplantation look-after and well-being of the ICD-receiving patient, the patient is required to report for outpatient visits at regular intervals. These visits are generally scheduled after every one month. In some cases, patients are called every three months. During these visits, the doctor examines whether there were any variations in rhythmic patterns of heart signals during the reporting period. If they happened, he/she studies the details of type of rhythmic disturbances that actually occurred. He/she also looks at the adequacy/redundancy of necessary electrical shocks that were delivered, whenever they were called for. It is important to know whether these shocks were actually useful or prodigally given? On the whole, the doctor evaluates the effectiveness of the treatment provided and tries to find whether the measures taken are sufficient. Depending on these inputs, the doctor makes suitable amendments/modifications in the programmed parameters to obtain the correct settings.

The above patient monitoring is performed in a noninvasive fashion. For this purpose, a programming wand is applied over the chest region of the patient. Application of the wand establishes wireless messaging from the ICD with a computer which is externally located. Information flow in both directions between the ICD and computer enables the monitoring.

During patient checkups, the battery condition is always determined. When the battery energy diminishes to a preset level, ICD replacement is compulsory. No chances of insufficient power to energize the ICD are allowed for dealing with any cardiac emergency.

The patients with ICDs have to adhere strictly to many restrictions in lifestyle as instructed to them by their doctors. Such patients are prohibited from undergoing MRI procedures. However, they are allowed to undertake X-ray scanning. Doorways of stores fitted with electronic theft detection devices should be avoided. The patients should not stand close to them. Nor they should stand in the vicinity of airport security chambers. Holding magnetic items next to their unit is forbidden. Stereo speakers are a common example of such items. Also, they should never hold cellular telephones against the device. Some ICD patients may be advised to avoid situations or tasks in which they or someone else could be injured in case they suffer from dizziness or happen to lose consciousness. On a case-by-case basis, sports requiring tiresome exercises such as diving, piloting, athletics, or similar activities may be allowed on a limited format. It is likely that even after ICD implantation, antiarrhythmic drugs may be prescribed to some patients.

Among the adverse effects of implantation may be mentioned the risks of infection at the implant site. The device may be visible from outside as a lump below the skin. So, cosmetic concerns arise. In addition, psychological apprehension and discomfort of electric shocks always persist. Multiple shocks may pose challenging medical and psychological management issues. These issues may be troublesome for the attending healthcare providers [14–16]. ICD-related suspicions and reservations, e.g., unnecessary worry, are the most familiar signs shown by ICD beneficiaries [17]. Patients must be imparted supplementary education and knowledge about further care after discharging from hospital [18]. At some places, support groups have been formed. In these groups, the ICD patients can share and discuss their problems with medical experts. These discussions allay many vague and imaginary fears which haunt the patients.

15.12 Discussion and Conclusions

An ICD prolongs life in patients who have either experienced or are prone to serious abnormal heart rhythms, often resulting from a damaged heart [19, 20]. The ICD is superior to conventional pharmacological therapy and has established its supremacy over medicines. There are several strong forecasters of death in ICD patients. Some of these are the age of the patient, renal dysfunction, and chronic obstructive pulmonary disease. Diabetes and peripheral vascular disease make the situation more murky [21]. The older age group is less active from physical standpoint. They are less pleased with its physical operation. They also show a little more fretfulness than their younger counterparts [22]. Among the aged patients, ICD intervention may not be found to be economical. It may not prove to be worth its cost. But the procedure may become cost-effective in those patients who are likely to live for > 5–7 years after implantation [23, 24]. Widespread use of ICDs for primary prevention will encumber and burden the financial blueprint of several healthcare systems. This is because of the prohibitively high price of each ICD device. Further, a sizable population of patients is potentially eligible to receive ICDs [25]. Over and above, there are extensive variations in implantation rates for ICDs. The main reason is that a common or uniform policy for sudden death prevention is not laid down. Even in a terrestrial region where the broad-spectrum level of health care is progressive and well esteemed by the people, gross variability in implant rates is encountered [26, 27].

Review Exercises

15.1 Differentiate between ventricular tachycardia and ventricular fibrillation. Which is more likely to lead to death if not treated immediately?

15.2 What is meant by cardioversion? How is it achieved? What is defibrillation? How does antitachycardia pacing differ from the above two modalities?

15.3 How does a defibrillator differ from a pacemaker? Which device caters to the need of a more life-threatening situation? Will a pacemaker help in such a circumstance?

15.4 What motivated Drs. Michel Mirowski and Morton Mower toward the development of a device that will keep a vigil on heart's activity and deliver a high-voltage shock in cardiac emergency. Describe the early phase of defibrillator progress.

15.5 Name the two main components of a defibrillator? Explain their functions. What is the name of the material from which the outer case of a defibrillator is made? What kind of battery is used in the defibrillator?

15.6 Which of the two kinds of leads, epicardial or transvenous, gives better electric field distribution? Which one needs less energy for defibrillation? Which one is implanted by thoracotomy?

15.7 Is the same kind of treatment always applicable to ventricular tachycardias of different rates? If not, what are the different possibilities?

15.8 What rhythm characteristics are used to detect ventricular arrhythmias? What is dynamic sensing? Why is it necessary? How is it done?

15.9 How does a single-chamber ICD differentiate among different arrhythmias? What kind of information is furnished by dual-chamber detection algorithms?

15.10 Describe the three ways to manage ventricular arrhythmias and the manner of their application under different conditions of the heart.

References

1. Hu K (2006) Prevention of sudden cardiac death. Internist (Berl) 47(10):1040–1049

2. Kadish A, Dyer A, Daubert JP et al (2004) Prophylactic defibrillator implantation in patients with nonischemic dilated cardiomyopathy. N Engl J Med 350:2151–2158
3. Bardy GH, Lee KL, Mark DB et al (2005) Amiodarone or an implantable cardioverter–defibrillator for congestive heart failure. N Engl J Med 352(3):225–237
4. Ghislandi S, Torbica A, Boriani G (2013) Assessing the outcomes of implantable cardioverter defibrillator treatment in a real world setting: results from hospital record data. BMC Health Serv Res 13:100–109
5. Josephson M, Wellens HJ (2004) Implantable defibrillators and sudden cardiac death. Circulation 109(22):2685–2691
6. Masoudi FA, Go AS, Magid DJ et al (2012) Longitudinal study of implantable cardioverter-defibrillators: methods and clinical characteristics of patients receiving implantable cardioverter defibrillators for primary prevention in contemporary practice. Circ Cardiovasc Qual Outcomes 5(6):e78–e85
7. Russo AM (2014) Are we improving care of Medicare patients undergoing primary prevention implantable cardioverter-defibrillator implantation? Circulation 130(10):808–810
8. Plombon WJ, Stroebel JC (2003) Implantable cardioverter/defibrillator. US 20030004552A1, 15 p
9. Huikuri HV, Castellanos A, Myerburg RJ (2001) Sudden death due to cardiac arrhythmias. N Engl J Med 345:1473–1482
10. DiMarco JP (2003) Implantable cardioverter–defibrillators. N Engl J Med 349(19):1836–1847
11. Deyell MW, Tung S, Ignaszewski A (2010) The implantable cardioverter-defibrillator: from Mirowski to its current use. B C Med J 52(5):248–253
12. Glikson M, Friedman PA (2001) The implantable cardioverter defibrillator. Lancet 357:1107–1117
13. De Maria E, Olaru A, Cappelli S (2015) The entirely subcutaneous defibrillator (s-icd): state of the art and selection of the ideal candidate. Curr Cardiol Rev 11(2):180–186
14. Sears SF Jr, Conti JB (2003) Understanding implantable cardioverter defibrillator shocks and storms: medical and psychosocial considerations for research and clinical care. Clin Cardiol 26(3):107–111
15. Stutts LA, Cross NJ, Conti JB et al (2007) Examination of research trends on patient factors in patients with implantable cardioverter defibrillators. Clin Cardiol 30:64–68
16. Sears SF, Matchett M, Conti JB (2009) Effective management of ICD patient psychosocial issues and patient critical events. J Cardiovasc Electrophysiol 20(11):1297–1304
17. Camm J, Sears SF Jr, Todaro JF et al (1999) Examining the psychosocial impact of implantable cardioverter defibrillators: a literature review. Clin Cardiol 22(7):481–489
18. Palacios-Ceña D, Losa-Iglesias ME, Alvarez-López C et al (2011) Patients, intimate partners and family experiences of implantable cardioverter defibrillators: qualitative systematic review. J Adv Nurs 67(12):2537–2550
19. Reiffel JA, Dizon J (2002) The implantable cardioverter-defibrillator: patient perspective. Circulation 105:1022–1024
20. Birnie DH, Sambell C, Johansen H et al (2007) Use of implantable cardioverter defibrillators in Canadian and US survivors of out-of-hospital cardiac arrest. Can Med Assoc J 177:41–46
21. Alba AC, Braga J, Gewarges M et al (2013) Predictors of mortality in patients with an implantable cardiac defibrillator: a systematic review and meta-analysis. Can J Cardiol 29(12):1729–1740
22. Hamilton GA, Carroll DL (2004) The effects of age on quality of life in implantable cardioverter defibrillator recipients. J Clin Nurs 13:194–200
23. Barra S, Providência R, Paiva L et al (2015) Implantable cardioverter-defibrillators in the elderly: rationale and specific age-related considerations. Eur J Pacing Arrhythm Card Electrophysiol 7(2):174–186
24. Groeneveld PW, Farmer SA, Suh JJ (2008) Outcomes and costs of implantable cardioverter-defibrillators for primary prevention of sudden cardiac death among the elderly. Heart Rhythm 5(5):646–653

25. Hlatky MA, Mark DB (2007) The high cost of implantable defibrillators. Eur Heart J 28(4):388–391
26. Boriani G, Berti E, Biffi M et al (2010) Implantable electrical devices for prevention of sudden cardiac death: data on implant rates from a 'real world' regional registry. Eur J Pacing Arrhythm Card Electrophysiol 12(9):1224–1230
27. Lubinski A, Bissinger A, Boersma L et al (2011) Determinants of geographic variations in implantation of cardiac defibrillators in the European Society of Cardiology member countries—data from the European Heart Rhythm Association White Book. Eur J Pacing Arrhythm Card Electrophysiol 13(5):654–662

Chapter 16
Deep Brain Stimulation

Abstract Deep brain stimulation is used as a substitute for permanent neuroablative procedures (destruction/inactivation of nerve tissue by surgery, injections, lasers, etc.) in the management of disorders associated with movement, notably Parkinson's disease, essential tremor, and dystonia. The technique involves stereotactic placement of an electrode into the identified area of the brain and delivering electrical pulses to that area. "Stereotactic surgery or stereotaxy" relates to "stereo-axis," a combination of "stereo" and "taxis," originating from "stereo" meaning "three-dimensional" + Greek "taxis" meaning "orientation," to direct the tip of an electrode in the brain. For the treatment of Parkinson's disease in sophisticated phase, the target area of the brain is subthalamic nucleus (STN). For medically refractory tremor, the focal area is ventral intermediate nucleus (Vim) of the thalamus. For both cervical and generalized dystonias and Parkinson's disease, the area of interest is the globus pallidus internus (GPi). Therapeutic stimulation parameters generally used are amplitude, pulse duration, and frequency. The high-frequency stimulation (100–200 Hz) given in this method impersonates the effects of surgical ablation. The method combines the advantages of adjustability of stimulus parameters with the reversibility of treatment. These merits have enabled deep brain stimulation to largely supersede the ablation practice.

Keywords DBS · Subthalamic nucleus · Globus pallidus · Parkinson's disease · Dystonia · Tremor · Lesioning · Depression · Obsessive–compulsive disorder

16.1 Introduction

Deep brain stimulation (DBS), sometimes called neuromodulation, is a reversible, stereotactic neurosurgical procedure (method for locating points using an external, three-dimensional Cartesian coordinate system as a reference frame), whereby multi-contact unilateral or bilateral electrodes (affecting only one side or both sides of an organ) are permanently implanted in specific anatomical locations of the brain (the hypothalamus, thalamus, globus pallidus, or subthalamic nucleus); the electrodes are

V. K. Khanna, *Implantable Medical Electronics*,
https://doi.org/10.1007/978-3-031-99868-3_16

connected to an implanted, battery-powered pulse generator in the chest; and with the help of these electrodes, continuous electrical stimulation of programmed intensity and frequency is administered to the specific, targeted sites into the deep structures of the brain [1, 2].

In action, the DBS implant is similar to a cardiac pacemaker. It seeks to do for the brain all that a pacemaker does for the heart. Hence, it parallels the cardiac pacemaker function in the brain. In size, the DBS implant is approximately the extent of a stopwatch. Figure 16.1, showing the parts of the human brain will aid in comprehending the deliberations in this chapter.

The precise location of electrodes varies with the type of disorder. Several parameters, including widths of electrical pulses and intensities of stimulating currents, are adjusted for every patient by skilled health professionals. Other options are mode of stimulation such as monopolar or bipolar stimulation and polarity of the electrode. The latter depends on which contact of the quadripolar lead is positive or negative. The intention is to optimize improvement in movement disorder. At the same time, any side effects must be reduced to the minimum proportions.

DBS is a well-established method to cure the debilitating motor symptoms of pharmacologically refractory movement disorders. Infamous among these is the Parkinson's disease. Especially in end-stage drug-resistant patients, Parkinson's disease is very difficult to cope with. Other prominent diseases for which DBS is a routine treatment method are essential tremor and dystonia. The DBS treatment is associated with minimal morbidity. It leads to striking improvements in the ability

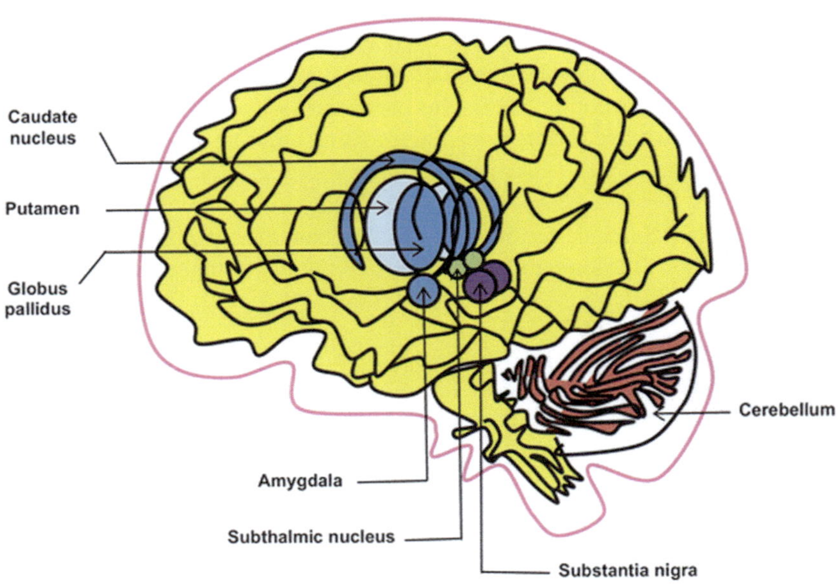

Fig. 16.1 Structure of the human brain

to carry out muscle-and-nerve acts (motor functions) and quality of life being led by the patient.

Apart from movement disorders, many have talked about the impending benefit of DBS of chosen regions of the brain to successfully alleviate various other neurological and psychiatric conditions. These include treatment-resistant depression (TRD) unresponsive to at least two antidepressant courses and obsessive–compulsive disorder (OCD) [3]. However, there is paucity of data to speculate that deep brain stimulation is out of harm's way and/or valuable for controlling these diseases. This is because of limited studies and small sample sizes used.

16.2 Movement Disorders

16.2.1 Parkinson's Disease (PD)

It is a chronic (long-term), progressive (gradually worsening), neurodegenerative (structure or function loss or even death of neurons) disorder caused by dying dopamine-generating cells, known as dopaminergic neurons in the *substantia nigra* located in midbrain. This death and disappearance of cells results from vascular (relating to blood vessels) or inflammatory changes taking place in the basal ganglia. The basal ganglia (Fig. 16.2) are a collection of nuclei on both sides of the thalamus, a large, dual-lobed symmetrical structure composed of gray matter and lying deep within the brain. The thalamus is involved in relaying sensory and motor signals. Consciousness and sleep regulation also fall under its dominion. Dopamine is the neurotransmitter responsible for conveying messages to the division of the brain in-charge of controlling the motion and coordination of muscles. As the disease advances, the quantity of dopamine formed in the brain declines.

Dopamine deficiency produces situations of hyperactivity and underactivity. The indirect basal ganglia pathways become hyperactive, while the direct basal ganglia pathways become overactive. The result is increase in reticence of the thalamus. As a result, activity of motor cortex is reduced. Due to these changes, the person is unable to control movement in a normal fashion. This leads to tremor of parts of the body, as seen not only in hands, arms, and legs but also in jaw and face. It also causes bradykinesia or reduction in and slothfulness of movement. Other symptoms are muscular rigidity. The limbs and trunk become stiff. The overall consequence is instability of posture or disabling of balance and coordination. Often an unusual, stooping or droopy posture and shuffling gait of the patient are seen. Apart from above, the patient also acquires a fixed, inexpressive facial expression. This is attributed to the poorer control over muscle movement. Additional disturbances are related to autonomic, sensory, sleep, cognitive, and psychiatric aspects. The disease degrades the patient's voice and sense of aroma. Interestingly, the disease is likely to occur to a person of any age group. But middle-aged and elderly people are the most sufferers. It afflicts 0.01 fraction of the population >55 years in age.

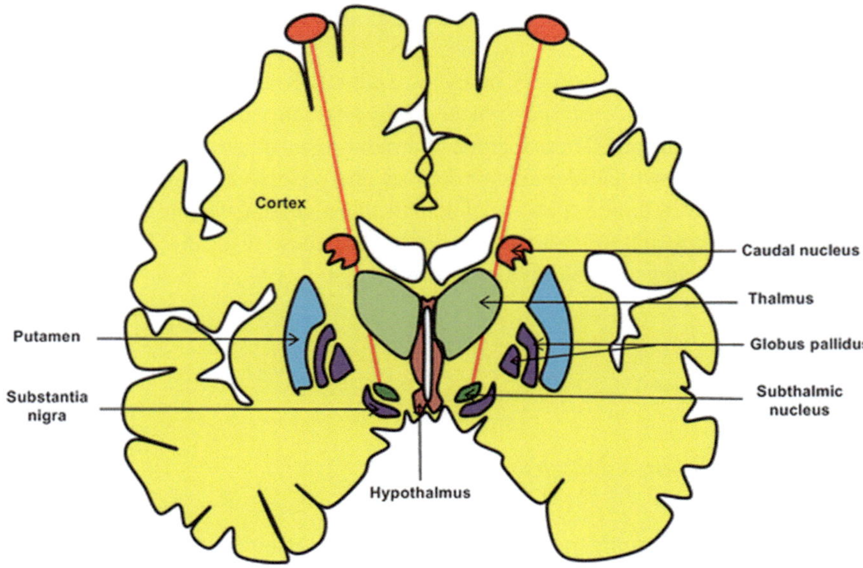

Fig. 16.2 The basal ganglia. Also shown in this diagram are the two electrodes inserted from the scalp and penetrating deep into the brain for electrical stimulation

During the first few years, the disease responds to replenishment of dopamine with the drug levodopa. However, in later years, its management with levodopa becomes difficult. The treatment is convoluted by drug intolerance and unwanted side effects. Stated in other words, usually, treatment for Parkinson's disease with levodopa is initially noticeably effective. But the same does not hold true subsequently and after long-term use.

16.2.2 Tremor

Tremor is an involuntary quivering movement. It is seen in the form of an uncontrollable trembling or shaking motion. This motion is usually rhythmic, back-and-forth, irregular, and unpredictable. Even normal persons experience a very trifling tremor when performing their daily chores. Sometimes, the everyday level of tremor becomes noticeable. It is particularly obvious in older people, e.g., it is clearly visible in people having trouble in writing or making a drawing. Difficulties may be noticed in holding utensils, e.g., a fork. Such manifest tremor is also normal, and it is often caused by emotional stress, anxiety, or anger. But in a few people, tremor is a warning sign of neurodegenerative diseases. Examples of these diseases are multiple sclerosis, stroke, etc. Tremor is seen in patients suffering from traumatic brain injury. This injury damages parts of the brain stem or the cerebellum. Tremor appears as an unwanted effect of certain drugs. Certain drugs such as corticosteroids and those used for treating a few psychiatric disorders are responsible for inducing tremor. Allied conditions are alcohol abuse or its withdrawal and poisoning by mercury. Overactive thyroid or failure of liver leads to similar conditions. Some forms of tremor are hereditary and run down generations. Causes of others are not known.

Essential tremor is the most common type of tremor. Up to 10 % of people are affected by it as they become old. People above 40 years of age are most commonly affected. But a remarkable increase in its prevalence is seen in the age group over 70. The tremor is more likely discerned in the hands, arms, head, or eyelids of the patient. But it occurs in the legs or feet infrequently. A tremor patient experiences difficulty in holding or using small objects such as silverware or a pen. A commonly used drug is propranolol, a beta blocker. Another drug is primidone, a drug for seizures.

16.2.3 Dystonia

The word "dystonia" is derived from Greek. It means altered muscle tone. Dystonia is a movement disorder. This disorder is observed as an irrepressible contraction of the muscles of the body. These contractions make the affected body part to twist involuntarily. So, recurring movements are produced. Often atypical or nonconforming postures are seen. About 1 % of the population is affected by dystonia. Women are more susceptible to dystonia than men. The causes of dystonia are not utterly known. However, they appear to be linked to changes in the area of the brain called the basal ganglia. This is a region in which Parkinson's disease shows its effect.

Dystonia can affect one muscle. It can also influence a group of muscles or the entire body of the patient. Depending on the portion(s) of body concerned, dystonias are classified into five types as (a) generalized dystonia in which mostly or all the body is affected; (b) focal dystonia in which a specific body part suffers, (c) multifocal dystonia when more than one unrelated body part is influenced, (d) segmental dystonia wherein adjoining body parts are involved, and (e) hemidystonia when the arm and leg on the same side of the body are affected.

The treatment for many focal dystonias is done by injection of botulinum toxin. This toxin is a neuromuscular protein. It is produced by a bacterium called *Clostridium botulinum.* It acts as a muscle relaxer and blocks the activity of nerves in the muscles. Botulinum toxin practically eradicates the erratic posturing of the head in torticollis. It eliminates the eye spasms in blepharospasm. Dramatic response to levodopa and related medications is observed in some children with generalized dystonia. These are known as DOPA-responsive.

16.3 Lesioning Procedures and the Need of DBS

16.3.1 Pallidotomy, Thalamotomy, and Subthalamotomy

Pioneer neurosurgeon Sir Victor A.H. Horsley with his colleague Robert Henry Clarke performed lesioning of affected parts in regions lying deep within the brain of animals. This lesioning dates back to early 1900s. The lesions were produced through holes formed by surgery. The holes were shaped by injecting corrosive chemicals. The creation of such lesions can be deemed as the keystone of surgical treatment of movement disorders. The treatment was based on carrying out lesioning of the various decisive targets within the basal ganglia (BG) region of the brain. In 1939, Dr. Russell Meyers in New York carried out surgery of deep brain region in the basal ganglia. He did this surgery for treatment of tremor and rigidity of Parkinson's disease. This led to introduction of further surgical techniques.

The stereotactic frame developed in the 1940s was the principal enabling instrument. Pallidotomy, thalamotomy, and subthalamotomy were the three main types of surgical procedures used. In these procedures, a small heated probe is accurately inserted into different regions of the brain. The heated probe destroys the tissue. The procedure is called pallidotomy if the globus pallidus internus region is lesioned. It is termed thalamotomy when lesioning of the thalamus in the brain is carried out. The procedure is referred to as subthalamotomy when the subthalamic nucleus is lesioned. Among these, pallidotomy has been the one most widely applied techniques to relieve symptoms of PD. Thalamotomy has been used for tremor treatment. Subthalamotomy has been applied for relief from dystonia. Despite the encouraging results obtained and the hopes raised, the mortality rates of these procedures were ~0.12. Their application often led to cognitive decline. Speech jumbling and other snags followed.

16.3.2 Advent of DBS

Postoperative complications almost always accompanied surgery. Further, morbidity (unhealthfulness) rates were relatively high in surgery. Therefore, lesioning did not receive any support, neither from the doctors nor from the patients. As a result, almost complete abandonment of surgical treatments was favored. In 1967, the long-term, oral levodopa schedule at high doses was launched in the treatment of Parkinson's disease. Soon Levodopa was confirmed to be a safer and more effective treatment. But success of Levodopa was short-lived. In the later years of 1970s, the complications of levodopa therapy were clearly proven. Predominantly, dyskinesias and motor fluctuations were annoying. The precincts of levodopa therapy were understood. Moreover, improved imaging techniques provided comprehensive information about the anatomy and physiology of the basal ganglia. Consequently, there was a shift in interest, and it was again diverted toward surgical treatments for PD.

In the later years of 1980s, Dr. Alim-Louis Benabid, French neurosurgeon, was preparing a patient for thalamotomy [4]. During this operation, he came across an interesting observation. It came to the notice of the surgeon that exclusive use of electrical stimulation (without lesioning) could avert tremor. The Grenoble group in France instituted for the first time, stimulation of the ventralis intermedius nucleus of the thalamus (Vim) at high frequencies. They came up with a path-breaking finding that electrical stimulation could be used as a substitute for thalamotomy for the treatment of tremor. The finding that identical clinical effects were produced by stimulation at high frequencies at the sites of the ablative surgeries and the ablation technique, played an important role in guiding future research. It promoted great interest in DBS technology.

In the concluding years of 1980s and early years of 1990s, there was a revitalization of attentiveness and commitment toward surgical techniques focused to new functional targets in the BG. Virtually a renaissance took place in this field. In 1987, Benabid reported that the treatment of tremor could be done by stimulating ventralis intermedius (VIM) nucleus. However, he was not the first to perform the same. Ohye and colleagues published a paper in 1964. This paper was about stimulation of ventrolateral and subventrolateral thalamus. The triumph of thalamic stimulation for relief from tremor led to the exploration of new targets, which could probably be stimulated to derive therapeutic effects.

Among the targets under consideration, those receiving focal notice were the subthalamic nucleus (STN) and globus pallidus interna (GPi). These were implicated to be causing the symptoms of Parkinson's disease. Deep brain stimulation was a subject of serious investigation as a treatment option to neuroablative procedures with interminable effects, notably thalamotomy and pallidotomy. It became apparent that in disparity to the everlasting effects of ablation that could never be reversed, unfavorable effects produced by stimulation were by and large revocable after removal of stimulation. Due to this reversibility of effects, DBS was found to offer a riskless and more agreeable alternative to the chancy and perilous ablation procedures. Ever since, the technique has been most painstakingly examined and judged as a replacement of thalamotomy. This is especially accepted for one-sided control of essential tremor. Tremor allied to Parkinson's disease also falls under its dominion.

Thalamic brain stimulation therapy became available in 1995 in many countries of Europe as well as in Canada and Australia for patients affected by ET and PD. This therapy was advised for patients who had disabling and medically refractory tremor. Afterward in 1997, DBS was approved for the treatment of unilateral tremor (tremor of ET or PD) by the US Food and Drug Administration (FDA). The VIM of the thalamus was the targeted region of the brain. Unilateral studies limited FDA approval to unilateral implants only.

The location of the implant determines the specific symptoms treated by DBS. In 1997, the US Food and Drug Administration (FDA) approved the use of unilateral DBS of the thalamus. This approval was granted for the treatment of tremor related to Parkinson's disease (PD) and essential tremor. Shortly thereafter, in 2002, FDA extended its approval to stimulation of other parts. The extension was done for inclusion of unilateral or bilateral stimulation of the subthalamic nucleus (STN) and the globus pallidus pars interna (GPi) for use in patients with PD. In 2003, DBS of unilateral or bilateral GPi or STN was approved by FDA for the management of selected types of ailments. Chronic, drug-refractory dystonia was covered under DBS. This dystonia included general or segmental dystonia, hemidystonia, and cervical dystonia.

DBS is done in thalamic ventrointermediate (VIM) nucleus for controlling pharmacologically refractory essential tremor (ET) or parkinsonian tremor [5]. It was found that in subjects suffering from Parkinson's disease at an advanced stage, the treatment is more effectual in gait improvement when DBS of pedunculopontine nucleus, i.e., (PPN)-DBS accompanies the standard STN-DBS [6]. This combination of therapeutic targets may also be helpful in managing extrapyramidal symptoms (EPS), which are drug-induced movement disorders. An example is progressive supranuclear palsy. The stimulation of two targets proved to be more useful than the isolated stimulation of each target. In twin-target stimulation, the use of two pulse generators is helpful to independently control the parameters for each individual target.

16.3.3 DBS Versus Lesioning

In so far as the degree of improvement of tremor and parkinsonism are concerned, the beneficial results of DBS and ablation exhibit striking resemblance [7]. Naturally, DBS has almost completely replaced lesioning procedures. Therefore, it will be good to describe the salient properties of DBS vis-à-vis lesioning to bring out their relative pros and cons.

Advantages of DBS over lesioning: (a) DBS differs from pallidotomy, thalamotomy, and subthalamotomy in the primary aspect related to destruction of brain tissue. It does not perpetually destroy the brain tissue. In contradistinction, during the creation of radio-frequency lesions, the influenced region may be unduly enlarged. By stretching, it may include neighboring benign areas. Besides inability of lesioning of a well-defined region, it is also possible that by mistake, lesioning may take place in a nearby site and not the actual targeted site. As a result of errors of this nature, vital structures adjacent to the intended target may be inadvertently entangled in lesioning, which may cause permanent damages in the brain. Compared with lesioning, effects of DBS electrode implantation are restricted to a smaller region causing reduced permanent neurological adverse effects. Only nominal or insignificant damage to brain tissue occurs during surgery for DBS implant. (b) DBS does not entail methodical, purposeful demolition of brain regions. Owing to its reversibility, it does not forbid the use of prospective therapies at a later date. In opposition, the effects of ablation are irreversible. They cannot be adjusted without additional surgery. (c) In order to improve efficacy and counter hostile effects, the DBS parameters are amenable to postoperative tunability. Lesioning does not provide such postoperative corrections. It is a one-time operation, either successful or unsuccessful. (d) In distinction to ablative procedures, DBS can be carefully performed bilaterally securing well-being of the individual. Hence, it is a comparatively safer technique than lesioning.

Disadvantages of DBS: (a) The operations of ablation and electrode implantation in DBS are habitually done keeping the patient conscious and awake. In DBS, an additional surgery is required which is performed with the patient under general anesthesia to introduce the IPG and the connecting wire. Furthermore, the IPG needs to be replaced by surgery after every 2–7 years because the life of the battery ends after this period. Extra surgeries required in DBS are inconvenient for the patient. (b) In DBS, several visits to the doctor are generally enforced after implantation of electrode to adjust the stimulation parameters. This is especially so for patients who have undergone GPi or STN-DBS. Contrarily, follow-up required after ablation is the least possible. More rigorous follow-up for DBS than lesioning is often awkward for the patient. (c) Almost 25–30 % of patients who undergo DBS are troubled by complications related to DBS hardware. Even more than 1 year postoperatively, such complications are likely to happen. The complications take place in the form of mechanical breakage of the hardware. Skin erosion may occur over the hardware. Often, this erosion may be accompanied by skin infection. The complications usually require that spoiled or contaminated hardware be replaced. Sometimes a course of antibiotics is necessary. In ablation, no such complications are possible. (d) When a patient falls down, DBS hardware may be damaged or displaced from its position. Hence, in patients suffering from severe postural instability and frequent falls, ablative surgery is advisable. For this selected group of patients, DBS is not a judicious choice. (e) DBS is an overpriced procedure compared with cheaper ablation. Hence, in several regions of the world, lesioning procedures represent the only reasonably priced surgical preference. Only when the cost of DBS is significantly lowered can it be provided to masses.

16.4 Patient Selection/Exclusion Criteria for DBS

The key to success of DBS surgery lies in the proper selection of the befitting candidates. For optimally selecting patients for surgery, a multidisciplinary committee is constituted. It consists mainly of a neurophysician specializing in movement disorders, along with a functional neurosurgeon. A neuropsychologist must be included in this team. The neurophysician ensures the correct diagnosis of the disease. He/she also makes sure that the entire suitable range of treatments that does not involve surgery has been practiced. The neurosurgeon has the responsibility of evaluating the general health of the patient apropos the risks of surgery. He/she forthrightly shares with the patient the impending difficulties with surgery. These complications must be inclusive of a 1–2 % probability of intracerebral (within the cerebrum) hemorrhage per operated side. Such complications can precipitate grave neurologic disability or even death of the patient.

Before deciding on the intervention, the patient's personal expectancy from surgery needs to be formally looked into. The treating neurophysician should frankly apprise the patient of the genuine panorama of DBS implantation. The neurophysician must dispel any unrealistic results expected from surgery. In some situations, it might be beneficial to counsel the patient psychotherapeutically before taking the final decision. The patient and the psycho-physician should outspokenly thrash out with each other the chances of significant improvement. In this dialogue, the unfortunate and unfriendly effects of more aggressive drug therapy must be brought out compared with surgery.

Many vital factors must be taken into consideration before arriving at the decision to perform DBS surgery. Personal, professional, and social situations of the patient are of topmost priority. Regardless of these factors, surgery should not be gratuitously deferred. Otherwise the patient may lose his or her profession. There may also be a marked diminution in liberty of the patient in carrying out routine activities accompanied by loss in quality of life. The neuropsychologist performs an exhaustive examination regarding the patient's mental abilities of perception and memory. The objective is to ascertain whether there are any symptoms of dementia, a severe decline in mental abilities. Interviews with the patient and caregiver can resolve any behavioral abnormalities. They also help to overcome psychiatric problems. Among these, anxiety, depression, psychosis, mania, etc., are the diseases that need exigent attention.

No steadfastly validated benchmarks exist to choose patients for surgery. Notwithstanding, some general basic strategies have been set out. These are applied by many surgical centers. The main aim of the qualifying process is to spot patients that show greater probability of improvement. These are the patients in whom the anticipated benefit far exceeds the probable danger linked with the surgical intercession. This means that the individual risk benefit profile of the patient must be evaluated. Generally, appropriate candidates qualified for surgery are those that show appreciable disability in performing everyday activities or tasks necessary for occupation/service despite having undergone best possible drug therapy.

The principal indication for deep brain stimulation in Parkinson's disease is highly developed idiopathic PD, a multisystemic synucleinopathy (related with abnormal deposition of synuclein) of the human nervous system. It produces complications of motor functions such as fluctuations and dyskinesias, characterized by distortions in voluntary movements. Consequently, pertinent disability or parkinsonian tremor resistant to therapy is produced. In order to forecast that DBS will be successful, the criteria are laid down for supreme candidate with advanced idiopathic PD. This candidate is one who has a conserved good levodopa response but is distressed due to the side effects accrued from enduring treatment, e.g., motor fluctuations and dyskinesias. A first-rate levodopa response of parkinsonian tremor is not essential. However, attempts of therapy with doses of levodopa as high as 1500 mg per day, dopamine agonists (substances that directly stimulate the receptors in nerves which normally would be stimulated by dopamine), and clozapine (an atypical antipsychotic medicine) are compulsory before deciding on surgery. Besides idiopathic PD, remaining parkinsonian syndromes include the multiple system atrophy, a progressive neurodegenerative disorder leading to symptoms affecting both the autonomic nervous system and movement. These do not receive benefit from DBS.

Some exclusion criteria must also be decided. These are severe brain atrophy (shrinkage), major depression, or acute psychosis. Patients must possess the mental and physical stamina to tolerate a long and challenging procedure. They must be healthy from cognitive viewpoint and show low anxiety levels. Then only they will be able to provide useful intraoperative feedback. Their assistance is essential during adjustment of stimulation parameters postoperatively. Deranged, frenzied, and wild patients tend to become more befuddled during the surgical operation. They are unable to give reliable reports of unsympathetic effects during stimulation that is done intraoperatively. For such patients, stimulation parameters cannot be programmed for best results. This is mainly because of their lack of awareness about their own motor status. Important preconditions for DBS surgery are good general health conditions in order that intraoperative and perioperative complications may be eliminated. They are also indispensable to warrant good doctor–patient cooperation during prolonged awake surgery.

16.5 DBS Surgical Methodology

An overnight before surgery, antiparkinson and antitremor medications are withheld. Drug suspension is done in order to capitalize on tremor and off-period parkinsonism. Most favorable appraisal of the payback of stimulation is thus allowed during the surgery.

The methodology of surgical operation for DBS is similar to that of ablative stereotactic neurosurgery. In this remark, the ultimate phase of implantation of electrode is excluded. Surgery is performed, as a rule, by keeping the patient awaken. In fact, the patient is an agile participant to facilitate useful feedback. The advantageous and harmful effects of microstimulation as well as macrostimulation are enquired from the dexterous patient while targeting the stimulation areas and during lead implantation. On the odd occasion, surgery is performed under general anesthesia. This is mandatory in patients who experience difficulty in bearing surgery. These are especially the children and grown persons stricken with acute dystonia.

The first step involves the fixing of a stereotactic head frame on the skull of the patient. This helps to establish a 3-D coordinate system for reference. The coordinate system allows interrelationship of the imaging prior to operation with corresponding points in the brain. It also intercepts movement of head during surgery. The stereotactic system stabilizes the head position during surgery while the neurosurgeon maps the brain. Some surgeons now perform deep brain stimulation with stereotaxis without frame.

Thereafter CT and/or MRI scanning of the brain are performed. These scanning operations are done using a stereotactic frame fixed on the head of the patient [8]. The purpose is identification of the anterior and posterior commissures. These commissures are white matter tracts crossing the midline of the brain and connecting the two cerebral hemispheres. Some groups also perform preoperative cerebral ventriculography with this intention. Cerebral ventriculography is a medical imaging test in which the X-ray radiography of the ventricles of the brain is performed. The ventricles consist of a communicating network of cavities that are filled with cerebrospinal fluid (CSF). The radiographic examination is done after withdrawing CSF from the ventricles. Either a contrast medium or radiopaque agent is then injected. At the currently available MRI resolution, individual thalamic nuclei remain invisible. Therefore, for imaging thalamic targets, indirect targeting is necessary. The indirect targeting is done using stereotactic arithmetic-based coordinates. However, MRI is used for viewing the STN and GPi. Thus, it is possible to directly target these structures. Of late, intraoperative MRI is being attempted together with guidance of electrode placement in real time. These attempts are in exploratory stage.

The coordinates of the targeted area in the brain are calculated. Then the probe carrier of the stereotactic frame is adjusted consistently. A small burr fissure in the form of a rough-edged hole is drilled at a distance of a few centimeters from the midline and near the coronal suture, which is the dense, fibrous connective tissue joining the frontal and parietal bones of the skull. A hole of roughly 4 mm size is punctured into the dura for the lead to reach the targeted area. Dura is the short form of dura mater, the outermost, toughest, and most fibrous of the three membranes (meninges) encasing the brain and the spinal cord.

The DBS electrode is roughly 1.27 mm in diameter. It is guided to the targeted area in the brain. The precision achieved ranges approximately within 1–2 mm. The DBS electrode is prepared with four contact points. These contacts are each 1.5–3 mm long. A gap of 0.5, 1.5, or 4 mm separates the contacts. This gap depends on the particular model. Each contact is separately stimulated and evaluated. Desired and undesired effects of each contact are observed. Acute effects are seen as very theatrical influences in motion disorders. Such effects involve instantaneous termination of tremor in Parkinson's disease. In psychiatric diseases, the effects may be more subtle or nonexistent. The neurosurgeon works with the patient in the operating room, and the two together evaluate various stimulation sites to make the most of the therapy. The aim is to provide maximum effectiveness with minimum side effects. The lead site is selected to the satisfaction of both the surgeon and the patient. In many cases, temporary microelectrodes are first used to identify the exact stimulation target. Permanent electrode implantation is done later. Only when the location of the electrode is found to be most advantageous, the electrode is firmly fastened to the skull. Stereotactic CT/MR imaging is performed immediately.

About 7–10 days afterward, the patient is administered full anesthesia. Below the skin of the scalp, neck, and shoulder of the patient, an insulated extension wire is tunneled. This wire connects the lead to the implanted pulse generator (IPG) or the neurostimulator. The IPG is implanted subcutaneously below the patient's clavicle or collarbone.

About 3–4 weeks after the leads have been implanted, the IPG is turned on. It is programmed by the neurosurgeon to set the stimulation parameters. Electrical stimulation of high frequency starts from the neurostimulator. It moves through the annexed wire and the lead into the brain.

16.6 The DBS System

An assemblage of three components constitutes the DBS system. The primary component of the DBS system is the implanted pulse generator (IPG) or the neurostimulator. The IPG is a small-size device, sealed and protected from the environment. It contains a battery and electronic circuit. It is implanted underneath the skin in the upper body near the chest of the patient. The IPG generates the electrical pulse waveforms that are discharged via the electrodes. The second component of DBS is the lead. It is an insulated coil of wire containing extremely thin electrodes, usually four in number. Each wire terminates in a 1.5 mm electrode located at the tip to be rooted in the brain. These electrodes are placed in one of three areas of the brain. The electrodes are arranged in the targeted site for stimulation. The extension wire connects the lead with the IPG. This extension forms the third component of the DBS system. The extension is a subcutaneously placed insulated wire shaped like a coil that connects the DBS lead to the neurostimulator.

For turning the battery-operated neurostimulator on and off, patients are supplied a handheld magnet. Battery life varies from patient to patient, depending on the usage of DBS system. Stimulation parameter setting and power requirements decide how long the battery will actually last. When the battery runs low, the IPG can is replaced in a simple outpatient procedure.

Based on the symptoms of a particular patient, the DBS may be used on one side of the brain (unilateral DBS), as shown in Fig. 16.3, or two sides (bilateral DBS), as illustrated in Fig. 16.4.

16.7 Mechanisms of DBS Action

During the course of the last decade, DBS has progressed to be the most hopeful healing choice for PD that has reached an advanced stage. Other movement disorders which are refractory to drug therapy have also come under its purview. Nevertheless, the cognizance of the root mechanisms that have made stimulation at frequencies >100 Hz a therapeutic tool has still remained intangible and elusive [9]. Not much is known about the fundamental principles of electrical stimulation at above frequencies. To date, the stimulation parameters used for DBS are determined by experiments, relying more on experience or observation than on theoretical grasp [10]. The unsettled understanding of the mechanism of interaction of the applied stimulus pulses with the contiguous neuronal elements is a burning research topic for neurologists. It is slowing down progress and expansion of this know-how [11].

The general stimulation parameters for therapy by monopolar cathodic configuration of DBS may be stated as follows: stimulus amplitude = 1–5 V; stimulus pulse duration = 60–200 ms; and stimulus frequency = 120–180 Hz. These parameters have been derived for the most part by trying out various values and observing their effects. The stimulation parameters obtained through the trial-and-error approach

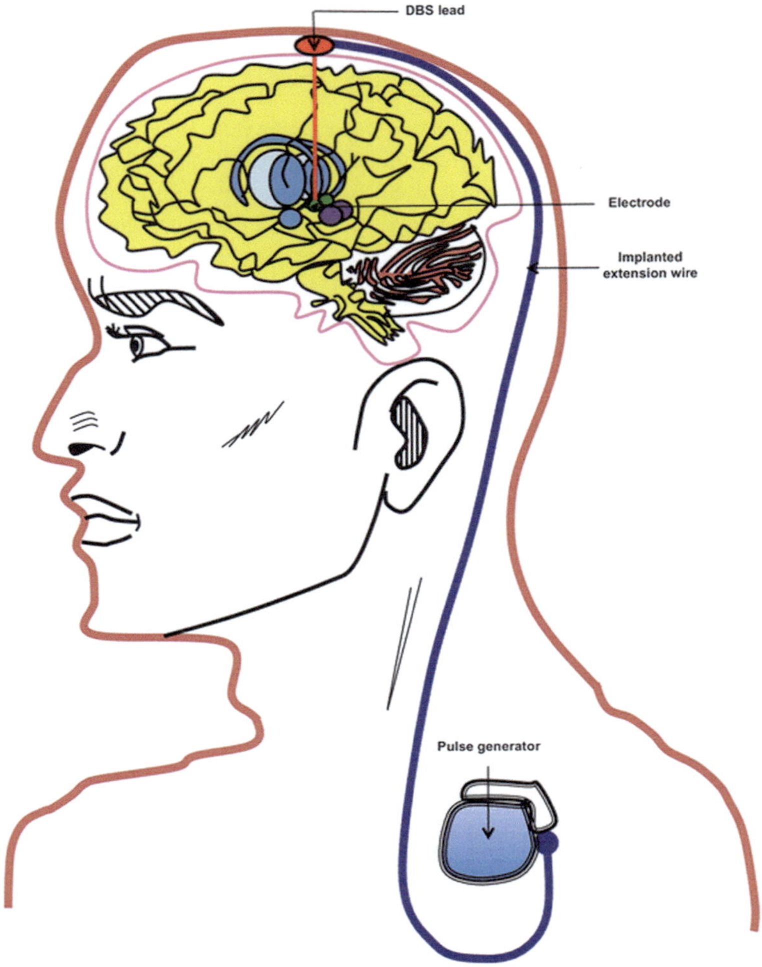

Fig. 16.3 Deep brain stimulation system using a single pulse generator for unilateral DBS

have proven to be effective, mainly due to the reason that the influence of DBS on the control of tremor and parkinsonian motor symptoms is found to be nearly taking place in period of time so short to be imperceptible. They are easily unequivocally noticeable. However, such ease of titration will not be allowed by the new therapies utilizing DBS technology.

For years, the belief was prevalent that when neuronal structures were stimulated, only excitation of the axons or cell bodies took place. This was evident from the rudimentary physiological principles. Astonishingly, beginning with the epoch of DBS, it was recognized that high-frequency stimulation impersonated the effects obtained previously by ablation in structures of the brain. From experimental and clinical findings, a vast amount of evidence has been amassed about the stimulation frequency.

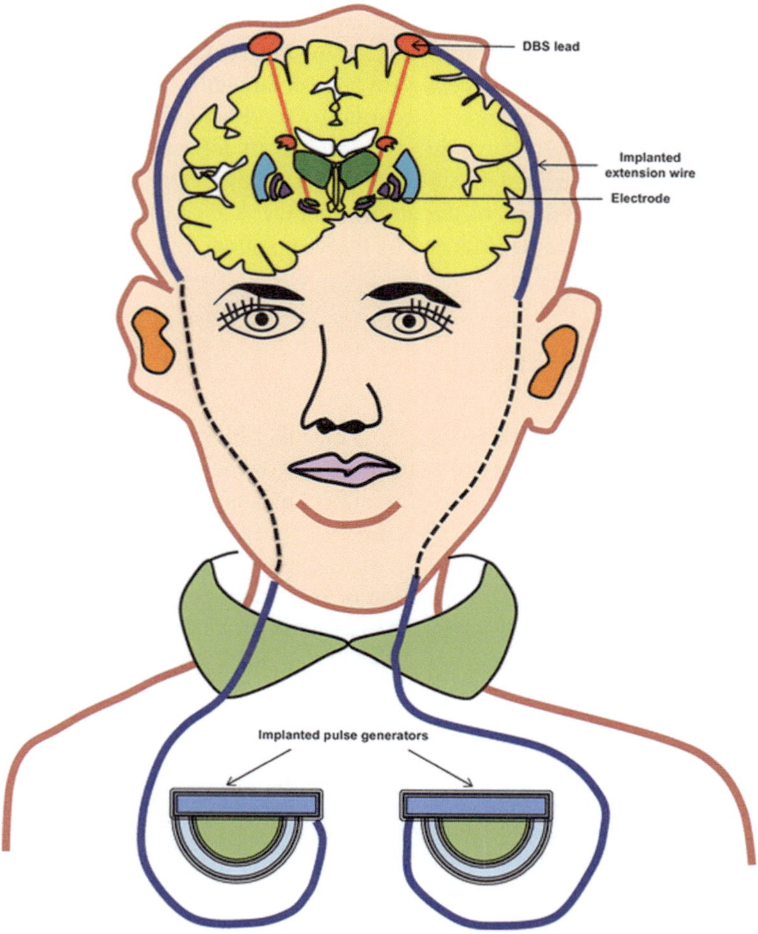

Fig. 16.4 Deep brain stimulation system using two pulse generators for bilateral DBS

This substantiation suggests that stimulation frequency is a primary factor influencing the irrefutable effect of DBS. Thus, there is comprehensive documentation of the frequency dependence of the effects of DBS. It has been found that the maximum relief from symptoms occurs at frequencies >100 Hz. Further, no therapeutic relief takes place at frequencies <50 Hz. Width of the electrical pulse decides which neural elements are impinged upon, by preference. Pulses of longer widths influence the cell soma or body. Pulses of shorter widths mainly affect axons. Furthermore, DBS provides highly focused stimulation confined to a small area because relatively low currents are used. This low value results in small spreads of current, on average about 2–3 mm at 2 mA intensity. The strength of stimulation decreases in proportion to the square of the distance from the electrode.

Based on experiments conducted on animals and after intraoperative revelations in human subjects, four major hypotheses elucidating the mechanism of high-frequency stimulation were proposed [12]. These were (a) depolarization cordoning of neuronal transmission in which this blockade occurred by rendering voltage-dependent ion channels inactive, (b) overcrowding or congestion of information by levying an efferent stimulation-controlled pattern at high frequency, (c) synaptic inhibition by prompting inhibitory afferents to the aimed nucleus, and (d) synaptic letdown by exhaustion of neurotransmitter induced by the stimulation.

16.8 Risks of DBS Surgery

The most serious danger of DBS surgery is bleeding into the brain (hemorrhage). This may lead to brain stroke resulting in abrupt death of brain cells. Less than 1 % of patients experience a stroke. The effects of a stroke can include paralysis causing impairment of body movement. Loss of speech, coma, or even death may result. There is also some chance (10 %) of less serious complications. Among the less serious ones is infection. Malfunction of the stimulator and movement of the electrode or generator are often vexing. Any of these problems may require removal of part or all of the DBS system.

16.9 DBS for Psychiatric and Neurological Disorders

16.9.1 Major Depression

Probably the most pervasive of all mental illnesses is major depression with symptoms of sadness, hopelessness, worthlessness, sleeplessness/too much sleep, restlessness, suicide thoughts, and so on. Long-established methods of conquering depression stem for the most part from fortuitous observations of effects of certain substances in countering depression. An example of such substances is iproniazid or imipramine. Both these drugs were accidentally discovered. It is known that nonsurgical therapies have relieved many patients from depression symptoms to a fairly acceptable level. The staple nonsurgical therapies are pharmacological treatment with drugs, psychotherapy using psychological methods and electroconvulsive therapy (ECT) by passing electric current through the brain to spark a seizure causing changes in chemistry of the brain. However, a sizable minority consists of patients in whom these treatments do not provide relief. Such tenacious, treatment-resistant depression is a ruthlessly incapacitating complaint. For this kind of stubborn depression, there are no verified treatment options once numerous medications, psychotherapy, and electroconvulsive therapy have bitterly flopped.

The mechanism of action of psychotropic drugs is by changing neurochemistry to a large degree in far-flung areas of the brain. It transpires that many of these regions in which neurochemistry has been changed may not be related to depression at all. Therefore, more dedicated approaches of treatment work on modulation of explicit networks in the brain. These comprise a more efficient method to assist patients who are treatment-resistant. One such approach employing DBS may help these patients to come out of depression. An open-label investigation testified a 60 % 6-month response percentage. This was achieved by lingering DBS of the subcallosal cingulate white matter (SCCwm). Grave and persistent effects counterpoising the depression of patients have also been described after stimulation of further white matter targets. These include the ventral most aspect of the anterior limb of the internal capsule (ALIC) and the inferior thalamic peduncle. They also embrace the gray matter targets. These targets are the globus pallidus internus and nucleus accumbens.

The mechanism by which DBS exerts antidepressant effects is intonation of activities taking place inside certain neural networks. These networks are accountable for controlling mood and attendant cognitive functions. They also regulate the circadian rhythms and motor functions. For delivery of continuous stimulation, DBS treatment of depression entails the two-sided stationing of electrodes at precise neuroanatomical sites [13–15]. More than half of patients were reported to return to work. Also, there were enhancements in quality of life past 1 year subsequent to DBS implantation. This indicates that DBS application to the subcallosal cingulate gyrus is beneficial for treatment-resistant depression. DBS provides both short-term and long-term benefits. Despite all encouraging and promising results, DBS for this malady is still an investigational area. Double-blind scientific trials are being pursued [16].

16.9.2 Obsessive–Compulsive Disorder

Obsessive–compulsive disorder (OCD) is a psychiatric malady which is over and over again prolonged, stern, and awfully unbearable. OCD is an anxiety disorder in which the affected individual is overcome by recurrent, unreasonable fears, thoughts, or impulses. In reaction to these obsessions, the individual performs certain acts, often repetitively. The repetitive acts are known as compulsions and are performed by the patient as a way to get out of the vicious cycle of stressful feelings. The patient develops a ritualistic behavior in which some rituals become part and parcel of activities like checking the door locks or the cooking heater several times to confirm that one has not forgotten to lock the door or switch off the heater when going out. In many cases, OCD is refractory to ideal treatments. A sizeable section of patients are either irresponsive to therapy or can obtain fragmentary reprieve.

In severe, refractory cases, lesions have been utilized as a "last alternative treatment." These were neurosurgical lesions. They were applied stereotaxically in the anterior limb of the internal capsules (anterior capsulotomy). Extensive case reporting of OCD counsels substantial benefit of lesions. However, controlled studies have been problematical. It has been observed that effects of lesioning can be simulated by stimulation with electrical pulses. These currents are supplied at intensities that are too stumpy to destroy tissues. This argument has shown the way toward the fabrication of implantable pulse generator systems to treat neurological illnesses for which lesionings have been endorsed to be successful as “last recourse solutions,” but for which safety trepidations disadvantaged extensive adoption.

The adjustment of control parameters such as pulse width and current crucially dictates the effectiveness of DBS for OCD mitigation. The values of these parameters vary in turn with the areas of the brain that are stimulated, as well as tissue types. Further complications are introduced by the fact that the effects of a particular stimulation setting do not appear at a level sufficient enough to be noticed until the patient has spent hours or days with that setting. It is agreed that DBS is a budding technique for treating psychiatric disorders which are refractory to traditional medicines. However, supplementary development work is called for before the procedure is utilized beyond the confinement of prudently regulated research proprieties [17].

16.9.3 Alzheimer’s Disease

Memory loss takes place and cognitive functions decline when brain cells die. This decline starts in a mild way but gets progressively worse. DBS can modulate the exclusive brain circuits in a changeable and reversible manner. Clinical benefits may be obtained in Alzheimer’s disease [18].

16.10 Discussion and Conclusions

DBS has reinforced and fortified its roots firmly. The prevailing ablative procedures have been pushed aside. For long, these procedures continued to be the backbone treatment by surgical route to get comfort from movement disorders [19]. But now they are left far behind. DBS has expressively bettered the quality of life of several patients. It has allowed them to convalesce and recuperate their individuality. For these patients, resumption of many normal activities has become possible. Very often, DBS has been found to decrease the daily medication dose required to manage symptoms. But it must clearly be remembered that DBS is not a cure for any disease. It has neither been shown to prevent the progression of disease. The goal of the DBS surgery is mainly to help in controlling the symptoms so that the patient can be comfortable.

DBS has become a standard therapeutic tool for some forms of severe movement disorders described in this chapter. For treating tremor and/or ataxia (difficulties in the coordination of muscles), DBS may prove to be a viable therapy in conjunction with or without using other remedial agents. These treatments are indicated in patients grieved with FXTAS (fragile X-associated tremor/ataxia syndrome) [20].

For dealing with other neurological and some psychiatric disorders, application of DBS is under investigation. Its utilization has been extended to indications, such as epilepsy and cluster headache [21]. Its domain has also been stretched to cover notorious psychiatric disorders. It has proven to be an effective sword cutting the symptoms of obsessive–compulsive disorder and depression [4]. Notwithstanding all these upbeat observations, several scientific, clinical, and ethical issues still remain unresolved [22]. These relate to the recognition and empathy of the targets for different diseases. The choice of the patients and the evaluation of the results are equally important [23].

Review Exercises

16.1 Where are the electrodes implanted for deep brain stimulation? Is the location of placement of electrodes the same in all kinds of disorders? Is the pulse generator placed in the brain?

16.2 What are the main disabilities for which deep brain stimulation is a well-recognized and accepted treatment? Name some other disorders in which the technique also helps.

16.3 What is the neurotransmitter whose degeneration in the brain leads to Parkinson's disease? What effects are produced by the deficiency of this neurotransmitter? What symptoms follow?

16.4 What is the drug levodopa used for? What happens after long-term use of this drug?

16.5 What is meant by tremor? Normal persons also experience trembling. How is this trembling different from that in the disease called tremor? What are the causative factors leading to this disease?

16.6 What is dystonia? With which part of the brain is it related? State the main features of the five classes of dystonia. What is the prevailing treatment for this disease?

16.7 What is meant by the following terms: pallidotomy, thalamotomy, and subthalamotomy? Name the diseases that are treated by performing these procedures.

16.8 What drug encouraged toward abandonment of surgical methods for Parkinson's disease? What difficulties were faced in long-term drug therapy?

16.9 What chance or accidental observation led to the use of ventralis intermedius nucleus stimulation for management of tremor?

16.10 In what ways is deep brain stimulation safer than neuroablative techniques for treatment of movement disorders?

16.11 Prepare a comparative report describing the relative merits and drawbacks of deep brain stimulation and lesioning techniques for the treatment of brain disorders affecting human movements.

16.12 What team of medical experts is formed to judge the suitability of patients for deep brain stimulation therapy? What are the roles of the different specialists in this team?

16.13 What are the criteria for selection of candidates for therapy by deep brain stimulation? What are the exclusion criteria?
16.14 Why is a stereotactic head frame fixed to the patient for DBS surgery? Why is the surgery performed with the patient awake? Why are CT and/or MRI scanning necessary? What is ventriculography?
16.15 How is the DBS electrode fitted into the brain? After how many days is the pulse generator implanted and how? When is it switched on?
16.16 What are the main components of the DBS system? What is bilateral DBS? How is the DBS system switched on or off?
16.17 What are the typical values of the parameters for controlling tremor and parkinsonian motor symptoms by deep brain stimulation? What frequencies provide the greatest relief from symptoms? At what frequencies the effects become imperceptible?
16.18 State the four main hypotheses for the mechanism of stimulation at a high frequency imitated from animal experiments and human studies?
16.19 Highlight the main risks and complications associated with DBS surgery?
16.20 What is the traditional approach to treatment of mental depression? What is treatment-resistant depression?
16.21 In what way is DBS a more focused technique to target the regions of brain related to depression?
16.22 What is OCD? Is it responsive to lesioning or DBS?

References

1. Garcia L, D'Alessandro G, Bioulac B et al (2005) High-frequency stimulation in Parkinson's disease: More or less? Trends Neurosci 28(4):209–216
2. Carron R, Chaillet A, Filipchuk A et al (2013) Closing the loop of deep brain stimulation. Front Syst Neurosci 7(112):1–18
3. Perlmutter JS, Mink JW (2006) Deep brain stimulation. Annu Rev Neurosci 29:229–257
4. Benabid AL (2007) What the future holds for deep brain stimulation. Expert Rev Med Devices 4(6):895–903
5. Xie T, Bernard J, Warnke P (2012) Post subthalamic area deep brain stimulation for tremors: a mini-review. Transl Neurodegener 1:20, 8 pages
6. Stefani A, Lozano AM, Peppe A et al (2007) Bilateral deep brain stimulation of the pedunculopontine and subthalamic nuclei in severe Parkinson's disease. Brain 130:1596–1607
7. Kern DS, Kumar R (2007) Deep brain stimulation. Neurologist 13(5):237–252
8. Burchiel KJ, Mccartney S, Lee A et al (2013) Accuracy of deep brain stimulation electrode placement using intraoperative computed tomography without microelectrode recording. J Neurosurg 119:301–306
9. Anderson T, Hu B, Pittman Q et al (2004) Mechanisms of deep brain stimulation: an intracellular study in rat thalamus. J Physiol 559(1):301–313
10. McIntyre CC, Savasta M, Kerkerian-Le Goff L et al (2004) Uncovering the mechanism(s) of action of deep brain stimulation: activation, inhibition, or both. Clin Neurophysiol 115:1239–1248
11. Yousif N, Purswani N, Bayford R et al (2010) Evaluating the impact of the deep brain stimulation induced electric field on subthalamic neurons: a computational modelling study. J Neurosci Methods 188(1):105–112

12. Breit S, Schulz JB, Benabid A-L (2004) Deep brain stimulation. Cell Tissue Res 318:275–288
13. Mayberg HS, Lozano AM, Voon V et al (2005) Deep brain stimulation for treatment-resistant depression. Neuron 45:651–660
14. Lozano AM, Mayberg HS, Giacobbe P (2008) Subcallosal cingulate gyrus deep brain stimulation for treatment-resistant depression. Biol Psychiatry 64:461–467
15. Kennedy SH, Giacobbe P, Rizvi SJ et al (2011) Deep brain stimulation for treatment-resistant depression: follow-up after 3 to 6 years. Am J Psychiatry 168:502–510
16. Cusin C, Dougherty DD (2012) Somatic therapies for treatment-resistant depression: ECT, TMS, VNS, DBS. Biol Mood Anxiety Disord 2:14, 9 pages
17. Abelson JL, Curtis GC, Sagher O et al (2005) Deep brain stimulation for refractory obsessive—compulsive disorder. Biol Psychiatry 57:510–516
18. Laxton AW, Tang-Wai DF, McAndrews MP et al (2010) A phase I trial of deep brain stimulation of memory circuits in Alzheimer's disease. Ann Neurol 68(4):521–534
19. Yu H, Neimat JS (2008) The treatment of movement disorders by deep brain stimulation. Neurotherapeutics 5(1):26–36
20. Hagerman RJ, Pak JS, Ortigas M et al (2012) Case series: deep brain stimulation in patients with FXTAS. Brain Disord Ther 1(2):104, 6 pages
21. Franzini A, Ferroli P, Leone M et al (2004) Hypothalamic deep brain stimulation for the treatment of chronic cluster headaches: a series report. Neuromodulation 7(1):1–8
22. Christen M, Müllerc S (2012) Current status and future challenges of deep brain stimulation in Switzerland. Swiss Med Wkly 142:w13570. https://doi.org/10.4414/smw.2012.13570, 8 pages
23. Sironi VA (2011) Origin and evolution of deep brain stimulation. Front Integr Neurosci 5(42):1–5

Chapter 17
Epidural Spinal Cord Stimulation

Abstract Spinal cord stimulation was first hosted in 1967 as a technique for treating chronic back pain. In this treatment, the nerves in the spinal column, also called the spine or backbone, are imparted mild electrical impulses or shocks. The impulses are supplied through leads that are implanted into the epidural space. The location of these leads is adjoining the lower facet of the spinal cord between T9 and L1. The implantation is carried out under fluoroscopic control in a relatively minor surgical procedure. The leads are supplied current from the pulse generator positioned between the skin and the fascial layers (connective tissue fibers, largely collagen). The electric impulses interfere with and modify the nerve activity to minimize the sensation of pain propagation to the brain. An additional available feature is programmability of electrode activation. Either constant-voltage or constant-current pulse trains can be chosen. Facility to minimize stimulus energy requirements is provided. The technique is vulnerable to migration of the lead. Shunting of the stimulus currents by the cerebrospinal fluid (CSF) and various other complications diminish its clinical efficacy.

Keywords SCS · Dorsal column stimulation · Back pain · Paresthesia · Dura mater · Epidural anesthesia · Gate control theory

17.1 Introduction and Historical Glimpses

Chronic pain is the principal cause of bodily and poignant anguish. It leads to domestic and societal commotion, frailty, and nonattendance of work [1]. The first clinical application of dorsal or posterior column stimulation (DCS) was documented by Shealy et al. [2]; "dorsal" means situated at the backside. They inserted the stimulator into patients suffering from cancer pain in 1967 [3]. This technique was portrayed as an original analgesic (pain relieving, palliative) method. It could alleviate pain in a multiplicity of chronic pain syndromes. Nevertheless, it was proved in later years that a greater number of structures in the nervous system might be activated by

V. K. Khanna, *Implantable Medical Electronics*,
https://doi.org/10.1007/978-3-031-99868-3_17

application of an electrical potential to the dorsal epidural space. By delivering electrical impulses of small amplitudes, unswervingly into the spinal cord in the region popularly known as epidural space, the straight spreading of pain signals traveling along the spinal cord toward the brain is interfered with and disturbed. Therefore, the term "dorsal column stimulation (DCS)" transformed to "spinal cord stimulation (SCS)."

The SCS is tactically designed to exchange the unpleasant sensory feeling of pain with a more gratifying tingling feeling. This tingling feeling is referred to as paresthesia. "Paresthesia" is a condition causing a patient to feel a sensation of numbness, burning, or tingling. It may appear like itching or prickling with no obvious durable effect on the body. This condition is vastly different from paralysis, which involves the loss of both movement and sensation. Paresthesia is only accompanied by loss of sensation. The perception of paresthesia in the region where pain is felt "masks" pain signals.

Pursuant to the first-time implantation of a stimulation system for spinal cord therapy wayback in 1967, techniques of neuromodulation have been widely practiced for treating patients with chronic pain. Evidence has mounted in support of the benefits of SCS in managing constant, obdurate pain of the torso and limbs. Many examples can be quoted. These include unilateral/bilateral pain insolent to orthodox and surgical treatments for chronic low back pain syndrome, radiculopathy (disease of spinal nerves/nerve roots), postsurgical pain, degenerative disk disease or herniated disk (slipped/ruptured disk), peripheral causalgia (burning pain due to peripheral nerve injury), epidural fibrosis (scar tissue formation near the nerve root), arachnoiditis (arachnoid inflammation), and complex regional pain syndrome. SCS has also been investigated to treat paralysis, spasticity (stiff/rigid muscles), gastrointestinal issues, and urinary system dysfunction. The stimulation site on the spine determines which nerve networks are targeted.

Continuing pain reduction and improvement in health-linked quality of life has been achieved by spinal cord stimulation in individuals suffering from chronic reflex sympathetic dystrophy (RSD) syndrome of unknown pathophysiology, a burning pain with swelling and tenderness [4, 5], and intractable cancer pain [6, 7]. In 1989, FDA approved SCS as a means to relieve pain from nerve damage in the trunk, arms, or legs. SCS now covers nearly 70 % of all neuromodulation treatments.

17.2 Epidural Space and Epidural Anesthesia

The term *dura mater*literally means "tough mother." It is the outermost, dense, leathery, and fibrous of the three membranes (dura mater, arachnoid membrane, and pia mater) that constitute the shielding envelope of the brain and spinal cord. The term "epidural space," also known as epidural cavity, extradural space, or peridural space, has its origin from ancient Greek, "on, upon" + dura mater. The epidural space (Fig. 17.1) is an anatomical area in the spinal cord confined between the walls and the dura mater of the vertebral canal. It is situated exterior to the dura mater, which enfolds the following: arachnoid mater, subarachnoid space, the cerebrospinal fluid, and the spinal cord. On the upper side, the epidural space terminates at the foramen magnum. It is the extremity at which the spine connects with the bottom of the skull. On the lower side, it ends at the sharp point of the sacrum at the sacrococcygeal membrane. The epidural space is composed of fat and small blood vessels. Its contents are mainly lymphatics (veinlike vessels carrying clear-to-white lymph fluid), roots of the spinal nerves, loose fatty tissue, small arteries, and internal vertebral venous plexuses (intraspinal veins).

Epidural anesthesia is the utmost universally practiced method for pain relief during labor. It is a local anesthesia. It occludes pain only in a specific region of the body. An epidural provides only analgesia or relief from pain. It does not aim to provide anesthesia, which causes total lack of feeling. The epidural obstructs the nerve impulses starting from the spinal segments on the lower side. The result is a decrease in responsiveness in the lower half of the body.

The procedure for providing epidural anesthesia is as follows: The middle back is washed in waistline area with an antiseptic solution to counter any chances of infection. A local anesthetic is infused by injecting into a small region on the back to create the feeling of numbness. A needle pierces the numbed region encircling the spinal cord in the lower back. A catheter is guided through this needle into the epidural space. The needle is then carefully withdrawn. But the catheter is left in its position. The catheter is securely fixed by adhesive tape to the back to check it from sliding off. Medicine is supplied by giving injections periodically. It may also be delivered in continuous infusion.

17.3 SCS Equipment

17.3.1 The Hardware and the Electrodes

The SCS hardware (Fig. 17.2) consists of a programmable pulse generator, an extension cord, and an electrode lead [8]. The pulse generator is usually implanted anteriorly (in front) and subcutaneously (beneath the skin) through a passage tunneled between the skin and fascial layers (fibrous connective tissues enveloping/separating/binding together muscles and organs). The lead with multiple contacts is positioned in the dorsal epidural space situated at the back or dorsum. It is connected via extension cables to the pulse generator.

By programming the system, the following parameters are set at definite values: pulse amplitude, width of the pulse, and frequency of repetition of pulses. The rostrocaudal position (between head and tail) of the multiple-contact lead is alterable. The position is changed to provide electrical stimulation at many levels of the spine.

The placement of the cathode is done at a location between the dorsal median sulcus (longitudinal groove marking the midline of the medulla oblongata) and the dorsal root entry zone (DREZ) area. Upon stimulation, the current starts flowing

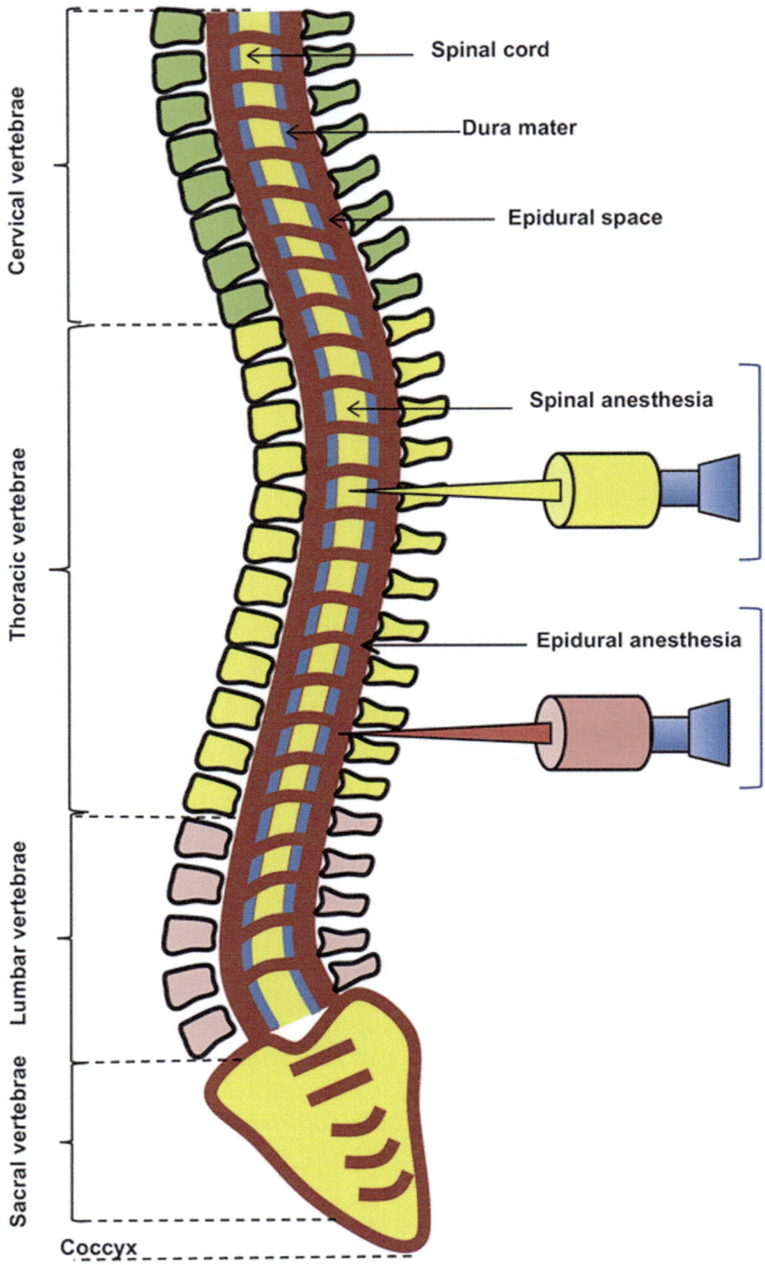

Fig. 17.1 Anatomy of the spinal cord and regions for spinal and epidural anesthesia

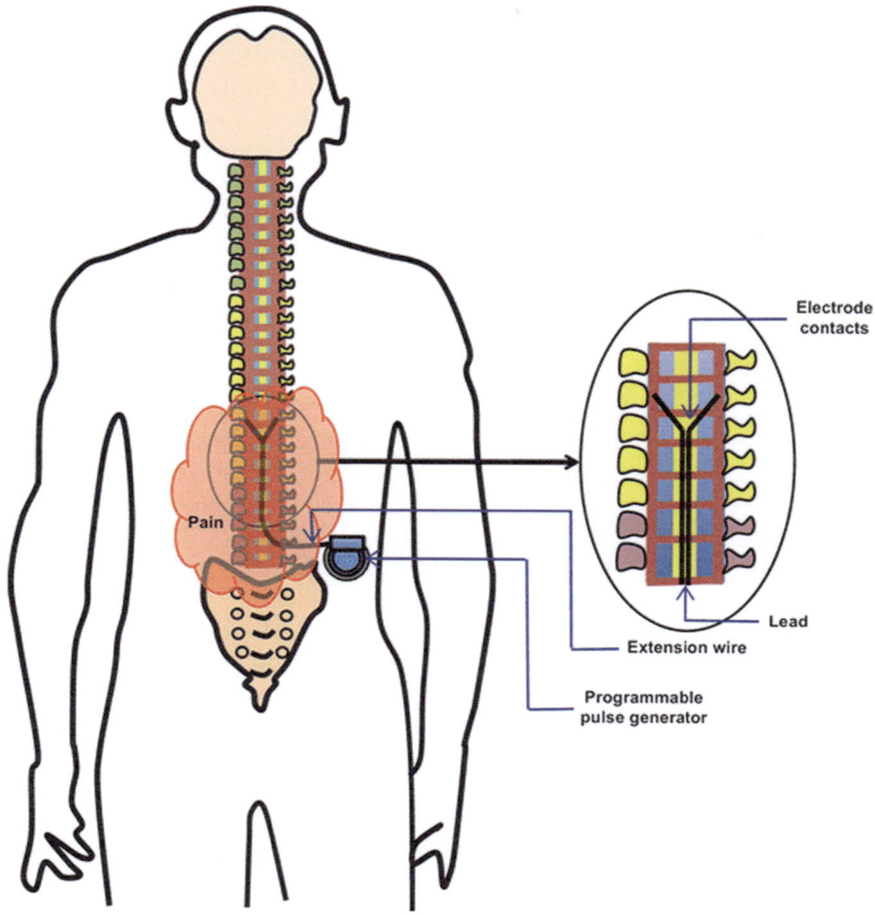

Fig. 17.2 Implanted spinal cord stimulator in a patient

from the cathode (negatively charged electrode) toward the anode (positively charged electrode). Fundamentally, every neural structure situated proximally to the cathode can receive activation at sufficiently high electrical stimulation. But current follows the path of least resistance. It therefore flows through anatomical structures having low electrical resistivity. Cerebrospinal fluid carries approximately 90 % of injected current because of its lowest electrical resistivity. Then the longitudinal white matter follows. Due to its anisotropic or directionally dependent characteristics, transverse white matter is more resistive than gray matter. Epidural fat and dura mater likewise possess high resistivity. Vertebral bone has the highest electrical resistivity. Therefore, it practically behaves as an insulator. Hence, the besieging tissues, e.g., the heart and structures of the pelvis, are protected from receiving stimulation.

The first-generation electrodes were unipolar. One lead was positioned at the stimulation site and the other at an area of zero potential (the earth or a large conducting body). Unipolar electrodes provided paresthesia in a restricted zone. Therefore, a new lead design was later developed. The number of electrodes in this design varied from four to eight. Two kinds of leads are presently used: the percutaneous lead and the paddle lead. The percutaneous electrode is inserted via Tuohy needles (hollow hypodermic needles). It is a perfect lead for both provisional and final implants. For placing the paddle lead, surgery is necessary by orthopedic neurosurgical procedures known as laminotomy or partial laminectomy. In laminotomy, a small part of the lamina is removed while in laminectomy, it is completely removed. The paddle lead offers greater stability and lesser tendency for migration. It is especially befitting for patients who have a history of lead relocation. It is also appropriate for the patients in whom assignment of the trial lead was previously demanding.

Programmable multiple-electrode arrays have shown superiority over the devices with a solitary channel. The reason is that they permit anode–cathode guarding and changes of polarity. In this manner, they enable optimal current steering.

Axonal activation and paresthesia (tingling/pricking sensation) distribution are determined by the comparative positions of the electrodes (cathodes and anodes). They are also impacted by their distances from the spinal cord. Using a two-channel pulse generator and applying pulses differing in simultaneity, a profounder dissemination of the pulses into the spinal cord is attained without creating a high-intensity electric field. The idea of electrical field routing through selectively engaging axonal nerve fiber tracts in the dorsal columns was publicized by a transverse tripole array (+, −, +) system. Electrical steering of the paresthesia across the axial back region is simplified. At the same time, the stimulation of the nerve roots was minimized.

17.3.2 Implantable Power Sources

There are three different types of implantable power sources to be discussed in the forthcoming subsections.

17.3.2.1 The Conventional Non-rechargeable Unit

The main component of the conventional unit is an implanted pulse generator. The pulse generator houses the integrated circuitry and batteries. It delivers low-voltage stimulation for pain relief. The pulse generator is accompanied by an external handheld, telemetry device. This is the external controller which activates the pulse generator and programs its parameters. It works transcutaneously. It is used to switch stimulation on or off and to attune the intensity of stimulation between the bounds set by the doctor. The patient too can control the stimulation parameters. The life of the battery is limited. The life expectancy of the battery is determined by the length of time over which it is used. Other decisive factors influencing battery life are the levels of the exploited parameters, viz., width of the pulse, the applied voltage, frequency, etc. This kind of unit generally lasts for 3–5 years before needing surgical replacement.

17.3.2.2 Rechargeable Unit

This unit is supplied to patients who require a high amplitude of stimulation for relieving pain. It lasts for a long period of 10–25 years. Recharging is done through an induction coil worn by the patient over the implanted device. During recharging, stimulation can be continued. The rechargeable unit is acquiring popularity among recipients due to its small dimensions and easy upkeep.

17.3.2.3 Radio-Frequency Unit

This power source is a radio-frequency device. Patients requiring high power adjustments are prescribed this kind of unit. These are the patients with pain in the back or legs. A radio-frequency unit consists of an implantable receiver and an external transmitter with antenna knotted on a belt or kept in a pocket by the patient. Energy emitted from the transmitter is picked up by the receiver. It travels along the lead and transmits impulses to stimulate the nerves along the spine near the implanted electrodes.

17.4 Mechanisms of Action

Spinal cord stimulation is an irrefutable outgrowth of the well-known gate control theory (GCT) of pain propounded by Melzack and Wall in [9]. This theory describes gracefully and succinctly the attenuation of spinal pain transmission by activation of afferent fibers. The technical context of the SCS trials is provided by GCT in terms of a draft outline for probing the exchanges between native and remote excitatory and inhibitory organizations in the dorsal horn. The dorsal horn is a longitudinal subdivision of gray matter from posterior spinal cord to dorsal roots.

The expected mechanisms of pain relief by SCS are predominantly described in terms of gating. SCS closes the gate to incoming signals of pain from the peripheral nerves and spinal cord. It was hypothesized in the GCT theory that the gate was opened by excessive small fiber activity. Also, an excessive large fiber activity closed the gate. It was proposed in the theory that when the large non-nociceptive (not pain-related) myelinated (with myelin sheath) fibers of the peripheral nerves were stimulated, the activities of the small nociceptive (pain-related) projections in the dorsal horn of the spinal cord were repressed. In comparison to small fibers, the activation threshold of large fibers for depolarization by an electric field was lower. Further, large fibers could be stimulated selectively.

The correct mechanisms of pain reprieve caused by stimulation of the spinal cord still continue to elude scientists and remain shrouded by mystery. Existence of other possible dominant mechanisms has been conjectured. Animal studies have shown that SCS props up the activation of chemical substances such as gamma-aminobutyric acid (GABA)-B and the adenosine A-1 receptors. These may modulate the pain. Explicitly, a faulty local GABAergic function partially uplifts the basal levels of exciting neurotransmitters in neuropathic pain states. It was shown that SCS prompted the liberation of neurotransmitters concerned with the variation and control of pain signals in the spinal cord, such as GABA, substance-P, and serotonin.

In the model of pain for animals with injury to sciatic nerve (a large nerve that runs through the buttock down the back of each leg), SCS inhibited the over-excitability of the extensive active span cells in the dorsal horn. This signifies that the major antinociceptive (anti-pain) effects of SCS occur via A-fibers, which are the nerve fibers in nerve trunks and peripheral nerves with the highest conduction velocity. Using SCS, it is possible to eliminate peripheral ischemic pain caused by inadequate local blood supply due to the blockage of blood vessels leading to that region. It may do so by rebalancing the oxygen supply/demand ratio, as evidenced by anti-ischemic and antianginal characteristics. At low stimulation intensities, SCS enfeebles over-excitation of the sympathetic nervous system. Antianginal effect might similarly spring up from the subdual of the central nervous system. Other reasons could be the steadying of the inherent cardiac nervous system. Liberation of adenosine is another possibility. With increase of stimulation level, the nitric oxide reliant on liberation of the calcitonin gene-related peptide could perhaps induce vasodilatation, i.e., the dilatation of blood vessels. This may produce anti-ischemic effects.

17.5 SCS Indications

To be an eligible candidate for SCS, the patient should have suffered from chronic pain for a period >1 year. Candidacy is decided through the individual's qualifying physical and intellectual tests. The status of a patient's health plays a vital role in determining whether the person is fit enough to draw healing advantage. It is abundantly clear that neuromodulation does not remove the underlying cause of pain. It is only a technique to provide relief from pain. Hence, the treatment is a part of a unified healthcare team approach. This approach seeks to provide long-term moral support and care to the patient. Only a collective effort by pain specialists, neurosurgeons, and rehabilitation physicians can offer effective SCS treatment.

A two-stage process is adopted for implanting the stimulator device. The first stage is a selection examination preceding the actual implantation to assess the suitability of the patient for permanent implantation. It is only when the patient is satisfied and convinced fully with the results of the trial that the SCS device is permanently implanted in the second stage.

In the first stage, the percutaneous method of electrode placement is applied. This method is less invasive. The electrode is inserted through a hollow needle.

The doctor carefully looks at the image on the monitor and guides a hollow needle inside the epidural space in the region above the spinal canal. This needle serves as a passage through which thin wires are threaded. In each wire, there are a small number of electrical contact points near the end. By attaching the leads to the power supply, a mild current is supplied. The patient is asked to understand and assess the exact feeling. After the doctor has listened to the patient's feedback, the positions of the electrodes are maneuvered for patient's satisfaction. This is done until the area of pain is covered by a tingling sensation. For realization of analgesia, paresthesia must overlap the site of pain. Then only the pain effects are overshadowed by the paresthesia. The trial period lasts from 3 to 15 days. As the patient carries out routine activities, the degree of relief is observed. After the patient is contented and responds affirmatively, it is inferred that the exact position of lead has been located. If the pain is reduced by more than a factor of half from the baseline value, the trial is deemed to be successful. Then the second stage of implantation is completed, wherein the SCS lead is strongly fixed.

17.6 Discussion and Conclusions

SCS stands out as an anodyne and relaxing treatment for various ingrained neuropathic pains, i.e., for management of lingering pain conditions accompanied by tissue injury [10]. It is useful to deal with assorted medical conditions of constant pain. One such condition is the failed back surgery syndrome (FBSS), namely, continued back pain after surgery. Another situation is multifarious provincial pain syndrome called complex regional pain syndrome(CRPS), a scarce form of prolonged pain typically affecting a limb such as an arm, a leg, a hand, or a foot, generally following an injury or trauma. Pain conditions associated with ischemic and coronary artery diseases [1, 11, 12], as well as cancer [13], are relieved using SCS. Time and again, the starting medical care acquirement expenditures of SCS gadgetry are compensated by a decrease in healthcare backup reinforcements and charges after implantation [14].

Review Exercises

17.1 Differentiate between dorsal column stimulation and spinal cord stimulation. Why was the term, "dorsal column stimulation" changed to, "spinal cord stimulation"?
17.2 What is paresthesia? How does it differ from paralysis?
17.3 Explain the following terms: (a) dura mater, and (b) epidural space.
17.4 Name a popular method of pain relief during labor. How is this anesthesia carried out?
17.5 Describe the hardware of SCS equipment. What is the path of current flowing during spinal cord stimulation?

17.6 What are the two types of electrode leads available for SCS? Which type of lead is less prone to migration?

17.7 What are the three types of implantable power sources for SCS? Indicate the typical life span of each class. Discuss their relative benefits and limitations.

17.8 What is gate control theory of pain? How does it explain the pain relief obtained by SCS? What other mechanisms are likely to participate, as suggested by animal studies?

17.9 What type of patients can benefit from implantation of SCS equipment? What are the two stages in which this implantation is carried out? Why does a trial implantation precede the permanent one?

References

1. Jeon YH (2012) Spinal cord stimulation in pain management: a review. Korean J Pain 25(3):143–150
2. Shealy CN, Mortimer JT, Reswick JB (1967) Electrical inhibition of pain by stimulation of the dorsal columns: preliminary clinical report. Anesth Analg 46(4):489–491
3. Raphael JH, Mutagi HS, Kapur S (2009) Spinal cord stimulation and its anaesthetic implications. Contin Educ Anaesth Crit Care Pain 9(3):78–81
4. Kemler MA, Barendse GAM, Kleef MV et al (2000) Spinal cord stimulation in patients with chronic reflex sympathetic dystrophy. N Engl J Med 343:618–624
5. Kemler MA, De Vet HCW, Barendse GAM et al (2002) The effect of spinal cord stimulation in patients with chronic reflex sympathetic dystrophy: Two years' follow-up of the randomized controlled trial. Ann Neurol 55(1):13–18
6. Yakovlev AE, Resch BE (2012) Spinal cord stimulation for cancer-related low back pain. Am J Hosp Palliat Care 29(2):93–97
7. Flagg A 2nd, McGreevy K, Williams K (2012) Spinal cord stimulation in the treatment of cancer-related pain: "back to the origins". Curr Pain Headache Rep 16(4):343–349
8. Krabbenbos IP, van Dongen EPA, Nijhuis HJA et al (2012) Mechanisms of spinal cord stimulation in neuropathic pain. In: Carrillo-Ruiz J (ed) Topics in neuromodulation treatment. InTech. http://www.intechopen.com/books/topics-in-neuromodulationtreatment/mechanisms-of-action-of-spinal-cord-stimulation-in-neuropathic-pain. Accessed 12 Feb 2015
9. Melzack, Patrick DW (1967) Pain mechanisms: a new theory. Surv Anesthesiol 11(2):89–90
10. Cameron T (2004) Safety and efficacy of spinal cord stimulation for the treatment of chronic pain: a 20-year literature review. J Neurosurg 100(3 Suppl Spine):254–267
11. Kumar K, Taylor RS, Jacques L et al (2007) Spinal cord stimulation versus conventional medical management for neuropathic pain: a multicentre randomized controlled trial in patients with failed back surgery syndrome. Pain 132(1–2):179–188
12. Frey ME, Manchikanti L, Benyamin RM et al (2009) Spinal cord stimulation for patients with failed back surgery syndrome: a systematic review. Pain Physician 12(2):379–397
13. Lihua P, Su M, Zejun Z et al (2013) Spinal cord stimulation for cancer-related pain in adults. Cochrane Database Syst Rev 2:CD009389
14. Taylor RS, Taylor RJ, Van Buyten JP et al (2004) The cost effectiveness of spinal cord stimulation in the treatment of pain: a systematic review of the literature. J Pain Symptom Manage 27(4):370–378

Chapter 18
Vagus Nerve Stimulation

Abstract Vagus nerve stimulation (VNS) therapy is used for halting those seizures of medically refractory epilepsy patients in which therapy by antiepileptic drugs (AEDs) has failed to provide any reasonable comfort. Another disease which can be treated by VNS is chronic or recurrent depression in adult patients that is unmanageable by antidepressant drugs. In VNS, the stimulator device is implanted in the upper region of the patient's chest. The electrodes are fastened to the vagus nerve in the neck. Although not yet confirmatively known, it is believed that constant and recurrent electrical stimulation of the vagus nerve causes the release of brain neurotransmitters that decrease seizure activity, as required for epilepsy control, or regulate the patient's mood, as needed for the treatment of depression. Actually, the therapy for clinical depression was started on the basis of improvements in cognition and mood that were observed in patients who were treated for epilepsy. Encouraging results have also been reported by VNS for the management of rheumatoid arthritis.

Keywords VNS · Vagus nerve · Epilepsy · Epileptic seizure · Depression · Obesity · Rheumatoid arthritis

18.1 Introduction

Vagus nerve stimulation (VNS) is a kind of neurostimulation, a technique involving the therapeutic activation of part of the nervous system by currents supplied through microelectrodes. It was developed in the 1980s. It is said to be an extracranial neuromodulation. The word "extracranium" means "outside the cranium." The cranium is the skull or the hard bony braincase that houses and protects the brain. VNS therapy is used to prevent epileptic seizures.

V. K. Khanna, *Implantable Medical Electronics*,
https://doi.org/10.1007/978-3-031-99868-3_18

18.2 Epileptic Seizures

Epilepsy is a set of neurological disorders (central nervous system illnesses) exemplified by repeated episodes of convulsive seizures (colloquially fits), sensory disorders, atypical behavior, loss of consciousness, or all of these symptoms [1]. Regardless of advances in the surgical and medical treatment of epilepsy, patients continue experiencing breakthrough seizures or suffering from incapacitating, unpleasant consequences perpetrated by antiepileptic drugs. A breakthrough seizure is one that befalls in spite of the fact that anticonvulsants have provided a sustained period of freedom from seizures. It may be provoked by withdrawal from excessive use of alcohol, by illness, due to forgetting to take the medication, or due to taking less than prescribed antiepileptic drugs over a period of time.

18.3 Vagus Nerve Stimulation for Medically Refractory Epilepsy

VNS has been explored as a treatment substitute for a select group of epilepsy patients. These are the patients suffering from partial-onset epileptic seizures that are medically unmanageable [2]. Partial-onset seizures refer to seizures that affect only a portion of the brain. They are placed under one of the two subdivisions, viz., simple or complex types. The former subdivision refers to seizures in which there is no loss of consciousness but changes take place in emotions or feelings. In the latter subdivision, there is loss of consciousness and awareness.

The VNS patients belong to the class for whom either surgery is not advised or those for whom surgery has been unsuccessfully tried. Looking from the perspective that VNS is an outlying or fringe intervention used to cure a disease that is wholly connected with phenomena taking place inside the brain [3], VNS appears to be a unique epilepsy treatment. This uniqueness is because of the matchless anatomical disposition of the vagus nerve. In fact, the vagus nerve affords a convenient peripheral medium whereby the brain can be influenced without the invasive intracranial surgery. It is widely believed that electrical stimulation of vagal afferent nerves conducting impulses inwards to the brain directly and indirectly influences well-defined seizure-related circuitry within the brain. But therapeutic mechanisms of VNS have not been fully elucidated.

In 1997, VNS therapy was approved in the USA. This approval was granted for use as a form of an additional subordinate therapy for reduction of seizure frequency in adults and teenagers aged more than 12 years suffering from partial-onset seizures [4]. It was stipulated that seizures of these patients must have proven to be irrepressible by antiepileptic medications. The VNS may be considered to be a medical necessity for dealing with medically refractory seizures. The refractory seizures are defined as those with the following properties: (1) seizures that occur despite the prescription of antiepileptic drugs in recommended therapeutic doses or (2) seizures that cannot be controlled with therapeutic doses of these medications because of intolerance of their pernicious effects by the patient. VNS has become a valuable treatment choice in the doctor's toolkit for patients distressed by refractory epilepsy. Currently, it is regularly available in epilepsy clinics all over the globe.

18.4 Promising Areas of VNS Therapy

For a long time, the use of a vagus nerve stimulator has been considered as an investigational treatment (not approved for general use but under study) for many conditions. Among these conditions, the following stand out prominently: heart failure and fibromyalgia (a rheumatic condition). Other conditions are depression, essential tremor, headaches, and obesity. The reason is that sufficient evidence was not available to arrive at a definite conclusion regarding health outcomes or benefits of this procedure for these sicknesses. But in 2005, VNS therapy received approval in the USA for patients suffering from chronic or recurrent depression with age $\geq$18 years [4]. Its use was permitted as a subordinate, long-standing treatment. It was also declared that such patients must have encountered a major depressive episode. In addition, four or more antidepressant treatments must have failed to give any solace to these patients.

18.5 Anatomical Basis of VNS

Parasympathetic nervous system, also dubbed as the "rest and digest system," is a part of the nervous system of involuntary character. This part acts to slow down the heart beat. It increases activities of the intestines and glands. It also causes relaxation of the sphincter muscles in the gastrointestinal tract. Vagus nerve refers to either one of the lengthiest duos of cranial nerves. These nerves exert parasympathetic control over the cardiac functions and the working of many other organs inside the body. Examples of these organs are thoracic and abdominal viscera. Among the 12 cranial nerves that originate in the brain, vagus nerve is the longest nerve.

As mentioned before, "cranium" means the skull. The vagus nerve represents the tenth number cranial nerve (Cranial nerve X). The 12 pairs of cranial nerves are shown in Fig. 18.1. They start from the bottom surface of the brain and the brain stem, not from the spinal cord. Particular nerves are information-fetching nerves bringing information from the sense organs to the brain. A few others control muscles. Another few are connected to the glands or internal organs. "Vagus" is a Latin word. It means "wandering." The name points to the complexity of connections formed within the body by the branches of this nerve. The vagus nerve begins in the region between the pons at the base of the brain stem and the spinal column. It comprises a large bundle of nerve fibers extending down the neck and connecting the brain to the visceral organs (Fig. 18.2). These are the organs lying within the chest and the gastrointestinal tract, notably the heart, lungs, liver, pancreas, and intestines. The vagus nerve transports impulses from the brain to influence the rate of heart beating and the gastric character. It also carries a large volume of information backward from the various organs of the body to the brain. This information from the body organs is transferred to miscellaneous regions of the brain. It influences the activity in the limbic system of the brain. The limbic system is a complicated group of structures. These structures are situated on the two sides of the thalamus and beneath the cerebrum. It is the section of the brain that exerts control over emotion. Hence, by stimulating the vagus nerve, it is possible to manipulate the brain function delicately, dexterously, and appreciably. Its stimulation is a valuable tool in the realm of neuromedicine.

The vagus nerve is a mixed nerve. By the "mixed nerve" is meant that it is a combination of afferent sensory and efferent motor fibers. These fibers move together in a common passageway. "Afferent" means conducting inward toward the nervous system; "efferent" implies conducting outward therefrom. Each fiber has its own origin, destination, and activation threshold. The vagus nerve consists of ~80 % afferent fibers. They start off from the heart and aorta. Some fibers are also connected to the lungs and gastrointestinal tract. The remaining ~20 % fibers in the vagus nerve are efferent fibers that parasympathetically innervate these structures. They also supply the voluntary, striped muscles of the larynx and pharynx with nerves. The sensory afferents far exceed the motor efferents in number. They constitute around 65–80 % of all vagal fibers.

Stimulation of the vagus nerve triggers the release of brain neurotransmitters. These are the chemical substances serving as messengers of information from a nerve cell to another part which could be a nerve, tissue, or muscle. It could be an organ too. These neurotransmitters decrease seizure activity contributing toward epilepsy control or regulation of the patient's mood. In this way, they help in the treatment of depression, hence neuropsychiatric disorders [5].

18.6 The VNS System

In VNS, a commercially available device is employed for the enduring, spasmodic stimulation of the vagus nerve located on the left side (Fig. 18.3). This device is an electrical signal generator. In size, it is similar to a pocket watch. As with a cardiac pacemaker, the device is placed in the chest region on the left side. The lead wires coming from the device are wound around the left vagus nerve. The electrodes are implanted on the left cervical vagal trunk, approximately 8 cm above the clavicle. Then current is delivered to the nerve occasionally. The electrical stimulation is provided without human intervention at prescribed intervals. The stimulation is done both during the waking and sleeping states of the patient. The electrical stimuli are high-frequency, short-duration pulses interchanging in polarity. They are supplied as trains of pulses with breaks or gaps between trains. From knowledge acquired through animal experiments, a computer program of the stimulation parameters is written confining between a range of magnitudes that are not injurious

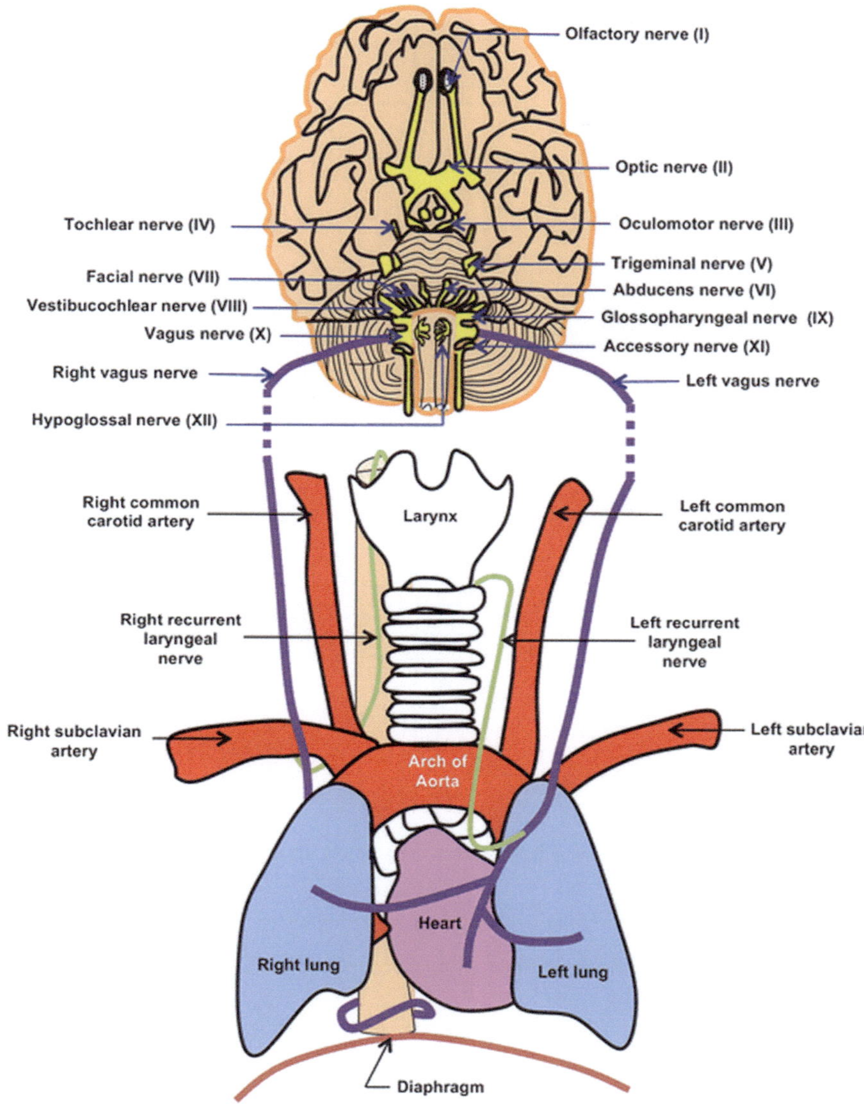

Fig. 18.1 The cranial nerves: origin and path of vagus nerves through the human body up to the diaphragm

to the nerves. Placing magnetic transducer (wand) over the implanted generator and using a laptop computer, programming is performed by a qualified medical professional. By bringing a magnet held in the hand over the pulse generator and then withdrawing the magnet away, the patient can also trigger a single train of stimuli. By holding or sticking the magnet with tape above the generator, the patient can

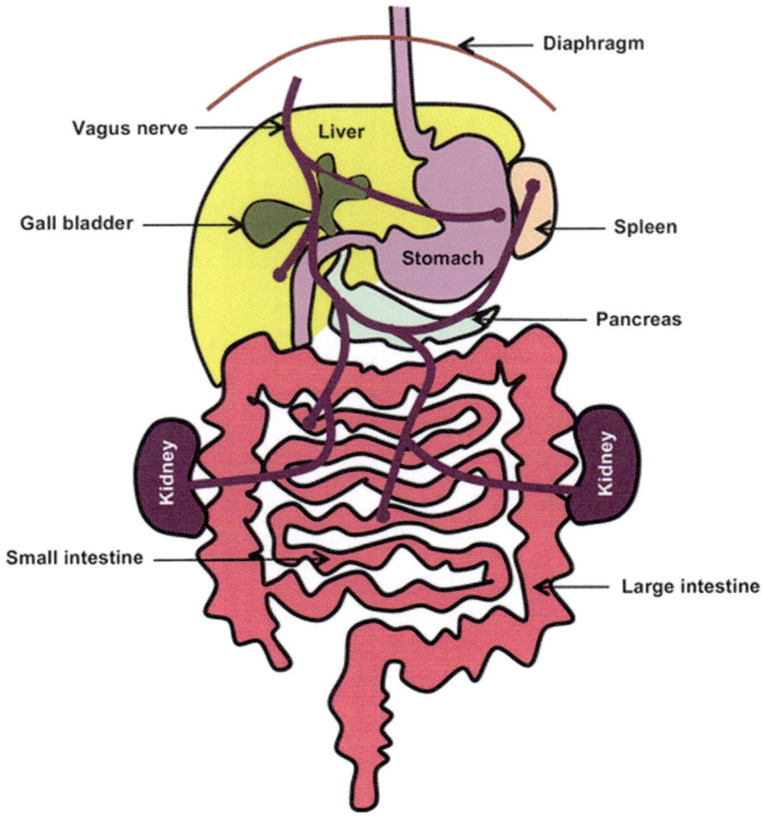

Fig. 18.2 Path of vagus nerve through the human body below the diaphragm showing the various organs that it touches

cease the stimulation for a period of time which can be as long as wished by the patient.

Further details of the VNS system are given as follows: The pulse generator is encased in a titanium enclosure which is a hermetically sealed. It is supplied power by a 3.3-V battery. It typically delivers current pulses of up to a few mA. The pulse width ranges from 100 to 1000 μs. The lead system consists of a set of bipolar electrodes and an anchor for fixation. These are enfolded around the left vagus nerve in the neck region, adjacent to the carotid artery. The VNS lead is several cm in length and a few mm in diameter. The inner conductor is shaped like a helix shape. Its constructional material is an alloy containing nickel, cobalt, chromium, and molybdenum. The lead impedance increases during the life of the implant. As a result, the projected battery life decreases. The battery life also decreases with increasing stimulation frequency, pulse width, and stimulation current. The stimulating electrodes are made of an alloy composed of platinum and iridium with center-to-center spacing of several mm. The insulation is made of silicone.

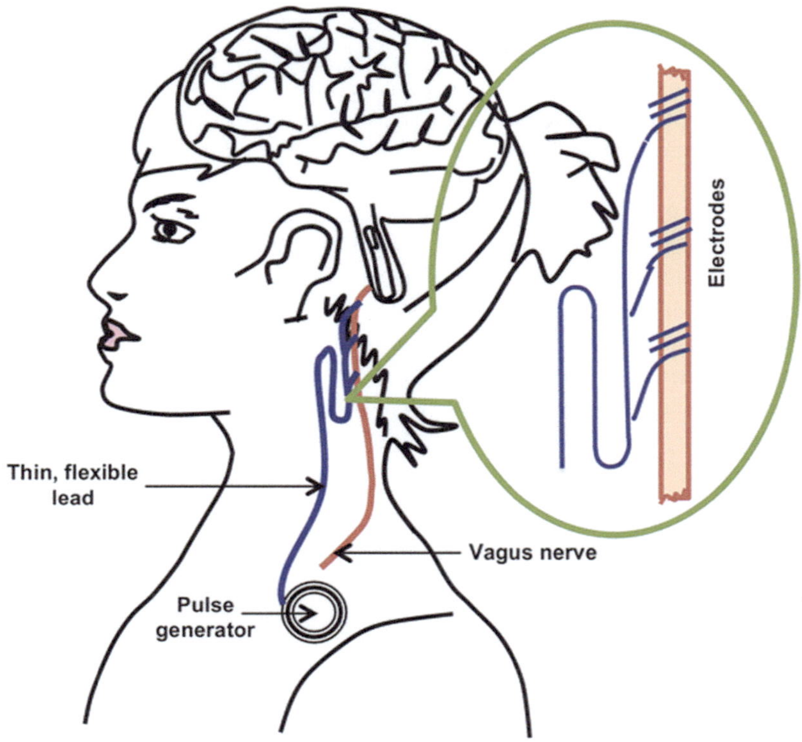

Fig. 18.3 Implanted vagus nerve stimulator, lead, and electrodes

18.7 Implantation of VNS System

A neurosurgeon performs the implantation of VNS system. Implantation for VNS is a lower-risk surgery. It is associated with very small incidence rate of surgical complications. These complications are much lower as compared to other neurosurgical interventions [6]. The implant procedure is a short outpatient procedure. It does not involve the brain, hence very simplified. The device is entrenched by keeping the patient under general anesthesia. Helical electrodes are placed on the leftward cervical vagus nerve. During this period, discontinuous stimulation is done using neurocybernetic prosthesis (Cyberonics, Inc.). The main component of this prosthesis system is a multi-programmable pulse generator. This pulse generator allows programming with a wide range of different parameter values. The pulse generator is implanted subcutaneously, i.e., under the skin in the upper region of the patient's chest. To carry out this implantation, a hypodermic pocket is formed in the anterior (in the front) wall of the chest. The pocket is located below the left collarbone or clavicle. The device is placed inside the pocket. After inserting the device, the lead wires whose tips are in the form of coils are arranged and wrapped around the vagus nerve. The complete surgical procedure for VNS is performed as a procedure completed in a single day stay at the hospital. Sometimes, in difficult situations, it may be essential for the patient to stay overnight in the hospital.

Stimulus parameters vary with patient. Notwithstanding, studies have suggested that maximum protection from seizures is achieved when stimuli are given periodically. The frequencies lie in the range 20–30 Hz [3]. Presently, a majority of patients are stimulated at a frequency of 30 Hz. A constant stimulation cycle of 35 s (30 s on and 5 s off) is followed. Clinical trials have established that VNS therapy decreases the attacks of complex partial seizures in most of the patients examined. About 20–40 % of patients have shown >50 % fall in the frequency of seizures.

On delivery of pulses, the patients experience a mild tingling sensation above the neck on its left side. These can be often ignored. Other feelings may include vocal alteration and heavy or forced breathing. Also observed are coughing, indigestion, and vomiting. Patient may experience swallowing difficulties. Side effects usually occur only during stimulation. They are found to diminish over time. Overall, they are mostly well tolerated by patients.

18.8 VNS in Depression

Major depressive disorder (MDD) stands as the fourth primary cause of ill-health globally [7]. Treatment has the aspiration of triumphing over and achieving complete remission from symptoms. Restoration of day-to-day activities naturally follows. It is equally important to prevent relapses and recurrences through medications that have antidepressant effects. Many types of psychotherapies that are time-limited may also be helpful. A combination of both may be gainfully exploited [8].

18.9 Reasons for Antidepressive Action of VNS

Antidepressant effects of VNS are suggested by the known anatomical protuberances of the vagus nerve to sundry brain regions. These regions have been incriminated in mood disorders [9]. Moreover, improvements were noticed in the mood and cognition of patients receiving VNS stimulation for epilepsy improvements. These hinted at the possible use of VNS for treatment of depression patients. It may be contended that mood improvements are explained by the decrease in frequency of seizures, whereby the patient feels comfortable and relaxed. Nonetheless, significant mood improvements were also detected in patients showing little or no decrease in seizures. This was a definite indicator of the positive effects of VNS on depression assuagement. Further, it is known that several anticonvulsant medications are used to treat mood disorders. Some of these drugs are carbamazepine, lamotrigine, gabapentin, and valproate. Hence, a logical expectation is that VNS too can bring about mood changes in analogy with these medications [8, 10].

Long-term medication trials have given useful results. To challenge these usual results, longer-term statistics concerning suggestive and practical results of patients undergoing VNS appear to be promising [11]. Particularly interesting is the observation that VNS may provide lasting and inexorable benefits. This is especially hailed enthusiastically, realizing the highly recurrent nature of MDD [7].

The influence of acute left cervical VNS on passionate grading of food in adult patients of major depression was examined. It was found that VNS device activation may be linked with severe change in provocation response to sugary foods in these patients. Serious VNS device activation induced a significant change in longing and urges for such foods [12, 13].

To some extent, antidepressant action of VNS is interpreted by the fact that afferent fibers extend to the nucleus tractus solitarius, which is the principal visceral sensory nucleus in the brain. This is further linked straightway and secondarily with several brain structures. These are the regions accountable for mood and anxiety regulation [7]. Notable examples of these regions are given below: The reticular formation is a set of interconnected nuclei located in the central core of the brain stem and consists of more than 100 small neural networks. It performs several autonomic functions: motor control (walking or running), sensory control (pain modulation), visceral control (breathing, heart rate, and blood pressure), and consciousness control (alertness and sleeping) [14]. The parabrachial nucleus (PBN) plays a prominent role in controlling the cardiovascular and respiratory functions [15]. The locus coeruleus handles the physiological responses to stress, decision-making, and panic. It is also associated with learning and memory. The amygdala has a role in generalized anxiety and is concerned with the processing of emotions such as fear conditioning and anger. The hypothalamus is responsible for the release of many essential hormones. The insula, also called the insular cortex, plays a crucial role in understanding what it feels like to be human. It is the fountainhead of social emotions such as sexual desire and abhorrence, gratification, and disgrace, wrongdoing, and reparation. It aids ethical intuitionism, understanding and sharing the feelings of others, and emotional responses to music. Thalamus is concerned with sensory perception and motor function regulation. The orbitofrontal cortex (OFC) is involved in the cognitive processing of decision-making. It is related to various functions such as sense of smell, emotions, and flexibility of behavior.

18.10 Drawbacks of VNS for Depression Treatment

Drawbacks of VNS for treating depression are the involvement of surgery and perioperative risks. The perioperative risks include the risks involved in preoperative, intraoperative, and postoperative phases of surgery, i.e., before, during, and after surgery. Potential side effects are hoarseness, throat pain, and coughing. Breathlessness, tingling or pricking sensation, etc., may accompany. Consequently, VNS treatment is advocated only for selected patients who have tried out the orthodox somatic treatments for MDD and were found to be either intolerant or lacking in response. "Somatic" indicates relating to the soma or body, as distinguished from the mind.

18.11 VNS for Obesity Treatment

"Obesity" is a term used to describe a medical condition of being extremely overweight. Generally, obese persons have a lot of fat in their bodies. Hence, they are said to be excessively or grossly fatty. Overweight and obesity conditions are defined from body mass index value of the individual. A person is classified as obese if the body weight is at least 20 % higher than normal value. People develop obesity gradually over a period of time. Obesity naturally grows if the energy expended by a person in pursuing daily chores, reckoned in calories, is less than the energy derived from foodstuff, also measured in calories. The (calorie input–calorie output) difference varies among people. A vital factor that might cause an increase in weight is the genetic makeup of a person. Following a sedentary lifestyle causes gain in weight. Eating high-fat foods and getting less sleep are also influential factors. The longer the time duration over which a person has remained overweight, the more difficult it is to lose weight. Obesity fosters further obesity. Dietary changes are necessary to prevent obesity. Also helpful is the reduction of daily calorie intake by eating more fruits, vegetables, and whole grains.

Regarding treating obesity with VNS, it is recalled that vagus nerve is an important nerve, transporting signals of satiety or absence-of-hunger from the digestive tract to the brain. Therefore, obesity treatment consists of stimulating the two branches of the vagus nerve, those moving leftwards and rightwards, of a patient at the same time with pulses of electrical current in a prearranged order. This sequence consists of two segments: a first segment characterized by the continuous application of pulses, alternating with a second segment with pulses absent [16]. Preferably, the location for application of electrical pulses to the vagus nerve is supradiaphragmatic, i.e., above the diaphragm. The applied pulses should have a magnitude <6 mA. However, this magnitude is chosen previously to be less than the intensity that can cause retching (heaving or gagging) in the person using the implant. The retching level is estimated during the preliminary implant. Its evaluation is essential to make sure that the patient does not suffer from vomiting sensation whenever he/she uses VNS. The pulse width is adjusted at a value <500 ms, and the pulse repetition frequency is kept at 20–30 Hz. Preferably, the magnitude of the second period is ~1.8× the first period in duration, i.e., duty cycle = on-period/off-period = 1/1.8. Using an external programmer, the physician varies the pulse parameters, including both on- and off-times.

18.12 VNS for Rheumatoid Arthritis

Rheumatoid arthritis is a disease affecting the autoimmune system. It degrades the coating of the joints. Chronic inflammation of the linkages, junctures, couplings, and other parts of the body is the consequence. It may eventually result in bone erosion. Joint deformity ensues. Current medications can reduce inflammation in joints. This relieves pain. It also prevents or slows down joint damage. The pain relief and joint prevention are achieved by overpowering the aggressive action of the immune cells with medicines. Antirheumatic drugs are substances like aspirin, corticosteroids, and other disease-modifying agents. In some patients, toxicity issues impose restriction on the use of these drugs. In others, treatment failure limits them.

In place of immunosuppressive drugs to block inflammation, the VNS approach uses a mini computer to generate a stimulating pulse. This pulse is imparted via an electrode joined to the vagus nerve of the patient [17]. The result is stimulation of the process of action potential generation in the vagus nerve. The action potentials thus produced propagate to the spleen. The spleen is a soft and cushioned abdominal organ filled with blood and lying neighboring to the stomach. It performs an important, critical function in the immune system of the body. The spleen produces and removes blood cells. The immune cells are a prime birthplace of tumor necrosis factor (TNF) in the body. TNF is a cell signaling protein tangled in general inflammation. TNF can induce fever because it is an endogenous pyrogen (cytokine). It can cause programmed cell death (PCD), cachexia (general weakness and weight loss), inflammation, etc.

VNS selectively stimulates the neural circuit feeding the spleen. This makes it possible to control the immune system regionally. Therefore, other body organs are not adversely affected. This stimulation enables the vagus nerve to shut down TNF release from the spleen. It thus applies a decelerating action counteracting destructive inflammation. In contrast, the effects of immunosuppressive drugs arriving in the body are not constrained merely to the spleen. They spread throughout the body. Hence, they produce unwanted effects in other organs of the body and are harmful to the patient at large.

18.13 Discussion and Conclusions

The inability to agreeably treat with medicine all patients of epilepsy has provided an incessant momentum to explore original types of treatment. VNS is an innocuous and effectual subordinate method of relief from epilepsy [18]. VNS has been applauded in epilepsy centers globally as a priceless and dependable non-pharmacological alternative, i.e., one not involving drugs for therapy. It is acknowledged as an anticonvulsant therapeutic technique to lessen the severity and frequency of seizures for patients who are not proper candidates for excision surgery [19].

Over the years, appreciation of the effects of VNS therapy related to the anatomy, physiology, and pathology of the nervous system has significantly improved. The role of the different operating parameters in therapy is being understood. Main parameters are frequency and intensity of the electrical pulses. Besides, the pulse width, duration, dose, etc., are also vital determinants of the effectiveness of therapy. The improved understanding is beginning to help in discovering other diseases that VNS might be able to treat besides epilepsy [20]. Looking at the anatomical disposition of vagus nerve, VNS is applicable in providing respite in conditions, such as depression, obesity, and rheumatoid arthritis.

Review Exercises

18.1 What is epilepsy? What is a breakthrough seizure? How is it triggered?

18.2 What is parasympathetic nervous system? Which important nerve exerts parasympathetic control over the heart and other organs?

18.3 What is a partial-onset seizure? What are the two types of partial-onset seizures?

18.4 For what type of seizure is vagus nerve stimulation advised? Define a medically refractory seizure.

18.5 Mention the names of two diseases in which VNS therapy may be considered apart from epilepsy.

18.6 "Vagus" is a Greek word. What does it mean? What does this meaning indicate about this nerve? Describe in what respects is vagus nerve a mixed cranial nerve?

18.7 What percentage of fibers of the vagus nerve is afferent? What fraction of fibers is of efferent type?

18.8 How big is a vagus nerve stimulator in size? In which part of the body is it implanted?

18.9 How is a VNS system implanted surgically? What is the typical stimulation frequency used in this therapy?

18.10 Besides epilepsy, in what major ailments the use of VNS is considered beneficial?

References

1. Ramani R (2008) Vagus nerve stimulation therapy for seizures. J Neurosurg Anesthesiol 20(1):29–35
2. Uthman BM, Reichl AM, Dean JC et al (2004) Effectiveness of vagus nerve stimulation in epilepsy patients. Neurology 63:1124–1126
3. Krahl SE (2012) Vagus nerve stimulation for epilepsy: a review of the peripheral mechanisms. Surg Neurol Int 3(Suppl 1):S47–S52
4. VNS Therapy for Epilepsy (2007) http://www.epilepsynorcal.org/docs/VNS.pdf. Accessed 15 Feb 2015

5. George MS, Sackeim HA, Rush AJ et al (2000) Vagus nerve stimulation (VNS): a new tool for brain research and therapy. Biol Psychiatry 47:287–295
6. Schachter SC, Schmidt D (2000) Vagus nerve stimulation. Martin Dunitz, London, p 101
7. Rong P-J, Fang J-L, Wang L-P et al (2012) Transcutaneous vagus nerve stimulation for the treatment of depression: a study protocol for a double blinded randomized clinical trial. BMC Complement Altern Med 12:255 (6 p)
8. Rush AJ, George MS, Sackeim HA et al (2000) Vagus nerve stimulation (VNS) for treatment-resistant depressions: a multicenter study. Biol Psychiatry 47:276–286
9. Walsh SP, Kling MA (2004) VNS and depression: current status and future directions. Expert Rev Med Devices 1(1):155–160
10. Panescu D (2005) Vagus nerve stimulation for the treatment of depression. IEEE Eng Med Biol Mag 24(6):68–72
11. Goodnick PJ, Rush AJ, George MS et al (2001) Vagus nerve stimulation in depression. Expert Opin Pharmacother 2(7):1061–1063
12. Bodenlos JS, Kose S, Borckardt JJ et al (2007) Vagus nerve stimulation and emotional responses to food among depressed patients. J Diabetes Sci Technol 1(5):771–779
13. Bodenlos JS, Kose S, Borckardt JJ et al (2007) Vagus nerve stimulation acutely alters food craving in adults with depression. Appetite 48(2):145–153
14. Reticular Formation (2014) Boundless anatomy and physiology. Boundless.com. https://www.boundless.com/physiology/textbooks/boundless-anatomy-and-physiology-textbook/central-nervous-system-cns-12/parts-of-the-brain-stem-117/reticular-formation-641-9022/. Accessed 16 Feb 2015
15. Mizusawa A, Ogawa H, Kikuchi Y et al (1995) Role of the parabrachial nucleus in ventilator. J Physiol 489(3):877–884
16. Barrett B, Reddy RK, Roslin MS (2003) Treatment of obesity by bilateral vagus nerve stimulation. US Patent US 6,587,719 B1, Jul 1
17. Andersson U, Tracey KJ (2012) A new approach to rheumatoid arthritis: treating inflammation with computerized nerve stimulation. Cerebrum 2012:3
18. Chayasirisobhon S, Koulouris S, Parker E et al (2001) Vagus nerve stimulation for refractory epilepsy. Perm J 5(4):21–26
19. Vonck K, Boon P (2008) The mechanism of action of vagus nerve stimulation therapy. Eur Neurol Rev 3(2):97–100
20. George MS, Nahas Z, Bohning DE et al (2002) Vagus nerve stimulation therapy: a research update. Neurology 59(6 Suppl 4):S56–S61

Chapter 19
Diaphragmatic/Phrenic Nerve Stimulation

Abstract Diaphragmatic/phrenic nerve stimulation is an alternate stand-in to mechanical ventilation for persons suffering from immedicable ventilatory insufficiency or failure. Suitable patients to receive benefit from this stimulation include those whose phrenic nerves and diaphragms are undamaged, and whose pulmonary function is satisfactory. The phrenic nerve begins from the cervical spine. It starts from the C3, C4 and C5 roots. It is the nerve that regulates and governs the movements of the diaphragm. The diaphragm is accountable for the volume of the air movement throughout natural breathing. The phrenic nerve stimulation device consists of an electrode surgically inserted and winding over the phrenic nerve. It is connected to a receiver operating at radio frequencies. This receiver is placed in the wall of the chest. Upon interception of radio-frequency signals from an external transmitter by an antenna that the patient wears over the receiver, regular electrical pulses are applied to the phrenic nerve. These pulses initiate contractions of the diaphragm, and the diaphragm contractions lead to the intake of air, similar to natural breathing. Hence, the implanted stimulator is called the breathing pacemaker. The respiratory rate is determined by the intensity, duration and rate of impulse. It is controlled by the external transmitter.

Keywords Respiration · Phrenic nerve pacing · Diaphragm pacing · Ventilation · Thoracotomy

19.1 Introduction

Diaphragmatic/phrenic nerve (D/P) stimulation is known by various names. The common names are phrenic pacing, diaphragm pacing, or electrophrenic respiration [1]. It consists of electrical stimulation of the diaphragm. This stimulation is carried out via the phrenic nerve. The phrenic nerve is a vital nerve constituting the foremost nerve supply to the diaphragm. The diaphragm is a respiratory muscle which handles and controls breathing [2]. The phrenic nerve stimulator is a system consisting of an electrode(s) implanted by surgical operation and wound around the phrenic nerve(s).

V. K. Khanna, *Implantable Medical Electronics*,
https://doi.org/10.1007/978-3-031-99868-3_19

It is connected to an implanted radio-frequency receiver placed in the chest wall. Radio-frequency signals are sent by an external transmitter to the device through an antenna which the patient wears over the receiver.

Some patients with damaged cervical spine still have an uninjured phrenic nerve. Implanted phrenic nerve stimulator supplies electrical pulses on a regular basis directly to this nerve causing the diaphragm to shrink and producing the inflow of air, which is of the same kind as natural breathing. For this intervention, unimpaired phrenic nerves and operational diaphragm muscles are indispensable [3, 4].

D/P nerve stimulation, either full- or part-time, is considered as a medically necessary option for patients suffering from hypoventilation (very shallow/very slow breathing) originating from various conditions, including severe, chronic pulmonary disease with ventilatory insufficiency, that requires invasive mechanical ventilation with pressure and rate support. It has been successfully used to treat respiratory paralysis caused by lesions of the brain stem and cervical spinal cord by providing ventilatory support with maximal patient mobility and quality of life [5]. Emphatically, it can only treat conditions affecting the upper motor neurons or the medullary respiratory control center but fails to take care of patients with primary neuromuscular disorder, phrenic nerve injury, or spinal cord injury at levels C3–C5 [6].

19.2 Respiration, Phrenic Nerves, and Diaphragm

Respiration is the action or process of inhaling and exhaling. It works predominantly under the supervisory instructions of the medullary respiratory control centers (RCs) located in the medulla oblongata and the pons of the brain. RCs are a group of nerve cells that control the tempo of breathing. They do so in rejoinder to variations in quantities of O_2, CO_2, and H^+ ions in the blood and cerebrospinal fluid. The RCs work in conjunction with voluntary cerebral influence [7]. The upper motor neurons (neurons starting in the motor cortex of the brain and ending within the medulla or the spinal cord) emanate from cell bodies in the medullary respiratory RCs. The axons of these neurons have synapses in the spinal cord in the neck at the levels C3–C5. Here, they exchange information with the lower motor neurons. The axons of these motor neurons form the phrenic nerves that innervate the diaphragm.

The diaphragm is sometimes called the thoracic diaphragm. It is a prime respiratory muscle, forming a vital part of the breathing system. It plays a central role in normal respiration. It is a sheet of muscle in the form of an egg-shaped cylindroid construction. This structure is crowned by the fibrous hemisphere. The diaphragm separates the chest from the abdomen. The central tendon is the uppermost component of the diaphragm and is made up of entwined collagenous fibers; collagen is the tough protein constituent of bones, cartilage, tendons, and other connective tissues. The cylindrical segment of the diaphragm is composed of an unbroken band of muscle fibers. The bulk of this band is straightway touching the internal facet of the lower ribs.

The lungs are confined in the chest or thoracic cavity. They are surrounded by the rib cage in which the ribs form the front, back, and sides with the upwardly arching diaphragm forming the floor of the cavity. During inhalation, the diaphragm undergoes contraction. During contraction, it pulls the central tendon down and is drawn interiorly into the abdominal cavity until it becomes flat. Simultaneously, the exterior intercostal muscles amid the ribs lift up the anterior rib cage. This enlarges the thoracic cavity. As the thoracic cavity becomes deeper and larger in size, space is provided for more air. To fill up this larger space, air is sucked in from the atmosphere. During exhalation, the rib cage falls back to its resting position. At the same time, the diaphragm relaxes and rises up to its dome-shaped position in the thorax. With the thoracic cavity decreasing in size, air in the lungs is expelled from the body.

As already mentioned, the diaphragm is supplied by the phrenic nerves. These nerves depend on the motoneurons of the cervical vertebrae C3–C5. The C4 neurological segment is responsible for initiation of a large amount of innervations. The third and fifth cervical nerves also contribute to stimulations. The phrenic nerves enclose sensory, motor, as well as sympathetic nerve fibers. The fibers are responsible for motor and sensory supplies. The motor supply is furnished by these fibers to the diaphragm. The sensory supply is provided by them to the middle tendon. The right and left phrenic nerves scurry along the scalene anterior muscle, the lateral muscle of inferior half of neck, deep to the carotid sheath. The carotid sheath is the dense tissue consisting of fibers and covering the carotid artery, internal jugular vein, and vagus nerve on either side (Fig. 19.1). The path followed by the right phrenic nerve is as follows: It passes over the brachiocephalic artery; this artery is the first branch of aortic arch. Then it moves rearward to the subclavian vein; this is the continuation of the axillary vein in the upper body. Thereafter, it moves over the right atrium, the upper right chamber of the heart. Finally, it pierces the diaphragm at the plane of T8 (eighth thoracic vertebrae). The trajectory of the left phrenic nerve is as under: It sprints rearward to the left subclavian vein. Then it flees over the pericardium (the membranous sac enclosing the heart) of the left ventricle (the lower chamber of the heart receiving blood from the left upper chamber or atrium). Eventually, it enters the left hemidiaphragm (half of the diaphragm) separately.

Any injury to any one of the three parts, namely, the central nervous system, phrenic nerve, or diaphragm, undermines the function of the diaphragm. Consequently, respiratory difficulties arise. These appear notably as a decrease in tidal volume and minute ventilation. Discernible abnormalities in gas exchange are found. Likewise, any damage to the spinal nerves or cord between the C0 and C4 vertebrae engenders quadriplegia or tetraplegia, paralysis of the full body below neck level [8]. It can harm breathing, necessitating lifelong ventilation support. Artificial respiratory support device is required for a long period of time in up to 4 % of cases of spinal cord injury for providing normal breathing. Quadriplegia decreases life expectancy. It influences motor and sensory functions and demeans the quality of life. To some extent, the extent of functional disablement and consequent self-reliance and self-sufficiency achievable are determined by the inclemency of the injury and the position where lesion is situated in the spinal cord.

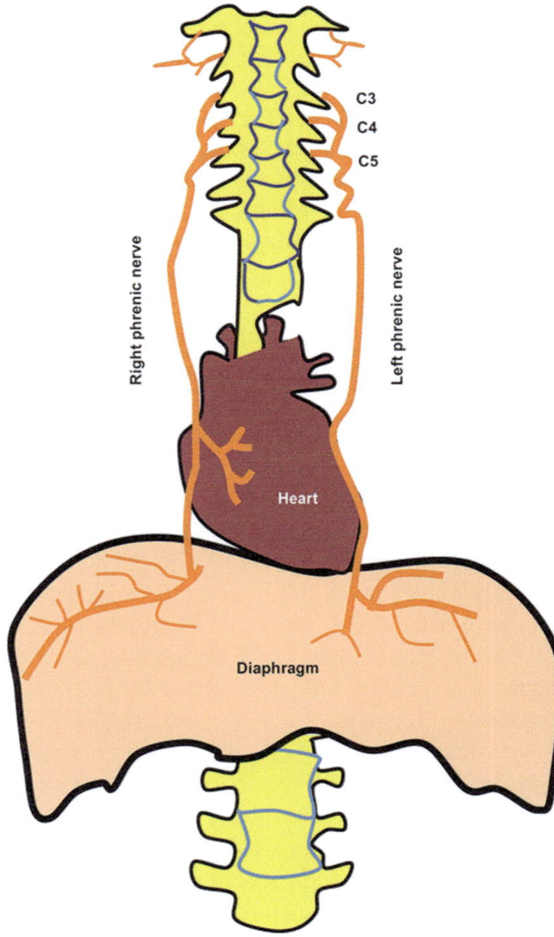

Fig. 19.1 Progression of the two phrenic nerves from spinal nerves to the diaphragm

19.3 Diaphragm Pacing Versus Mechanical Ventilation

The prevailing treatment to manage respiratory failure patients involves the use of artificial mechanical ventilators [9]. In mechanical ventilation, air is pumped into the lungs. Air forcing is done under positive pressure. In this manner, function of the lung is executed. But this type of ventilation messes up the coughing ability of the patient. The unwanted result is occurrence of respiratory infections. It also limits speech. Infections are avoidable by the regular suction of secretions. But these secretions are very disturbing and troublesome to patients. Due to these comorbidities and complications, ventilated patients can execute limited activities. Thus chronic use of ventilators reduces independence of the patient. It causes increased death rate in comparison to similar patients who are not supported by ventilator. D/P nerve stimulation provides physiological respiratory function far superior to that offered by mechanical ventilators. It allows the patient to have increased mobility and a more normal lifestyle.

Physiologically, pacing is more accurate and comfortable for the receiving patient. The main reason for superiority of pacing over mechanical ventilation is that in pacing, the inhaled air is pulled inward into the lungs. The diaphragm exerts the pulling action under negative pressure, hence by the musculature (the system or arrangement of muscles). Improved venous return, i.e., negative, not positive pressure, is beneficial. In ventilation, air is thrust into the chest under positive mechanical pressure.

Pacing allows for normal breathing and speech patterns. It eases eating and drinking. Also, it improves sense of smell [10]. This is because pacing enables flow of air through the passage in the nose. Thereby the sense of smell is restored.

The pacing device is small in size. It weighs around a pound (453.59 g). It does not require the massive batteries and discomfited tubing of a mechanical ventilator. Thus it greatly enhances the mobility of patients.

The pacing device operates noiselessly and is unobtrusive because the external components are smaller in size as compared to ventilator. This feature improves the ability of patients to keenly participate in community and learning activities.

One primary reason why many patients prefer to choose pacing over mechanical ventilation is the total cost difference (initial + running cost). Diaphragm pacing costs about 90 % less than the comparable costs for keeping a patient on a positive-pressure ventilator. It is cost-effective because patients need not be admitted in hospitals. The outlay on a ventilator and related throwaway materials is cut down. Of course, the start-up cost for a breathing pacemaker system may be discouraging. But one should not be pessimistic. Ongoing operating costs are very low. Typically in the first year following implantation, a positive return on investment (ROI) is obtained.

Compared to ventilator-supported patients, phrenic pacing patients have shown a lower infection rate. This is because the patient has a much lower probability of infections in the upper airway. These infections include pneumonia associated with ventilator use. The infection rate is also lowered by the cutback in evacuation. By getting rid of external circuits for humidifier and ventilator, the likelihood of infections decreases. Moreover, it is possible that the tracheostomy (or tracheotomy) tube may be removed; tracheostomy is an incision in windpipe. Some patients have their tracheostomy closed. When a tracheostomy is not required, the patient can breathe, speak, eat, and drink like a normal person. Hence, pacing reduces hospital readmissions.

Thus on the whole, when compared to mechanical ventilation, pacing noticeably improves the quality of life of the patient. It significantly reduces upper airway infections and improves the quality of speech. It reduces costs for disposable equipment. On the whole, it reduces mortality and prolongs life span (Table 19.1).

Table 19.1 Mechanical ventilation and diaphragm pacing

Sl. No.	Mechanical ventilation	Diaphragm pacing
1.	Phrenic nerve of the patient need not be intact	Phrenic nerve of the patient must be intact
2.	An external machine is used to deliver air and oxygen to the patient. This air/O_2 mixture is supplied through a tube placed in the windpipe	It uses the patient's own diaphragm as the ventilator
3.	The patient is tied to the ventilator through the tracheostomy tube; tracheostomy is a surgical procedure in which an opening is incised through the neck into the trachea (windpipe)	Patient is not tethered to the short length of ventilator tubing
4.	It permits limited mobility to the patient	It provides freedom to the patient from being constantly tied up to the ventilator machine
5.	Normal life activities cannot be performed by the patient	The patient can participate in normal activities, with ability to smell, eat, and talk
6.	Quality of life of the patient is low	It improves the quality of life of the patient
7.	Daily incremental cost for ICU patients is high	It is far less expensive when matched with the long-term cost of the ventilator facility, disposables, and caregiver support requirements
8.	The patient needs frequent suctioning and hospitalizations	Less need of suctioning and fewer hospitalizations of the patient are required
9.	It introduces complications and gives decreased life expectancy due to respiratory infections	It introduces less complications and increases life expectancy

19.4 Indications for D/P Nerve Stimulator

As repeatedly mentioned, for phrenic pacing to be effective, it is mandatory that patient's phrenic nerves should be intact. Moreover, the diaphragm should be functional. Also, uncompromised lung function is essential. Besides these indispensable requirements, the patient ought to be mentally alert, competent, and motivated. Then only the patient will be able to reap the benefits of guidance required for restoration to normalcy. Before implantation, the patient may go through a few tests. These tests are electromyography (EMG) to gauge conduction by phrenic nerve, investigation of pulmonary function (noninvasive diagnostic tests to obtain measurable feedback on the functioning of the lungs), and a polysomnography (sleep study).

Candidates to be managed by diaphragm pacing can be subdivided into several groups of respiratory insufficiency. The first group of patients is those who require unceasing support by ventilator arising from central alveolar hypoventilation (shallow breaths, especially during sleep). Also included in this group are the patients who are dependent upon the therapy of sporadic or enduring use of a mechanical ventilator along with maintenance of a permanent tracheotomy aperture. The second group belongs to patients suffering from reduced ventilator drive during day or night, i.e., sleep apnea (pauses in breathing or shallow breaths during sleep). The third group pertains to patients suffering from Ondine's curse (failure of central nervous system by birth in exercising control over breathing while slumbering). The therapy should also be thought out for patients in whom the central nervous system is implicated for apnea (suspension of breathing). The location of lesion in the central nervous system must be higher than the third cervical level. This is because phrenic nerve stimulation can enable pacing of the diaphragm only when the nerve cell bodies situated in the frontward horns of C3–C5 are usable. In the fourth group are stroke victim patients. Under the fifth group fall the patients having injury or disease in the brain stem or spinal cord, i.e., patients in whom neuronal conduction is broken up at the upper cervical level. Characteristically befitting aspirants for phrenic pacing are patients having cervical injuries at C3 or higher level.

19.5 The Pacing System

Internal components of the pacing system consist of a generator/receiver, electrodes, and leads. External components are an antenna and a transmitter (Fig. 19.2). The electrodes are sutured to the subcutaneous tissue of the phrenic nerves in the neck or in the chest. The generator/radio receiver is protected in a receptacle made under the skin in the chest wall. The external transmitter/antenna assembly oversees the entire functioning of the system. It supplies power to the system. It produces the stimuli and controls the intensity of pulses, their duration, and rate of generation. Thus the rate of respiration can be adjusted by tuning the relevant parameters in the transmitter/antenna section.

The transmitter/antenna device is small and light. The antenna is secured externally over the receiver. The transmitter contains the batteries. There are no batteries inside the implanted components. Since the external transmitter supplies the power, and no batteries are present inside the body of the patient, an obvious expectation is that the implanted components will continue to work in a satisfactory manner during the full lifetime of the patient.

A smooth, rhythmic contraction of the diaphragm is achieved upon applying repetitive stimulus patterns produced by the pacer to the phrenic nerves. This results in a regular breathing cycle. The breathing cycle consists of inhalation when phrenic nerves are stimulated electrically and exhalation when such stimulation pauses. The degree of diaphragmatic contraction which determines the tidal volume is controlled by the magnitude of electrical voltage that has been applied. Two types of pacing are used: unilateral or bilateral. This selection is based on the condition of the patient.

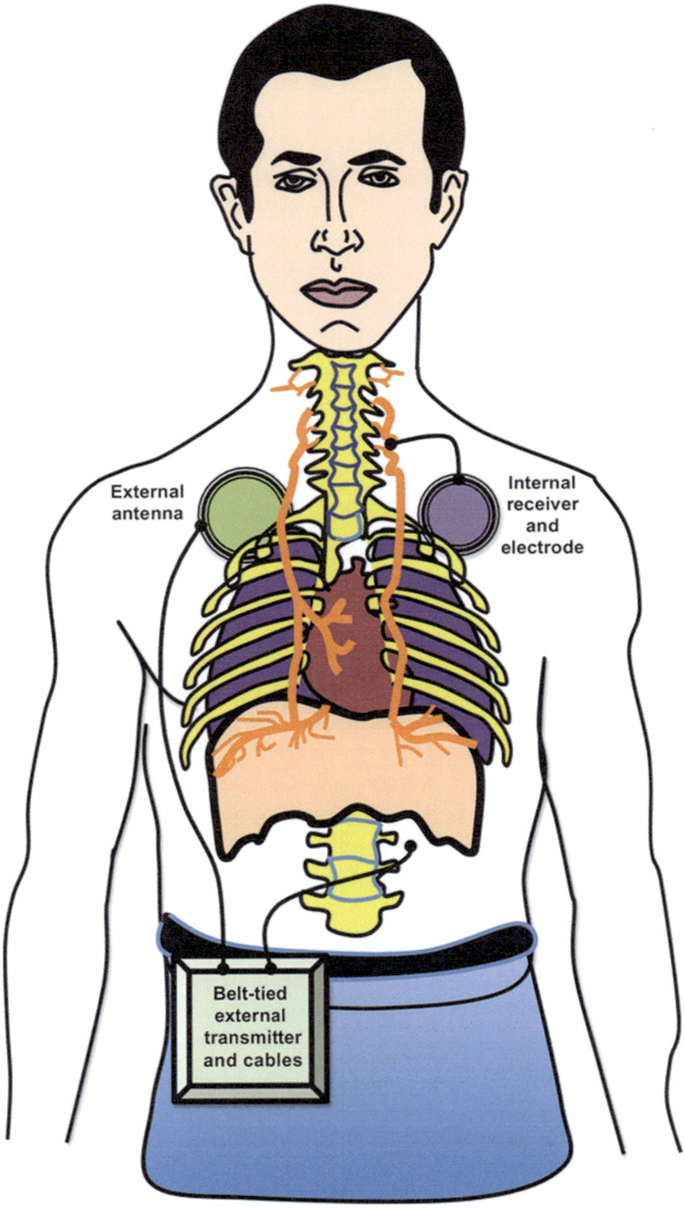

Fig. 19.2 External and internal components of the phrenic nerve stimulator

19.6 Surgical Procedure

After the patient has suffered injury, a period >6 months is allowed to elapse. During this time span, it is expected that the patient has fully recovered from injury to spinal cord and/or phrenic nerves. Then a surgical operation can be considered to implant the phrenic nerve pacing device. However, previous to the pacing device is implanted, phrenic nerve conduction studies are necessary. Hemidiaphragm responses after phrenic nerve stimulation are studied to make certain that the phrenic nerves and diaphragm of the patient are functional [11].

The surgical procedure used to insert the radio-frequency receiver is called bilateral anterior thoracotomy (clamshell incision). It is done through the second or third rib approach, via an incision in the skin measuring about 5–6 cm in length [12]. Thoracotomy is the process of making a cut into the chest wall. Electrode placement around the phrenic nerves is carried out followed by their stabilization. In turn, the electrodes are joined to a passive receiver. They are embedded under the skin. This is done bilaterally to complete the surgery. The recovery after the operation is rapid.

The implanted electrodes are sturdy and long-lasting. Their lifetime is typically >20 years in some instances. The battery-powered external transmitter is robust. It may need to be substituted after a span ranging between 5 and 10 years. The cables running from the external transmitter to the antenna are encrusted with silicone. But careless handling may damage them and should be avoided. Notwithstanding, they are a readily available, low-cost item. Hence, the patient need not worry about them.

19.7 Training, Rehabilitation, and Precautions

Stimulation is not started immediately after implantation surgery. It commences 4–6 weeks after implantation. This time delay is necessary to allow curing and fibrosis (formation of fibrous connective tissue in reparative response to injury or damage) around the leads. Initially, it is generally done for 1–2 h per day. As the patient adapts to the pacing, the pacing period is gradually increased. Fluoroscopy is used to determine the current that produces maximal diaphragm excursion. It is an X-ray procedure to see internal organs in motion through real-time video images.

During the training and rehabilitation period, the doctor tries to obtain optimal contraction of the diaphragm using the lowest applied current. If this contraction is optimum, the tidal volume (the lung volume = the volume of air sucked in or expelled out in quiet breathing at rest) has maximum value. For maximizing the tidal volume, the intensity and duration of the stimulus are adjusted. The stimulation rate is fixed at the most clinically appropriate rate. This rate is ~8–10 breaths per minute for adults. For up to 3–6 months, full pacing may not be attained.

There are two modes in which phrenic pacing may be applied: intermittent or continuous. For patients who use D/P pacing on a 24-h basis, a caregiver must be present at all times. Sometimes emergency situations may arise. To cope with these circumstances, a backup transmitter, mechanical ventilator, or other respiratory support device should be available at hand. As a precautionary measure, external stimuli that might interfere with the pacemaker must be avoided. These are magnetic resonance imaging, lithotripsy (a procedure to break stones in the kidney, bladder, or ureter into smaller pieces with the help of ultrasonic shock waves so that these pieces can pass out of the body), diathermy (a physical therapy in which tissues are heated by high-frequency electric current), metal detectors, etc.

19.8 Discussion and Conclusions

Diaphragm pacing bestows on the patient the ability to speak. It increases the patient's mobility. The breathing by diaphragm pacing is like normal breathing. The patient has the ability to smell, i.e., olfactory sensation. He/she has good social interaction. Participation of the patient in rehabilitation activities is extended. More occupational opportunities are provided at a lower cost than mechanical ventilation [13].

Review Exercises

19.1 What are the different names by which diaphragmatic/phrenic nerve stimulation is called?

19.2 For what disease conditions is D/P nerve stimulation advised? Under what conditions this treatment fails to provide comfort?

19.3 Explain how the thoracic diaphragm takes part in the breathing system. How do phrenic nerves control the movements of the diaphragm?

19.4 What is meant by hypoventilation? What is the conventional treatment offered to respiratory failure patients?

19.5 In what ways is diaphragm pacing superior to mechanical ventilation? Mention one shortcoming of diaphragm pacing.

19.6 What type of patients is suitable for receiving diaphragm pacing? For which type of patients this technique will not work?

19.7 Describe a typical diaphragm pacing system, mentioning its main parts and describing their functions. Does this system have batteries inside?

19.8 How much period after injury should be typically allowed for planning the implantation of a phrenic nerve pacing system? What is thoracotomy?

19.9 What is the time frame allowed after implantation before starting pacing? What imaging technique is used to find the current for maximum excursion of the diaphragm?

19.10 Name some forms of external stimuli that are likely to interfere with the operation of phrenic nerve pacing system.

References

1. Surani S (2011) Diaphragmatic pacemaker, modern pacemakers—present and future. In: Das MR (ed) InTech. http://www.intechopen.com/books/modernpacemakers-present-and-future/diaphragmatic-pacemaker. Accessed 11 Feb 2015
2. Sardarzadeh S, Pooyan M (2012) An implantable electrical stimulator for phrenic nerve stimulation. J Biomed Sci Eng 5:141–145
3. Taira T, Takeda N, Itoh K et al (2003) Phrenic nerve stimulation for diaphragm pacing with a spinal cord stimulator: technical note. Surg Neurol 59(2):128–132
4. Mitsuyama T, Taira T, Takeda N (2003) Diaphragm pacing with the spinal cord stimulator. Acta Neurochir Suppl 87:89–92
5. Sitthinamsuwan B, Nunta-aree S (2009) Phrenic nerve stimulation for diaphragmatic pacing in a patient with high cervical spinal cord injury. J Med Assoc Thai 92(12):1691–1695
6. Taira T, Hori T (2007) Diaphragm pacing with a spinal cord stimulator: current state and future directions. Acta Neurochir Suppl 97(Pt 1):289–292
7. Creasey G, Elefteriades J, DiMarco A et al (1996) Electrical stimulation to restore respiration. J Rehabil Res Dev 33(2):123–132
8. Son BC, Kim DR, Kim IS et al (2013) Phrenic nerve stimulation for diaphragm pacing in a quadriplegic patient. J Korean Neurosurg Soc 54(4):359–362
9. Hall JB, McShane PJ (2013) Overview of mechanical ventilation. http://www.merckmanuals.com/professional/critical_care_medicine/respiatory_failure_and_mechanical_ventilation/overview_of_mechanical_ventilation.html. Accessed 11 Feb 2015
10. Adler D, Gonzalez-Bermejo J, Duguet A et al (2009) Diaphragm pacing restores olfaction in tetraplegia. Eur Respir J 34(2):365–370
11. Wetstein L (1987) Technique for implantation of phrenic nerve electrodes. Ann Thorac Surg 43(3):335–336
12. Bhimji S, Mosenifar Z (2013) Diaphragm pacing technique. Medscape. http://emedicine.medscape.com/article/1970348-technique. Accessed 11 Feb 2015
13. Stuban S (2006) The latest facts about the diaphragm pacing stimulation system. Ventilator-assisted Living 20(3):1–3. http://www.synapsebiomedical.com/publications/images/VAL_20-3p1-3.pdf. Accessed 11 Feb 2015

Chapter 20
Sacral Nerve Stimulation

Abstract Urinary and fecal incontinence are the two principal disorders for which sacral nerve stimulation is an effective treatment. This treatment involves the application of electrical stimulation to the sacral nerves via an implantable system. The system consists of an electrode placed extradurally (outside the dura mater), close to the third sacral anterior nerve root (S3). Along with the electrode is stationed an implantable pulse generator (IPG). An extension cord connects the electrode to the generator. The rationale of the therapy is that the functioning of pelvic floor muscles supporting the bladder and bowel can be controlled by stimulating the sacral nerves electrically. It has been reported that moderately low amplitudes of the signal ~0–3.0 V suffice to provide relief. Within the advocated parameter limits (210 μs, 10–16 Hz), uninterrupted stimulation is practicable without evoking any sensation of pain.

Keywords Sacral nerve · SNS · SNM · Urinary incontinence · Sphincter · Overactive bladder · Fecal incontinence · Tined lead

20.1 Introduction

Urinary retention and overactive bladder are serious bladder control problems. They lead to a lot of shame, awkwardness, and discomfiture to the sufferers. Overactive bladder syndrome disapprovingly impacts the everyday life of anguished people [1]. The anxiety and restlessness aroused by bladder control issues can often be frustrating, embarrassing, and sometimes devastating. These concerns prevent the patients from controlling when and how much to urinate. For such patients, easy routine chores and social activities may become puzzling. The patients may have to trim down their interests and hobbies. In extreme cases, they may even discontinue working. They are always preoccupied with bladder menace and trapped by a constant apprehension of urine trickling mishaps. They may develop the habit to stay nearby a rest room always. Patients with bladder control tribulations can be of any age, child, adolescent, or old. For long, incontinence has been treated using pharmaceuticals and surgery. Pharmaceuticals are often inadequate to resolve the matter and also

V. K. Khanna, *Implantable Medical Electronics*,
https://doi.org/10.1007/978-3-031-99868-3_20

cause unwanted side effects. Several surgical procedures have a low success rate and are irreversible [2]. They may have probably undergone fruitless treatment attempts with drugs, counseling on modification of behavior, changes in diet, and exercises for strengthening the muscles of the pelvic floor. They may have sometimes used a catheter to empty the bladder without any satisfying relief from symptoms. In the past, the patients unresponsive to conventional therapies had only few options. But now the doctors advise the patients to consider a therapy called sacral nerve stimulation (SNS) [3].

The sacral nerves are a subset of the spinal nerves. Looking at the spinal nerves, these nerves comprise a mixed set of 31 pairs of nerves connecting the spinal cord with different parts of the body. They are the carriers of three types of signals between the spinal cord and related part of the body. These are the motor signals for coordinating movements, sensory signals for detection, and autonomic or involuntary signals. Starting from the top end of the spine and moving downwards, the spinal nerves consist of the 8 cervical nerves, the 12 thoracic nerves, the 5 lumbar nerves, the 5 sacral nerves, and the single coccygeal nerve, as shown in Fig. 20.1.

The sacral nerves are the nerves at the base of the spine that control the bladder, the bowel, the pelvic floor, and allied muscles, as depicted in Fig. 20.2.

SNS, also branded as sacral neuromodulation (SNM), is a minimally invasive procedure. In this procedure, a programmable pulse generator is implanted for delivering low-amplitude electrical current pulses to stimulate the sacral nerves [4]. These pulses are delivered via quadripolar tined leads. The pulse delivery is done through the S3 foramen. SNS is the treatment for the management of pelvi-perineal dysfunction in adult male and female patients. These patients show detrusor overactivity, non-obstructive urinary retention, or painful bladder syndrome. To be eligible for SNS, the patients must have faced one of these conditions for at least a year. They must have shown nonresponsiveness to applicable orthodox pharmaceutical, behavioral, and medical treatments. Then only their urinary dysfunction can be qualified as refractory to prevailing treatment.

Since the late 1990s, the Food and Drug Administration (FDA) has granted approval to SNS for providing relief from voiding dysfunction that is refractory in nature. Since 1997, its usage for urge incontinence (UI) has been approved. From 1999, its deployment for urgency–frequency syndrome (U/F) has been accepted and so also for idiopathic, non-obstructive urinary retention (UR) [5]. Even though the mechanism of action has evaded understanding, it has emerged as a valuable treatment for voiding dysfunction. This is true at both extremes of urgency–frequency and urinary retention. It is successful in the long run and safe too. This success has been reported by copious findings [6, 7]. SNS treatment has been applied to the above types of pelvic disorder symptoms based on increasing awareness regarding the effects of pelvic floor muscle dysfunction on disordering of regular sensory afferent signaling and reflex pathways. This disorganization is connected with bladder and bowel functioning, as also with the feeling of chronic pain [8].

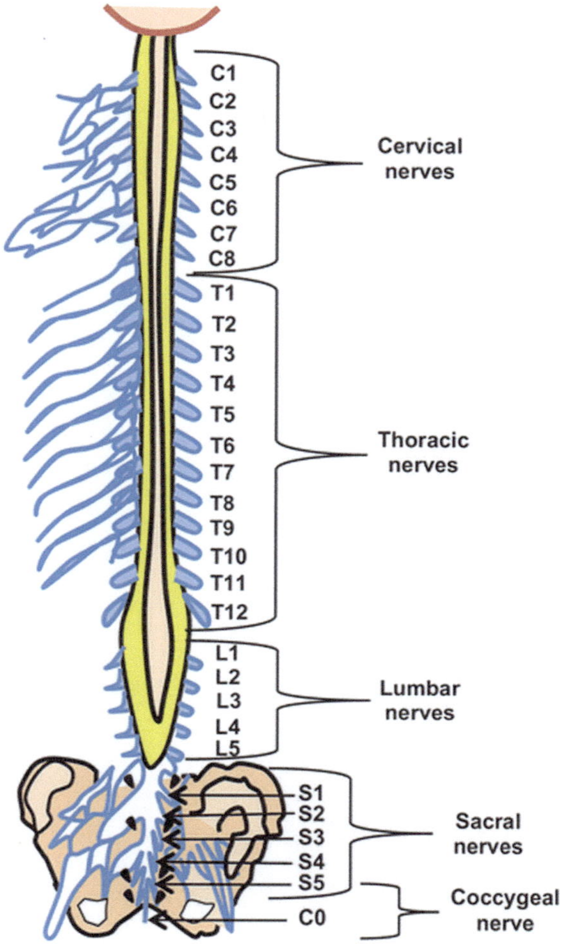

Fig. 20.1 Posterior view of the spinal nerves

20.2 The Urinary System and Bladder Control Problems

As a prelude to understanding the SNS, it is rewarding to learn the functioning of the urinary system [9]. The urinary system comprises six main components: the two kidneys, two ureters, the bladder, and the urethra. The kidneys are organs in the abdominal cavity. These organs resemble the familiar beans, edible seeds of leguminous plants. They are responsible for rejection of surplus fluid and unusable/unwanted substances from the blood to produce urine. The ureters are 25–30 cm long tubes with diameters of 0.3–0.4 cm. They are made of smooth muscle fibers. These fibers impel the urine from the kidneys toward the bladder. The bladder accumulates the urine exuded by kidneys. The urethra is a duct connecting the urinary bladder to the genitals. The urine is drained out through urethra. The opening and closing of the urethra are controlled by a cylindrical muscle called sphincter muscle.

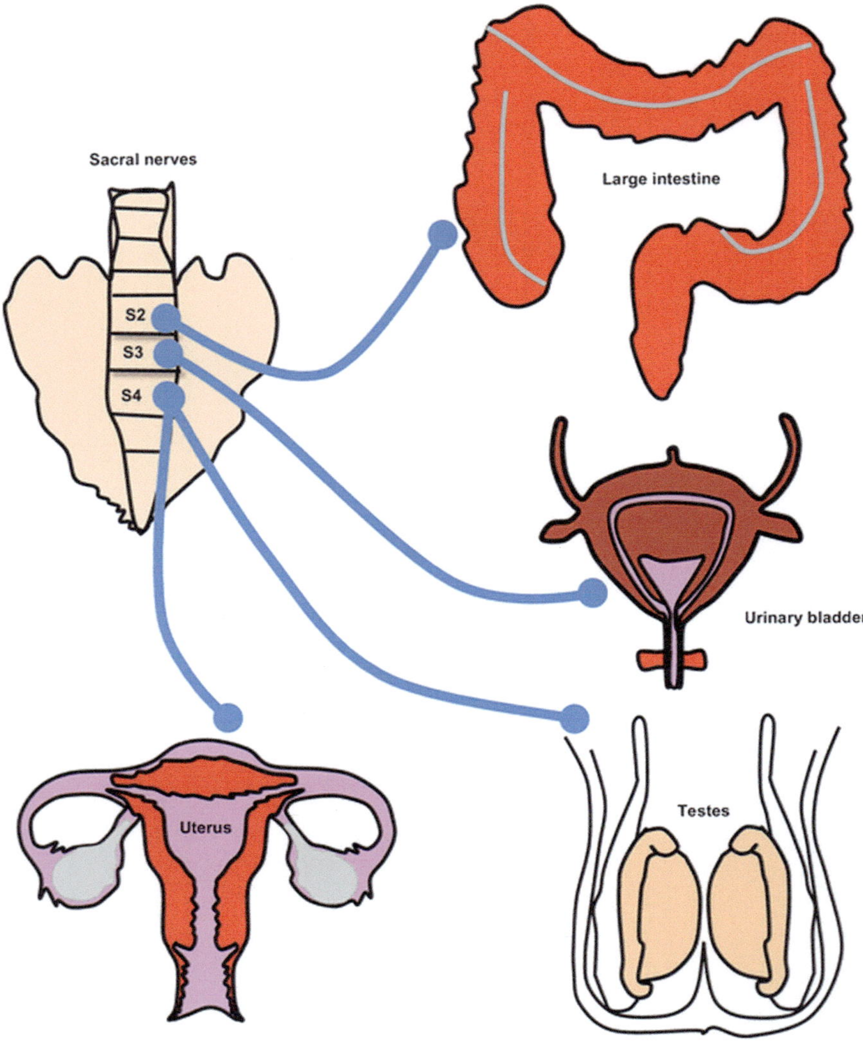

Fig. 20.2 Organs controlled by the sacral nerves

When the bladder starts to fill with urine, the information about "bladder filling" is sent to the brain through the sacral nerves. The brain is able to know that the bladder is being filled up. As the bladder becomes full, this message intensifies. With the strengthening of the message, one decides to urinate. Then the brain dispatches a communiqué to the bladder through the agency of sacral nerves, commanding the bladder muscle to undergo contraction. On receiving the message, the pelvic muscles experience relaxation, allowing the urine to trickle out from the bladder. This is the process of urination, which is usually under voluntary control. One can decide when and where one wants to urinate.

Upon disruption of the communication in two directions between the brain and bladder, as happens in disease condition, patients begin to show symptoms of problems faced in bladder control. SNS augments and extends the brain–bladder communication, thereby alleviating these symptoms.

20.3 Indications for SNS

Three body functions are vital decisive factors for ability of a person to withhold urine/feces: (1) a storage function performed by the urethra/bladder or colon, (2) a doorkeeper function executed by the urethral or anal sphincter, and (3) the ability of the brain and nerve sensitivity to allow/disallow the discharge of urine or feces.

Incontinence is the consequence of a dysfunction or deficiency in any of these factors. Urinary incontinence manifests itself in two forms. These forms are stress and urge incontinences. Stress incontinence arises from an unsteady and unreliable detrusor muscle (three-layered smooth muscular component of the urinary bladder wall) controlling the urinary sphincter (ring-shaped circular band of muscle, surrounding and able to contract and close a body orifice). Weakening of the detrusor muscle causes urine leakage from the bladder. This could arise from any pressure on the abdomen produced by movements resulting from sneezing, coughing, etc. Urge incontinence is represented by an abrupt impulsive call to micturate and helplessness to restrain urination for a time span; hyperactivity of the urinary sphincter accompanies the trouble. Both ailments are treated by SNS through nonstop stimulation of the sacral nerves which exercise dominating influence over the urinary sphincter.

To clarify a few terms, urge incontinence means the unintentional loss of urine, as without one's conscious wishes. It is characterized by an abrupt, intense desire for voiding. Urgency–frequency is observed as frequent, uncontrollable urge to urinate. Frequent urination means needing to pass urine more often than usual. Urgent urination is developing a sudden, strong desire to urinate. Urinary retention or ischuria is the inability to empty the bladder. Overactive bladder or spastic bladder, a problem in which the bladder squeezes urine out at the mistaken time, includes urge incontinence and urgency–frequency. Either one of these is present or both exist in combination.

SNS may be the only therapy available in some pathological conditions of the pelvis. Clean intermittent/sporadic self-catheterization (CISC) represents the solo treatment of chronic urinary retention. Catheters implanted in the bladder, either suprapubic (above the pubis) catheter or transurethral (via urethra) catheter, are the types of catheters used. The CISC is an excruciating and tormenting process; therefore, SNS is an effective alternative. Sacral nerves control and induce several bodily functions. Among these are pelvic floor disorders such as urinary incontinence; urinary urge/frequency; chronic urinary retention; pelvic voiding dysfunction (lack of coordination between the bladder muscle and the urethra); bowel dysfunction, e.g., constipation/obstipation, diarrhea, and irritable bowel syndrome (gastrointestinal functional disorder); erectile dysfunction (impotence or sexual dysfunction);

and lingering pain syndromes, a chronic neurological disorder. Disorders of these functions can be treated by electrical stimulation.

SNS has grown to be a reputable treatment for two classes of patients within urological practice: (1) patients with detrusor overactivity of unknown origin, manifesting in UI, U/F, and UR, and (2) women ailing from urinary retention with unidentified causes or defects in the distal urethral sphincter, called Fowler's syndrome in honor of Prof. Clare J. Fowler, who first chronicled it in 1985 [10].

20.4 Diagnosis and Suitability

Incontinence or lack of control expressively impacts the quality of life of a person. In the initial stages, a patient is often reluctant to discuss the problem. A patient visits a physician's clinic when the urinary problems aggravate to the extent of posing troubles in the workplace or in social events. Usually the family physician refers the patient to a physician specializing in urinary tract/urogenital system called urologist, for diagnosing the detailed genesis of incontinence. Patient history is reviewed and a physical examination is performed. To determine the root cause or source of the incontinence, the doctor assesses the capacity of the bladder. Functioning of the urethral/sphincter also comes under scrutiny.

Cystoscopy or cystourethroscopy is the endoscopy of the urinary bladder via the urethra. It is a nonsurgical procedure using a flexible tube with camera and light arrangement. Local anesthesia is adequate to carry out this procedure. The urologist can see the interior lining of urinary bladder and urethra with a scope (cystoscope). The examination helps to diagnose bladder tumors. Any obstruction of the bladder is identified. Moreover, any abnormalities of the bladder and its lining are explored. Such exhaustive tests are necessary to rule out all additional reasons that are likely to create urinary urgency and frequency, before prescribing SNS. After confirmation by the doctor that the reason of urinary incontinence is sphincter insufficiency, SNS is attempted on a temporary basis. For patients complaining of fecal incontinence, diagnosis and treatment proceed on the same lines as in the case of urinary incontinence [11–15]. If a patient does not show any improvement with a temporary implant, he/she is not considered for the permanent implant.

Bladder control problems that may be corrected by SNS therapy are:

1. Patient must be irresponsive to conformist therapy. The same must be evidenced by proper documentation. The documents should show that behavioral therapy was tried without success. Evidence for pharmacological therapy and similarly for therapy by surgical correction must be provided.
2. Patient must be physically fit for surgery. Then only implantation with anesthesia can be done.
3. Patient must have gone through temporary stimulation, shown betterment in this phase, and adapted well to the same. Permanent implantation will then be beneficial. Only a patient demonstrating a 50 % or superior improvement during

temporary stimulation is eligible for permanent implantation. The improvement is measurable. Voiding diaries must be maintained as proof.
4. Patient must possess adequate ability to record voiding diary data. This is essential to enable proper evaluation of the scientific outcomes of the implantation.

20.5 The SNS System and Implantation Procedure

20.5.1 The SNS System

SNS is a kind of electrical stimulation therapy. This therapy is conducted through an implanted device (Fig. 20.3). The implanted device has two components. These are a programmable pulse generator called the neurostimulator and a thin insulated wire known as a lead. The neurostimulator is the core controlling section of this therapy. Appearance-wise, the neurostimulator looks like a cardiac pacemaker. Dimensionally, its size equals that of a pocket stopwatch. By surgery, the two components are inserted and placed in the lower abdomen of the patient [16–18]. The implanted device directs placid electrical pulses to nerves that are situated a little above the coccyx or tail bone. These nerves are called sacral nerves, recognized by the alphanumeric characters S2, S3, and S4. They turn on or slow down muscles and organs which take part in controlling urine ejection. Among these are pelvic floor muscles, viz., those of the bladder and sphincter.

The electrical stimulation may eradicate or lessen definite disturbances about controlling the bladder in a few people. By facilitating the communication between the brain and bladder, it helps to relieve the symptoms of excessively active bladder or retention of urine. The relieved symptoms include UI and those of U/F in some patients [19].

20.5.2 Stages of the Implantation Procedure

The general approach is that a test lead is temporarily implanted in a preoperative screening test or trial. The trial is carried out in surgeon's clinic. The procedure is minimally invasive. Medical care is given to the candidate as an outdoor patient. The procedure is least painful to the patient and assures faster recovery and less discomfort. The patient's symptoms are closely watched to find if there is any noticeable improvement. If this screening test tip-offs that the treatment is viable for the individual patient, then only the patient is qualified for receiving the permanent implant. After the permanent implantation is deemed necessary, inpatient surgery for permanent implantation is scheduled and the patient is called for this operation. The two stages of the implantation procedure are discussed in detail in the following subsections.

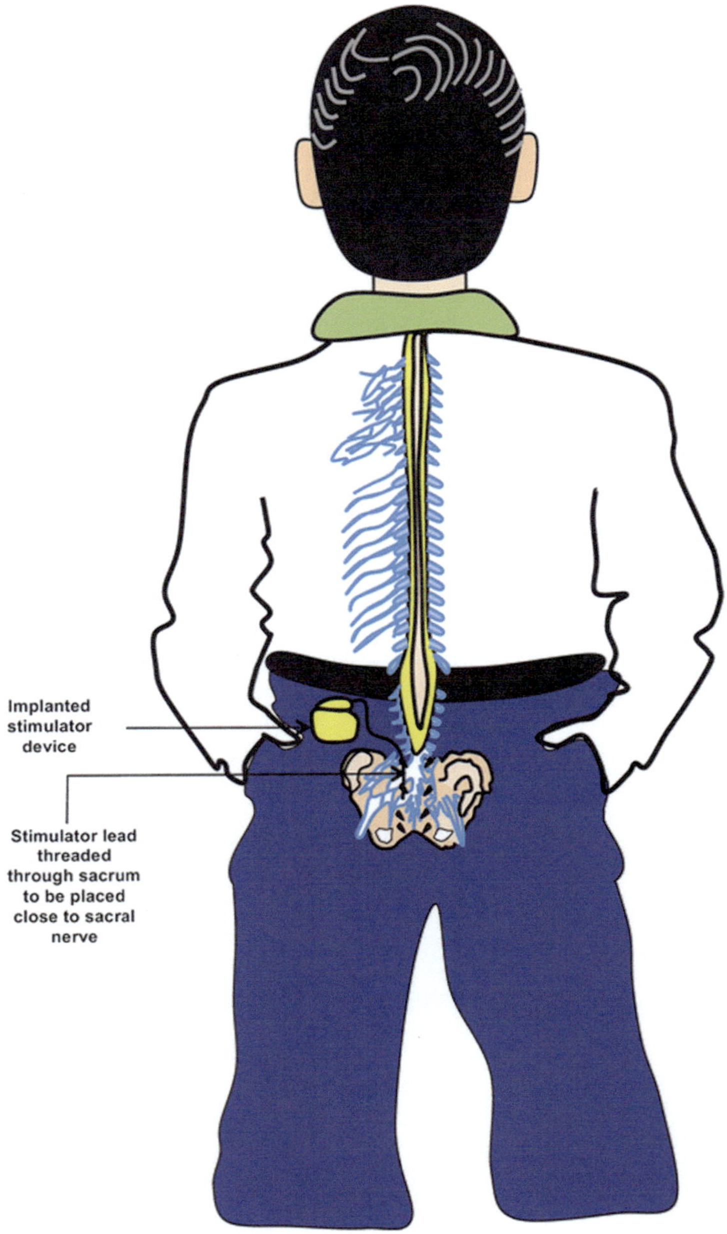

Fig. 20.3 Implanted sacral nerve stimulator

20.5.2.1 Stage I: Minimally Invasive Screening Test

This is done as an outpatient procedure. Local anesthesia is used. A needle is tucked in to pierce the sacral foramen (generally S3). Typical nerve responses of the patient are evaluated. If these responses are found to be proper, electrode lead implantation can be done either as a temporary lead or as a tined lead. Temporary lead fixation is a one-phase procedure. Tined lead involves a two-phase procedure.

A temporary lead is placed in the sacrum and fixed to the outer surface of the sacral region. Small electrical pulses from an external pulse generator are applied to check if a satisfactory response is obtained. The optimal site for the electrode is pinpointed. This percutaneous nerve evaluation procedure is done by a minimally invasive procedure. It is performed via needle-puncture of the skin. It is usually carried out in the operating theater under local anesthesia with the patient fully conscious. Hence, the patient can provide feedback about the location where the test stimulation is sensed. The procedure takes up to 1 h. The patient is subjected to overnight observation.

The tined lead has tines, i.e., slender pointed projecting parts, as in a fork. These tines are designed to be deployed as a fixation system [20, 21]. Classically, the lead is positioned in the S3 foramen. By tunneling, it is connected to an extension made through the skin. This connection is made through a small slit at the probable site of neurostimulator placement. The lead then contralaterally (referring to the opposite side) exits the skin. After the lead placement has been finalized, subchronic neurostimulation (repeating over a short period) is performed. This stimulation is done by linking the electrode lead with a pulse generator placed outside. The maximum comfortable level of stimulation is identified and maintained over a period of 3–7 days. The symptoms and voiding function are also recorded in order to assess the effectiveness of the stimulation.

20.5.2.2 Stage II: Permanent Implantation

Patients showing improvement in at least two major symptoms by >50 % during the screening test are considered for permanent implantation. Formerly, the pulse generator was implanted in a compartment below the skin in the lower part of the abdominal wall. This procedure required a long operating time and three incisions. It often yielded suboptimal results. Postoperatively, the patients reported displacement or pain at the pulse generator site. They also complained about interference from magnetic fields. The current practice involves positioning the pulse generator in a pouch situated under the skin in the lateral superior quadrant of the buttock. The situation is approximately 5–10 cm caudal (the hind part) to the iliac crest or crest of the ilium (the curved superior border or margin of the ilium), allowing for a shorter operation and less invasive placement.

Either general or local anesthesia is employed during the permanent surgical implantation of an SNS system. The patient is admitted in the hospital and required to stay overnight. Following anesthetization of the patient, the surgeon cuts open and

Table 20.1 Two stages of SNS implantation

Stage I: screening test	Stage II: permanent implantation
A transitory electrode is placed to the left or right of the S3 hind foramen. Neuromodulation is done for 3–7 days using a pulse generator located outside. If the results are positive, everlasting implantation is arranged	A pulse generator is inserted in a pouch below the skin. Its location is in the upper, outer quarter circle of the buttock or the lower abdomen. The electrical pulses are transmitted through a thin lead wire carrying a small electrode tip hooked to the sacral nerve

places the neurostimulator beneath the skin of the abdomen. Electrical pulses run along thin wires or leads running from the stimulator to the sacral nerves situated in the lower back. After stimulator placement and lead fixation, the surgeon stitches the cut made in the abdomen.

Presently, a minimally invasive percutaneous approach has outmoded the original open surgical approach to the sacrum. Through imaging, a quad electrode is positioned carefully. This electrode contains four self-sufficient electrodes for stimulation. After the quad electrode is in position, its outer covering is separated. Then plastic tines bounce out to fix the parts in correct positions. The electrode is tunneled to a sack in the upper outer quarter circle of the buttock. This is the region where the pulse generator is implanted.

Table 20.1 summarizes the main aspects of the two-stage SNS implantation protocol.

20.5.3 Stimulation Parameters

After stationing of leads, stimulation parameters are attuned to optimize functioning of the device. Low amplitudes in the range 0–3.0 V can produce needful effects. Continuous stimulation is feasible without producing any pain sensation, for operation confined within the advocated parameters for stimulation (210 μs, 10–16 Hz). Most widely acknowledged form of SNS is unilateral stimulation. Bilateral stimulation has been advised to allow lower stimulation intensities, for prolonging battery life and for reducing the chances of nerve damage.

20.6 Discussion and Conclusions

SNS has been successful in curing pelvic dysfunction. These abnormalities include urinary symptoms, fecal incontinence, and refractory idiopathic constipation [22, 23]. Fecal incontinence is a complex and multifactorial health problem. The treatment for this sickness has to be individualized. The causes and seriousness need to be analyzed separately in each case [24]. Application of neuromodulation to different urologic indications of infants, children, and adolescents has also met with success [25].

SNS was found to be an extremely economical therapy for fecal incontinence [26]. Costs can be further lowered if the doctor is careful about patient selection, treatment is carried out in an outpatient setting, and cheaper devices are used. The technique has been improved by introducing a minimally invasive percutaneous two-stage implantation protocol. This modality permits the patients to undergo an extended test stimulation employing a permanent lead [27].

Review Exercises

20.1 Name the nerve that controls the bladder and the bowel.
20.2 What do the acronyms SNS and SNM stand for?
20.3 What are the main parts of the urinary system? What functions do they perform? Explain how bladder control problems arise on disruption of the communication between the brain and bladder?
20.4 What are the main deciding factors for a person's ability to withhold urine? Define the terms: urge incontinence, urgency–frequency, urinary retention, and overactive bladder.
20.5 Explain stress incontinence and urge incontinence.
20.6 Prepare a list of the bladder control problems that can be treated by SNS therapy.
20.7 Describe the sacral nerve stimulation system and explain how it is used to solve urinary control problems.
20.8 Why is SNS implantation done as a two-stage procedure?
20.9 Explain the terms "temporary lead" and "tined lead" with reference to SNS implantation.
20.10 What class of patients is considered for permanent implantation after the minimally invasive screening test? Describe the open surgical and percutaneous approaches to the implantation of a sacral nerve stimulator.

References

1. Banakhar MA, Al-Shaiji T, Hassouna M (2012) Sacral neuromodulation and refractory overactive bladder: an emerging tool for an old problem. Ther Adv Urol 4(4):179–185
2. Gerber MT (2000) Single and multi-polar implantable lead for sacral nerve electrical stimulation. United States Patent US006055456A
3. Brazzelli M, Murray A, Fraser C (2006) Efficacy and safety of sacral nerve stimulation for urinary urge incontinence: a systematic review. J Urol 175(3 Pt 1):835–841
4. Laviana A, Jellison F, Kim JH (2014) Sacral neuromodulation for refractory overactive bladder, interstitial cystitis, and painful bladder syndrome. Neurosurg Clin N Am 25(1):33–46
5. Sutherland SE, Lavers A, Carlson A et al (2007) Sacral nerve stimulation for voiding dysfunction: one institution's 11-year experience. NeurourolUrodyn 26:19–28

6. Hassouna MM, Siegel SW, Lycklama Ànÿeholt AAB et al (2000) Sacral neuromodulation in the treatment of urgency-frequency symptoms: a multicenter study on efficacy and safety. J Urol 163(6):1849–1854
7. van Kerrebroeck PEV, van Voskuilen AC, Heesakkers JPFA et al (2007) Results of sacral neuromodulation therapy for urinary voiding dysfunction: outcomes of a prospective, worldwide clinical study. J Urol 178:2029–2034
8. Elneil S (2012) Sacral neuromodulation in pelvic floor disorders: therapy helps normalize organ function. http://www.neuromodulation.com/assets/documents/Fact_Sheets/fact_sheet_pelvic_floor_disorders.pdf. Accessed 11 Feb 2015
9. Urology Team PA (2007–2015) Interstim or sacral nerve stimulation. http://www.urologyteam.com/male-urology/interstim-or-sacral-nerve-stimulation
10. Symons S, Barnecott J, Harrison J (2005) Sacral nerve stimulation (neuromodulation) for the treatment of lower urinary tract symptoms in adult patients. Adv Clin Neurosci Rehabil 5(1):35–37
11. Matzel KE, Kamm MA, Stosser M et al (2004) Sacral spinal nerve stimulation for faecal incontinence: multicentre study. Lancet 363:1270–1276
12. Tjandra JJ, Lim JF, Matzel K (2004) Sacral nerve stimulation: an emerging treatment for faecal incontinence. ANZ J Surg 74:1098–1106
13. Michelsen HB, Thompson-Fawcett M, Lundby L et al (2010) Six years of experience with sacral nerve stimulation for fecal incontinence. Dis Colon Rectum 53:414–421
14. Wexner S, Coller JA, Devroede G et al (2010) Sacral nerve stimulation for fecal incontinence: results of a 120-patient prospective multicenter study. Ann Surg 251:441–449
15. Mellgren A, Wexner SD, Coller JA et al (2011) Long-term efficacy and safety of sacral nerve stimulation for fecal incontinence. Dis Colon Rectum 54:1065–1075
16. Rittenmeyer H (2008) Sacral nerve neuromodulation (InterStim). Part I: review of the InterStim system. Urol Nurs 28(1):15–20
17. Rittenmeyer H (2008) Sacral nerve neuromodulation (InterStim). Part II: review of programming. Urol Nurs 28(1):21–25
18. Rittenmeyer H (2008) Sacral nerve neuromodulation (InterStim). Part III: case studies. Urol Nurs 28(1):26–28
19. Groen J, Bosch JLHR (2001) Neuromodulation techniques in the treatment of the overactive bladder. Br J Urol Int 87:723–731
20. Spinelli M, Weil E, Ostardo E et al (2005) New tined lead electrode in sacral neuromodulation: experience from a multicenter European study. World J Urol 23:225–229
21. Marcelissen TA, Leong RK, de Bie RA et al (2010) Long-term results of sacral neuromodulation with the tined lead procedure. J Urol 184(5):1997–2000
22. Baeten CG (2011) Status of sacral neuromodulation for refractory constipation. Colorectal Dis 13(Suppl 2):19–22
23. van Wunnik BP, Peeters B, Govaert B et al (2012) Sacral neuromodulation therapy: a promising treatment for adolescents with refractory functional constipation. Dis Colon Rectum 55(3):278–285
24. Moya P, Arroyo A, Lacueva J et al (2014) Sacral nerve stimulation in the treatment of severe faecal incontinence: long-term clinical, manometric and quality of life results. Tech Coloproctol 18(2):179–185
25. Loret N (2012) Application of sacral and pudendal nerve stimulation in children. Dissertation for Master of Medicine in Medicine. Universiteit Gent, Belgium, 49 p. http://lib.ugent.be/fulltxt/RUG01/001/892/849/RUG01-001892849_2012_0001_AC.pdf. Accessed 03 Aug 2015
26. Hetzer FH, Bieler A, Hahnloser D et al (2006) Outcome and cost analysis of sacral nerve stimulation for faecal incontinence. Br J Surg 93:1411–1417
27. Spinelli M (2007) The future of the sacral nerve stimulation. Pelviperineology 26:17–18

Chapter 21
Cochlear Implants

Abstract Cochlear implantation is a multidisciplinary therapy capable of treating acute-to-overwhelming sensorineural hearing loss in children and adults. The cochlear implant system comprises a two-piece equipment. The equipment consists of an internal part that requires surgical placement and an external part generally worn behind the ear. The internal part of the implant consists of a receiver–stimulator containing the electronic circuitry, the receiving antenna, a magnet, and an electrode array. Direct electrical stimulation is provided to the auditory nerves by inserting the electrode array inside the cochlea. The external part is battery-powered. It consists of a microphone to pick up sound, a speech processor with manual controls, and a transmitting coil to convey information to the internal part of the implant. The two parts work in tandem. Following this implantation, most adults can converse on the telephone, while children can pursue mainstream classrooms.

Keywords Bionic ear · Sensorineural hearing loss · Cochlea · Hearing aid · Electrical hearing · Tonotopy · Microphone · Speech processor · Microelectrode array

21.1 Introduction

"Bionic ear" is another name for the cochlear implant. The word "bionic" originates from "bio- + (electro)nic." "Bionic" indicates the application of electronic engineering and technology to biological systems. It aims at the augmentation or replacement of physiological functions by electronic or mechanical components. As research progresses to link machine and mind, bionic body parts are acquiring many capabilities of natural human ones.

The cochlea is a spiral-shaped cavity in the inner ear. It is usually spiraled like the shell of a snail and accommodates nerve endings, which convert sound vibrations into nerve impulses. The cochlear implant (CI) is an electronic device containing a microelectrode array implanted in the cochlea to support its proper functioning. This implant uses electrical impulses for direct stimulation of the auditory nerve.

V. K. Khanna, *Implantable Medical Electronics*,
https://doi.org/10.1007/978-3-031-99868-3_21

The stimulation enables a totally deaf person to perceive acoustic signals. CI can provide partial hearing to individuals with varying degrees of sensorineural hearing loss (SNHL). This is a type of hearing loss caused by injury to or deterioration of the inner ear. Its origin lies in the vestibulocochlear nerve (VIIIth cranial nerve). This nerve is sometimes called the auditory nerve. Beginning from the ear, it runs to the central processing centers of the brain. In SNHL, the patient may experience difficulty in understanding sound, even when speech is sufficiently loud. Usually, the amount of loss varies with the frequency of the sound. Hence, the audiogram (plotted with sound volume on the ordinate and sound frequency on the abscissa) is not horizontal. It is sloping or dipping at different frequencies.

CI is the only means available presently to break through the ruthless silence disconnecting and estranging the deaf from society [1]. Many individuals with profound deafness have reaped the benefits of this device. Over the past 25 years, and more than ever during the preceding 5 years, it has emerged as a greeted instrument in the quality-of-life enrichment for many persons who were formerly deaf and hard of hearing. The trends indicate that more children are receiving CIs than ever before. The age at which they receive implants is declining from school age to infancy, 12 months, or below [2]. The assistance provided by cochlear implants for development of speaking and communication skills, talking and exchanging ideas in community get-togethers, and scholastic triumphs is proven beyond doubt. It has been found that both speech and lexicon benefit greatly when cochlear implantation is done in children at an age less than 2.5 years [3]. It is expected that soon general education teachers will be instructing a larger number of children with high-tech ears [4].

An extended, acclaimed, fascinating, and captivating history antedates the development of the cochlear implant. This encompasses dynamic reciprocity and teamwork between specialists in engineering and medicine. An intricate balance is struck between tests carried out under designed conditions and moral principles.

21.2 Causes of Hearing Loss

Many dissimilar causes are responsible for partial or total hearing loss [5]. Hearing loss is subdivided into three principal categories. These are known as conductive, sensorineural, and mixed hearing losses.

21.2.1 Conductive Hearing Loss

This loss is related to defects in either of the two parts of the ear: the outer or middle ear (Fig. 21.1). The main parts involved are tympanic membrane or eardrum vibrating with sound and three small bones called auditory ossicles transmitting sound from the tympanic membrane to the inner ear. Some types of conductive hearing loss are attributed to congenital defects. It may so happen that the ear canal is absent or

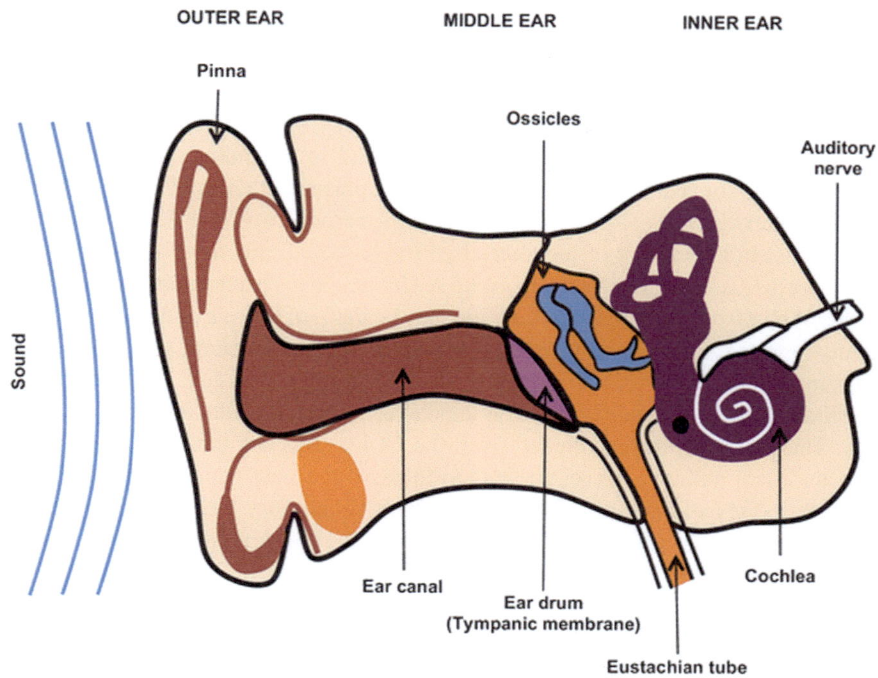

Fig. 21.1 Parts of the ear

closed in a child at the time of birth. Also, middle ear structures existing at birth may be altogether absent, malformed, or dysfunctional. Some or all these defects may possibly be corrected by surgery. Defects that are not amenable to surgical correction can generally be corrected by wearing a hearing aid, which is just a sound amplifier.

21.2.2 Sensorineural Hearing Loss

This is a hearing loss related to nerves. It arises from troubles of the inner ear. The main reason for this type of loss lies in the inner ear or cochlea. The cochlea has the shape of a snail, a mollusk with a hard, coiled outer shell looking like a spiral. It is filled with fluid and contains the sensory hair cells. These cells may be destroyed by injury or in some other way. A natural outcome is profound deafness because it is these hair cells which convert sound waves into electrical impulses. The impulses produced are transmitted to the brain for analysis. For further clarification, the aforesaid electrical impulses move from the hair cells to auditory nerve fibers for interpretation. As the hair cells have been lost, the impulses become ineffective [6]. Therefore, the nerve fibers are not stimulated. Consequently, no hearing response is produced. Individuals suffering from this loss but still possessing an undamaged auditory nerve have profited from the advancements in cochlear implant technology.

Sensorineural damages arise from aging. Other factors that are responsible for these damages consist of before-birth and accompanying-birth defects arising from the infections transported by a woman to her child in the uterus. Some infectious diseases are toxoplasmosis due to the parasite toxoplasma, rubella or German measles owing to rubella virus (RuV), and herpes from herpes viruses. Infections such as meningitis (inflammation of meninges by viral/bacterial infection), mumps caused by mumps virus, scarlet fever due to group A *Streptococcus* bacteria too are damaging agents. Apart from these reasons, mention may be made of the role played by heredity, trauma, exposure to loud noise, and tumors in the auditory system such as acoustic neuroma.

Mild hearing loss conditions can be satisfactorily supplemented by simple hearing aids. These are therefore excluded from the CI list. But if a deaf patient does not have an intact auditory nerve, one has to think of the next generation of auditory prosthesis. This will be the superficial or intrusive auditory brain stem implant. It will circumvent the auditory nerve. It will try straightway stimulation of auditory processing centers residing in the brainstem.

21.2.3 Mixed Hearing Loss

A blend of hearing losses of conductive and sensorineural types constitutes mixed hearing loss.

Table 21.1 presents a comparison of conductive with sensorineural loss.

21.3 CI Versus Hearing Aid

A cochlear implant is strikingly dissimilar from a hearing aid [7].

1. A hearing aid is merely a sound amplifier. A CI is a complex arrangement serving as an electrical stimulator of nerves. A hearing aid only amplifies the sound level to boost its perception. This makes the sound detectable by the inner ear for a patient who has a damaged middle and/or inner ear. A CI totally bypasses two parts of ear: the outer and middle ones. It electrically stimulates acoustic nerves.
2. A hearing aid is worn externally. It is an external device requiring no surgery. Unlike a typical hearing aid, the cochlear instrument consists of two parts: an externally worn device and an internal implant. Signals produced by the implant are transmitted via the auditory nerve passageway to the brain. The brain distinguishes and identifies these signals as sound.

Table 21.1 Conductive and sensorineural hearing losses

Sl. No.	Conductive hearing loss	Sensorineural hearing loss
1.	It arises from any obstruction that inhibits sound vibrations from reaching the inner ear	It is instigated by mutilation of the infinitesimally small hair cells in the cochlea of the ear
2.	It affects the following parts of the ear: middle ear (three ossicles), eardrum (tympanic membrane), or inner ear	It attacks the ear parts given below: inner ear, vestibulocochlear nerve, or central processing centers
3.	Its root causes are a blockage in the outer or middle ear, such as by buildup of excess earwax, otosclerosis (abnormal bone growth near middle ear), infections in the middle ear, tiny apertures in the tympanic membrane, etc.	Its main causes are aging, exposure to very loud noises, genetic predisposition or susceptibility, infections of the inner ear due to different viruses, chemotherapy, radiation exposures, and head injury
4.	It can be either temporary or permanent depending on its cause	It is always irreversible and permanent
5.	Its influence is usually trivial or medium in gradation, in the limits between 25 and 65 dB	Its impact can be mild, moderate, severe, or profound
6.	It can often be corrected with medical management or minor surgery	It cannot be remedied medically, but a hearing aid or cochlear implant is virtually always helpful

3. Listening through a hearing aid is like habitual or customary hearing, whereas hearing from CI differs from normal hearing. In CI, less sound information is received and processed by the brain in distinction to normal hearing. This explains why the quality of sound perceived with a CI is different from that of the natural acoustic hearing.
4. Little or no learning process is involved in hearing from the hearing aid. But hearing from CI entails a time-consuming process of learning or relearning and acceptance. Frequently, patient participation in a rigorous preimplantation protocol is necessary. This is required to help the doctor to take decision regarding the suitability of the patient for candidacy. The ensuing step is surgery to insert a part of the device and then follows an activation process of the external portion of the device that the patient normally wears. In this process, the external portion of the device is programmed. The patient has to go through an intensive auditory training program. Finally, an appropriate educational program is mandated to realize maximum benefit from the device.
5. In comparison to the ancestral hearing aids, a greater cognizance to a wider range of sounds has been brought about by cochlear implants. Experience, exposure, and observation have corroborated this finding. CI allows several patients to distinguish cautionary signals. These patients can comprehend or make out other types of sounds in the ecosystem. Their lip-reading capacity is also enriched. With time and intensive training, the patients can take pleasure and satisfaction from dialogues such as a face-to-face or even a telephonic conversation. Newer

devices and processing strategies have enabled CI recipients to receive more benefits. They can hear better in noise or perform conversation over cell phones. They can learn foreign languages too. They can enjoy and appreciate a variety of music. Notwithstanding, these optimistic remarks, downplaying of any undue expectations to regain normal hearing, are essential. This moderation must be done during preoperative counseling and postimplant training of patients.

6. Hearing aids are low-priced. CIs are currently too expensive. As a result, they are virtually unavailable in developing countries.

Table 21.2 presents the main differences between hearing aids and cochlear implants at a glance.

Table 21.2 Hearing aid and cochlear implant

Sl. No.	Hearing aid	Cochlear implant
1.	It makes sound louder by amplification	It helps in sound perception by electrically stimulating the auditory nerve
2.	It does not bypass normal hearing	It bypasses the sensory cells in the cochlea
3.	It does not require surgical operation	It requires surgical operation. For surgery, an inevitable prerequisite is the removal of the hair cells within the cochlea. Perpetual loss of some or all residual natural hearing may occur from such hair shaving. Fortunately, by using flexible electrodes and applying correct surgical methods, preservation of hair cells is achievable
4.	It is externally placed	It consists of one external and one implanted component
5.	It is helpful for people suffering from mild-to-moderate hearing loss. These losses may have arisen from age-related damage to sensory cells, by exposure to loud noise, reactions to drugs, etc.	It is used by people with severe-to-profound hearing loss, mainly, sensorineural hearing loss. A reason for this loss is damage to the hair cells in the inner ear. Nerve pathways of the patient from the inner ear to the brain may also be mutilated
6.	It provides acoustic hearing	It provides electrical hearing
7.	It does not require learning	It needs learning to interpret the sounds produced by the implant
8.	It is available at affordable cost	It is expensive
9.	It provides more flexibility, i.e., it can be manually adjusted, repaired, replaced, and removed	It offers less flexibility, i.e., once a company's device is implanted, the patient has no option unless the device fails or the patient chooses another surgery/device

21.4 Acoustic Versus Electrical Hearing

Let us consider a person with a normal functioning cochlea. In such a person, frequencies of sinusoidal vibrations for the different regions of the basilar membrane in the *organ of Corti* do not match [7]. The reason for this frequency mismatching is not far to seek. It is the outcome of disparities in thickness and width along the length of the membrane. Therefore, it is said that frequency encoding is performed "tonotopically" by the cell bodies of the cochlear nerve (spiral ganglion). These cell bodies broadcast information from diverse areas of the basilar membrane. The term "tonotopy" needs explanation. It means a structural organization in the auditory pathway. This arrangement facilitates the transmission of diverse tone frequencies unconnectedly. This transmission takes place along individual parts of the structure. Like normal hearing, the principle of tonotopy is also endorsed for electrical hearing.

In a normal hearing person, sound stimuli generate patterns of electrical excitation. These patterns propagate along the nervous system pathways to the auditory nerve. Upon electrical stimulation of the auditory nerve, an action potential or spike is produced. This action potential is transmitted to the auditory cortex of the brain. But it must be emphasized that in a patient with severe-to-profound SNHL, the situation is different. This is because the hair cells are nonfunctional. Consequently, there is no production of spikes along the auditory nerve by the auditory system as observed in a response to sound signal by the normal ear. Noting this failure, the CI electrode is placed in the *scala tympani* neighboring the spiral ganglion of the auditory nerve. This electrode directly stimulates the auditory nerve electrically. It bypasses the middle ear and the part of the inner ear where the *organ of Corti* is located. As the brain is unable to discriminate whether the spikes have been generated by hair cells or a cochlear implant, it interprets the CI spikes as sound similar to those from hair cells.

21.5 Components of the Device

As already indicated, the cochlear implant is a two-piece device composed of two basic components. Of these, one component is worn externally. The other component is surgically implanted [8]. The externally worn component of the CI comprises a microphone and a speech processor coupled to a transmitting coil with cables and magnet. The surgically implanted component consists of the receiver/stimulator and electrode (Fig. 21.2).

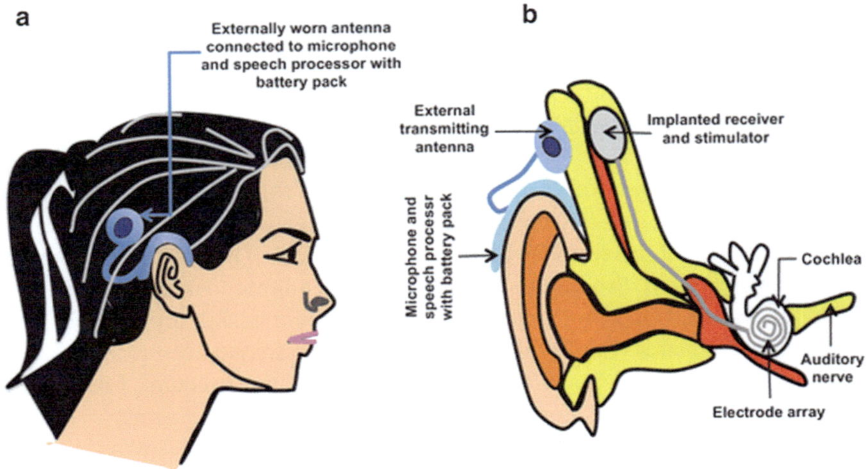

Fig. 21.2 Components of the cochlear implant system: **a** externally worn and **b** internal

21.5.1 External Functionality

The tiny microphone is enclosed in a headpiece that is worn at the ear level. It resembles the microphone of a hearing aid. Its role is to detect and pick up the sound vibrations. After picking up the vibrations, it converts them into electrical signals. The electrical signals thus obtained are transmitted to the speech processor.

The patient can wear the speech processor on the body in the style of a pager. A cable connects the speech processor to the headpiece. Sometimes, it is placed behind the ear similar to a hearing aid. It consists of three main parts: a digital signal processing (DSP) section to mathematically manipulate the signal, a power amplifier stage to boost the power level, and a radio-frequency transmitter for telecommunication.

The DSP is the brain or information-processing center of the cochlear implant system. It receives electrical signals pertaining to sound. It extracts distinctive features in the signal. These features are converted into a rivulet of bits. The bits are transmitted by the radio-frequency link. In other words, the DSP performs signal processing for conversion of the received signal into electrical stimuli. These stimuli are conveyed to the implanted component for electrical stimulation.

For preprocessing, the signal is supplied to an amplifier. Here, it undergoes two operations, viz., filtering and compression. The aim of signal processing unit is to analyze and split the received signal into channels. The signal splitting is done in accordance with its frequency content. This is achieved using a number of bandpass filters. Short-term spectra of the signals are estimated. The signal is subdivided into audio spectral bands. In the next step, the spectral bands are converted into pulsatile signals using envelope detectors.

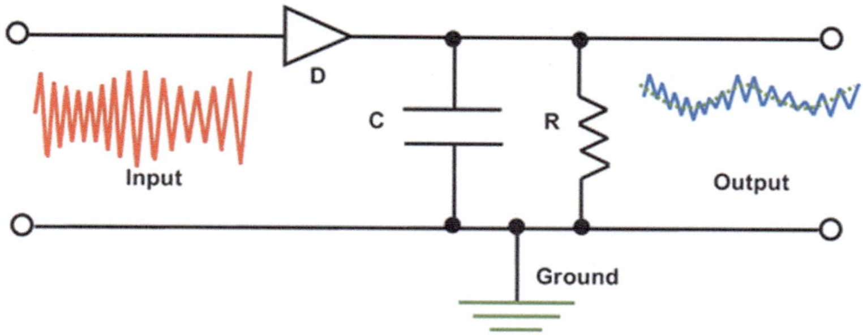

Fig. 21.3 Envelope detector circuit

An envelope detector, also called amplitude demodulator or detector, is a simple and cheap electronic circuit consisting of a diode, a capacitor, and a resistor (Fig. 21.3). It takes a high-frequency signal as the input and delivers an output signal, which represents the envelope of the input signal. This circuit is essentially a half-wave rectifier. When the input signal rises, a capacitor is charged to the peak value of the input waveform. The capacitor voltage is increased through the rectifying diode. As the signal falls, the capacitor is discharged through the bleeder resistor, and the capacitor voltage decreases. Thus the outline of the incoming signal is produced.

Following conversion of spectral bands into digital signals by envelope detectors, shaping of digital impulses is done. The pulses are controlled in parameters such as amplitude, pulse widths, and application rates. Calculations of the electrical stimulation parameters are performed for appropriate representation of the spectrum.

The DSP contains memory units or maps. These maps are used to store information specific to the patients. The maps as well as speech processing parameters are amenable to modification or adjustments. A PC-based fitting program is used for this modification. The stimulation parameters depend on an exclusive set of values. This unique set is determined for each implant user separately when the device is programmed. The digitized output of the DSP is a digital code. This digital code is routed via a cable to the primary windings of an electromagnetic transmitter coil.

The transmitting coil is a small disk. This disk is comparable in size to a quarter. It sticks to the skin behind the ear. This sticking is not by any adhesive but by utilizing the attractive force of a magnet. This magnet is also necessary for securing alignment between the internal and external implant components.

A small cable joins the transmitting coil to the microphone. The main role of transmitting coil is transmission of the encoded pulsatile signals to the implant. A secondary but no less important role of this coil is delivery of power. Power is delivered across the skin of the patient to the internal components of the implant via radio-frequency signals. For ensuring high signal quality and power transmission efficiency, the abovementioned sticking magnet aligns the external device to the internal implant.

There are some variations among different CI models. In some models, the microphone and transmitter are included in a single body. In other models, the microphone is located in a segment placed behind the ear appearing as a standard hearing aid.

21.5.2 Internal Functionality

The internal functionality of a CI system (Fig. 21.4) is based on two surgically implanted components: (1) RF receiver-cum-stimulator circuit receives RF signals and supplies stimulating electrical pulses in accordance with the sound signals as provided by the external unit. Besides, responsibility of this circuit also includes supervising all the activities that take place internally because these activities have to be relayed to the external unit. Through a feedback loop, the internal unit of the implant monitors critical electric and neural activities inside. In return, it transfers these happenings to the external unit. For properly executing all the above functions, the receiver-cum-stimulator assembly is housed in a hermetically sealed biocompatible package. This is done to ensure airtight enclosure and absence of any biological rejection possibility. The assembly is implanted subcutaneously by surgery behind the ear. It encloses a magnet. Coupling of this magnet with the one in the transmitter worn externally makes flawless alignment between them possible. (2) Microelectrode array is an electrode array that is introduced inside the cochlea. Its function is provision of direct electrical stimulation to residual nerve fibers of the patient.

It must be noted that the internal unit is battery-less. Therefore, it has to derive power from outside. The stimulator obtains power from the RF signal.

During operation of CI, the signals from the external transmitter are electromagnetically captured. The charged-up receiver-cum-stimulator performs the operations of decoding and conversion of the RF bit stream into electric currents for feeding the applicable electrodes. With this intent, the internal receiver first deciphers the data transferred by the external processor. The parameters of the required pattern of stimulation are obtained. The extracted parameter information is then used to control a stimulator circuit. This circuit supplies impulses of electric currents to the

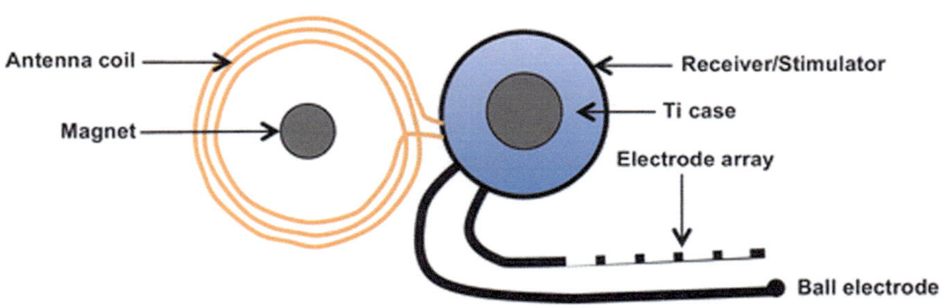

Fig. 21.4 Internal components of the cochlear implant

assemblage of internal electrodes that have been inserted into the cochlea. The electrodes directly cause stimulation of the cochlear or auditory nerve. They sidestep the transducer cells, which are either nonexistent or nonfunctional. Upon receipt of the signals by the brain as impulses in the form of intelligent wave patterns, the signals are recognized or interpreted as sounds or hearing. The external components are held in their respective positions by the magnet, as repeatedly pointed out. Obviously, the cochlear implant does not magnify sound. So, none of its components is considered as a hearing aid.

21.6 Candidacy for Cochlear Implantation

Primary categories of candidates for CI are the post-lingually deaf and pre-lingually deaf patients. Post-lingual deafness develops after learning to speak and acquiring knowledge of language, usually at > 6 years of age. Pre-lingual deafness refers to people who were born with a hearing loss. It also includes those who lost hearing before they began to speak. Main decisive considerations for suitability of a patient for CI are medical history of the patient, health condition, and inner ear structure. Most essentially, the presence of functional auditory nerve fibers to receive electrical stimuli inside the cochlea is necessary for the cochlear implant success. A clear appreciation and strong enticement to accept the long-term commitment and cooperation are demanded from the patient as well as the caregivers.

21.6.1 Presurgery

The surgery is preceded by a series of evaluations and examinations, e.g., computerized tomography (CT) and magnetic resonance imaging (MRI), audiogram scans, and experiments with hearing aids. These investigations are designed to judge the well-being and viability aspects. Both corporal and psychosomatic appraisals are planned to make the candidates ready for cochlear implantation. By the CT or MRI scans, the doctor examines the outer, middle, and inner ear structure for infections with microorganisms, aberrations, or malformations. The audiogram provides the audiologist a quantitative assessment of hearing levels of the patient. A physical evaluation is necessary to prepare the patient for surgical anesthesia. The psychological examination and counseling are done for knowledge acquisition and general attentiveness to provide the patient a realistic expectation from the surgery.

21.6.2 Surgical Procedure

For this surgery, the candidate is usually kept as an outdoor patient. Rarely does this surgery require an overnight hospital stay. Carried out under general anesthesia, the surgery is typically 2–4 h long. A trifling quantity of hair behind the ear is smooth-shaven. A cut is made adjoining the cavity behind the ear. Drilling through the mastoid bone forms a cradle of 3–4 mm size. Having done so, the internal device is installed and secured subcutaneously. Opening up the mastoid bone to gain access to the inner ear, a small orifice is made on the cochlea. Through this hole, the electrodes are threaded into the helices of the cochlea.

21.6.3 Postsurgery

Two to six weeks after the surgery, the implant recipient returns to the doctor for matching the speech processor to the implant. After activating the implant, the processor is instructed for scheduled operations through a computer program, and parameters are adjusted to appease hearing needs of the individual. Noncompulsory rehabilitation program includes therapy for auditory and speech improvement.

21.7 Discussion and Conclusions

Reported designs of cochlear implants include both marketable and investigational systems under development, and all proposed designs share several common features [9]. They detect audio signals with a microphone sealed inside a package worn by the user, much like the hearing aids tied behind the ear. From this microphone, an electric signal is conveyed to an electronic signal processor. This signal corresponds to the pressure variations due to airborne sound waves. The processor converts distinctive features of sound signals into patterns of electric stimuli that excite the nerves to elicit appropriate sensations of hearing from the patient. Considerable manipulability is exercised by the engineers designing the speech processing circuitry and algorithms. This has led to the formulation and realization of many idiosyncratic schemes.

The deliberations and criticism over the resolution to use cochlear implants in children who were deaf before learning to speak have brought forward several legal and ethical issues. These are notified permission, agreement, and parental decision-making, the cost–benefit analysis of surgery and therapy, danger valuation, etc. But the most remarkable amelioration in hearing, speech, and language has been noticed in children. Please remember, the lower the age of a child when the surgery for implantation was carried out, the more is the benefit obtained from CI [6, 10].

Review Exercises

21.1 Explain the meaning of the term, "bionic." Hence, justify why is the cochlear implant referred to as a "bionic ear."
21.2 What is meant by sensorineural hearing loss? What are its main causes?
21.3 Describe six important features that differentiate a cochlear implant from a hearing aid.
21.4 In what ways hearing from a cochlear implant is dissimilar to that from a hearing aid?
21.5 What is tonotopy? Does the principle of tonotopy apply to electrical hearing? Explain.
21.6 What are the roles of externally worn and surgically implanted components of a cochlear implant? How do these two components work together to provide the perception of sound to the patient?
21.7 What are the functions of the following parts in generating the output signal of the external component of a cochlear implant: (1) the microphone, (2) the speech processor, and (3) RF transmitter?
21.8 Does the internal unit of a cochlear plant have a battery? How is it powered? How does it deliver electrical stimulation to the cochlear or auditory nerve?
21.9 What are the main candidates of cochlear implantation? What characteristics qualify a patient to receive this implant?
21.10 Before planning CI surgery, what tests and examinations are typically performed? How is the implantation surgery done? What are the follow-up steps to surgery?

References

1. Zeng FG, Rebscher S, Harrison W et al (2008) Cochlear implants: system design, integration, and evaluation. IEEE Rev Biomed Eng 1:115–141
2. Fagan MK, Pisoni DB, Horn DL et al (2007) Neuropsychological correlates of vocabulary, reading, and working memory in deaf children with cochlear implants. J Deaf Stud Deaf Educ 12(4):461–471
3. Connor CM, Craig HK, Raudenbush SW (2006) The age at which young deaf children receive cochlear implants and their vocabulary and speech-production growth: is there an added value for early implantation? Ear Hearing 27(6):628–644
4. Stith JL, Drasgow E (2005) Including children with cochlear implants in general education elementary classrooms. Teach Except Child Plus 2(1):13 p. http://escholarship.bc.edu/education/tecplus/vol2/iss1/2. Accessed 10 Feb 2015
5. Tierney J, Zissman MA, Eddington DK (1994) Digital signal processing applications in cochlear implant research. Lincoln Lab J 7(1):31–62
6. Weinberg A (2005) Pediatric cochlear implants: the great debate. Penn Bioethics J 1(1):1–4
7. Roux TL, Laurent C (2015) Cochlear implants in developing countries. In: Swanepoel DW, Fagan J (eds) Open access guide to audiology and hearing aids for otolaryngologists, ©University of Cape Town. https://vula.uct.ac.za/access/content/group/27b5cb1b-1b65-4280-9437-a9898ddd4c40/Cochlear%20Implants%20in%20Developing%20Countries.pdf. Accessed 10 Feb 2015

8. Shpizner BA, Holliday RA, Roland JT et al (1995) Postoperative imaging of the multichannel cochlear implant. Am J Neuroradiol 16:1517–1524
9. McDermott HJ (2004) Music perception with cochlear implants: a review. Trends Amplif 8(2):49–82
10. Ali W, O'Connell R (2007) The effectiveness of early cochlear implantation for infants and young children with hearing loss. Nzhta Tech Brief 6(5):1–63

Chapter 22
Retinal Prostheses

Abstract Severe visual impairment up to the level of blindness is caused by either age-related macular degeneration or retinitis pigmentosa. These are the two usual diseases that lead to degeneration of the outer part of the retina. But even after cellular degeneration in these diseases, i.e., degradation of light-sensing photoreceptors, the remaining visual system of neural networks in the retina may not be damaged in many patients. For such cases, a subretinal implant containing microphotodiodes is placed beneath the retina. Currents produced in the photodiodes by the incoming light energize microelectrodes which stimulate sensory neurons in the retina. Otherwise, an epiretinal implant placed on the surface of the retina is employed along with a video camera. The camera captures the light signal and translates the data into an electrical signal through a microprocessor. This signal is transduced across the nerve cells, through the optic nerve, and eventually to the brain for the conception of an image.

Keywords Retina · Age-related macular degeneration · Retinitis pigmentosa · Subretinal implant · Epiretinal implant · Microphotodiode · Argus II retinal system · Alpha IMS retinal implant

22.1 Introduction

Retinal prosthesis is a groundbreaking medical technology of applying microelectronics to restore partial eyesight to patients whose retina has undergone degeneration. Thisrestoration is done through currents supplied to microelectrodes that are surgically implanted either above or below the surface of the retina [1]. Fortunately, in patients with degenerated photoreceptors of the retina, the nerve cells that carry the signal to the brain often remain intact. As the nerve cells are left unimpaired, it is possible to electrically stimulate them and thereby produce signals that help the patient in image perception. It transpires that 50 % of blindness cases are attributed to retinal damage. Hence, this prosthesis has been the focus of attention. Both private companies and research laboratories are working on it all over the globe.

V. K. Khanna, *Implantable Medical Electronics*,
https://doi.org/10.1007/978-3-031-99868-3_22

22.2 Role of Retina in Vision

Vision is an indispensable ingredient in the quality of life. It brings brightness and color to the world. It also adds manifold dimensions to everything that is circumjacent to us. It is an outstandingly complex form of information processing system. The function of this system depends on an amazing neuroprocessor located at the back of the eyeball. Thisneuroprocessor is named as the retina [2]. The retina therefore plays a central role in the vision of a person.

For visualization of a picture, the image of an object is formed in the eye and transmitted to the brain. For this image construction and transmission, light has to travel across several layers. "Seeing" is commenced when the light crossing the black opening called the pupil in the center of the iris is concentrated onto the sensory neuroepithelium of the retina by the eye lens. As a result, a reduced, inverted image of the object is projected onto the outermost layer of the retina. This layer is composed of photoreceptor cells,therods and cones (Fig. 22.1).

The retina is anerve layer with a complex structure (Fig. 22.2). It comprises ten different layers of cells, including ganglion cells, amacrine cells, bipolar cells, horizontal cells, photoreceptor cells (the rods and cones), and nerve fibers.

The*ganglion cells* represent the first link in the chain of neurons in the retina.Ganglion cells are neurons positioned adjacent to the inward surface of the retina. They gather pictorial information in the form of statistical data in their

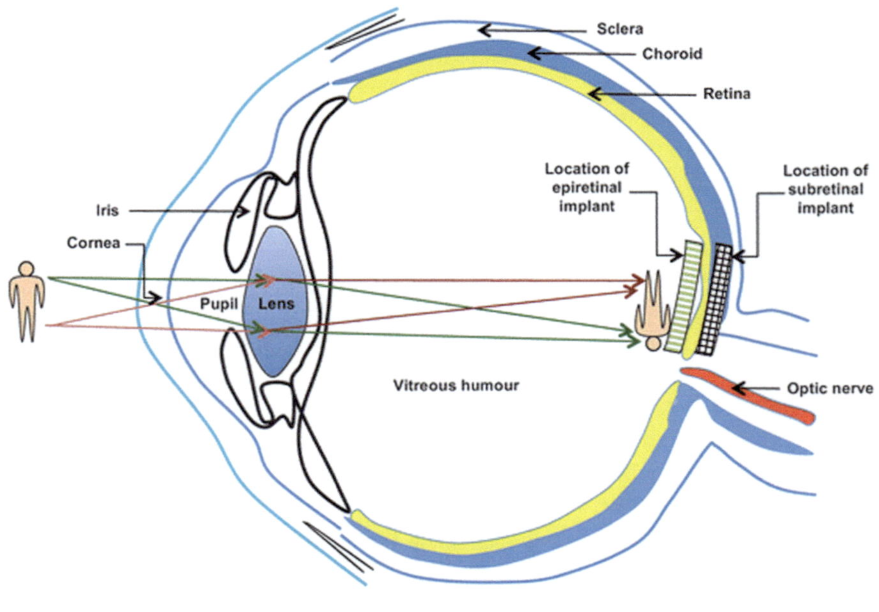

Fig. 22.1 Inverted image construction onthe retina. The diagram also shows the locations of epi- and subretinal implants to be described in this chapter

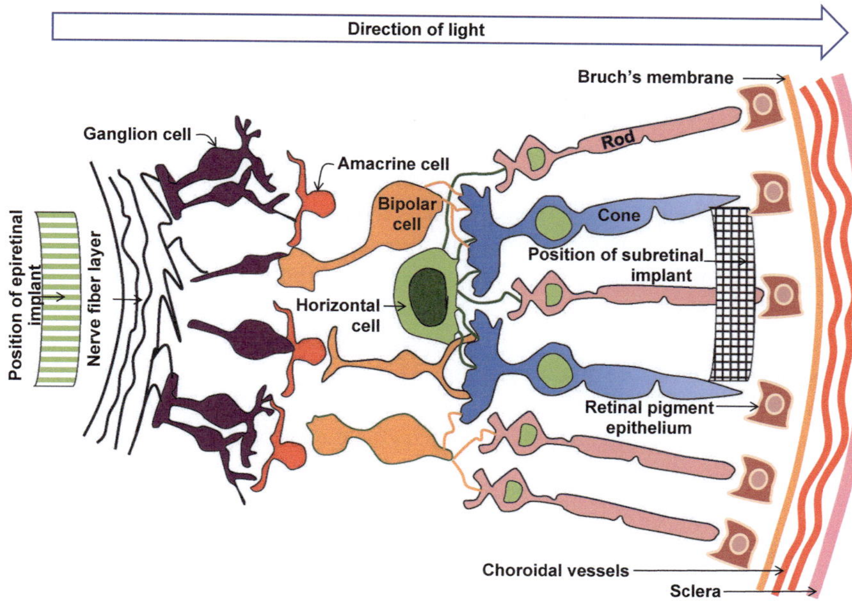

Fig. 22.2 Structure oftheretina showing its constituent layers. Also shown are the positions for placement of epiretinal and subretinal implants to be discussed further on

dendrites from bipolar cells and amacrine cells. The information collected is communicated through their axons to the brain. The ganglion cells differ expressively in regard to their dimensions, connections, associations and feedbacks to ophthalmic stimulations.

Amacrine cells are interneurons in the retina. An interneuron is also called a relay neuron. It is a neuron that carries impulses between a sensory and a motor neuron. The amacrine cells are small nerve cells which have dendrites but no axon.

Bipolar cells are signal couriers between the photoreceptors and the ganglion cells. They are so-called because they have two polar extensions protruding from opposite ends of their cell bodies. One extension is toward the photoreceptors and the other toward the ganglion cells for delivery of the processed signal to ganglion cells.

Horizontal cells are the sideways interlocking neurons. They are found in the outer plexiform layer of the retina. Their role is to assimilate/control the inputs received from numerous photoreceptor cells. They allow the eyes to perform adjustments to be able to view correctly, both when the illumination is bright and faint.

The *rods* spread along a much broader extent in the retina. They are used for achromatic vision (without colors). They are necessary for dark-adapted or scotopic vision under low-light conditions. They are also required for peripheral or side vision (seeing objects and motion outside the direct line of vision). They convert the local luminous intensity per unit area and color variations of the image projected on the retina into signals, both chemical and electrical. In dim light conditions, they provide a smaller amount of spatial resolution.

Cones are mainly found in the middle region of the retina, the fovea. They are essential for color vision and high-resolution vision. They generate chromatic or color images of high clarity in the expanse of location of objects, calledspatial resolution.

The signals generated by the photoreceptor rods and cones excite the complex circuitry of retinal neurons encompassing the four types mentioned above: horizontal cells, bipolar cells, amacrine cells, and ganglion cells. Gigantic visual information is accumulated from the innumerable photoreceptors of the retina. Hence, it has to be compressed into electrical signals. These signals are transported by highly dedicated ganglion neurons. The axons of these neurons form the optic nerve. By way of the optic nerve, the signals are conveyed to the primary visual cortex of the brain, which processes all the information relating to vision. The signal transmission takes place via a route that passes through the lateral geniculate nucleus.

22.3 Vision Impairment and Its Remedial Schemes

As we understand, the optical pathway for vision is a long journey starting from the eye optics and the lens, then the retina, to the optic nerve, visual cortex, or other cortical areas. These are the multiple regions that diligently carry out the needed signal processing which enables a person to see objects. Destruction or injury to any single step in the above series or chain may cause blindness. About 50 % of all blindness patients have suffered some form of insufficiency of the retina. In diseases causing blindness, a gradual erosion of the exterior retina takes place. Two such diseases are age-related macular degeneration (ARMD) and retinitis pigmentosa (RP) [3]. The former is a common cause of blindness in the elderly people.

22.3.1 Age-Related Macular Degeneration

As the age of apersonadvances,the danger of degenerative diseases affecting the retinal cells begins to haunt. For adult patients with age more than 65 years, ARMD stands as the principal reason of vision deprivation in grown-ups. This disease is characterized by loss of cells in the macula close to the center of the retina. Initially, the symptoms observed are the loss of fine vision. These are followed by dwindling central vision with eyes focusing straight ahead, e.g., to drive or read. Eventually, legal blindness may ensue in many patients with central visual acuity measurement of 20/200 or a reduced amount, in superior eye with correction. The central visual acuity is an indicator of sharpness or clarity of vision. Examination of the retinal status of these patients reveals that they can be deprived of as high as 70 % of photoreceptors. Still they may not incur any loss of other types of retinal cells.

22.3.2 *Retinitis Pigmentosa*

The cardinal basisofhereditary blindness, RP isagenetic disorder that primarily affects photoreceptors in the retina causing incurable blindness. Its symptoms are a reduction in night vision, peripheral vision, and in more acute stages, central vision. Randomized experiments have evidenced that increased vitamin A intake serves to put on the brakes against the symptoms of photoreceptor degeneration. This means that the cells undergo apoptosis (cell death as a normal part of growth of an organism) or necrosis (cell death by disease or injury). However, vitamin A over dosages may harm the liver. RP is accompanied by the exhaustion of up to 0.95 fraction of the photoreceptor layer. Nonetheless,up to 80 % oftheinner nuclear layer and ~30 % of the ganglion cell layer are left undamaged.

22.3.3 *Evolution of the Concept of Electrical Stimulation of the Retina*

Electrical stimulation of theretina and related technological approaches have been attempted to artificially distribute ocular information (in imitation of natural approach) to the outlasting portion of the retina that continues to remain in existence withstanding all the degeneration processes. This has been done as a countermeasure for restoring at least low-resolution vision to such patients. This vision may be sufficient enough to provide at least light perception and object recognition [4]. During electrical stimulation, the information is transmitted to the brain through neurons in the eye using a series of energized electrodes targeting the retina, which communicates with the visual cortex of the brain. Experimental studies and trials have been conducted on such multielectrode devices calledmicroelectrode arrayscontainingfrom only 16 electrodes to >1000 electrodes.

22.4 Two Kinds of Retinal Implant

Presently, research and development efforts are being made on two different varieties of retinal implants [5]: subretinalimplants and epiretinal implants. In the first type of retinal prosthesis, namely, the subretinal type, the implant is placed beneath the retina. In the second type, viz., the epiretinal type, the implanted device is lodged on the exterior surface of the retina. Both types of devices rely on the presupposition that even during degeneration of cells in ARMD or RP, the neural network of the retina remains undamaged, i.e., regardless of the non-operation of the light-sensing photoreceptors, the remaining visual system is still functional. While some retinal implant devices are approved, rigorous clinical trials are under way to ascertain the safety, effectiveness, and potential performances of others.

The first approach seems to be more elegant and reminiscent of the actual working of the eye. But it requires implantation of more sophisticated electronics into the eye, increasing the risk of failure of components that are neither easily fixable nor replaceable [6].

22.4.1 Subretinal Implant

As obvious from its name, thesubretinal device is implanted underneath the retina. The implant is squeezed between the pigment epithelial layer and the outer layer of the retina. Thephotoreceptor cells are seated in this outer layer [7]. During surgery for implantation, access must be gained to the subretinal space. This is done ab externo by incision in sclera, the white of the eye. Other surgical procedure is ab interno, which is done through the vitreous cavity and retina. The vitreous cavity is the cavity in the eye which lies posterior to the lens but in front of the retina.

Practically, subretinal prosthetic device for restoring vision consists of a microphotodiode array (Fig. 22.3).

Themicrophotodiode array contains thousands of small-area, light-sensitive photodiodes, laid out in a well-defined geometrical pattern such that all the photodiodes in the array are equipped with microelectrodes of gold (Au) and titanium nitride (TiN). The complete circuit assembly is fabricated in a compact, low-dimensional footprint on a very thin plate that can be housed in the subretinal space.

Figure 22.4 shows the construction and working of asilicon photodiode. The photodiode has a P–N–N$^+$ structure. The P–N junction is a shallow diffused junction. Incident light having energy greater than silicon energy gap is absorbed. It dislodges electrons from the atoms of the crystal lattice creating electron–hole pairs. In zero-bias conditions, the generated electrons and holes drift under the electric field of the depletion region to produce a current flow similar to that obtained in a solar cell. This phenomenon is called thephotovoltaic effect. Subretinal implants normally do not useabattery, but application of a reverse bias to a photodiode enlarges the depletion region width and hence increases the effective area in which electron–hole pair generation effectively contributes to current because the carriers produced far away from the depletion region may be lost by recombination before reaching the depletion region and may not be able to participate in conducting current. Thus, the reverse voltage enhances the sensitivity of the photodiode.

To understand how vision is helped by subretinal implantation, it is necessary to point out that this implant lying between two vital layers, the bipolar cell layer and the retinal pigment epithelium, is essentially meant to substitute the function of photoreceptors. Effectively in the subretinal approach, the place of rods and cones is usurped by a silicon chip carrying a vast number, in the range of thousands, of optically sensitive microphotodiodes, each furnished with a stimulation electrode. Light coming from the image that is incident on the retina modulates the action ofmicrophotodiodes. Currents are produced in the photodiodes in accordance with light variations. The generated currents activate the microelectrodes. Thus triggered

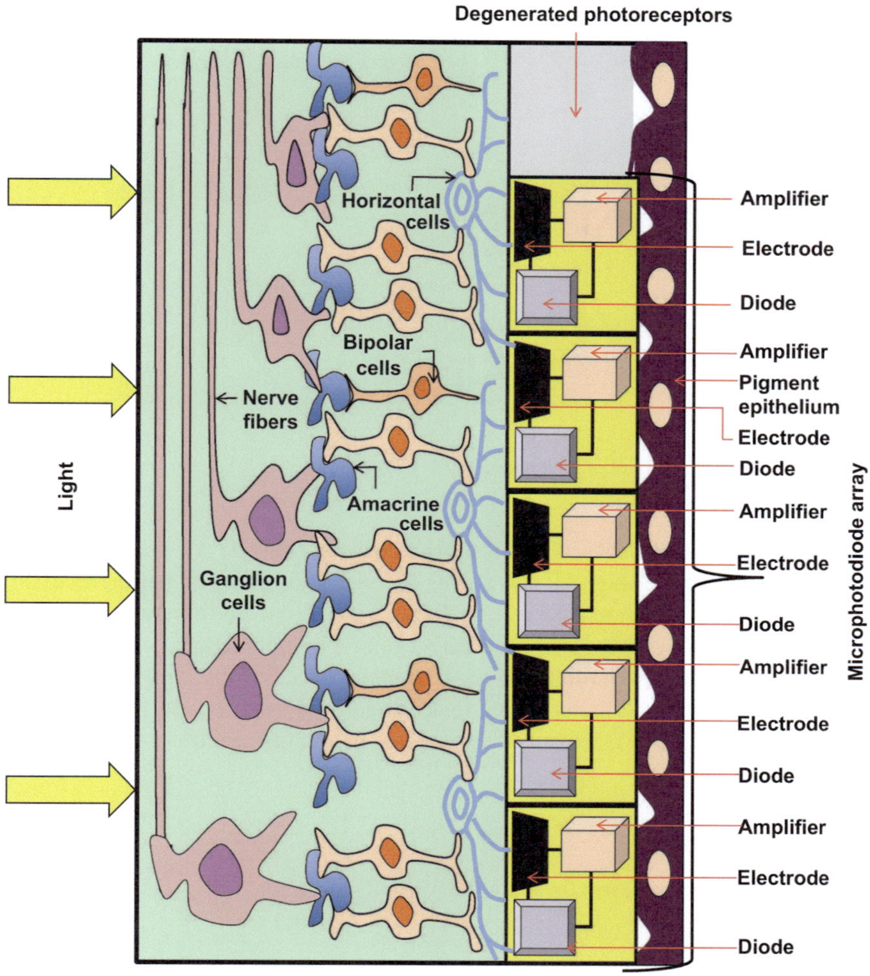

Fig. 22.3 Diagrammatic representation of a subretinal implant

by the microphotodiodes, the microelectrodes instill minuscule currents into the residual neural cells left behind, viz., horizontal cells, bipolar cells, amacrine cells, and ganglion cells, of the retinal inner layer. These currents are able to stimulate retinal sensory neurons. Consequently, a visual perception is created in the brain of the patient, representative of the original incident image.

There are manyadvantages of subretinal prostheses. The microphotodiode array directly replaces the lost or degenerated cells. The remaining cells of the retina are still capable of processing electrical signals and can participate in vision process. Fixing the high density microphotodiode array in the subretinal position is easy. Moreover, there is no need of any external camera or external image processing equipment. Also, eye movement to locate the objects is not restricted.

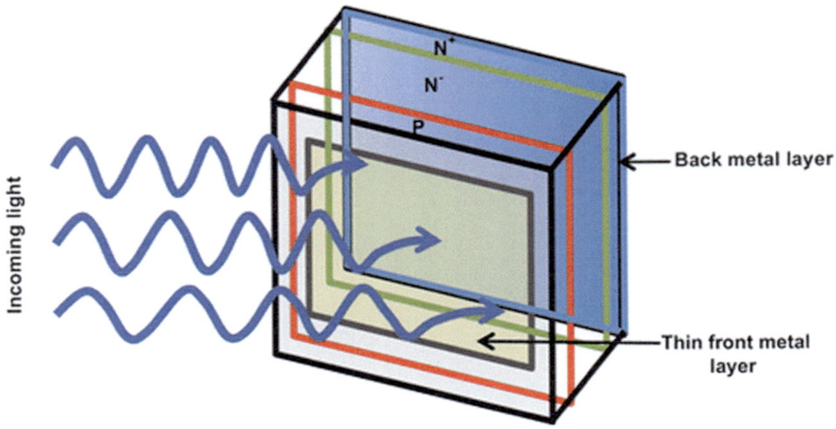

Fig. 22.4 Schematic diagram ofaphotodiode

In principle, a subretinal implant does not require any outside paraphernalia except the microphotodiode array. All the processes, viz.,acquisition of light, processing of signals, and stimulation of nerve cells, are carried out by this array. However, alimitation to this implant is that the single microphotodiode array cannot supply adequate current for stimulation. Hence, the implant must be supported by an external energy source, which is included in some versions of this implant. Also, flexible substrates are needed to take care of the delicate nature of the retina and to decrease the light intensity [8].

22.4.2 *Epiretinal Implant*

There are twomaindifferences between the subretinal and epiretinal implants [9]. Firstly, the epiretinal device is implanted on the surface of the retina and functions with healthy ganglion and bipolar cells. In opposition, the subretinal implant was placed beneath the retina. Secondly, the epiretinal implant has no areas sensitive to light. In contrast, the subretinal approach utilized the light sensitivity of microphotodiodes. Therefore, the epiretinal implant acceptselectrical signals from a remote camera and processing unit located outside the body, whereas the subretinal implant received them from the microphotodiode array inside the eye.

In the epiretinal prosthesis, substantial power and data telemetry mechanisms are involved. The typical epiretinal prosthesis system works as a two-component arrangement that consists of an external part attached to the arm of the patient's eyeglasses (Fig. 22.5), consisting of a small video camera, an image processor, and a transmitter coil; and an internal part carrying the decoding circuitry and microelectrode array. The video camera seizes the visual images from the external world. The image processor digitizes the images after reducing their resolution [10, 11]. It transforms the visual images into control signals through amicroprocessor. In this manner, the images are transfigured into templates of electrical stimulation. These patterns are fed to the transmitter. Radio-frequency links are used to transfer these signals to the internal part.

The signals are identified and picked up by the internal part. This part consists of a receiver for the relayed information signals and power for the implant. The signals for themicroelectrodes are decoded by the integrated circuitry. The obtained spatial and temporal stimulation patterns are used for excitation of the remaining, viable inner retinal neurons that have escaped the onslaught of disease, by controlling the stimulation microelectrode array. Upon stimulation, the electrodes produce action potentials in the upper ganglion cell layers. They directly excite the axons of the inner-layer ganglion cells. These axons form the optic nerve. In this way, the electric signals are transduced across the nerve cells, through the optic nerve, and ultimately to the brain for causing visual sensations and the creation of an image.

Thus, the internal part works by receiving two types ofinput: (1) statistical information about the visible objects/scenery known as the visual signal and (2) electrical power for feeding the electronics and the electrodes stimulating the retina for generating the visual image akin to the object. The above functions are executed by wireless communication and radio-frequency transmission through a transmitting coil attached to the arm of special glasses; this coil transmits the data and power to the coils kept on the prosthetic device.

Epiretinal implants of various designs have been developed. These designs differ with regard to the implanted and outwardcomponents constituting the devices and their operation to facilitate vision. The guiding principle in these designs is topreservethe normal anatomy/physiology of the eye to the maximum extent possible. At the same time, the proportion of implanted electronics required to power the device must be curtailed. In addition, this device must be engineered in such a way that it must exist stably in the saline environment of the vitreous humor without disturbing the rest of the tissue in the eye.

In essence, the epiretinal implant is a readout chip. It receives electrical signals containing information about the image from an outlying camera and processing unit. Theinformation-bearing signals are used to generate electrical impulses. The impulses are coupled to the ganglion cells and their axons. Through this coupling, the impulses can propagate via the ganglion cell axons of the optic nerve to the brain.

A comprehensivecomparisonbetween subretinal and epiretinal implants is presented in Table 22.1.

22.5 Argus II Retinal System

It is thefirstretinalimplantsystem designed for adult patients who are aged $\geq$25 years. This system is meant for patients afflicted with advanced retinitis pigmentosa. These patients could be suffering from plain or no light perception vision. The system consists of three main components. Of these, one component is internal and remaining

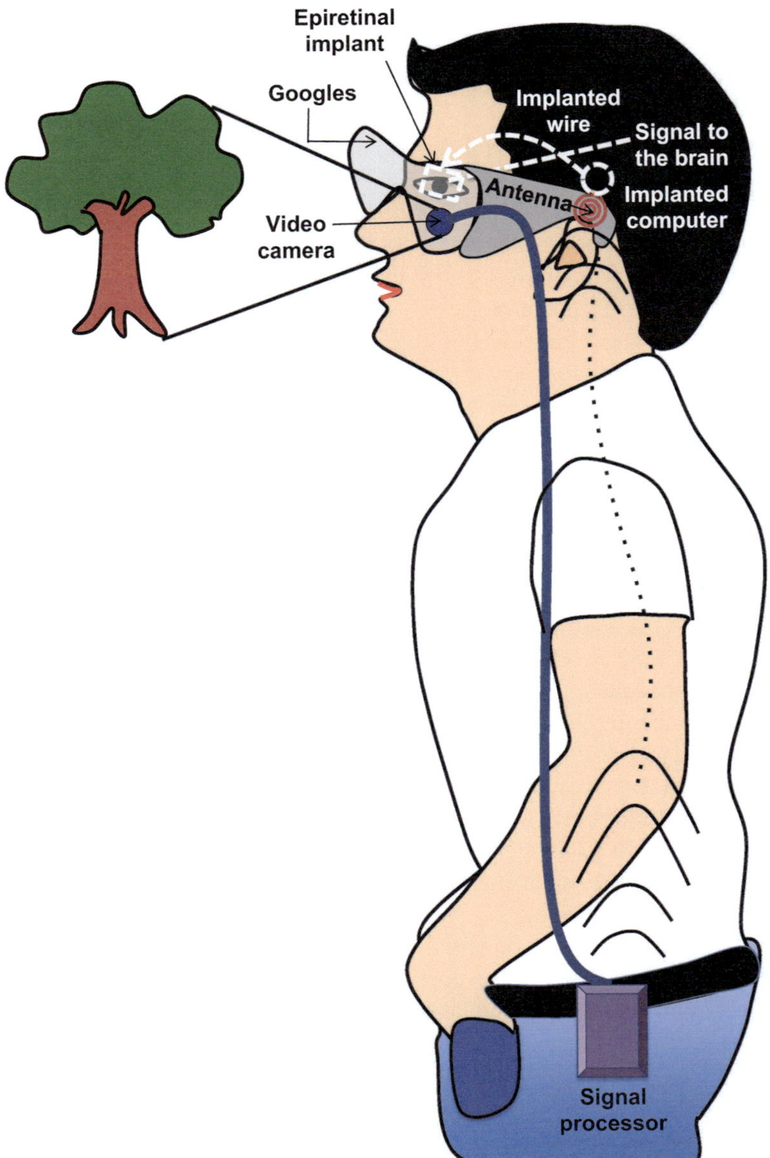

Fig. 22.5 One version ofanepiretinal implant in which a video camera in glasses captures the image of an object and sends it to a signal processor from where it is wirelessly transmitted to an antenna over the glasses, then to an implanted microcomputer, and finally to the epiretinal implant which energizes the electrodes to send signals to the brain

Table 22.1 Subretinalandepiretinal implants

Sl. No.	Subretinal implant	Epiretinal implant
1.	It is placed behind theretina	It is placed on or above the retina
2.	It sits on the outer surface of the retina. It resides as a layer sandwiched between and touching the photoreceptor layer and the retinal pigmentepithelium	It sits inside the retina. It resides adjoining the inner surface of the retina
3.	Its implantation is difficult	Its implantation is easier
4.	Its placement is more clear-cut, as the electrode array for stimulation is positioned directly contiguous to the spoiled photoreceptors	Its mechanical fixation is done using micro-tacking to steady the implant on top of the inner retinal layer
5.	It directly stimulates the photoreceptor layer and depends on the regular processing capability of the retinal layers located on the inner side and in the middle	It directly stimulates the ganglion cells, finding a way around all other retinal layers
6.	It is simpler in design with no external components	It is complex in design with components located internally as well as externally. It works as a two-component device consisting of an extraocular and intraocular part
7.	It allows for normal inner retinal processing, including amplification, by depending on the function of the residual retinal layers. Therefore, when taken as a whole, it requires a smaller threshold to incite a visualresponse	It requires more sophisticated image processing techniques because stimulation is done at the ganglion cell layer and care has to be taken for the processing that would normally have been done with the bypassed retinal layers
8.	It provides relief to patients having retinal diseases within the photoreceptor layer of theretina	It provides visual discernment to individuals whose retinal infirmity stretches further than the photoreceptor layer
9.	It does not impose the requirement of any external camera or transmitter. Microphotodiodes mounted on a single chip perform all the operations including light acquisition, processing, and stimulation	It entails the use of an external video camera, signal processing chip, and transmitter along with implanted decoding circuit and array of microelectrodes
10.	It enables the patients to move their gaze to locate objects through normal eyemovements	It requires patients to make head movements for changing their gaze, as the camera is externally located
11.	Its stimulation is inherently more precise and accurate. This happens because the blueprint of light falling on the microphotodiodes is an undeviating manifestation of the imagerequired	Its stimulation is ill-defined. The stimulation field embraces the ganglion cell bodies but it may overstretch to include close-by axons, related with other areas of the retina. As a result, a slightly imprecise and malformed stimulation pattern is obtained. This pattern ought to be amended by processing electronically

(continued)

Table 22.1 (continued)

Sl. No.	Subretinal implant	Epiretinal implant
12.	Its mounting calls for minimal fixation effort. The underlying reason is that the subretinal space is mechanically held back in position and negative pressure is created by the retinal pigment epithelium within the subretinal space	It is made stable or steadfast by the force exerted by the vitreous humor
13.	It is larger in size but mechanically restricted by the nominal distance separating the outer retina from the retinal pigmentepithelium	It allows more miniaturization allowing for a smaller implant because the main portion of electronics is incorporated into the external components
14.	It does not permit upgrading of electronics software/hardware because it is embedded in the eye	It allows simple upgrades to electronics without additional surgery
15.	It does not allow the doctor to have any control over image processing or be able to adapt the processing to differentpatients	It allows the doctor to exercise complete freedom over processing of images and adaptation of this processing to suit individual patient because of the external electronics
16.	It suffers from higher risk of temperature-induced damage to the retina from heat produced by the implant during operation. This is due to nearness between the implant and the retina	It imposes less risk of thermal damage because the vitreous humor in the vitreous cavity serves as a heat sink and dissipates the warmth produced by the electronic circuits under operation
17.	It does not require external electrical power, but in some cases, the implant may draw power from external segment for enhancement of the imagesignal	Its external components are battery powered. The implant is supplied power via radio-frequency induction coils
18.	Its performance may be affected due to insufficiency of incident light. As a consequence, the microphotodiodes are unable to spawn tolerable currents. Therefore, often an outside power supply is incorporated in the system for amplifying the effect of availablelight	It faces no such problem because electronics part lies outside

two components are externally placed [12]. The internal component is a small electronic device working as an epiretinal implant. One external component is a petite video camera fixed to a pair of spectacles of the patient. The other external component is a video processing unit that the patient wears or carries, may be in a pocket [12].

The glasses of the patient together with video camera capture the image of the surroundings. The captured image is in the form of electrical signals. The video processing unit performs signal processing. The processed signals are wirelessly transmitted to the internal component of the system for electrical stimulation of the retina. Upon retinal stimulation, the brain recognizes the image as an array of spots of light. It has been shown by medical studies that the system could help patients in identifying the location or movement of objects and people. It helped them to read large letters, words, or sentences. Moreover, they became more mobile in day-to-day activities. They could detect waysides and walk on footpaths without trudging off. This system was granted approval in EuropeinFebruary 2011 and by the FDA in the USA in February 2013.

22.6 Alpha IMS Retinal Implant

Afteramulticenterclinical trial, European regulatory approval was granted in 2013 to Retina Implant AG, Germany. The company has begun offering its Alpha IMS implant, a subretinal implant. This implant is meant for restoration of moderate sight to RP-blinded people [13]. The system performs the job of the retina from reception of light on a microchip of size 3×3 mm. The chip has a resolution of 1500 pixel. The optic nerve is stimulated on the basis of the image seen by the chip. As no external camera is used, viewing around is possible in natural way as one would expect, using the eyes rather than by head movements. Hence, the patient can look around by eye movement, without disturbing the head. The implant is reported to provide a realistic solution to restore purposeful vision. The implant-receiving patients were competent to distinguish numbers on doors, identify faces, and perceive expressions made by face.

A clear advantage offered by the Alpha IMS is that it does not use an external camera [14]. Instead, light detection takes place within the eye. Another exclusive feature is its higher-resolution framework. Being implanted underneath the retina, it allows the processing of input by the middle layer of the retina before sending the signal to the visual cortex. In a study, functional vision of most of the participants was reinstalled. Some subjects developed substantial visual ability. Postimplantation, a few patients were able to read large printed letters impromptu.Theseresultsare supportive and give confidence.

22.7 Optimal Candidates for Retinal Implants

The candidates mustbepatients affected by ARMD or RP. To qualify as a candidate for receiving a retinal implant, it is essential that at least the ganglion cell layer must be intact, as assessed in a noninvasive manner by a technique known as optical coherence tomography (OCT) imaging. Other vital factors that must be considered when determining suitability of patients include the overall health of the patient and family commitment toward rehabilitation. The latter is no less important for success of the implant.

22.8 Discussion and Conclusions

During the past few decades, treatment of extreme vision impairment by artificial means has reached closer to realization [15]. This research aims to fabricate an implantable medical device that provides helpful vision to those unfortunate sufferers who have exhausted all alternatives. An analogy can be drawn with the cochlear implants used for hearing losses of certain categories. Like cochlear implants, the retinal devices can reestablish useful vision by translating visual information into prototypes of electrical stimulation. These prototypes are used to excite the surviving inner retinal neurons in patients who are grievously affected by ARMD or retinitis pigmentosa.

Although many strides have been taken, the field of artificial vision is comparatively young. Aided by the ongoing progress in microelectronics technology, surgical instruments, and treatment possibilities, there has been a large leap toward restoring partial vision to AMD and RP patients. People who have used this technology have found it to be extremely useful. Compared to the millions of photoreceptors in the normal eye, there are only a small number of electrodes. Yet people can make out a general sense of their surroundings. Partial visual function can be restored to patients with advanced photoreceptor degeneration. This can give them back the possibility of recognizing or localizing objects. They can achieve self-sustained mobility. They can differentiate a cup from a plate. They can know where a door is located in their home. They can tell where a table is placed in the dining room. Some of the missing pieces of information can be filled in by the brain particularly when memory is taken into account.

The various retinal prostheses that have been reported have given many reassurances in limited clinical trials. Each type of prosthesis has distinct advantages and disadvantages. A relevant research topic is the study of likely performance fluctuations of microelectronic chip stuck inside the briny atmosphere prevalent in the eye over an extended period of time. This especially concerns the airtight and waterproof packaging of the micro-fabricated electrode arrays. Research will also need to address the minimization of heat produced by the chip, and its proper dissipation. A paramount biocompatibility issue is raised by the effects that persistent electrical stimulation on the retina can exert and about which not much is known.

Review Exercises

22.1 List the different processes that take place during transmission of an image to the brain? What are the functions of rods and cones in the retina?

22.2 What is the commonest cause of blindness in the elderly people? What kind of blindness arises from inherited reasons?

22.3 Are retinal implants able to correct all forms of blindness? If not, what are the types of blindness amenable to correction within limits by these implants?

22.4 Name the implant, which is placed beneath the retina. Which implant is placed above the surface of the retina?

22.5 How is the function of rods and cones in the eye performed by a subretinal implant?

22.6 Draw the cross-sectional diagram and describe the operation of a microphotodiode.

22.7 Does the epiretinal plant have any light-sensitive region? If not, how does it capture images?

22.8 Which retinal implant has no external components?

22.9 Which retinal implant may not need a battery? Some of the retinal implants may require a battery? What is the function of the battery, if included?

22.10 Which implant is easier to install: subretinal or epiretinal? Explain.

22.11 Which implant is less likely to cause thermal damage to the retina: subretinal or epiretinal? Why?

22.12 Which implant needs a video camera for capturing images: subretinal or epiretinal?

22.13 What are the primary considerations for a candidate's eligibility for undergoing retinal implant surgery?

22.14 Explain the statement, "The epiretinal implant is a readout chip."

22.15 Highlight some research problems in the area of retinal implants.

References

1. Chuang AT, Margo CE, Greenberg PB (2014) Retinal implants: a systematic review. Br J Ophthalmol 98(7):852–856
2. Zrenner E (2002) Will retinal implants restore vision? Science 295:1022–1025
3. Kien TT, Maul T, Bargiela A et al (2012) A review of retinal prosthesis approaches. Int J Mod Phys 9:209–231
4. Asher A, Segal WA, Baccus SA et al (2007) Image processing for a high-resolution optoelectronic retinal prosthesis. IEEE Trans Biomed Eng 54(6):993–1004
5. Zureick A (2011) Optoelectronics and retinal prosthesis: the revival of vision. Dartmouth Undergraduate J Sci 13(2):17–18
6. Rizzo JF (2011) Update on retinal prosthetic research: the Boston retinal implant project. J Neuroophthalmol 31:160–168
7. Stingl K, Bartz-Schmidt KU, Besch D et al (2013) Artificial vision with wirelessly powered subretinal electronic implant alpha-IMS. Proc Biol Sci 280:20130077, 8 p
8. Maynard EM (2001) Visual prostheses. Annu Rev Biomed Eng 3:145–168
9. Rao C, Yuan X-H, Zhang S-J et al (2008) Epiretinal prosthesis for outer retinal degenerative diseases. Int J Ophthalmol 1(3):273–276
10. Javaheri M, Hahn DS, Lakhanpal RR et al (2006) Retinal prostheses for the blind. Ann Acad Med 35(3):137–144
11. Butterwick A, Vankov A, Huie P et al (2007) Progress towards a high-resolution retinal prosthesis. Ophthalmic Technol XVII, SPIE 6426A:9 p
12. Argus II retinal prosthesis system—H110002. Last updated 13 Jan 2015. http://www.fda.gov/medicaldevices/productsandmedicalprocedures/deviceapprovalsandclearances/recently-approveddevices/ucm343162.htm. Accessed 7 Feb 2015
13. Alpha IMS vision restoring wireless retinal implant now cleared in Europe. http://www.medgadget.com/2013/07/alpha-ims-vision-restoring-wireless-retinal-implant-now-cleared-in-europe-video.html. Accessed 7 Feb 2015

14. Retinal implant alpha IMS brings sight to blind in new study, © Medgadget, LLC. 2004–2015. http://www.medgadget.com/2013/02/retinal-implant-alpha-ims.html. Accessed 7 Feb 2015
15. Karmel M (2012) Retinal prostheses: progress and problems. EyeNet 41–43. http://development.aao.org/publications/eyenet/201203/upload/Clinical-Update-Retina-PDF.pdf. Accessed 3 Aug 2015

Chapter 23
Smart Contact Lenses and Eye Implants

Abstract Lenses worn or implanted in the eye have garnered increasing attention in recent years due to their ability to wirelessly provide real-time data on intraocular pressure, glucose levels in tears, and biomarkers of various diseases. This chapter provides an overview of the progress in this field. A few examples discussed include contact lenses with integrated glucose sensors and displays, theranostic contact lenses for glaucoma treatment, intraocular lens (IOL)-based sensors for detecting aqueous humor (AH) biomarkers, implantable IOL pressure sensors, and contact lenses for human–machine interaction (HMI), among others.

Keywords CL · ICL · IOL · Glucose · Lactate · Cholesterol · Lysozyme · MMP-9 · BACE1 · Alzheimer's disease · Intraocular pressure · Glaucoma · HMI

23.1 Introduction

This chapter describes the electronic applications of different lenses, either worn on the eye surface as contact lenses (CLs) or implanted in the eye as implantable contact lenses (ICLs) with the natural lens intact or implanted in the eye as a replacement for natural lenses, the so-called intraocular lenses (IOLs) with the damaged lens removed. Therefore, we will be concerned with CLs, ICLs, or IOLs equipped with electronic capabilities to perform medical functions, such as diagnostics, therapy, or human–machine interaction. A comprehensive review of recent developments in this field will be presented.

V. K. Khanna, *Implantable Medical Electronics*,
https://doi.org/10.1007/978-3-031-99868-3_23

23.2 Types of Wearable and Implantable Lenses for the Eye

As already cursorily mentioned, lenses worn or implanted in the eye belong to one of the three types: traditional contact lenses (TCLs), implantable contact lenses (ICLs), and intraocular lenses (IOLs) [1]. Tables 23.1 and 23.2 present the salient features of these lenses.

Table 23.1 TCLs and ICLs

Sl. No.	Point of comparison	TCLs	ICLs
1	Purpose of the lens	Correction of refractive errors, e.g., nearsightedness, farsightedness, and astigmatism	Correction of vision problems as done by TCLs
2	Placement site of the lens	It is placed on the cornea's surface; the cornea is the eye's outer layer	It is surgically placed in the eye between the iris and the natural lens
3	Time duration for which the lens is placed	Temporary	Long-term use
4	Maintenance requirements of the lens	Cleaned daily, stored, and replaced	Cleaning and replacement of the lens are not necessary
5	Lens installation cost	Recurring replacement costs, including those of the cleaning solutions	A one-time expense

Table 23.2 ICLs and IOLs

Sl. No.	Point of comparison	ICLs	IOLs
1	Purpose of the lens	Correction of refractive errors, e.g., nearsightedness, farsightedness, and astigmatism, without replacing the natural lens	Restoration of vision by replacement of the clouded natural lens during cataract surgery
2	Placement site of the lens	The lens is surgically placed in the eye between the iris and the natural lens	Surgically implanted in the eye to replace the natural lens
3	Time duration for which the lens is placed	Long-term use	Permanent
4	Maintenance requirements of the lens	No cleaning or replacement requirement	Maintenance-free
5	Lens installation cost	A one-time expense	A one-time expense

Table 23.3 Regular eye lenses (TCLs, ICLs, and IOLs) and smart or electronic eye lenses

Sl. No.	Point of comparison	Regular eye lens	Electronic eye lens
1	Purpose of the lens	Correction of vision defects	Beyond correcting vision defects, electronic components are incorporated to provide additional functions
2	Applications of the lens	Mono- or multifocal, and toric (for astigmatism)	Health monitoring by tracking glucose and disease biomarkers in tear fluid and performing intraocular pressure measurement, drug delivery, and augmented reality operations

Contact lenses were initially designed for vision correction, but they have gradually evolved to be used in detecting ocular manifestations of systemic diseases in the human body. Regular eye lenses (TCLs, ICLs, and IOLs) are simply for vision correction. Smart or electronic eye lenses are reinforced with in-built devices and circuits to perform health care [2]. Table 23.3 compares the smart electronic lenses with the regular lenses. These electronic lenses are the focus of attention here.

Figures 23.1, 23.2 illustrate the placement of the three types of lenses mentioned above. Figure 23.1a shows a traditional contact lens positioned on the pupil of the eye. Figure 23.1b shows an implantable contact lens fixed between the iris and the eye's natural lens. Figure 23.2a shows the clouding of the natural lens, leading to cataract. In Fig. 23.2b, the clouded natural lens is removed, and an intraocular lens is installed during cataract surgery.

23.3 Monitoring of Glucose, Lactate, and Cholesterol Levels in Tear Fluid Using Eye Lenses

Eye lenses have been fabricated using electrochemical glucose, lactate, and cholesterol biosensors. Figure 23.3 shows the working mechanisms of these sensors. In each case, an electrode is modified with an enzyme that catalyzes the reaction of the corresponding analyte with oxygen present in the air. The chemical equations representing these reactions are written down. The production of hydrogen peroxide or oxygen consumption provides the sensing signal. The resulting current generated by the electrochemical oxidation of hydrogen peroxide is proportional to the analyte concentration. Hence, its measurement reveals the concentration of the analyte in each experiment. Among optical sensors, fluorescence properties enable colorimetric changes in the contact lens color to indicate the concentration of the analyte.

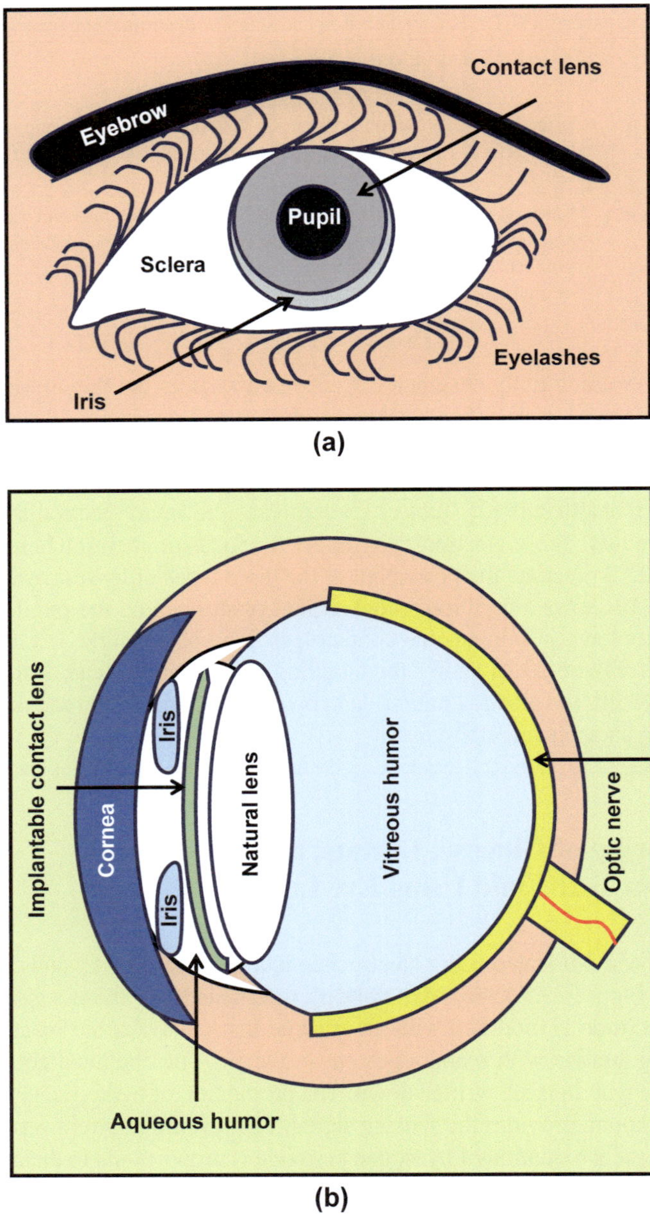

Fig. 23.1 Lenses fixed in the eye: **a** traditional contact lens, and **b** implantable contact lens

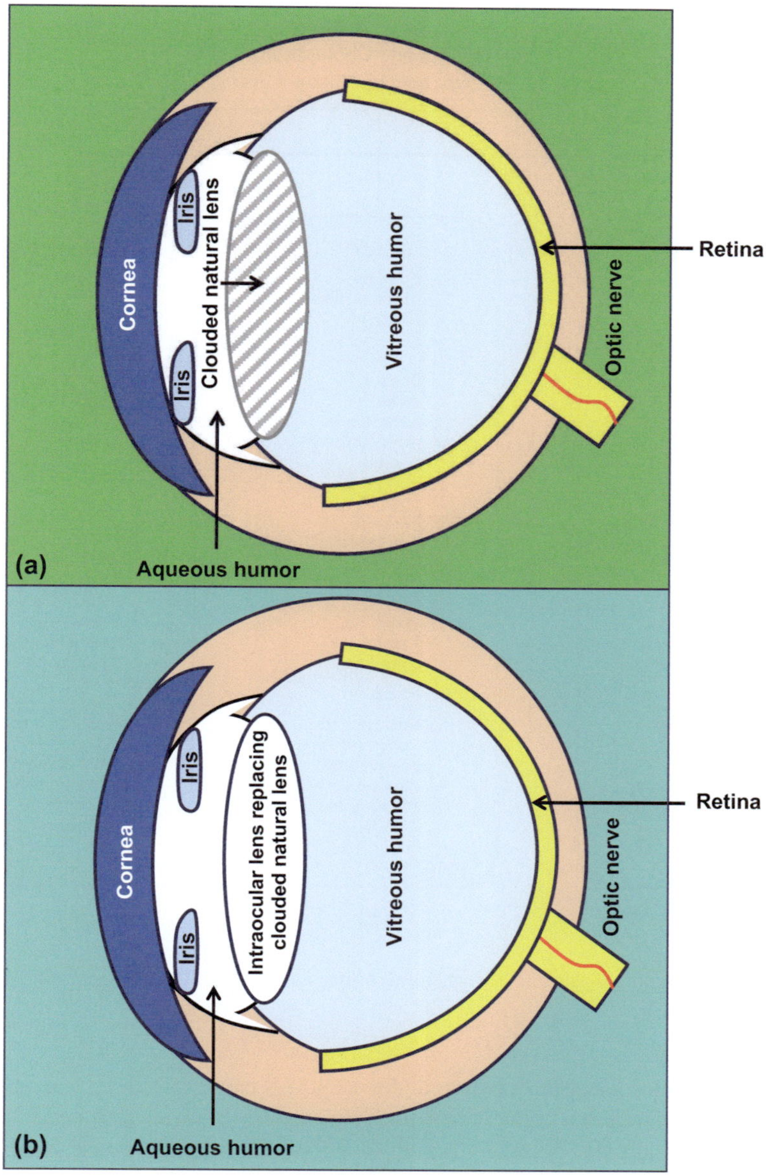

Fig. 23.2 Cataract surgery for replacing **a** a clouded natural lens with **b** an intraocular lens

23.3.1 Contact Lens with Integrated Glucose Sensor

Diabetes mellitus is a metabolic disorder characterized by elevated blood sugar levels greater than 126 mg/dL (milligram per deciliter) or 7.0 mmol/L (millimole per liter) in

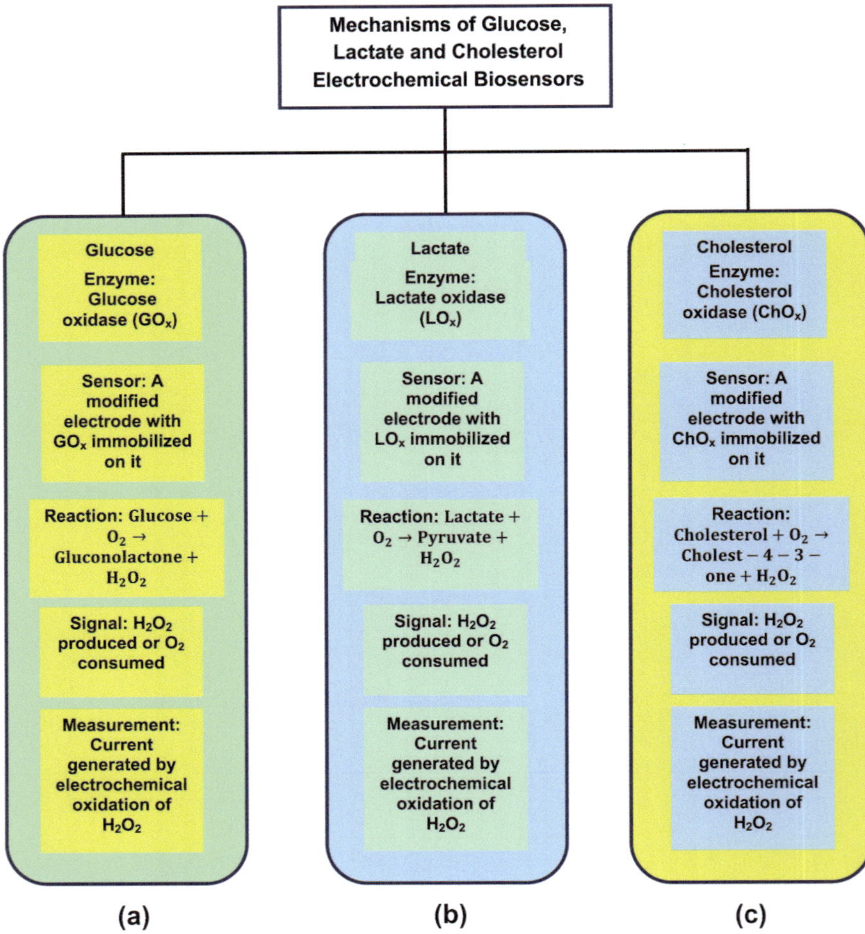

Fig. 23.3 Working principles of electrochemical biosensors for detection of: **a** glucose, **b** lactate, and **c** cholesterol

fasting conditions. Long-term elevated glucose levels cause microvascular or small-vessel disease. There is a higher risk of ischemic stroke (brain stroke), heart disease, and peripheral vascular disease, also called peripheral arterial disease. Additional complications include retinopathy, nephropathy, and neuropathy affecting the eyes, kidneys, and nerves. As hypo- and hyperglycemic episodes threaten homeostasis, the self-regulating process that ensures the survival of organisms, a quick reversal of any abnormalities in glucose levels is necessary. For this, continuous monitoring with glucose sensors is essential. The glucose sensors include optical (fluorescent and light diffractive sensors) and electrochemical devices [3].

23.3.2 Contact Lens with Glucose Sensor, Circuit, and Wireless Display

The glucose sensor, wireless power transfer circuit, and display pixels for real-time visualization of sensing signals are embedded within a contact lens using transparent and stretchable nanostructures, enabling direct glucose measurement without the need for extra equipment [4]. Figure 23.4 shows the contact lens with rectifier, light-emitting diode (LED), and glucose sensor as the functional devices, along with the antenna and interconnects made on a transparent stretchable conductor. The glucose sensor is fabricated on a silicon (Si) wafer with a 300-nm-thick thermal oxide layer. Evaporated and photolithographically patterned chromium/gold (Cr/Au) electrodes, with a thickness ratio of 3:100 nm, were used as source and drain connections. A graphene layer synthesized by chemical vapor deposition was transferred onto the sample and patterned by photolithography to form the channel. Then, the SU-8 (a trademarked term for an epoxy resin used as a photoresist) layer was patterned to open channel regions to immobilize glucose oxidase (GOD) and catalase (CAT) on the graphene channel. For functionalization with GOD and CAT, the fabricated sensor was placed in 1-pyrenebutanoic acid succinimidyl ester in methanol for 2 h at room temperature, followed by cleaning with methanol. The cleaned sensor was immersed in GOD and CAT in deionized (DI) water for 18 h at room temperature, rinsed with clean DI water, and dried with nitrogen gas. CAT decomposes hydrogen peroxide (H_2O_2) into water (H_2O) and oxygen (O_2), thereby improving the sensor's sensitivity by increasing the enzymatic activity of GOD.

23.3.3 Contact Lens for Glucose Monitoring and Drug Delivery in Diabetic Retinopathy

A multifunctional contact lens device is fabricated on a biocompatible polymer for continuous, noninvasive diagnosis of diabetes and simultaneous delivery of necessary therapy [5]. It contains ultrathin, flexible electrical circuits and a microcontroller chip for real-time electrochemical glucose biosensing. An on-demand flexible drug delivery system, wireless power management, and a remote radio frequency communication system are included. Wireless power is transmitted from an external power source using a transmitter coil, achieving resonant inductive coupling with a copper (Cu) receiver coil. Figure 23.5 illustrates the diabetes diagnosis and therapy lens, which incorporates a glucose biosensor, a drug delivery system, a wireless power transmission system that connects a transmitter coil to a receiver coil, an application-specific integrated circuit (ASIC) chip, and a remote communication system.

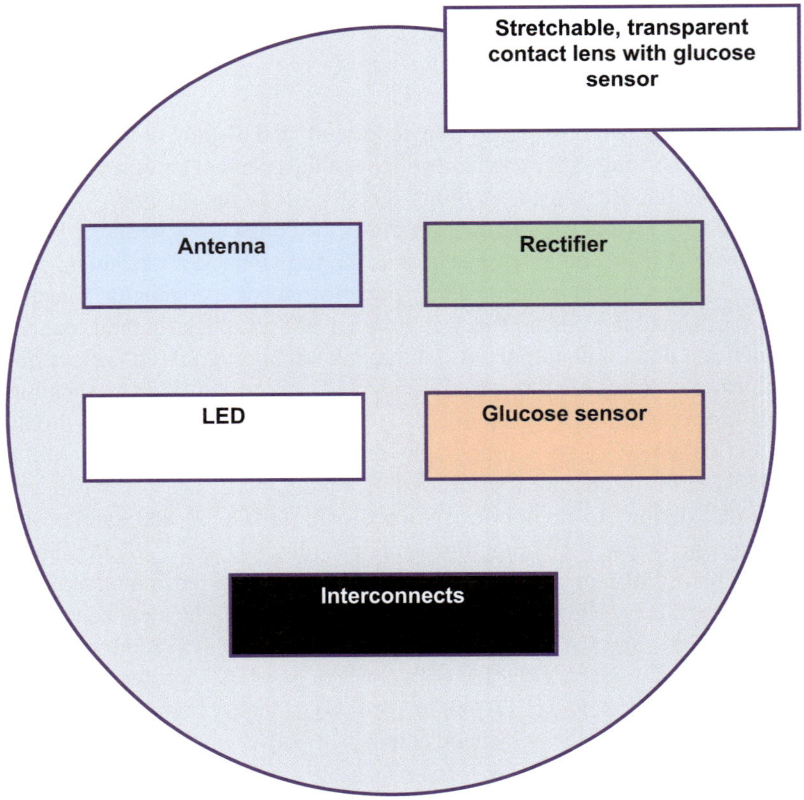

Fig. 23.4 Contact lens with the glucose sensor, wireless circuit, and display

23.3.4 Fluorescent Contact Lens for Monitoring Glucose via a Smartphone

The optical sensing principle is sometimes applied for glucose testing. Figure 23.6 illustrates a contact lens that utilizes the fluorescence phenomenon for glucose sensing. Soft and transparent contact lenses are prepared by grafting a glucose fluorescent probe and a reference fluorescent dye into the poly(2-hydroxyethyl methacrylate) (pHEMA) hydrogel network of the lens. The lens's fluorescent color changes from pink to blue with increasing glucose concentration. Fluorescent images captured by a smartphone can measure glucose concentrations in the range of 23 micromolar (μM) to 1.0 mM (mM) in tears [6].

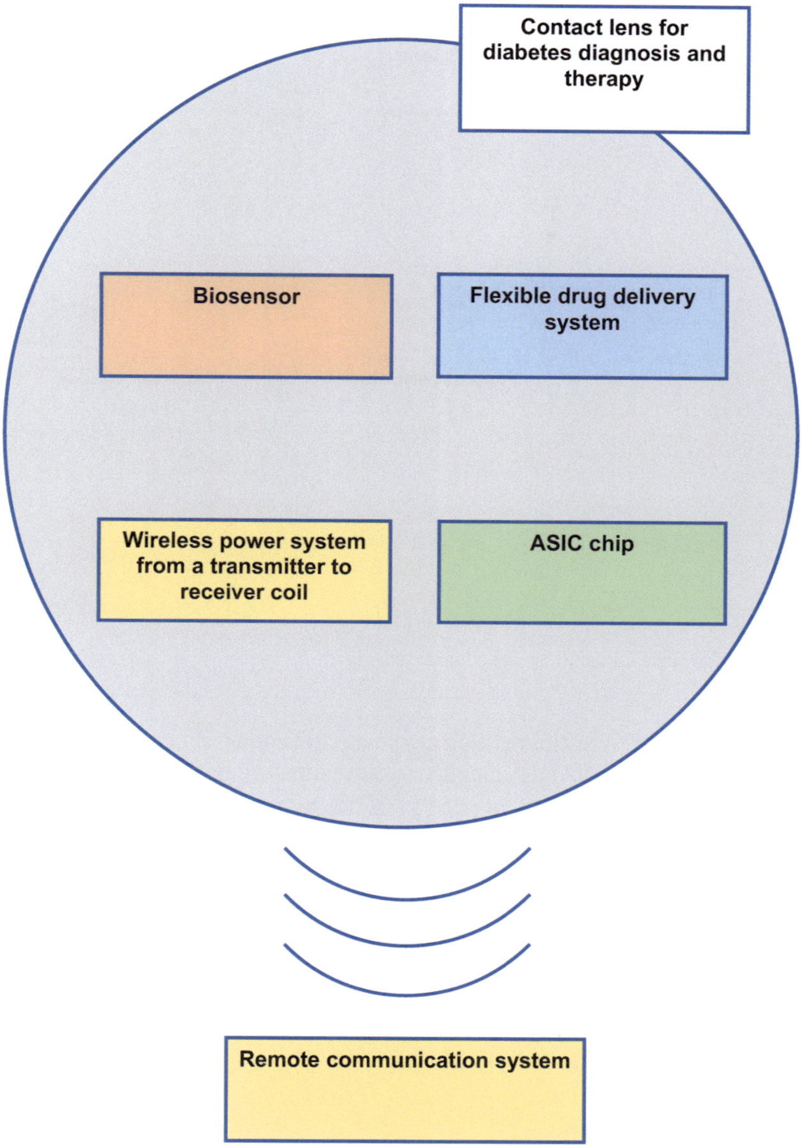

Fig. 23.5 Contact lens for diabetes diagnosis and treatment

23.3.5 Contact Lens with Integrated Lactate Sensor

Lactate is a vital metabolite playing a central role in the anaerobic glycolytic pathway in the human body. Anaerobic glycolysis works through glycolysis in which glucose

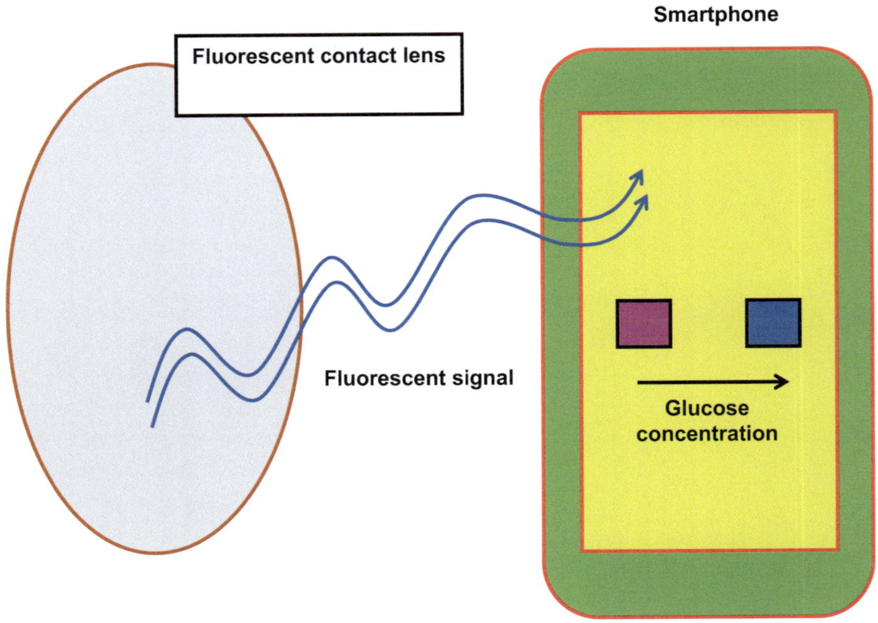

Fig. 23.6 Fluorescent content lens for glucose sensing

is broken down into pyruvate through enzymatic reactions, generating ATP (adenosine triphosphate) and NADH (nicotinamide adenine dinucleotide, reduced form), lactate formation pyruvate is converted to lactate by the enzyme lactate dehydrogenase, regenerating NAD^+ (nicotinamide adenine dinucleotide, oxidized form) from NADH and ATP regeneration allowing glycolysis to continue producing ATP even without oxygen. High lactate levels in the body, a condition known as hyperlactatemia, indicate either oxygen deficiency (hypoxia), overproduction of lactate, or impairment of its clearance. An amperometric, mono-enzymatic L-lactate sensor is developed on a polymer substrate molded into a contact lens shape. It has titanium/palladium/platinum (Ti/Pd/Pt) electrodes. The platinum sensing structures are functionalized by cross-linking lactate oxidase with glutaraldehyde and bovine serum albumin, and then coated with polyurethane. The sensor has a quick response time of 35 s, with an average sensitivity of ~approximately 53 microamperes millimolar $^{-1}$ cm $^{-2}$ (μA mM^{-1} cm^{-2}) in the linear range [7].

23.3.6 *Contact Lens with Cholesterol Sensor*

Cardiovascular diseases constitute a significant cause of early death worldwide. Genetic or environmental reasons, including unbalanced diets, cause them. Hyperlipidemia is a significant risk factor for these diseases. It refers to elevated cholesterol levels in the blood, which can lead to excessive fat accumulation and the formation of plaque in the arteries. A soft, wireless contact lens is demonstrated for measuring free cholesterol levels in tear fluids [8]. An electrochemical biosensor embedded in the lens provides continuous measurement of cholesterol concentrations. This biosensor uses cholesterol oxidase enzyme to detect cholesterol selectively. The lens also features a stretchable wireless antenna, designed with serpentine geometry, and an integrated circuit chip for near-field communication. A smartphone transfers power and data wirelessly without obstructing the wearer's vision.

The cholesterol biosensor consists of a two-electrode system featuring working and reference electrodes. Cholesterol oxidase (ChOx) enzyme is immobilized using Nafion polymer. Prussian blue printed on the working electrode works as an artificial peroxidase to reduce H_2O_2. The reaction is translated into an electrochemical response using chronoamperometry by biasing at − 0.1 V versus a silver/silver chloride (Ag/AgCl) electrode. The reduction current of H_2O_2 is proportional to the concentration of cholesterol. The biosensor detects free cholesterol from 0.4 to 46.4 mg per deciliter (mg dL^{-1}) (0.01 to 1.2 mM) [8].

23.4 Contact Lens-Based Lysozyme Detection in Tear Fluid

Lysozyme is a protein in tears, a biomarker for dry eye disease, dry eye syndrome, or keratoconjunctivitis sicca where the eyes do not produce enough tears, or the tears evaporate too quickly, leading to visual disturbances induced by insufficient lubrication of the ocular surface. It is also a biomarker for Sjögren's syndrome, an autoimmune disorder in which the moisture-producing glands are attacked.

Commercially available contact lenses are used to collect biomarkers in the tear fluid [9]. Time-lapse imaging captures an increasing fluorescent signal produced by an enzymatic assay performed in a 96-well plate. The enzyme activity is indirectly inferred from the calculated rate of change in the detected fluorescence compared to a standard curve. When the contact lens wearers played a mobile phone, causing digital eye-strain, a mean increase in lysozyme concentration measurements from 6.89 ± 2.02 μg per milliliter (μg/mL) to 10.72 ± 3.22 μg/mL (mean ± standard deviation) was observed [9].

23.5 Implantable IOL for Aqueous Humor Biomarkers for Central Nervous System Disorders

The eye, directly connected to the brain via the optic nerve, acts as a visual extension of the central nervous system (CNS). The aqueous humor in the eye is a clear fluid, secreted by the ciliary body. It contains a variety of biomolecules that can reflect changes in the CNS. Consequently, the variations in the central nervous system are reflected early in any changes in this fluid. Its analysis, therefore, provides a non-invasive method of detecting such disorders. This has given impetus to the development of lenses for intraocular monitoring.

In this context, matrix metalloproteinase-9 (MMP-9) is an important biomarker. In the CNS, elevated MMP-9 levels have been linked to tick-borne encephalitis (TBE). MMP-9 is implicated in dry eye disease (DED) and age-related macular degeneration (AMD). To fabricate MMP-9-responsive fluorogenic PEGDAAm hydrogel sensors, cleavable fluorogenic peptide-probes were conjugated with MMP-9. The fluorogenic hydrogel sensor was attached to an IOL to make the fluorogenic IOL (FIOL). The FIOL was implanted in rabbits by cataract surgery. Changes in the fluorescence signal in response to increasing intraocular MMP-9 concentrations were measured [10].

23.6 Implantable IOL Fluorescence Sensor for Alzheimer's Disease

Beta-site amyloid precursor protein cleaving enzyme 1 (BACE1), or beta-secretase, is an enzyme that cleaves amyloid precursor protein (APP) to form Amyloid-β (Aβ), a short chain of amino acids called a peptide. The Aβ aggregates to form insoluble plaques in the brains of Alzheimer's disease patients. Hence, the BACE1 is a biomarker and therapeutic target for neurogenerative brain diseases [11]. The BACE1 is found within the intraocular space, particularly in the retina. A pioneering platform for detecting BACE1 is the fluorescent aptamer-conjugated intraocular lens (FAIOL).

The fluorescent aptamer-conjugated intraocular lens (FAIOL) is an innovative platform for the detection of BACE1, offering a two-fold advantage:

(i) It allows for ongoing surveillance of degenerative brain illnesses, detection of minuscule patient changes over time, and expediting a correct diagnosis.
(ii) It features a user-friendly program that reliably detects and presents variance in the sensor signal. Therefore, user data interpretation is appreciably simplified.

The FAIOL is fabricated by integrating a dsDNA probe with an aptamer in an inverse opal hydrogel (IOH) and combining it with an IOL. The IOH, coupled with the specifically engineered aptamer, provides the targeted detection of BACE1. The signals are examined using mobile phone applications [12].

23.7 Theranostic Contact Lens for Glaucoma Treatment

Glaucoma is a chronic eye disease characterized by increased intraocular pressure that damages the optic nerve, resulting in progressive vision loss, which typically begins with peripheral vision and eventually affects central vision. As high intraocular pressure is a sign of glaucoma, its continuous monitoring and restriction to a lower acceptable range are recommended. Therefore, intraocular pressure (IOP) control is an effective medical care for glaucoma patients.

23.7.1 Contact Lens with Gold Hollow Nanowire-Based Pressure Sensor

Figure 23.7 illustrates a theranostic contact lens that combines diagnostic and therapeutic capabilities to monitor eye conditions and deliver medications as needed. The contact lens includes a sensitive, flexible gold hollow nanowire (AuHNW)-based intraocular pressure sensor and a flexible drug delivery system loaded with IOP-lowering drug timolol. A biocompatible layer protects the drug delivery system. IOP control is achieved by releasing timolol from the reservoir by electrical triggering. The lens also has an antenna for wireless communication and an application-specific integrated circuit chip for monitoring and controlling intraocular pressure in glaucoma. The signal received from IOP sensing enables timolol release via a wireless power and communication feedback system [13, 14].

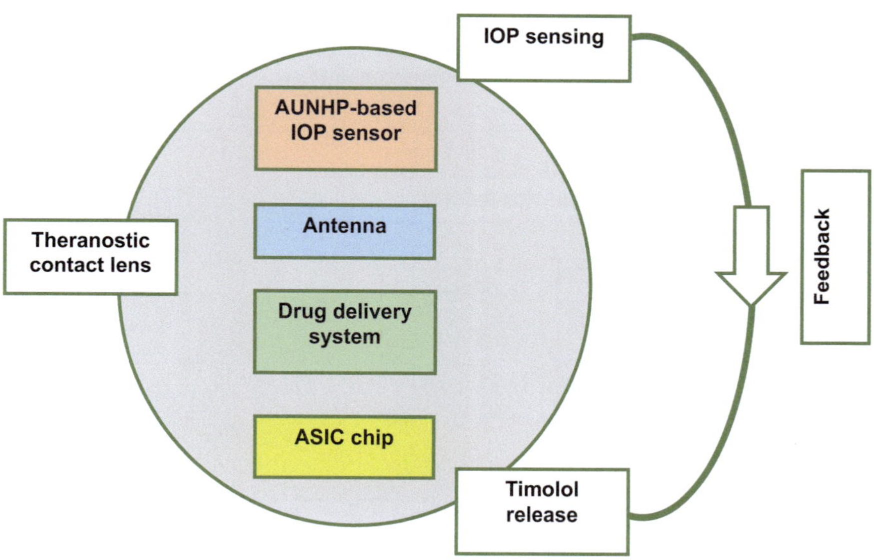

Fig. 23.7 Theranostic contact lens for personalized glaucoma management

23.7.2 Contact Lens with a Capacitive Pressure Sensor

Figure 23.8 shows a minimally invasive theranostic contact lens comprising pressure-sensing and drug-delivery modules. The pressure sensing module has a capacitive sensing circuit and inductance coils. An air film is sandwiched between the upper and lower lenses, acting as a dielectric material for the capacitor formed between them. The main feature of the lens is an inductance–capacitance–resistance (LCR) circuit designed with a cantilever structure. The capacitive sensing and reference plates are folded into the cantilever construction to form a beam-like structure, enhancing measurement sensitivity, repeatability, and reliability. The air film acts as the dielectric. Any change in IOP deforms the air film, altering the capacitance and hence the resonant frequency of the LCR circuit. Thus, the IOP transducer detects pressure fluctuations and wirelessly transmits them.

The wireless intraocular pressure sensing module detects intraocular pressure fluctuations ultra-sensitively. The drug delivery module employs an efficient wireless power transfer circuit to trigger the delivery of anti-glaucoma drug into the aqueous chamber via iontophoresis electrodes. At high-risk IOP conditions (IOP > 21 mmHg), wireless power transfer-activated iontophoresis effectively administers the drug *in situ* [15].

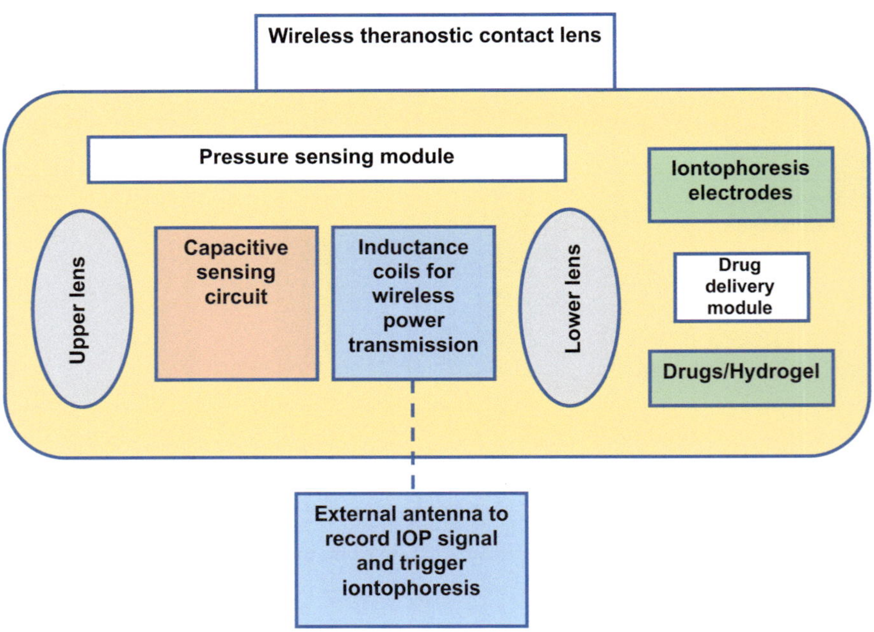

Fig. 23.8 Glaucoma theranostic lens with LCR circuit

23.8 Contact Lens for Human–Machine Interaction

Apart from detecting metabolites and biomarkers in tear fluid, contact lenses are utilized in human–machine interaction (HMI) applications, including eye tracking, vision assistance, and displays. Xia et al. [16] present a novel way of categorizing smart contact lenses into six core building blocks: sensing, energy management, driver electronics, communications, substrate, and input/output interface modules.

Eye tracking techniques enable highly efficient, natural, and effortless human–machine interaction. They detect users' eye movements and decode their attention and intentions. A tiny, invisible, and biocompatible smart contact lens is proposed for in situ eye tracking and wireless eye-machine interaction. The contact lens enhances the approaches to eye tracking techniques and encourages the development of human–machine interaction technology [17].

Employing the frequency encoding strategy and an advanced spherical conformal preparation technique, the chip-free and battery-free lens successfully detects eye movement and closure. The well-designed smart contact lens comprises four chipless passive RF tags with different working frequencies. A portable sweeping-frequency reader is installed at the framed glasses and opposite the user's eyeball. This reader wirelessly collects the tags' signal. The received signal strengths of multiple tags change because of variable coupling coefficients caused by eye movement. These changing signal strengths help to track the gazing point and to input the eye commands. A time-sequential eye-tracking algorithm is used. The lens has an angular accuracy of $< 0.5°$. This is even less than the vision range of the central fovea.

23.9 Discussion and Conclusions

This chapter surveyed the wide range of uses of electronic eye lenses in medical applications. Table 23.4 summarizes the lenses discussed.

With its tear fluid, the eye is a treasure trove of physiological information, providing an ideal avenue for non-invasive disease monitoring using smart contact lens sensors [18–21]. Eye lenses display promising capabilities for providing continuous health monitoring and delivering therapy through the chemical analysis of tear fluid. They can also act as helpful aids in human–machine interaction.

Review Exercises

23.1. Differentiate between the following terms: (a) contact lens and implantable contact lens, (b) implantable contact lens and intraocular lens, and (c) regular eye lens and electronic eye lens.

23.2 Explain the working mechanisms of electrochemical biosensors for: (a) glucose, (b) lactate, and (c) cholesterol.

Table 23.4 Multiple uses of electronic contact/implantable eye lenses

Electronic eye lens capabilities and applications	Reference
Electrochemical glucose sensor for tear fluid, wireless circuit, and display	[4]
Electrochemical glucose monitoring in tear fluid and drug delivery in diabetic retinopathy	[5]
Optical glucose monitoring in tear fluid via a smartphone	[6]
Electrochemical lactate sensor for tear fluid	[7]
Electrochemical cholesterol sensor for tear fluid	[8]
Lysozyme measurement in tear fluid; lysozyme is a biomarker for dry eye disease	[9]
MMP-9 measurement in tear fluid; MMP-9 is a biomarker for disorders of the central nervous system	[10]
BACE1 measurement in tear fluid; BACE1 is a biomarker for Alzheimer's disease	[11, 12]
Intraocular pressure measurement for glaucoma detection and drug delivery	[13–15]
Human–machine interaction	[16, 17]

23.3 What is diabetes mellitus? What are its harms? Why is continuous glucose monitoring advised for diabetic patients?

23.4 Draw a diagram of a contact lens with a glucose sensor, a wireless circuit, and a display, explaining the roles of its components. How is the glucose biosensor fabricated and functionalized for selective glucose detection?

23.5 Explain with a diagram the working of a contact lens for glucose monitoring and drug delivery in diabetic retinopathy.

23.6 How does a fluorescent contact lens provide glucose monitoring via a smartphone? Explain with the help of a diagram.

23.7 What role does lactate play in the anaerobic glycolytic pathway in the human body? What do elevated lactate levels in the human body indicate? How does the contact lens with an amperometric, mono-enzymatic L-lactate sensor work?

23.8 What does Hyperlipidemia refer to? How does a cholesterol biosensor incorporated in a contact lens detect cholesterol in a patient's tear fluid?

23.9 Elucidate the statement, "The eye is a visual extension of the central nervous system of an individual". How are central nervous disorders diagnosed with contact lenses for intraocular monitoring?

23.10 With what disease are high MMP-9 levels interrelated? What is a fluorogenic IOL used for?

23.11 What is the biomarker for neurogenerative brain diseases? How does a fluorescent aptamer-conjugated intraocular lens detect this biomarker?

23.12 What is meant by a theranostic contact lens? Describe a contact lens with a gold hollow nanowire-based pressure sensor for glaucoma monitoring. How is the anti-glaucoma drug released when the intraocular pressure rises?

23.13 What is the advantage of using the cantilever structure in the theranostic contact lens with a capacitive pressure sensor? What is the mechanism of delivery of anti-glucoma drug?

23.14 How can contact lenses improve human–machine interaction?
23.15 Describe the use of a contact lens to monitor eye movements through frequency encoding.

References

1. Inland eye specialists 2018 implantable contact lens—pros and cons. https://www.inlandeyespecialists.com/eye-surgery/implantable-contact-lens/#:~:text=Implantable%20contact%20lens%20(ICL)%20are,any%20side%20effects%20or%20discomfort
2. Abdulamier AA, Shaker LM, Al-Amiery AA, Qasim MT, Isahak WNRW, Luthfi AAI (2024) Advancements and applications of smart contact lenses: a comprehensive review. Results Eng 24:103268. https://doi.org/10.1016/j.rineng.2024.103268
3. Elsherif M, Moreddu R, Alam F, Salih AE, Ahmed I, Butt H (2022) Wearable smart contact lenses for continual glucose monitoring: a review. Front Med (Lausanne) 9(858784):1–15. https://doi.org/10.3389/fmed.2022.858784
4. Park J, Kim J, Kim S-Y, Cheong WH, Jang J, Park Y-G, Na K, Kim Y-T, Heo JH et al (2018) Soft, smart contact lenses with integrations of wireless circuits, glucose sensors, and displays. Sci Adv 4(1):1–11. https://doi.org/10.1126/sciadv.aap9841
5. Keum DH, Kim S-K, Koo J, Lee G-H, Jeon C, Mok JW, Mun BH, Lee KJ, Kamrani E et al (2020) Wireless smart contact lens for diabetic diagnosis and therapy. Sci Adv 6(37):1–12. https://doi.org/10.1126/sciadv.aba3252
6. Deng M, Song G, Zhong K, Wang Z, Xia X, Tian Y (2022) Wearable fluorescent contact lenses for monitoring glucose via a smartphone. Sens Actuators B: Chem Part 2(352):131067. https://doi.org/10.1016/j.snb.2021.131067
7. Thomas N, Lähdesmäki I, Parviz BA (2012) A contact lens with an integrated lactate sensor. Sens Actuators, B Chem 162:128–134. https://doi.org/10.1016/j.snb.2011.12.049
8. Song H, Shin H, Seo H, Park W, Joo BJ, Kim J, Kim J, Kim HK, Kim J, Park JU (2022) Wireless non-invasive monitoring of cholesterol using a smart contact lens. Adv Sci (Weinheim) 9(28):e2203597. https://doi.org/10.1002/advs.202203597
9. Ballard Z, Bazargan S, Jung D, Sathianathan S, Clemens A, Shir D, Al-Hashimi S, Ozcan A (2020) Contact lens-based lysozyme detection in tear using a mobile sensor. Lab Chip 20(8):1493–1502. https://doi.org/10.1039/c9lc01039d
10. Shin M-K, Ji YW, Moon C-E, Lee H, Kang B, Jinn W-S, Ki J, Mun B, Kim M-H, Lee HK, Haam S (2020) Matrix metalloproteinase 9-activatable peptide-conjugated hydrogel-based fluorogenic intraocular-lens sensor. Biosens Bioelectron 162:112254. https://doi.org/10.1016/j.bios.2020.112254
11. Hampel H, Vassar R, De Strooper B, Hardy J, Willem M, Singh N, Zhou J, Yan R et al (2021) The β-Secretase BACE1 in Alzheimer's disease. Biol Psychiat 89(8):745–756. https://doi.org/10.1016/j.biopsych.2020.02.001
12. Lim J, Lee H, Kim J, Moon C-E, Lee J-K, Kang Y, Kim R, Mun B, Ji YW, Haam S (2025) Implantable intraocular fluorescence sensor for visualized monitoring of Alzheimer's disease 35(3):2413723. https://doi.org/10.1002/adfm.202413723
13. Kim TY, Mok JW, Hong SH, Jeong SH, Choi H, Shin S, Joo C-K, Hahn SK (2022) Wireless theranostic smart contact lens for monitoring and control of intraocular pressure in glaucoma. Nat Commun 13(1):6801. https://doi.org/10.1038/s41467-022-34597-8
14. Ran M, Yan J, Zhang H (2023) Precisely integrated contact lens: an intraocular pressure guard for glaucoma patients. Biomed Technol 4:39–40. https://doi.org/10.1016/j.bmt.2023.03.001
15. Yang C, Wu Q, Liu J, Mo J, Li X, Yang C, Liu Z, Yang J, Jiang L et al (2022) Intelligent wireless theranostic contact lens for electrical sensing and regulation of intraocular pressure. Nat Commun 13:2556. https://doi.org/10.1038/s41467-022-29860-x

16. Xia Y, Khamis M, Fernadez FA, Heidari H, Butt H, Ahmed Z, Wilkinson T, Ghannam R (2023) State-of-the-art in smart contact lenses for human–machine interaction. IEEE Trans Hum-Mach Syst 53(1):187–200. https://doi.org/10.1109/THMS.2022.3224683
17. Zhu H, Yang H, Xu S, Ma Y, Zhu S, Mao Z, Chen W, Hu Z, Pan R, Xu Y, Xiong Y, Chen Y, Lu Y, Ning X, Jiang D, Yuan S, Xu F (2024) Frequency-encoded eye tracking smart contact lens for human-machine interaction. Nat Commun 15(1):3588. https://doi.org/10.1038/s41467-024-47851-y
18. Mertz L (2023) Smart contact lenses keep an eye on health. IEEE Pulse 14(3):20–23. https://doi.org/10.1109/MPULS.2023.3294087
19. Seo H, Chung WG, Kwon YW, Kim S, Hong Y-M, Park W, Kim E, Lee J et al (2023) Smart contact lenses as wearable ophthalmic devices for disease monitoring and health management. Chem Rev 123(19):11488–11558. https://doi.org/10.1021/acs.chemrev.3c00290
20. Kazanskiy NL, Khonina SN, Butt MA (2023) Smart contact lenses-A step towards non-invasive continuous eye health monitoring. Biosens (Basel) 13(10):933, 1–28. https://doi.org/10.3390/bios13100933
21. Liu X, Ye Y, Ge Y, Qu J, Liedberg B, Zhang Q, Wang Y (2024) Smart contact lenses for healthcare monitoring and therapy. ACS Nano 18(9):6817–6844. https://doi.org/10.1021/acsnano.3c12072

Chapter 24
Wireless Capsule Endoscopy

Abstract The concept of using noninvasive cameras swallowed in capsules to image the gastrointestinal tract has led to the development of a commercial diagnostic instrument called capsule endoscopy. This technique is a valuable alternative to the standard diagnostic modality for gastric lesions known as esophagogastroduodenoscopy (EGD). Advances in technology enabling the realization of capsule endoscopy are described, notably those in computer hardware, imaging techniques, and artificial intelligence. Diseases for which capsule endoscopy diagnosis is indicated are discussed, e.g., gastroesophageal reflux disease, obscure gastrointestinal bleeding, cirrhosis and portal hypertensive bleeding, coeliac disease, small bowel tumors, and colorectal polyps. Capsule endoscopy is limited by the passive movement of capsules via bowel peristalsis. Magnetic manipulation systems have been developed for capsule locomotion, orientation, localization, and manipulation. These maneuverable capsules provide a more detailed evaluation and accurate diagnosis of affected regions. Many research groups work on detecting suspected lesions and delivering drugs to specific sites for targeted therapeutics. This makes it an efficient tool for treating diverse pathologies without requiring sedation. The present status and research trends of capsule endoscopy are summarized.

Keywords Capsule · Endoscopy · EGD · Magnetically guided endoscopy · Gastrointestinal bleeding · Crohn's disease · Celiac disease · CT enteroclysis · MRI · Push enteroscopy

24.1 Preliminary Background

Capsule endoscopy is a noninvasive medical procedure, particularly useful for examining the small intestine [1]. Since it is noninvasive, there is no need to physically insert any diagnostic equipment into the body, avoiding any discomfort to the patient. It overcomes the limitations of traditional endoscopy methods conducted to examine

In this chapter, we describe a painless and harmless technique for taking pictures of the organs through which food and liquids pass after being consumed orally.

V. K. Khanna, *Implantable Medical Electronics*,
https://doi.org/10.1007/978-3-031-99868-3_24

the lining of the food pipe, esophagus, stomach, duodenum, or the first part of the small intestine. The conventional test is called esophagogastroduodenoscopy (EGD). The EGD helps investigate symptoms such as stomach pain, trouble swallowing, or stomach bleeding to identify diseases like ulcers and gastroesophageal reflux disease (GERD). This test is typically performed as an outpatient procedure under sedation, during which a thin, flexible tube known as an endoscope is inserted through the mouth and down into the patient's stomach. While it is generally a safe procedure, there are potential risks that include bleeding, sore throat, upper gastrointestinal tract perforation, and side effects of the sedative [2].

Various names are used to describe the capsule endoscopy technique, all highlighting a vital process feature. Of course, the first is "capsule endoscopy," which involves swallowing a tiny capsule during the test, although it is not necessarily wireless. The second is "wireless capsule endoscopy," which emphasizes the nonuse of wires in the method, although it does not necessarily contain a camera or image sensor. It is a radio telemetry system. The third is "video capsule endoscopy," which focuses on using a camera to capture and transmit video images of the body's internal parts, not necessarily wirelessly. The fourth is "small bowel capsule endoscopy," pointing toward its utilization for small bowel visualization. The fifth is "capsule enteroscopy," which examines the enteron or small intestine. The sixth is "PillCam," which suggests using a small, pill-sized capsule containing a camera. The seventh is "M2A" meaning mouth to anus capsule endoscopy [3]. Other names are: Endocapsule (Olympus) and MiroCam (IntroMedic) for obvious reasons. Ignoring all the distinctions among the names, the various names are used interchangeably in the literature. So, we shall incorporate the manifold features in the all-inclusive name "capsule endoscopy."

Figure 24.1 illustrates the differences between conventional and capsule endoscopies. In traditional endoscopy, shown on the left-hand side, the flexible endoscope tube is inserted through the patient's mouth to reach the stomach. In capsule endoscopy, shown on the right-hand side, the capsule, taken by the patient, reaches the stomach, from which it transmits images to a recorder tied to a belt worn by the patient.

24.2 Core Components of the Capsule Used in Endoscopy

The capsule (typically 2.5 cm long by 1.3 cm wide and weighing 3.0 to 4.7 g) consists of the following parts [4]:

(i) Source of Light: Light-emitting diodes illuminate the inside of the digestive tract.
(ii) Camera for Imaging: A high-resolution camera captures clear and sharp images of the digestive tract. It is based on either charge-coupled device (CCD) or complementary metal-oxide semiconductor (CMOS) technology.

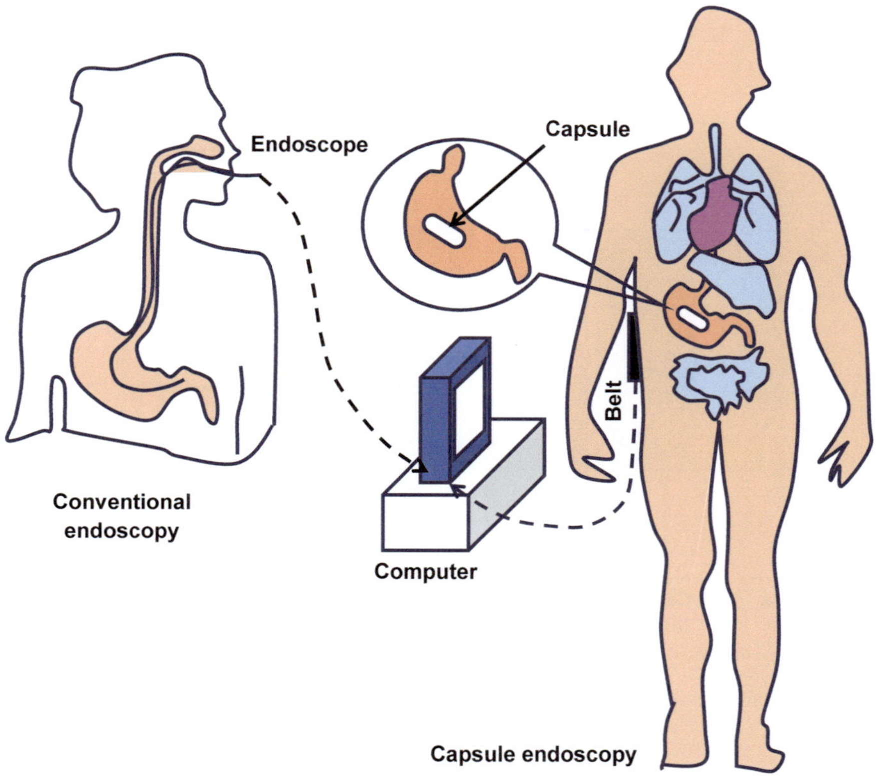

Fig. 24.1 Comparison between conventional and capsule endoscopies

(iii) Focusing Lens: A compact lens focuses light and images on the camera's sensor.
(iv) Wireless Transmitter: A small radio-frequency transmitter conveys the images acquired by the camera to a recording device fitted in the patient's belt. The recorder downloads the received images to a computer for analysis. A medical expert reviews the images to diagnose gastrointestinal disorders.
(v) Antenna: An antenna sends signals from the transmitter to the recorder.
(vi) Battery: A battery of cells powers the light-emitting diodes (LEDs), camera, and transmitter.
(vii) Microprocessor: It controls the camera's functioning and the LEDs' timing of image transmission to the recorder.
(viii) Biocompatible Package: The capsule's parts are assembled inside a small, disposable plastic housing for easy movement through the tract.

Figure 24.2 illustrates the components of a typical capsule used in capsule endoscopy without a magnet.

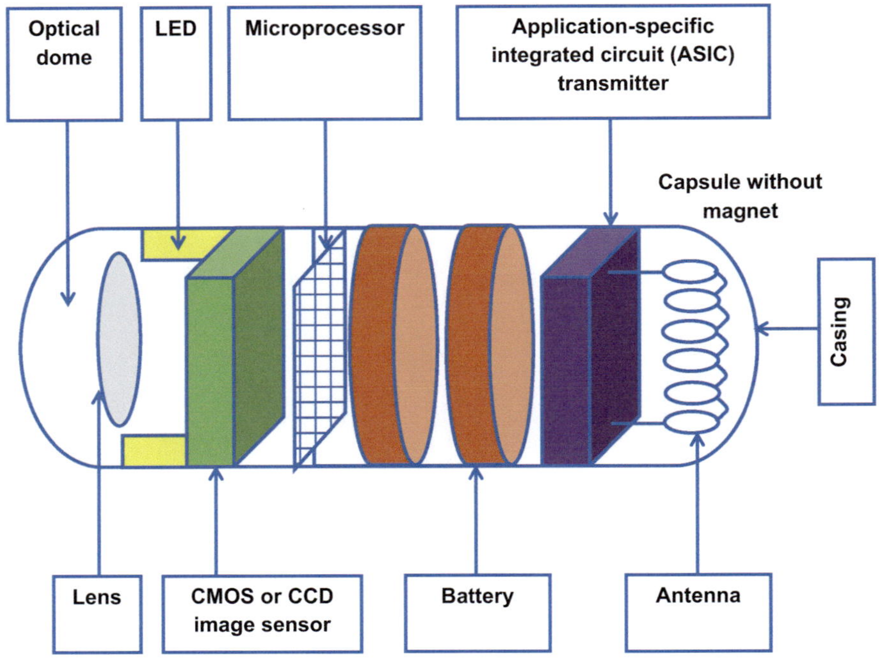

Fig. 24.2 Components of a capsule without a magnet

The components shown are: optical dome, lens, LED, CMOS or CCD image sensor, microprocessor, battery, application-specific integrated circuit (ASIC) transmitter, antenna, and casing.

24.3 Procedure of Capsule Endoscopy

(i) Pre-procedure Steps:

(a) Preparatory: The patient must follow a clear liquid diet the day before, taking a laxative to clear the digestive tract. Fasting from food or drinks is necessary for a certain period before the procedure.

(b) Advisory Instructions: The doctor issues specific instructions to the patient about the food and drinks to be taken before the procedure, including medications to be taken or avoided.

(ii) Steps Taken During the Procedure:

(a) Swallowing the Capsule: On the decided day of the procedure, the patient swallows the capsule, which is about the size of a large vitamin pill.

(b) Wearing a Small Data Recorder: The patient wears a small data recorder attached to a belt tied around the waist. The capsule wirelessly transmits the images to the recorder.
(c) Resumption of Normal Activity: The patient usually resumes normal activities after swallowing the capsule. The doctor's instructions should not be forgotten.
(d) Passage of the Capsule Passage Through the Patient's Digestive System: The capsule moves through the patient's digestive system, including the esophagus, stomach, and small intestine. During transit through the system, it takes an innumerable number of pictures as it travels. Ultimately, it is naturally excreted in the stool during the process of bowel emptying.

(iii) Post-procedure Steps:

(a) Review of Collected Patient Data: After the capsule finishes its journey, the data recorder is sent to the doctor's clinic for review, analysis of the image, and writing of the report.
(b) Results of the Procedure: The doctor reveals the findings to the patient, explaining any abnormalities or issues detected and the subsequent treatment that must be undertaken.

24.4 Applications of Capsule Endoscopy

Capsule endoscopy aids in the diagnosis of several conditions, including gastrointestinal bleeding, Crohn's disease, and celiac disease. A few applications are mentioned below:

(i) Pinpointing Gastrointestinal Bleeding Sources: This method pinpoints the source of bleeding in the small intestine, which is often difficult to locate with traditional endoscopic techniques.
(ii) Detection of Crohn's Disease and Inflammatory Bowel Disease (IBD): Inflammation and other abnormalities in the small intestine associated with these conditions are found.
(iii) Finding Signs of Celiac Disease: This method helps assess the small intestine for signs of celiac disease, an autoimmune disorder triggered by gluten.
(iv) Identification of Small Bowel Tumors: Tumors or polyps in the small intestine are identified.
(v) Resolving Unexplained Abdominal Pain and Diarrhea Issues: The clearer picture of the small intestine obtained, where other techniques failed, explains such issues.
(vi) Finding Sources of Obscure Gastrointestinal Bleeding: Recurrent and obscure gastrointestinal bleeding sources become detectable by capsule endoscopy after traditional endoscopic procedures have been unsuccessful.

24.5 Advantages of Capsule Endoscopy

Capsule endoscopy is a minimally invasive, safe, painless procedure that does not need sedation. Its principal advantages are as follows:

(i) Minimally Invasive and Safe Nature of the Procedure: Swallowing a small camera capsule makes it significantly less invasive than traditional endoscopy, which sends the endoscope down the food pipe.
(ii) Non-requirement of Patient's Sedation: This advantage makes the method more comfortable and convenient for patients. Sedation is required during traditional endoscopy.
(iii) Capability of Visualization of the Entire Small Intestine: Such examination is not achieved with traditional endoscopy.
(iv) High Diagnostic Yield for Detection of Diseases, Gastrointestinal Bleeding, and Other Problems: A high yield is assured for detecting conditions like Crohn's disease, ulcerative colitis, and other small bowel and digestive tract problems.
(v) Convenient and Easy Procedure for the Patient: It allows the patient to resume normal activities immediately after the procedure without waiting for a recovery period.
(vi) Cost-Effectiveness: Avoiding the need for sedation and staying in the recovery room after the procedure minimizes hospitalization expenses.

24.6 Shortcomings of Capsule Endoscopy

Capsule endoscopy has limitations, including its inability to perform specific procedures, such as biopsies or therapeutic interventions, the risk of capsule retention, and the time required for image interpretation. Its drawbacks include:

(i) Lack of Control of Capsule Locomotion: The capsule's movement through the gastrointestinal (GI) tract is passive, with the operator unable to control its position or movement.
(ii) Restrictions in Optical Technology Provisions: Owing to the capsule's small size, effective optical technology and techniques like air insufflation or flushing cannot be provided.
(iii) Issues of Image Quality: The presence of bile, mucus, air bubbles, and rapid transit of the capsule sometimes results in blurred or incomplete views.
(iv) Life of the Battery in the Capsule: As the battery fails after a certain period, complete examination of the small bowel is prevented if the capsule's transit through the digestive track is inordinately slow.
(v) Time-Consuming Interpretation of Images: Physicians may take a lot of time in analyzing the enormous number of images captured during the test.

(vi) Non-availability of Biopsy or Therapeutic Capabilities: It is purely a visual technique. It cannot be used to obtain tissue samples for biopsy examination or to perform therapeutic procedures.
(vii) Risk of Retention of the Capsule in Narrow Region of the Intestine: Surgical or endoscopic retrieval becomes necessary if a capsule is lodged in a constricted area.
(viii) Missing Crucial Lesions: Lesions in the duodenum and proximal jejunum may be missed during rapid movement of the capsule in some regions of the small bowel.
(ix) Incomplete Examinations: Incomplete examinations may occur due to the running out of the battery during delayed gastric emptying or prolonged small bowel transit
(x) Costly and Expensive Procedure: The cost of the procedure, inclusive of the capsule, equipment, and interpretation, can be prohibitively high.

Figure 24.3 presents a brief view of the diagnostic applications of capsule endoscopy in gastrointestinal bleeding, Crohn's and Celiac diseases, and small bowel tumors. Its advantages are its noninvasiveness and cost-effectiveness, and its disadvantages are the lack of control over capsule motion and the inability to perform biopsies. To summarize, we recite the poem.

Upsides and Downsides of Capsule Endoscopy

To start a diagnostic examination

A capsule is swallowed without any sedation,

The capsule transmits digestive tract information,

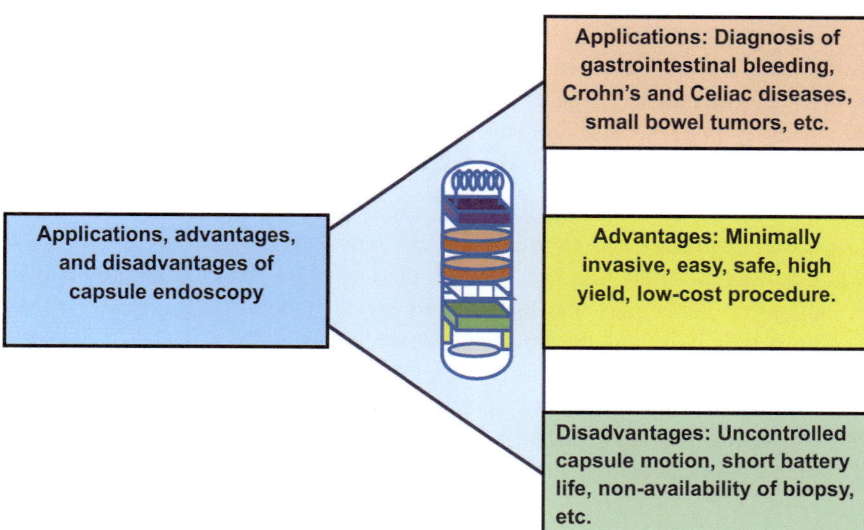

Fig. 24.3 Uses, merits, and demerits of capsule endoscopy

The patient feels no pain, no body invasion,

Capsule-sent images enable visualization of any bleeding in the stomach or inflammation,

Doctors scrutinize the pictures

And find out tumors and ulcers,

Through this easy intestinal scan

To formulate an effective treatment plan,

But the technique has limitations

About the inability to perform biopsies or therapeutic interventions.

24.7 Magnetically Guided Capsule Endoscopy

24.7.1 Working of Magnetically Guided Capsule Endoscopy

Magnetically guided capsule endoscopy (MGCE) is an improved version of the traditional capsule endoscopy [5, 6]. Figure 24.4 shows the components of a typical capsule containing a magnet used in magnetically guided capsule endoscopy. The components shown are: optical dome, lens, LED, CMOS or CCD image sensor, microprocessor, magnet, battery, ASIC transmitter, antenna, and casing. These components are essentially the same as those shown in Fig. 24.2, except for the inclusion of a magnet in the capsule for MGCE.

External magnetic fields produced by magnets or electromagnetic coils are applied to guide, navigate, and manipulate a capsule containing magnets inside the gastrointestinal tract. Neodymium-iron-boron (NdFeB) permanent magnets are used within the capsule. The NdFeB magnets are preferred due to their high magnetic strength and relatively compact size, because miniaturization of the capsule is essential. Magnetically guided capsule endoscopy proceeds as follows:

(i) Swallowing a Magnet-Containing Capsule: The patient swallows a capsule containing magnets, a camera, and other sensors.
(ii) Locomotion of the Capsule Through the Digestive Track: Externally applied magnetic fields guide the capsule through the patient's gastrointestinal tract, propelling it through the stomach and small intestine.
(iii) Adjusting the Orientation of the Capsule: The magnetic fields can orient the capsule so that it is positioned correctly for imaging.
(iv) Delivering Therapy: Magnetic fields can be used for therapeutic purposes in the form of removing metallic objects or delivering medication.

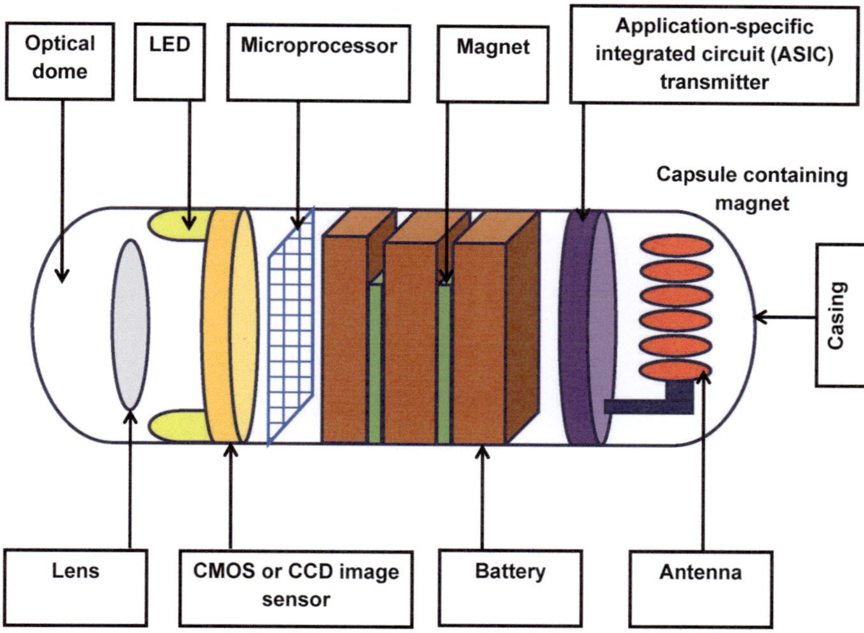

Fig. 24.4 Components of a capsule containing a magnet

24.7.2 Types of Magnetic Actuation Systems

Three major magnetic capsule endoscopy (MCE) systems are used: handheld, MRI-based, and robotic [7].

(i) Handheld Magnetic Capsule Endoscopy: This method uses a handheld magnetic device akin to a magnet paddle to guide the capsule through the gastrointestinal tract. A first-in-human study tested a new magnetic, maneuverable, wireless capsule, in which the capsule was manipulated in the esophagus for 10 min in a volunteer, causing it to turn somersaults, and remain angulated [8]. The study demonstrated the feasibility of remotely manipulating a capsule in the human esophagus and stomach to enhance diagnostic and therapeutic wireless capsule endoscopy.

(ii) MRI-Based Magnetic Capsule Endoscopy: The capsule is relocated using magnetic resonance imaging (MRI) technology. Like MRI systems, the system comprises a controller, an examination bed, a magnetic head that produces a magnetic field, and a translational and rotational platform that allows the external magnet to move linearly and rotationally. After swallowing the capsule, the patient lies down on the examination bed. Then, the magnetic head, operated by joystick movements, enables the capsule to move within the stomach.

In an observational cohort study, 3182 asymptomatic individuals undergoing magnetically controlled capsule gastroscopy (MCCG) were enrolled. Patients with ulcers

Table 24.1 Comparison among the handheld, MRI-based, and robotic capsule endoscopies

Type of MCE	Handheld system	MRI-based system	Robotic system
Advantages	Simplicity, cost-effectiveness, portability, and ease of operation	Provision of dynamic magnetic fields, a high control accuracy, and usability for a broader range of applications, including the stomach	Offering more precise and reliable control than manual manipulation, it is less expensive than MRI-based systems
Disadvantages	Only skilled operators can handle this system	It is expensive because it uses specialized equipment. Additionally, potential side effects related to MRI contrast agents are also a concern	A dedicated robotic system is more complex and expensive than a handheld system

and suspected malignancies were referred for gastroscopy and biopsy. Seven patients (0.22%) were diagnosed with MCCG among the 3182 enrolled individuals, indicating that MCCG can detect both cancerous and benign lesions, confirming its safety and clinical practicability in a large population [9].

(iii) Robotic Magnetic Capsule Endoscopy: This method employs a robotic arm or system to direct the capsule's travel through the GI tract. The interaction of a robot-assisted, magnetically driven, wireless capsule endoscope (WCE) with colonic tissue was investigated as it traversed the colorectal bends in both the dorsal and ventral directions, relying solely on three-dimensional (3D) accelerometer data [10]. Tilting the WCE by an angle exceeding the angle of the incline improved the WCE locomotion. An elliptical capsule shell geometry and water insufflation also improved the locomotion of WCE. The three types of systems are compared in Table 24.1.

24.7.3 Applications, Benefits, and Limitations of MGCE

A variety of gastrointestinal conditions, such as Crohn's disease, small bowel tumors, and polyposis syndrome, are diagnosed. Further, the effectiveness of treatment for gastrointestinal conditions is easily monitored. As in the original nonmagnetic embodiment of the technique, all this is done noninvasively without sedation, making the method well tolerated by the patient and a patient-friendly procedure that provides access to previously difficult areas. But the technique lacks therapeutic functionality, primarily due to constraints of capsular volume and limited energy storage in batteries. Poor bowel preparation or air bubbles deteriorate the quality of images. Further, there is a small, albeit rare, risk that the capsule may remain in the body.

Table 24.2 Comparison of capsule endoscopy with conventional endoscopy

Sl. No.	Point of comparison	Capsule endoscopy	Conventional endoscopy
1	Access	It allows access to the small intestine	The small intestine is difficult to reach with traditional endoscopes
2	Invasiveness	It is a noninvasive procedure that eliminates the need for sedation and reduces patient discomfort	It is an invasive procedure
3	Diagnostic yield	It has a higher diagnostic yield for small bowel pathology than other methods (push enteroscopy, barium contrast studies, CT enteroclysis, and MRI)	It has a lower yield
4	Limitations	It lacks therapeutic capabilities, e.g., biopsies or interventions during the procedure	It provides therapeutic capabilities
5	Esophagus and stomach	It can be used to examine the esophagus and stomach	It is the gold standard for diagnosing these regions

24.8 Comparison of Capsule Endoscopy with Related Techniques

Compared to conventional endoscopy and other imaging techniques, such as computerized tomography (CT) enteroclysis or magnetic resonance imaging (MRI), capsule endoscopy offers a unique perspective for examining the gastrointestinal tract. The customary procedures are less sensitive to mucosal disease.

24.8.1 Capsule Endoscopy Versus Conventional Endoscopy

A comparison of capsule endoscopy with conventional endoscopy is made in Table 24.2.

24.8.2 Capsule Endoscopy Versus Radiographic Techniques

Table 24.3 compares capsule endoscopy with techniques such as barium swallow/enteroclysis.

Table 24.3 Capsule endoscopy and radiographic methods

Sl. No.	Point of comparison	Capsule endoscopy	Radiographic methods
1	Visualization	It allows direct visualization of the small bowel mucosa	They provide indirect views
2	Sensitivity	It is superior in detecting mucosal disease and angiodysplasia	They are inferior, being less sensitive
3	Diagnostic yield	It has a higher diagnostic yield for small bowel pathology	They have a lower diagnostic yield

Table 24.4 Capsule endoscopy and CT enteroclysis/MRI

Sl. No.	Point of comparison	Capsule endoscopy	CT enteroclysis/MRI
1	Mucosal evaluation	It is better for evaluating the small bowel mucosa	They provide better visualization of the entire bowel wall and extra-enteric complications
2	Diagnostic yield	It has a higher diagnostic yield for small bowel pathology	They have a lower yield

24.8.3 Capsule Endoscopy Versus CT Enteroclysis/MRI

Table 24.4 provides a comparative overview of capsule endoscopy, computerized tomographic (CT) enteroclysis and magnetic resonance imaging (MRI).

24.8.4 Capsule Endoscopy Versus Push Enteroscopy

A comparison of capsule endoscopy and push enteroscopy is drawn in Table 24.5.

Table 24.5 Capsule endoscopy and push enteroscopy

Sl. No.	Point of comparison	Capsule endoscopy	Push enteroscopy
1	Invasiveness	It is less invasive	It is more invasive, requiring sedation and a longer procedure time
2	Diagnostic yield	It is superior in identifying bleeding sources and other small bowel pathologies	It is inferior in this aspect
3	Therapeutic capabilities	It does not allow for biopsies and interventions	It enables biopsies and other interventions

24.9 Future Directions and Prospects of Capsule Endoscopy

The future of capsule endoscopy (CE) holds promise for advancements in its diagnostic and therapeutic capabilities [11–13]. Artificial intelligence (AI)-powered image analysis will be a boon to the method. Active locomotion of the capsule will be an additional beneficial aspect. Therapeutic interventions like drug delivery are likely to be incorporated in capsule endoscopy. All these ameliorations and amendments are expected to be achieved without compromising the noninvasive nature of CE. A more detailed look at the potential promises of capsule endoscopy, together with expected challenges, is given as follows:

(i) Enhancement of Diagnostic Capabilities:

  (a) AI-Powered Analysis of Acquired Images: Artificial intelligence is poised to revolutionize capsule endoscopy by automating and accelerating image analysis. With its application, a faster and more accurate diagnosis of small bowel abnormalities will become possible.
  (b) Provision of Improved Visualization of the Gastrointestinal Tract: Higher resolution cameras, better lighting equipment, and broader fields of view, enabling clearer picturization, will facilitate more comprehensive visualization of the gastrointestinal tract.
  (c) Pan-Endoscopy: Advanced capsules will be developed. These will provide the ability to visualize the entire GI tract, from the esophagus to the colon, in a single procedure

(ii) Building Active Locomotion and Interventional Capabilities:

  (a) Introduction of Maneuverable Capsules: These capsules will be endowed with active locomotion mechanisms. They will allow for targeted investigations. Therapeutic interventions, such as drug delivery or biopsy, will become realities.
  (b) Magnetic Actuation of Capsules: Magnetic actuation for manipulating capsules by external magnets, a propitious technology for controlling capsule movement and orientation within the gastrointestinal tract, will gain widespread adoption and popularity. Magnetically controlled capsule endoscopy allows better maneuverability and visualization of the stomach [14–20].
  (c) Robotic Wireless Capsule Endoscopy: This technique improves organ visibility and technological success rates.
  (d) Taking Biopsies and Delivering Targeted Therapies with Drugs or other Treatments: This is a thrust area of research and development for future applications of capsule endoscopy. Capsules capable of performing biopsies, analyzing and coagulating lesions, and treating them are likely to emerge.

(iii) Expansion of Clinical Applications:

(a) Playing a Larger Role in Screening for Various Gastrointestinal Diseases: Crohn's disease, small bowel tumors, and polyposis syndromes are some of the diseases that are likely to receive increasing attention.
(b) Evaluation of Obscure GI Bleeding: Although capsule endoscopy is a valuable tool for investigating obscure gastrointestinal bleeding, its diagnostic yield in these cases is expected to improve with future advancements.
(c) Progress in Therapeutic Capsule Endoscopy (TCE): Research into using capsules for therapeutic interventions is gaining momentum. This has potential applications in treating conditions like bleeding lesions or polyps.

(iv) Challenges and Considerations:

(a) Capsule Retention: While rare, capsule retention remains a potential concern. Patients with known gastrointestinal strictures or inflammation are at a higher risk of such incidents.
(b) Cost and Accessibility: The cost of CE procedures and the need for specialized equipment could limit its widespread adoption. Technological advancements may help reduce costs and make the method more economical.
(c) Data Interpretation: Human interpretation of CE findings will remain crucial. AI can assist in image analysis. It is essential to develop AI-driven tools that can accurately and efficiently assist clinicians in this process.

Review Exercises

24.1 Explain the statement, "Capsule endoscopy is a non-invasive technique." How does capsule endoscopy overcome the limitations of traditional endoscopy methods?
24.2 What is the conventional test for examining the food pipe called? How is it performed? What are the potential risks involved in this test?
24.3 Write the different names used for capsule endoscopy and explain their meanings.
24.4 Name the core components of the capsule used for endoscopy, and explain the function of each component.
24.5 What technologies are used in the camera inside the capsule?
24.6 Explain the capsule endoscopy procedure, clearly mentioning the pre- and post-procedure steps and those followed during the procedure.
24.7 List five applications of capsule endoscopy.
24.8 Briefly describe the principal advantages and disadvantages of capsule endoscopy.
24.9 Name the three types of systems used in magnetic capsule endoscopy, and explain their working principles.

24.10 Which is the least expensive of the three magnetic capsule endoscopy systems: handheld, MRI-based, and robotic?
24.11 What difficulty is faced in using a handheld magnetic capsule endoscopy system to examine a patient with a high body mass index? Discuss giving reasons.
24.12 Perform a comparative analysis of capsule endoscopy with each of the following: (a) conventional endoscopy, (b) radiographic methods, (c) CT Enteroclysis/MRI, and (d) push enteroscopy.
24.13 Explain the drawbacks of magnetically guided capsule endoscopy, and suggest necessary improvements.
24.14 Present a detailed analysis of the future promises of capsule endoscopy, mentioning the chief challenges.
24.15 What is the likely impact of artificial intelligence in augmenting the performance of capsule endoscopy?

References

1. Mata A, Llach J, Bordas JM (2008) Wireless capsule endoscopy. World J Gastroenterol 14(13):1969–1971. https://doi.org/10.3748/wjg.14.1969
2. Underwood C (2021) EGD test. https://www.healthline.com/health/egd-esophagogastroduodenoscopy
3. Nam JH, Lee KH, Lim YJ (2021) Examination of entire gastrointestinal tract: a perspective of mouth to anus (M2A) capsule endoscopy. Diagn (Basel) 11(8):1367, 1–10. https://doi.org/10.3390/diagnostics11081367
4. Koprowski R (2015) Overview of technical solutions and assessment of clinical usefulness of capsule endoscopy. Biomed Eng Online 14(111):1–23. https://doi.org/10.1186/s12938-015-0108-3
5. Shamsudhin N, Zverev VI, Keller H, Pane S, Egolf PW, Nelson BJ, Tishin AM (2017) Magnetically guided capsule endoscopy. Med Phys 4(8):e91–e111. https://doi.org/10.1002/mp.12299
6. Hanscom M (2024) Magnetically guided capsule endoscopy similar to EGD for detecting esophagogastric varices in patients with cirrhosis. Gastroenterology 167(5):1046. https://doi.org/10.1053/j.gastro.2024.05.024
7. Kim JH, Nam SJ (2021) Capsule endoscopy for gastric evaluation. Diagn (Basel) 11(10):1792, 1–15. https://doi.org/10.3390/diagnostics11101792
8. Swain P, Toor A, Volke F, Keller J, Gerber J, Rabinovitz E, Rothstein RI (2010) Remote magnetic manipulation of a wireless capsule endoscope in the esophagus and stomach of humans (with videos). Gastrointest Endosc 71(7):1290–1293. https://doi.org/10.1016/j.gie.2010.01.064
9. Zhao A-J, Qian Y-Y, Sun H, Hou X, Pan J, Liu X, Zhou W, Chen Y-Z, Jiang X, Li Z-S, Liao Z (2018) Screening for gastric cancer with magnetically controlled capsule gastroscopy in asymptomatic individuals. Gastrointest Endosc 88(3):466-474.e1. https://doi.org/10.1016/j.gie.2018.05.003
10. Mahmood S, Schostek S, Schurr MO, Bergsland J, Balasingham I, Fosse E (2022) Robot-assisted magnetic capsule endoscopy; navigating colorectal inclinations. Minim Invasive Ther Allied Technol 31(6):930–938. https://doi.org/10.1080/13645706.2022.2032181

11. Slawinski PR, Obstein KL, Valdastri P (2015) Capsule endoscopy of the future: what's on the horizon? World J Gastroenterol 21(37):10528–10541. https://doi.org/10.3748/wjg.v21.i37.10528
12. Yang YJ (2020) The future of capsule endoscopy: the role of artificial intelligence and other technical advancements. Clin Endosc 53(4):387–394. https://doi.org/10.5946/ce.2020.133
13. Cao Q, Deng R, Pan Y, Liu R, Chen Y, Gong G, Zou J, Yang H, Han D (2024) Robotic wireless capsule endoscopy: recent advances and upcoming technologies. Nat Commun 15:4597. https://doi.org/10.1038/s41467-024-49019-0
14. Xu Y, Li K, Zhao Z, Meng MQ-H (2021) On reciprocally rotating magnetic actuation of a robotic capsule in unknown tubular environments. IEEE Trans Med Robot Bionics 3(4):919–927. https://doi.org/10.1109/TMRB.2021.3123407
15. Chen W, Sui J, Wang C (2022) Magnetically actuated capsule robots: a review. IEEE Access 10:88398–88420. https://doi.org/10.1109/ACCESS.2022.3197632
16. Du X, Wang Q, Jin D, Chiu PW, Pang CP, Chong KL, Zhang L (2022) Real-time navigation of an untethered miniature robot using mobile ultrasound imaging and magnetic actuation systems. IEEE Robot Autom Lett 7:7668–7675. https://doi.org/10.1109/LRA.2022.3184445
17. Song L, Dai Y, Wang L, Zhang W, Ji Y, Cao Y, Wei J, Wang F, Zhong J, Yang J, Feng L (2022) Motion control of capsule robot based on adaptive magnetic levitation using electromagnetic coil. IEEE Trans Autom Sci Eng 20(4):2720–2731
18. Zheng Y, Zheng G, Li YY, Gong X, Chen Z, Zhu L, Xu Y, Xie X, Wu S, Jiang L (2023) Implantable magnetically-actuated capsule for on-demand delivery. J Control Release 364:576–588. https://doi.org/10.1016/j.jconrel.2023.11.009
19. Li X, Zeng D, Xu H, Zhang Q, Liao B (2024) Magnetic actuation for wireless capsule endoscopy in a large workspace using a mobile-coil system. Micromachines 15(11):1373, 1–23. https://doi.org/10.3390/mi15111373
20. Rosa B, Cotter J (2024) Capsule endoscopy and panendoscopy: a journey to the future of gastrointestinal endoscopy. World J Gastroenterol 30(10):1270–1279. https://doi.org/10.3748/wjg.v30.i10.1270

Chapter 25
Drug Delivery Implants

Abstract Implantable drug delivery devices, nondegradable reservoir or biodegradable types, have shown great prospects. These devices have revealed ostensible possibilities of advancement in several applications demanding onerous efforts in controlled and precise, highly localized liberation of decisive doses of drugs with fewer side effects and without direct medical intervention. Actively controlled devices are more propitious than passive release devices. This greater potentiality of active devices is because the drug delivery process can be controlled post-implantation and even by telemetry, involving automatic measurements and telecommunication. Dissenting from passive devices, they do not rely on the chemistry of degradation of specific materials in the premeditated region of the implant. Numerous implantable drug delivery devices have been reconnoitered for use in chronic and terminal diseases. Diabetes, osteoporosis, and cancer are a few such examples. Robotic capsules are small, pill-sized devices that can be swallowed to perform diagnosis and drug delivery.

Keywords Drug delivery · Oral · Nasal · Pulmonary · Transdermal · Intravenous · Intramuscular drug delivery · IDDS · Zero-order controlled release · Sterilization · Biodegradation · Immunoisolating capsule · Microreservoir · Micropump · Osmotic pump · MEMS · Piezoelectric pump

25.1 Introduction

Drug delivery systems are contrivances, which are used to vouch for easy and safe introduction of drugs into the human body and their accurately traveling to the areas where they are called for [1, 2]. These systems strive, in the first place, to dole out the needed quantity of drug benignly and efficaciously to specific regions in the body and, secondly, to adjust the chronological profile of drug to extract uttermost curative action. Such systems have to take several aspects into cogitation, with ambit ranging from simplicity of delivery of drugs to the extent of fruitfulness achieved by the drugs administered by the selected system. Among the various types of drug

V. K. Khanna, *Implantable Medical Electronics*,
https://doi.org/10.1007/978-3-031-99868-3_25

delivery systems, the greatest expectation of success is assured by implantable drug delivery systems (IDDSs) in a mammoth number of applications. These critical applications inescapably require the carefully planned, organized, and error-free delivery of precise drug doses without any medical intermediation.

25.2 Conventional Drug Delivery Systems

The time-honored and often preferred routes for drug delivery (Fig. 25.1) are represented by the oral, nasal, pulmonary, transdermal, and intravenous or subcutaneous injection or infusion methods [3].

Largely speaking, the trump card of the oral method is its comparative simplicity. But the other methods have been improvised to assist more speedy delivery of drugs and to aim at targeting of specific organs in a more appropriate manner. They are invariably useful in circumstances where oral drug delivery is either nonoptimal or unfeasible (Table 25.1).

25.2.1 Oral Method

It represents the most habitually used dosage form for many products. Practically, it involves swallowing or gulping down a pharmaceutical compound with the intent of releasing it into the gastrointestinal tract (the mouth, stomach, small and large intestines) [4]. Many macromolecules undergo digestion in the gastrointestinal tract. Others suffer from improper absorption into the bloodstream. This loss of drug molecules is an important handicap offsetting the utility of this method. Moreover, a rapid onset of drug action is required in some situations. In these instances, oral administration stands malapropos.

25.2.2 Nasal Method

Over the surface of the nasal cavity lies a thin mucosa [5]. The mucosa is a mucus membrane or a sheath that exudes mucus, a greasy and adhesive fluid. The mucosa membrane is vascularized with blood vessels. Supported by this vascularization, i.e., foundation of blood vessels, a drug molecule is conveyed across the single epithelial cell layer promptly to the systemic blood circulation; epithelium is the cellular wrapper of inner and outward body surfaces. Thus in the nasal method, the first-pass hepatic and intestinal metabolism is circumvented by the drug molecules. Therefore, no gastrointestinal degradation is experienced.

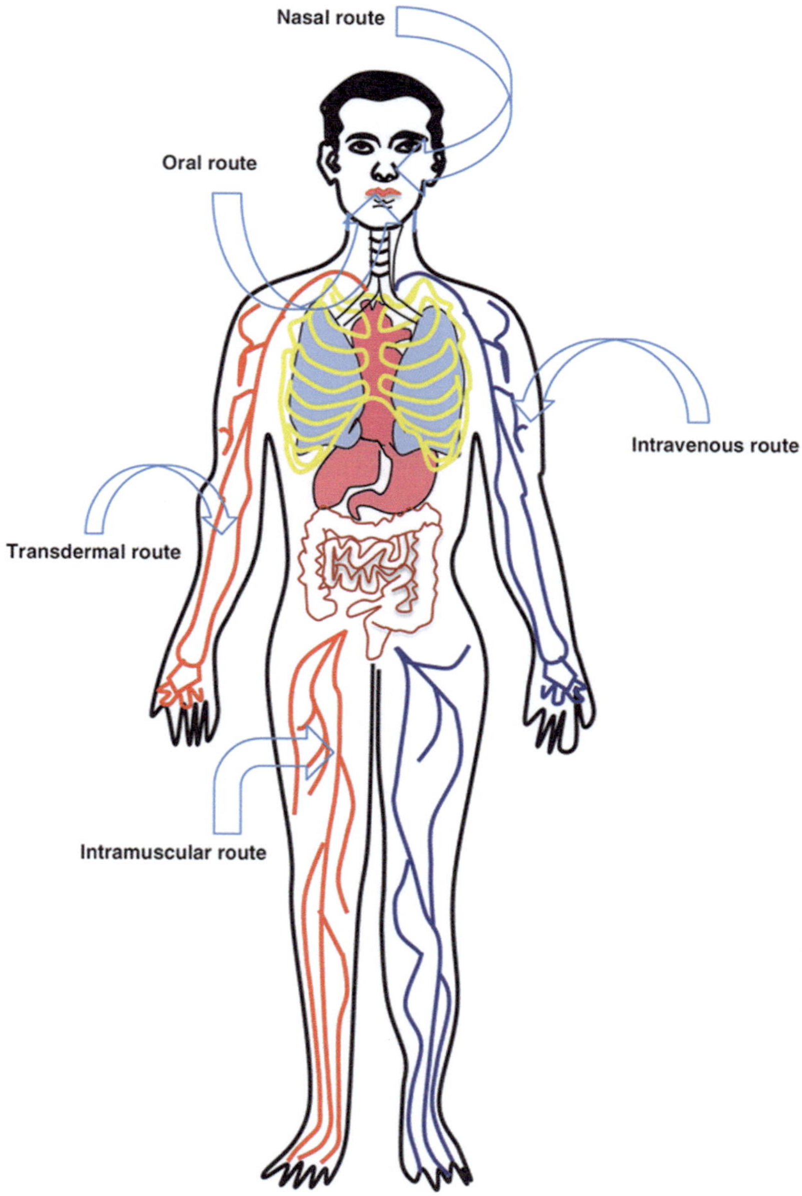

Fig. 25.1 Few routes of drug delivery in human body

Table 25.1 Traditional drug delivery routes

Sl. No.	Route	Features			
		Entry point	Gastrointestinal degradation	Delivery speed	Effectiveness
1	Oral	Mouth	Yes	Slow	Less
2	Nasal	Nose	No	Medium	Medium
3	Pulmonary	Lungs	No	Medium	Medium
4	Transdermal	Skin	No	Medium	Medium
5	Intravenous	Vein	No	Fastest	Most

25.2.3 Pulmonary Method

The lungs can drench up pharmaceuticals, and this method utilizes such soaking by lungs. The lungs are saturated by the pharmaceuticals, either for the purpose of local deposition or for systemic delivery [6]. A wide assortment of devices is on hand for pulmonary drug delivery. These include the dry powder inhalers, both of the passive breath-driven type and the active power-driven single-/multiple-dose variety. From the lungs, the inhaled drugs are absorbed into the bloodstream and supplied to the needy recipient places. The redeeming feature is that inhaled drugs require a low fraction of oral dose. A total of 40 metered doses of drug content are equivalent to one 4 mg tablet of salbutamol. The benefits accrued are the inconsequential side effects. Nonexposure of rest of the body to drug is responsible for lesser side effects. In people suffering from asthma and diabetes, long-term treatment by pulmonary drug delivery pledges consummate recovery.

Also, outbreak of action by the drug is very quick, and there is no degradation of drug by the liver. The main demerits are the relatively low efficiency of inhalation system and the smaller quantity of drug mass delivered per puff. Additionally, poor stability of drug formulation and lack of reproducibility in dosing are also problematic.

25.2.4 Transdermal Method

The word "transdermal" means the process of giving forth a drug through the skin. Transdermal drug delivery systems (TDDS) are also known as transdermal patches. These are medicated adhesive patches. By placing these patches on the skin of the patient, a specific, therapeutically effectual dose of medication is transported across the skin into the bloodstream [7].

25.2.4.1 Advantages of Transdermal Method

In this medication route, first-pass metabolism of drugs is a nonissue. By particular tribulations linked with the drug, e.g., gastrointestinal infuriation, inferior absorption, and breakdown during "hepatic first-pass" effect can be eschewed. To explain "hepatic first-pass" effect, the ingested drug absorbed in the gastrointestinal tract is transported to the liver, wherein it undergoes metabolism. During metabolism, degradation of some of the active drug material occurs. As a result, the active drug content entering the blood circulation is only a small fraction of the drug orally taken. For this reason, more therapeutic value is derived from many drugs taken via transdermal route than by oral delivery. Hence, a corresponding therapeutic benefit is received through transdermal means from a smaller daily dose than is required in oral intake. A decreased plasma concentration level of drug produces reduced side effects. Further, in the event of toxicity, the drug is easily eliminated. Following this simplified medication regimen, patient conformity is improved. Moreover, inter- or intrapatient variability is subdued. Fluctuations in plasma level of drugs are also diminished. Another encouraging feature is that drug delivery by this route permits the exploitation of drug contestants having small half-life (half-life is the time in which drug concentration inside the body is halved) as well as low therapeutic index (therapeutic index is the ratio of amount of drug causing therapeutic action to the amount of drug causing toxicity). Other advantages are reduction of dosing frequency, unwavering deliverance of drug over a long period of time, and prevention of loathsome effects or therapeutic failure normally related with inconsistent, unmethodical dosing.

25.2.4.2 Limitations of Transdermal Method

Drugs, which have high melting temperatures but low solubility in water and fat, can be given by this method. But the heavy drugs, molecules (> 500 Da) find difficulty in penetrating the stratum corneum, the outermost of the five layers of the epidermis made up of several layers of flat, keratinized (filled with keratin protein), nonnucleated, dead, or peeling cells. Also, the drugs exhibiting very low or very high partition coefficient do not succeed in reaching blood circulation. Partition or distribution coefficient is the ratio of the amounts of a substance distributed between two heterogeneous phases that are in equilibrium with each other.

25.2.5 Intravenous Method

The term "intravenous" means "into the vein." Intravenous drug administration entails the transportation of liquid substances straight into a vein [8]. Inserting a needle into a vein is the first step of this method. The insertion is done usually at a site proximate to the elbow, the wrist, or on the flipside of the hand. Then the medication is guided through that needle.

25.2.5.1 Intravenous Method for Fast Drug Delivery

Comparative to other routes, the intravenous route is the most expeditious way to deliver flowing substances and medicines all through the body. Naturally, in an emergency situation when medicine needs to be absorbed quickly, this method comes to rescue. In such circumstances, the drug has to be supplied to the systemic circulation by the shortest route. Then oral intake of pills or liquids may not be sufficiently quick to provide medication inside the body speedily. By breaking down delicate medication, enzymes in the stomach may present a serious obstacle to timely delivery if the medicine is taken by mouth.

25.2.5.2 Intravenous Method for Slow Drug Delivery

Nevertheless, the intravenous method should not be assumed as essential only when the patient must receive medication very rapidly. It is also helpful when the medication is to be given gradually but incessantly.

25.2.5.3 IV Push and IV Infusion

Based on the above, there are two ways to administer medication by intravenous (IV) method: IV push and IV infusion. An IV push means a fast insertion of medicine into the flowing blood in one attempt. An IV infusion provides an unhurried trickle of medicine into a vein over a stipulated length of time to convey a fixed allocation of drug. For easy treatment, an intravenous or IV line is created with a cannula or catheter placed into a vein. The medicine is administered through it frequently.

25.2.5.4 Risks of Intravenous Medication

A prone risk of intravenous medication may be that of infection at the injection site. For this reason, trained experts should manage the IV medication procedure using sterile equipment and techniques. Also, it must be borne in mind that the intravenous drugs are potentially dangerous. Hence, only registered nurses, doctors, or other skilled medical practitioners are allowed to administer IV therapy. Another possible risk of this medication is *phlebitis*. This is an inflammation of the veins because some drugs given as gradual infusions in due course of time can inadvertently be supplied too hastily in one go. The rapid supply harms or devastates the vein close to the site of entry of drug during injection. In such damage to a vein or that by employing an intravenous catheter line, often referred to as "infiltration," the medicine dribbles into neighboring tissue instead of the blood. A grave condition is possible if the line runs dry. Then severe problems in circulation can arise due to air bubbles in the syringe or the IV bag. Air bubbles entering the vein can migrate to the heart and lungs. This risk is called *air embolism*. This is a pathological condition having potential for severe morbidity and mortality. Air bubbles are trapped in a vein or artery, thereby blocking it. Further, due to IV therapy, blood clot formation and deep vein thrombosis are perilous. Clots get stuck in blood vessels causing tissue damage or death. It is known that the intravenously delivered medications have a very fast effect on the human body. Hence, a patient under this kind of medication must be kept under watchful attention at all stages. Toxic effects, unwanted side effects, and allergic reactions take place fast.

25.2.5.5 Limitations of Intravenous Medication

Unfortunately, for most of therapeutically active agents, the duration of drug action in intravenous drug administration is small. Therefore, recurrent injections must be given, and the patient has to choose between two viable options. The patient has to either travel frequently to a treatment site to receive injections or has to hoard a large supply of injections at home and call the nurse when necessary. Combined with the embarrassment of regular injections, poor patient fulfillment is achieved. Irrevocably, a drug injection schedule involving multiplicity of times is difficult to provide, necessitating professional help. A silver lining is that unaided drug administration is possible by portable infusion systems.

25.2.6 Intramuscular Method

It is a technique for delivering several drugs and almost all inactivated vaccines deep into the muscles. Besides deltoid muscle on the upper part of the arm covering the shoulder joint or vastus lateralis muscle, the largest part of the quadriceps femoris muscle of the thigh, intramuscular delivery may be done through ventrogluteal and dorsogluteal muscles in the buttocks. In the event of difficulty to locate veins or when a drug is irritating to the veins, this method is employed in place of intravenous injection. Also, intramuscular injection is superior to oral delivery by swallowing for the drugs that are destroyed by the digestive system. Common complications of this method are abscess formation at the injection site, hematoma (a localized swelling filled with blood), infection, bleeding, allergic reaction, etc.

25.3 Advantages of IDDSs Over Existing Methods

Physician or patient uses IDDSs to deliver a drug at a specific rate without taking recourse to habitual intervention. Such systems are exceptionally prudent whenever obedience to a preset drug program is indispensable [9]. A commendable advantage of these systems is the ability of provision of aimed regional delivery of medicines at a continuous rate. As a consequence of this targeted delivery, fewer drugs in lesser quantity may be required to heal the malady. Another advantage is the reduction of side effects to the smallest possible degree. Obviously, a higher overall worth of treatment is the outcome.

In preliminary studies using the controlled drug delivery systems, better efficacy over old-fashioned methods of cure has been demonstrated. Looking at these desirable features, many sustained release drug formulations have been developed by persistent efforts. These systems pave the way toward dispensing unstable drugs one time in a week to one time in a year. Previously, such drugs required frequent daily dosing. Prominent advantages offered are explained below (Fig. 25.2):

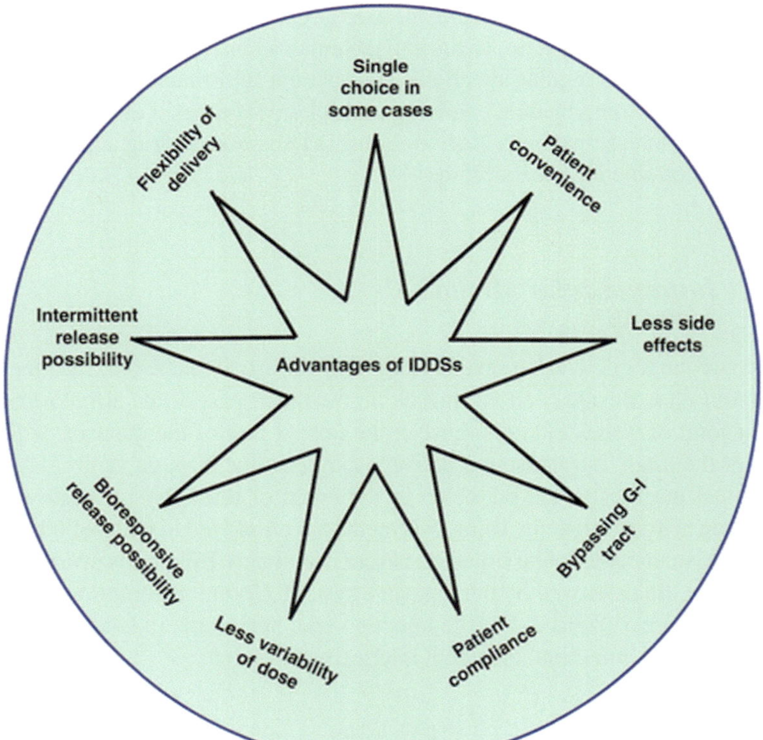

Fig. 25.2 Several ways in which IDDSs are superior to conventional drug administration methods

1. *Solo Practicable Choice on Some Occasions*: Every so often, pharmaceutical agents, notably drugs having short half-life or brisk metabolism or degradation rates, may be either ineffective or unavailable for oral or exterior infusion methods. Furthermore, the presence of physiological barriers might cause approachability of targeted affected places by these tracks to be infeasible. When unremitted administration is needed to these far-flung targets, e.g., in intrathecal supply of medicines for implacable chronic pain, implantable infusion pumps remain the single realistic option. Another application area of implantable devices lies in achieving and sustaining higher drug concentrations within restricted areas of the body. These areas could be either inaccessible or accessible with difficulty to peripheral drug administration.
2. *Convenience of the Patient*: IDDSs permit the patients to receive medication without going to the hospital, with trivial medical attention and interference. They are more comfortable to patients and cause less infection problems in comparison to chronic transcutaneous, catheter-based infusion systems fixed in patient's body for a sustained period of time. In conventional treatments, unrelenting IV infusion or periodical injections are used to confirm that an effective concentration of drug in the blood is maintained for long periods of time. In these treatments, all through drug administration, the patients must visit the hospital regularly for continual monitoring. The condition is further worsened for a short-acting medicine because the amount of injections or the rate of infusion needs to be increased to preserve a level of drug, which is effective for therapy.
3. *Side Effect Minimization*: The side effects are reduced because the required amount of drug is released at the targeted site in a dedicated, focused manner without vitiating the surroundings.
4. *Improved Drug Delivery*: Usage of IDDSs allows the drug to be distributed within a confined region or in blood circulatory system with slightest hindrance by metabolic or biological barricades. Bypassing of the gastrointestinal tract and the liver by the drug moiety is helpful to those medicines, which are either deactivated readily or absorbed scantily in the tract and/or the liver previous to systemic distribution.
5. *Compliance*: Compliance is largely increased by reduction or complete exclusion of role of the patient in dosing. Drug outpouring from an implanted system is independent of any input from the patient. Hence, it is not vulnerable to mistakes by a patient such as forgetting to take the dose of a medicine, as per schedule. Despite the drawback that some IDDSs require periodical refilling, the patient has less participation in dispensing the requisite medicine.
6. *Capability for Controlled Release*: Implants have been made for delivering drugs by zero-order controlled release kinetics. Zero-order or constant rate release of drug minimizes fluctuations in drug concentration in the blood. It improves the concentration of drug in the body with respect to time and space as compared to conventional therapy [10]. Such deviations may cause unsolicited periods of less exposure or more exposure for patients than required. Peaks in drug concentration cause toxicity. Troughs of concentration are periods of ineffectiveness. By avoidance of such variations, a reduction of dosing frequency may be allowed.

Together with automation of dosing operation, an increased patient compliance is obtained.
7. *Promise of Bio-responsive Drug Liberation*: Bio-responsive liberation from implantable devices is an interesting research area under way. Degradation of bio-responsive materials occurs in response to biochemical characteristics of a disease. Specially designed polymers restrict drug delivery to inflammation regions by hurriedly degrading in the presence of high concentrations of the hydrogen peroxide (solubilizing the polymer) and low pH (hydrolyzation of the polymer) [11].
8. *Possibility of Occasional Drug Liberation*: Occasional liberation can be assisted by pumps which can be programmed from outside according to factors such as circadian rhythms, unstable biochemical, metabolic necessities, and rhythmically pulsating liberation of several peptides and proteins. The circadian rhythm of a person is an internal biological clock that regulates physical, mental, and behavioral changes in a 24-h period, in response to light and darkness in the environment. Some examples are the cycle of sleeping and waking up, the system of regulation of temperature, and the endocrine system consisting of hormone-releasing glands.
9. *Flexibility*: The choice of materials, fabrication technologies, degree of drug loading, drug release rate, and other vital parameters permits considerable resilience in IDDSs. From a regulatory perspective, the IDDS is deemed as a new product. For a new drug entering the market, it can extend the protection of the drug in the market for an extra 5 years. In case of existing drugs, it can do so for 3 years.

25.4 Disadvantages of IDDSs Over Existing Methods

One disheartening weakness of these drug delivery systems is their presently exorbitantly high cost-to-benefit ratio. The cost factor owes the responsibility for preclusion of their adoption over conventional methods. Moreover, a few lately devised implants are in preliminary stages. Rigorous and elaborate scientific testing is in progress preceding their deployment in normal practice. Main limitations of IDDSs are (Fig. 25.3):

1. *Higher Cost*: Presently the total cost of implantation, i.e., the expenditure for surgery together with the cost of implantable device, is excessive and beyond the reach of the common man. Availability of devices at affordable costs and reduction of surgical and postsurgery expenses will boost the use of IDDSs in healthcare for masses.
2. *Surgical Placement and Invasiveness*: To start the therapy with IDDSs, a surgical procedure is essential. It may be minor or major in nature. Only professional surgical personnel are allowed to undertake this operation. Hence, specialized training is required to ensure patient safety and proper implementation of implant follow-ups.

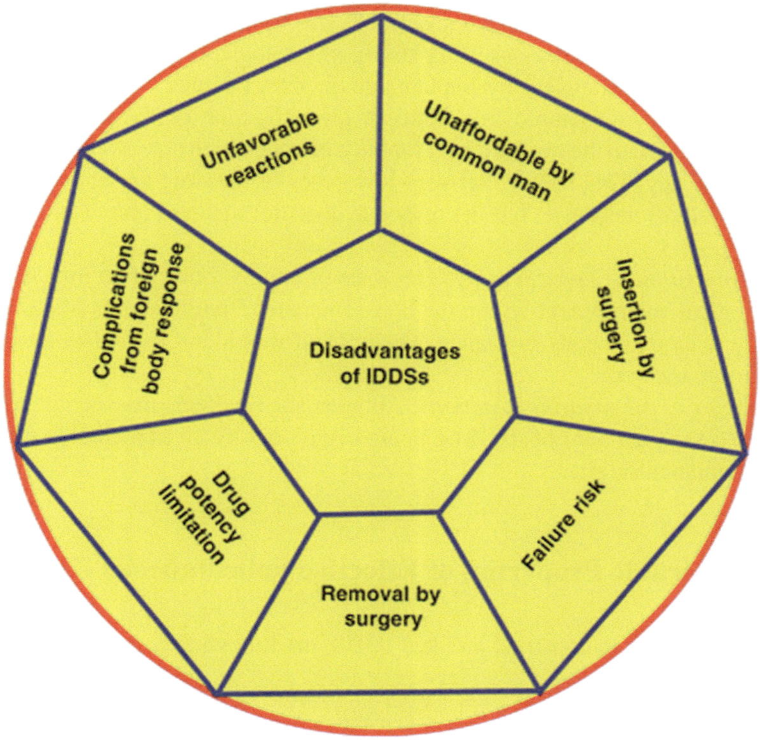

Fig. 25.3 Shortcomings of implantable drug delivery systems (IDDSs) over routine methods

The surgical procedure may be time-consuming as well as traumatic. In a few patients, some scars are formed at the site of implantation due to surgery. Also complications are found related to surgery. Further, some uncomfortable feeling may persist in the patient receiving the implant.

3. *Imperilment from Nonperformance by the Implanted Device*: Fear lurks in the minds of patients receiving treatment through implanted device regarding the likelihood of device failing to work for some reason after implantation, requiring corrective surgery. As such fear looms large, the patient may remain worried.
4. *Cessation of Therapy*: After the therapy has been completed, the inserted osmotic pumps and nonbiodegradable polymeric devices are no longer required. It is essential that they are taken out from the body. Hence, at the termination of therapy, implant removal needs surgery. Of course, surgical recovery is not mandated for polymeric implants that are biodegradable. But these implants too suffer from one shortcoming that their biodegradation is continuous. During the last phase of the lifetime of implant, sustenance of correct dose may be impractical due to the ongoing degradation.
5. *Confinement to Potent Drugs*: Usually, the physical dimensions of an implant are kept minimal to lessen annoyance and soreness to the patient. Consequently, a

majority of implants have a circumscribed capacity of loading the drug. Normally, only powerful medicines exerting strong physiological effects, e.g., hormones, are amenable to delivery by implantable devices. Potency of a medicine is the amount of drug required to produce a given fraction of the maximal effect, without regard to the magnitude of topmost effect. A high potency drug may have poor efficacy. This means that very low doses of the drug are able to produce a satisfactory response, but the response does not increase appreciably at high doses.

6. *Biocompatibility Apprehension*: The issue of compatibility of an implant in the biological environment within the body emanates from reactions of the human body to an extraneous substance. The implant should not cause any damage or injury to the body.
7. *Chances of Adversarial Reactions*: Antagonistic reactions may be activated by an excessive amount of the drug being dished out by an implantable device at the implantation site.

25.5 Desirable Properties of Effective Subcutaneous IDDSs

If one is interested to construct an ideal IDDS, the following features are desirable [12]:

1. *Patient Acquiescence*: By decreasing the repetitions of drug supply over the inclusive period of treatment, it ought to improve patient compliance, i.e., the degree of adherence of a patient to the prescribed treatment.
2. *Surgery*: Hypothetically, it should be readily administered and be uninhibited from any major surgical procedure. Implantation procedure should follow easy, quick, outdoor patient recipe. Nobody likes surgery. Who wants the body to be cut open?
3. *Drug Discharge Rate*: The drug liberation should occur in a rate-controlled manner, enhancing its effectiveness toward the ailment and reducing side effects [13]. The term, "controlled release" is different from, "sustained release" because controlled release is an ideal release of zero order, i.e., the drug is liberated evenly over time without regard to its concentration. In distinction, sustained release implies slow liberation of the medicine over a period of time, which is not necessarily controlled.
4. *Post-therapy Implant Retrieval*: When the time of terminating medication is reached, the implant retrievable should be carried out via a painless, trouble-free, outpatient procedure in which the patient is comfortable and so willingly participates. This will increase patient cooperation.
5. *Sterilization of the Implant*: The sterilization process of the implantable device should be simple and easy. This process can be subdivided into two main branches. These are terminal sterilization and aseptic processing. Terminal sterilization is an out-and-out, thoroughgoing sterilization, meaning that implanted objects have been subjected to a credible, validated process for annihilation of

any living microorganisms before use. These objects are expected to remain sterile in their packaging wrapper until the same is opened for use. In aseptic processing, the sterile components and packaging are assembled in environments as free from disease-causing germs as achievable [14]. Four methods of sterilization have found widespread acceptability in healthcare today. These are steam sterilization, peracetic acid (peroxyacetic acid or PAA), CH_3CO_3H liquid sterilization, ethylene oxide (oxirane, C_2H_4O) sterilization, and hydrogen peroxide (H_2O_2) sterilization [15].

6. *Manufacturability of the Implant*: The implant should be easily receptive to bulk production in large numbers at low outlay in a manner assuring uniformity and reproducibility to enable economical viability.
7. *Complications*: The implant should never be the cause of introduction of any potential medical complication of surgery, such as pain, infection, etc.
8. *Risks*: Safety, stability, and proper mechanical strength are essential qualities that an implant should possess.

25.6 Biodegradation-Based IDDS Classification

A biodegradable material is a substance that decomposes through the actions of living organisms, e.g., bacteria, fungi, etc., i.e., one which breaks down in natural environment into raw materials. From degradation viewpoint, the IDDSs are of two types [16]: biodegradable and nonbiodegradable. Many of these systems are polymer-based. The solubility and diffusion coefficient of the drug in the polymer play vital roles in the release kinetics of drugs. The incipient drug load is another determinant. Apart from these, material properties in the biological conditions can influence the drug release.

25.6.1 Biodegradable IDDSs

This category of IDDSs is very attractive if it can perform in the desired manner. As the device disappears over time, the need of a second surgical procedure for its retrieval is eliminated [16].

25.6.2 Nonbiodegradable IDDSs

After termination of dose, a trivial surgery must be performed to extricate the system from the body.

25.7 Passive and Active IDDSs

Based on the manner of drug delivery and control exercised thereupon, present IDDSs are arranged into two classes: passive or active. The strikingly conspicuous characteristics of the two groupings of systems are given below with reference to some principal aspects:

1. *Drug Release Mechanisms*: Passive drug delivery systems are based on chemical release mechanisms that begin instantaneously after implantation. The operation of passive IDDSs is dependent on the chemistry of degradation of the particular materials chosen for device fabrication in the intended implantation region. Mechanisms of drug release include diffusive processes, degradation of membrane caps, or actuation of valves. The drug outflow perpetuates until the committed dose is depleted.

   On the other hand, the active drug delivery systems are electrically activated. They can be supervised by the patient or doctor. This supervisory control allows much superior adjustments of the dosing of the drug than possible with passive systems.
2. *Controllability*: Controllability is the exercise of authoritative or dominating influence on the delivery of very precise doses. These doses are given, either intermittently or in response to the signal received from a sensor. The dosing is effected autonomously, i.e., in a self-governed manner [17]. In passive systems, the only control achievable is that over the total dose. Neither the physician nor the patient nor the system itself can alter the drug delivery timing or its delivery rate. Once implanted, drug release continues until the system is removed or the drug supply is drained out. In active systems, the drug delivery process allows dynamical manipulation after implantation and even monitoring from remote through telemetry.
3. *Case Studies*: The multitude of passive drug delivery systems uses the polymer depot structure. One approach of design of these systems is to maintain a steady rate of dispersal of the drug from the polymer through the mechanism of diffusion. Another approach is to make the system from a selected material, which degrades in the body at a certain rate, thus emancipating the drug at that rate.

The implants for malevolent malignant tumors of the brain and prostate cancer constitute illustrative commercial products employing this method. In the implants for prostate cancer, uninterrupted release of leuprolide acetate is accomplished by using osmotic pressure-generating agents [18].

The most widespread configuration used in active delivery systems is the mechanical pump. This pump can deliver the drug in a rhythmically pulsatile or continuous manner. But the implanted pumps are large. Moreover, being able to deliver liquid drug formulations only, a noticeable impediment to the broad use of this technology is the inadequate stability of most proteins when stowed in a liquid form at the temperature of the human body.

In the experimental demonstration of an active drug delivery described by Santini et al. [19] and Chung et al. [17], the system consisted of a collection of reservoirs fabricated on a microchip made of silicon. These reservoirs were filled with a sequence of different chemical preparations and sealed individually. After implantation, an externally applied electric potential was applied to discharge the contents of each reservoir. This potential produced the discharge by bringing about dissolution of the Au-made fastening membrane via an electrochemical mechanism. The method provides active control over the timing and dose of drug delivered, which greatly ameliorates therapeutic efficacy achieved by the system.

The efficacy is improved because it is possible to release the exclusive drug in its proper dose very fast in accordance with the instantaneous state of the patient. Apart from a reasonably simple construction, such systems are actuated electrically using low voltages in dissimilarity to mechanical pumping.

25.8 Micro- and Nanoscale IDDSs

Micro- and nanotechnologies have ushered a new era of IDDSs. Integration of these technologies into IDDS design and fabrication has brought about revolutionary changes in the dimensional and layout design features of implantable systems [20, 21]. Most importantly, the spatial extent of separate components and that of the complete systems has shrunk to a small form factor, which allows easy implantation and makes the systems only marginally invasive [22]. Additionally, a stable, sturdy, chemically nonreactive surface separating the implant and the body has been obtained by employing proper techniques for modifying surfaces and by selecting correct biocompatible materials. Even further, stricter regulation of drug release profiles has been exercised by using minutely adjustable drug release mechanisms. Possible profiles can be programmable, cyclic, pulsatile, or continuous. The idea of a smart drug delivery system is sketched in Fig. 25.4. Expectedly, it will contain the pump, the controlling circuit, and the sensor as essential components.

One version of an external insulin pump with indwelling catheter/cannula needle is shown in Fig. 25.5. Although an external device, it is interesting to venture into a short digression to learn about its working. This pump is a small, computerized device. It can be programmed according to a patient's condition to supply insulin at a constant rate for 24 h a day, 7 days a week (24/7). This constant rate is called the basal rate. The sensor keeps a watch on patient's blood sugar level and monitors the insulin rate continuously. If a patient takes more carbohydrates in food or drink, he/she must manually alter insulin supply rate to deliver a heavier dose known as the bolus rate to meet the body's requirement. The pump has to be removed/taken care of while performing activities like bathing, swimming, and other sports.

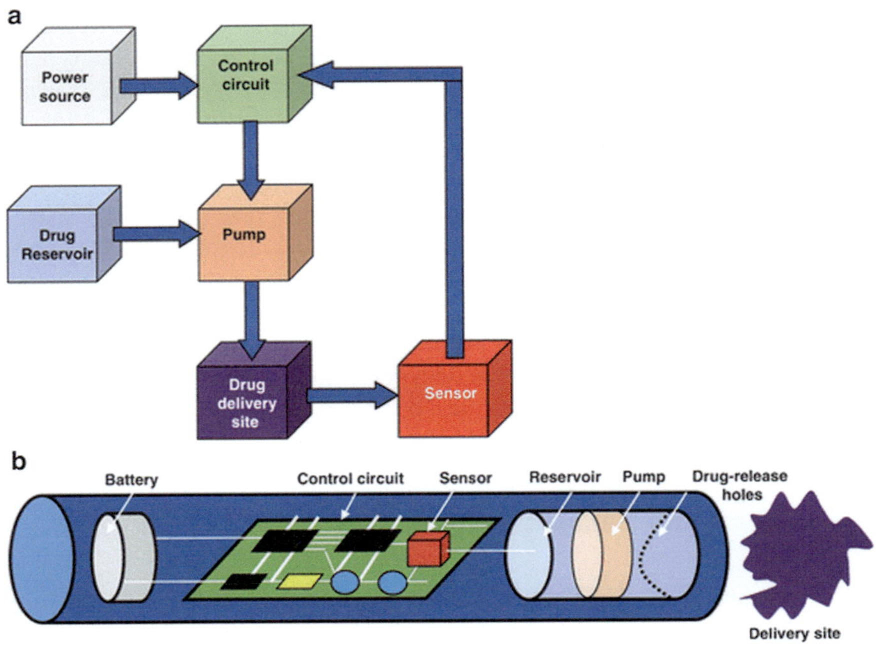

Fig. 25.4 Idea of a smart implantable drug delivery system: **a** Flow diagram and **b** schematic layout

25.8.1 Microreservoir-Based IDDSs

Microfabricated reservoirs called microreservoirs are drug receptacles for transitory storage of drug consignments before releasing them in the body. The drug liberation takes place in two modes: (1) moderately over a period of time through orifices of membranes in the reservoir wall, in which these membranes have a nanoporous structure [23], and (2) at predefined times by removal of an impervious seal, in which the drug is ejected by diffusion or other transference mechanisms. Microreservoir-based drug delivery devices have advantages of simple blueprint and easy realization. However, the curative lifetime of such drug-releasing implants is presently limited because drug payloads cannot be restocked.

25.8.1.1 Passive Devices

The working principle that was utilized in many micro- and nanofabricated drug delivery systems of reservoir type, which were developed in the beginning stages, was simple transport mechanism mediated by diffusion of drug molecules. The drugs were stockpiled or produced within the reservoirs [24]. As an example, silicon-based immunoisolating capsules (repositories sheltering drug release carriers from immune reactions) consisted of a nanoporous membrane integrated into a reservoir.

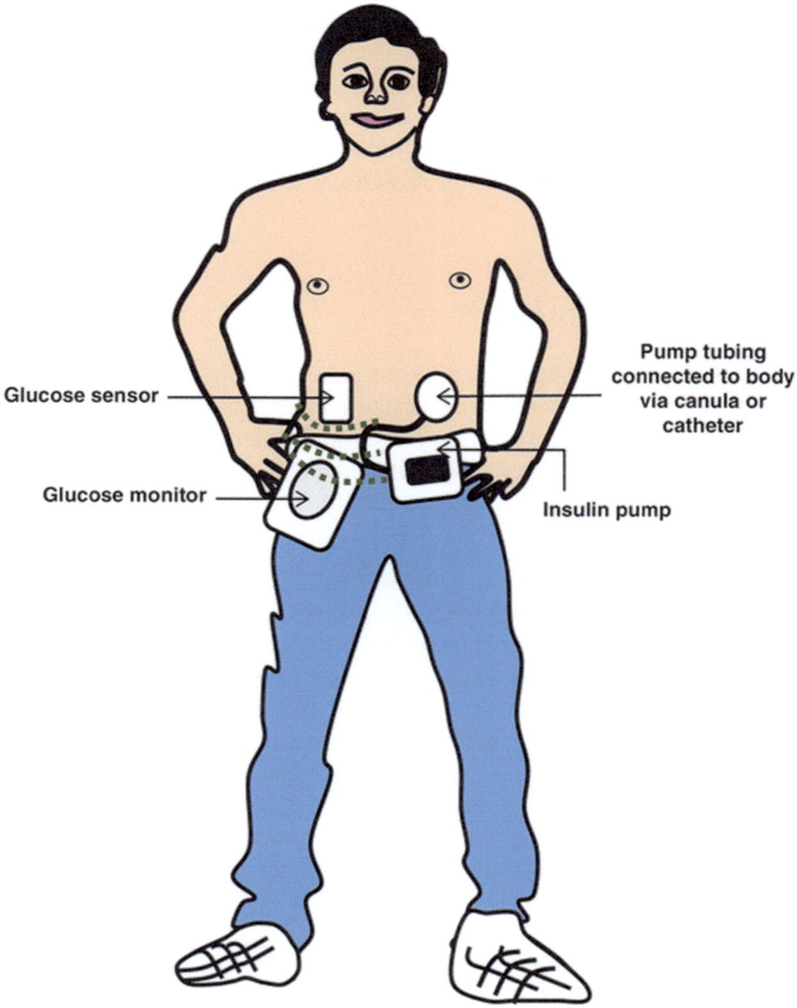

Fig. 25.5 Man with external insulin pump

Their operation was based on the clear-cut controlled feature dimensions [25]. In another example, drug reservoirs microfabricated by etching into silicon substrates were crowned with thin transitory metallic membranes. After implantation, the lyophilized drug (medicine dried by freezing and succeeding evaporation of water in high vacuum) in the reservoir was selectively uncovered for diffusive liberation. The idea used for uncovering was either electrochemical dissolution or deterioration of the membrane cap caused by electrothermal mechanism [19].

One more structure introduced consisted of microreservoirs gated with temperature-responsive hydrogel valves. These microreservoirs allowed time-based variation of drug liberation through valves. Wireless actuation of the valves was triggered through inductive coupling between planar microfabricated pairs of metal coils incorporated with the reservoirs. To allow drug release by heat dissipation, reversible contraction of valves made of hydrogel (gel with water as the liquid component) was carried out [26]. A hydrogel is a cross-linked polymeric network. This network can accommodate water in the obtainable or accessible spaces among its polymeric chains [27].

25.8.1.2 Actively Driven Devices

Drug-liberating reservoirs based on diffusion process gained popularity by virtue of their straightforwardness but also became notorious due to several built-in restraining factors. They were restricted in usage by their small volumes of doses, low speed of release of drug, and dependency on opportune biological situations. Hence, several suggestions of active contrivances to exorcise drugs after elimination of the seal of membrane were promulgated.

1. *Drug Expulsion by Electrolysis*: This takes the clue from the technique of uncovering the drug filled into wells made of silicon by electrochemically dissolving a metallic plugging film. It further improved upon the method by integrating electrodes into the interior of the well. These electrodes enabled the spewing out of the drug solution by electrolysis, the well-known electrochemical decomposition process that takes place on passing electric current through an ionized solution. By this scheme, ejaculation of vasopressin could be done for crisis handling of shock accompanied or produced by hemorrhage [28]. However, only 0.37 fraction of the solution of drug was recuperated from the tanks measuring 15 μL in volume. Moreover, it was found that the drug underwent chemical degradation during electrolysis of the solution.
2. *By Modulating Aperture Size*: Instead of removing the membrane covering of the medicine tank, another conception is based on liberation of the medicine as controlled by dimensional adjustment of a microscopic opening from which the medicine is ejected. Using a device made in this way, intraocular administration (supply by entering the eyeball) of docetaxel, an antimitotic chemotherapy medicine having low solubility, could be done for treating diabetes-induced retinopathy. In this disease, damage to the blood vessels in the retina is caused by recurrently high blood glucose levels. To implement the idea, the drug was enclosed in a polymer reservoir [29]. This reservoir was closed on one side using flexible membrane, in which there was a slit for releasing the drug. The membrane was magnetically sensitive. An external magnetic field was applied wirelessly to deform the membrane. Thus, modulation of the size of the slit could be done. Distinctly separate volumes of medicine having mass $= 171 \pm 16.7$ ng could be thereby released.
3. *By Thermal Energy*: In an alternate concept, heat energy supplied by resistively heating a metal film at the flooring of the tank was used for drug release. By

heating, bubbles were generated. The bubbles mechanically ruptured and blew out the fastened membrane. They also rapidly drove out the medicine from the tank having 20 μL volume. This method could salvage 0.85 fraction of active drug with negligible wastage caused by the action of heat on the drug molecules.

4. *By Pressurizing*: In yet another scheme, reservoir contents were quickly released by pre-pressurizing it, i.e., by putting a drug under a greater than normal pressure. Membranes composed of polymeric material having resistive structures made of metal were used to fabricate short-lived valves. These valves were sufficiently stout to securely hold reservoir contents. By the application of a brief pulse of current to the resistor made of metal, the polymer membrane was enfeebled by heating action, allowing the liberation of drug contents.

25.9 Infusion Micropumps for Drug Delivery

Based on the mechanism of displacing and delivering the drug, micropumps are classified into passive and active groups (Table 25.2). Active pumps need energy in electrical form for their functioning. Passive pumps apply other forms of energy such as mechanical or chemical to provide the difference of pressures for pumping the medicine. While they are more complex as compared to simple tanks, drug infusion micropumps can sometimes be refillable for long-lasting dosing.

25.9.1 Principles of Passive Micropumps

Passive micropumps operate on two common mechanisms, namely, osmosis and pre-pressurizing reservoirs. The overall size of such pumps is small because no electrical power source is necessary.

Table 25.2 Passive and active micropumps

Sl. No.	Passive micropump	Active micropump
1	It does not use electrical power	It is electrically powered
2	Its design is simple	Its design is complex
3	It provides limited control over drug delivery	It offers the ability to control the rate and temporal behavior of drug delivery
4	It affords considerable miniaturization	It gives only limited miniaturization
5	Its operating principles are: osmosis, spring-powering, etc.	Its working principles are: electrostatic, piezoelectric, electrochemical, thermal, etc.

25.9.1.1 Osmotic Principle

Osmotic mechanism is an inherently simple one (Fig. 25.6). It requires no electrical energy. Naturally, it provides increased robustness as well as opportunities for miniaturization. Osmosis is described as a procedure by which molecules of a solvent (a medium in which another substance is dissolved to form a solution) spontaneously cross through a semipermeable or selectively permeable membrane from a region having a low concentration of solute into one in which there is a high concentration of the same to equalize the solute concentrations in the two regions. This mechanism affords nonstop transport of substances. The delivery does not rely on the exact conformation of molecules, neither on their physical nor chemical properties.

Basically, osmotic pumps comprise three constructional elements. These are the osmotic agent, the solvent, and the medicine [18]. For delivering a drug that dissolves in water, a single-chamber osmotic pump is used (Fig. 25.7).

If a drug is insoluble in water, a dual-chamber pump is required, as illustrated in Fig. 25.8.

In osmotic pumps, drug is stored either in liquid or solid state. The solid state allows efficient storage of drug in a concentrated manner requiring minimal space. But drug delivery is done as a liquid solution after dissolving the drug in water. In this manner, osmotic systems constitute the most volume-saving technologies for supplying medicines. The constant availability of water in all fluids inside the human body makes tremendously shrunken implantable devices possible [18]. For drugs,

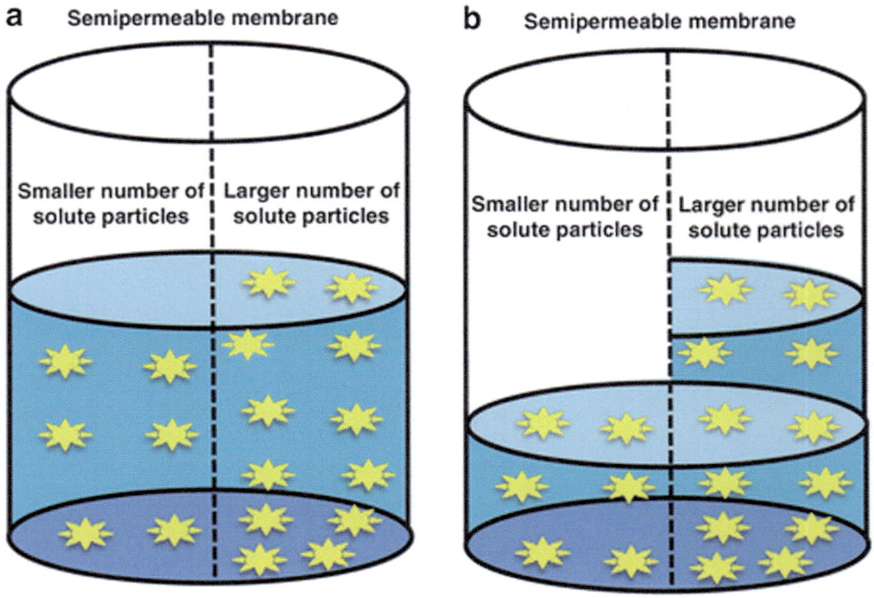

Fig. 25.6 Osmosis: Solvent level rises on the side of semipermeable membrane containing larger number of solute particles by diffusion of solvent across the membrane

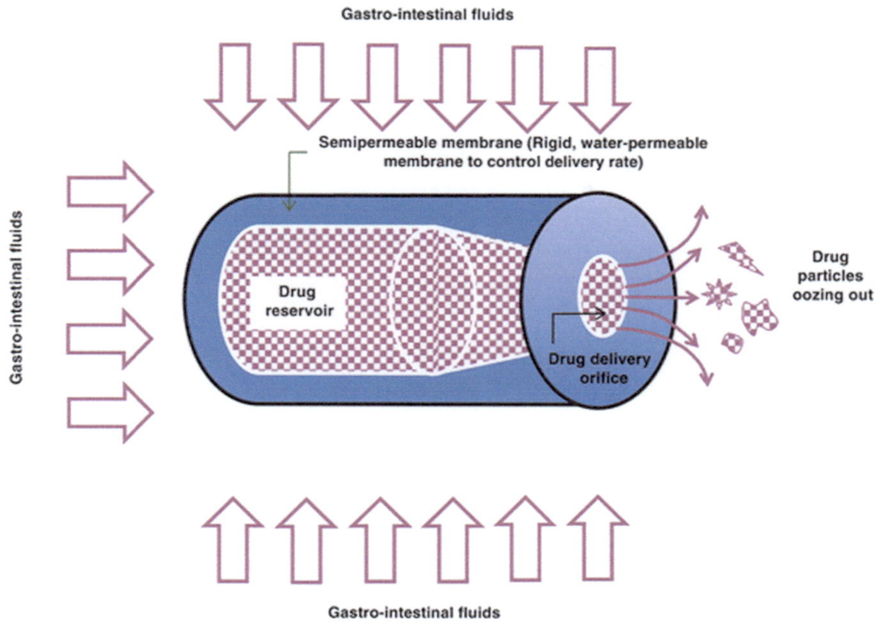

Fig. 25.7 Single-chamber osmotic pump for delivery of a water-soluble drug. Inflow of gastrointestinal fluids through the semipermeable membrane causes ejection of drug particles from the orifice. Drug delivery rate is controlled by the membrane

which are not water-soluble, a dual-chamber osmotic pump is required, as already indicated.

However, the simplicity of this mechanism also imposes restrictions on the control of dosing profile and timing. Once the osmotic gradient is initiated, the osmotic pumps continue to work without interruption. They release the drug at an unvarying rate. Further, they cannot be turned off. Moreover, as the osmotically active agent cannot be easily restored to its original state, osmotic pumps are naturally single use.

25.9.1.2 Spring-Powering Principle

Mechanical mechanisms based on spring loading as a means to force displacement of the medicine have not been the subject of much investigation. However, these pumps provide distinctive benefits over osmotic variety and other variants that are actively driven. They need to be looked into as a potential research area.

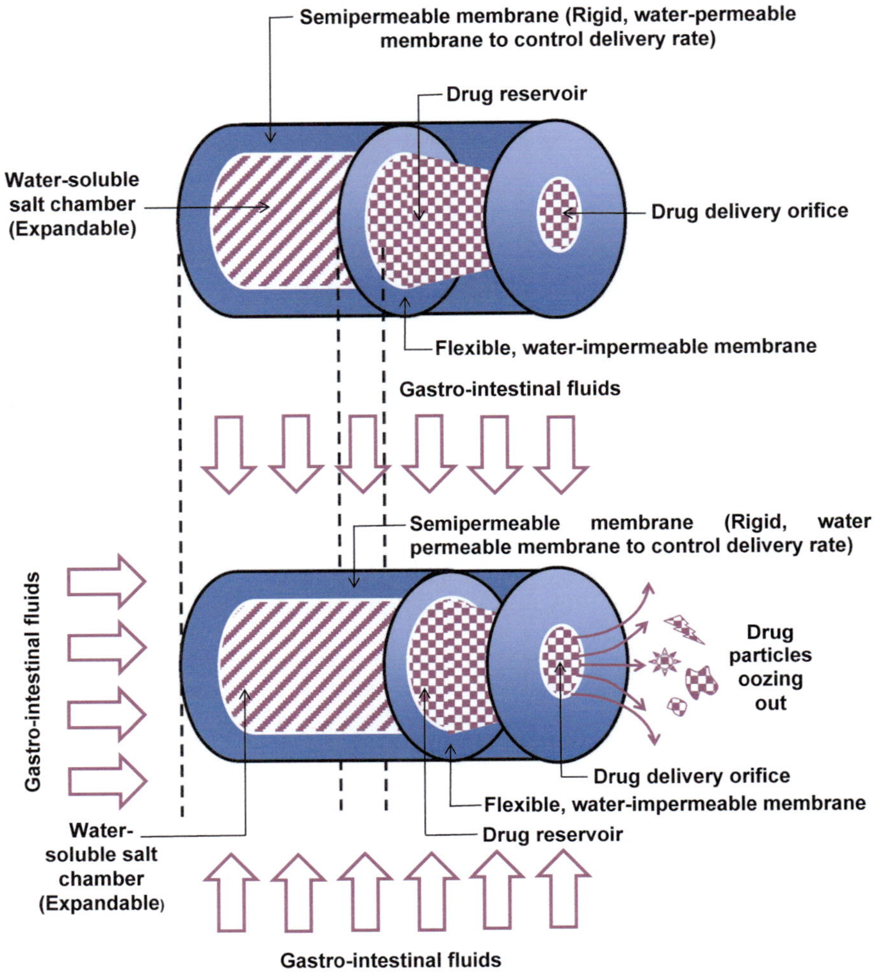

Fig. 25.8 Dual-chamber osmotic pump for delivery of a drug that is insoluble in water. Diffusion of gastrointestinal fluids into the salt chamber across the semipermeable membrane dilates the salt chamber pushing it into the drug tank and forcing drug particles out via the orifice

25.9.2 Principles of Active Micropumps

Microelectromechanical system (MEMS) technologies can downscale practical embodiments of electronically controllable pumping mechanisms into miniature sizes that are conducive to implantable or external transdermal patch pumps. MEMS consist of mechanical components, sensors, actuators, and electronic circuits, all on a single substrate made of silicon. This integration is achieved by utilization of the microfabrication technology with additional fabrication processes. MEMS contain moving or vibrating parts like diaphragms or membranes, beams, cantilevers, and so on. Dimensionally, complete MEMS measure a few millimeters × few millimeters. But individual features/devices are ~ 1–100 μm. In MEMS, the sensors collect knowledge from the ecosystem via diverse phenomena of mechanical, biological, chemical, thermal, optical, and magnetic nature. The microelectronic circuits process the pieces of information gathered from the sensors. Then, through some capability of arriving at timely decisions, the actuators are directed to take necessary action by movement, placement and positioning, regulation, pumping, etc. Thus MEMS are able to control a process or situation to achieve a looked-for outcome or to fulfill a well thought-out purpose. The microelectronics acts as the brains of MEMS. The sensors and actuators augment this decision-making capability by acting as the eyes and arms, for controlling the environment.

MEMS technology consists of the microelectronic fabrication processes such as silicon oxidation/diffusion, low-pressure chemical vapor deposition (LPCVD), plasma-enhanced chemical vapor deposition (PECVD), epitaxy, photolithography, sputtering, etc. Along with these standard microelectronic fabrication techniques, some more processes are included such as surface and bulk micromachining, silicon–silicon and silicon–glass wafer bonding, deep reactive ion etching (DRIE), LIGA, and micromolding. LIGA is an acronym for a sequence of process steps. This acronym represents the main process steps involved in its execution, viz., deep X-ray lithography, electroforming, and plastic molding. German words for these processes are *lithografie*, *galvanoformung*, and *abformung*, respectively, hence called LIGA. Thus, MEMS technologies are an extension of microelectronic processes. All the fabrication processes mentioned above are useful for the fabrication of drug delivery systems and can be utilized depending on the IDDS design.

On the microlevel, the mechanisms examined for making pumps are [30] *mechanical mechanisms* such as electrostatic, piezoelectric, electrochemical, thermal, shape memory alloy [31], bimetallic; and *nonmechanical mechanisms,* e.g., electrohydrodynamic, electroosmotic, ultrasonic and thermo-capillary mechanisms, as well as electromagnetic mechanisms.

Frequently commissioned mechanical pumps primarily use a pliable membrane. This supple membrane is put into action by using one of the aforesaid driving mechanisms of mechanical nature. This mechanism is either electrostatic or piezoelectric or electrochemical or thermal. Hence, the medicine kept in liquid state is driven out from a tank. The mechanisms used for driving the pumps clearly swerve from those employed in devices, which are fabricated on intermediate and milli-scales. These mechanisms are preferential because they enable the benefits of miniaturization. A tighter management of the inception, profile, and stoppage of medicine supply is feasible with these electronically controlled micropumps. However, these micropumps require components, which can move and bend, e.g., microvalves and thin, flexible diaphragms. Such components introduce complications in their manufacturing processes. Nonetheless, the lowering in power consumption obtained with downsizing encourages the opportunity of making pumps, which are both powered and operated wirelessly. These pumps do not need implantable batteries.

25.9.2.1 Electrostatic Principle

From elementary electrostatics, we know that a significant mechanical force is produced by electrostatic attraction between two oppositely charged metallic plates placed in vicinity at the micrometer scale. The magnitude and direction of this force are determined from Coulomb's inverse square law [32]. To apply this electrostatic interaction to pulsatile drug delivery, two sets of plates are used. These sets of plates act as the doors of a valve of a drug reservoir. One set of plates is movable, and the other is stationary in position. The drug delivery is controlled by carefully modulating the charges on the two sets of plates. Consequently, the distance between the plates changes as the plates come closer or move apart. By such control of intervening distance between the plates, the drug movement from the reservoir is stopped by constricting or narrowing the outlet (smaller distance between plates). Thereafter, drug supply is restored by dilating the outlet (larger distance between plates). Effectively, the drug flow from the reservoir takes place in a sequential style. Thus, this arrangement of electrostatically controlled plates delivers packets of drug in a throbbing manner.

Two familiar formats of electrostatic micropump configurations are the reciprocating diaphragm and peristaltic (multiple-reciprocating diaphragms) types [33]. Both these pumps are micromachines of positive-displacement variety. They work by compelling a fixed quantity of drug withdrawn from the intake pressure zone of the pump toward its discharge section.

Reciprocating pumps require a membrane and mechanical valves to regulate fluid flow [34]. The membrane serves as a piston or plunger. The pumping action is a two-stroke process: a suction stroke and a delivery stroke. In the suction stroke, the membrane moves rearward. The valve on the inlet side unlocks and the drug is sucked in from the reservoir into the pump chamber. As the exit valve is fastened, the drug is retained in the chamber. During the delivery stroke, the membrane moves forward to force the drug out from the chamber of the pump through the valve on the outlet side for delivery to the required site. As the inlet valve remains closed, the drug cannot move into the inlet side.

The peristaltic pump operates by squeezing a bendable tube in a fashion resembling our throat and intestines. Peristalsis is radially symmetrical contraction and expansion of muscles. A peristaltic pump is based on alternating compression and relaxation of the flexible tube drawing the contents into itself. In this pump, the reciprocating diaphragms are linearly arranged. These diaphragms are controlled individually. By activating the diaphragms in a recommended sequence to harmonize their action, aggregate pumping in one direction is achieved. In opposition to reciprocating pumps, each diaphragm works like a valve, which is controlled by an electronic mechanism. This valve closes the path of flow of the medicine to prevent it from flowing backward, i.e., in the reverse direction into the reservoir.

Despite providing a speedy mechanical response and capable of operating at a low power level, electrostatic mechanisms have the shortcoming that their operation is limited to small distances. This restricts the volume of fluid, which is transferred during every stroke of the pump and imposes the requirement of high voltages for operation of the pump. Hence, the utilization of electrostatic pumps for delivering drugs is curbed down.

25.9.2.2 Piezoelectric Principle

Piezoelectricity (electricity from pressure) is the ability of certain materials to produce an electric charge in response to mechanical deformation [35]. By fixing a thin film or sheet made of a piezoelectric material on an elastic diaphragm and subjecting it to the controlling electronic signal, the diaphragm is brought into reciprocatory motion or reciprocation (Fig. 25.9). When voltage is applied in one direction, the consequential upward bending produces an upstroke of the pump (suppose). The inlet valve opens, the drug is sucked inward, and the pump compartment is filled with drug. The outlet valve remains closed. When the voltage is applied in reverse direction, the membrane is deformed downward. By this downstroke, the valve on the outlet side opens, and drug is displaced out of the pump chamber into the body. The inlet valve is closed during this period. Thus, the valves on the two sides of the pump compartment define the direction of flow of the drug. As the membrane lifts up and down, the drug is pumped in a pulsatile manner similar to the electrostatic micropump.

Piezoelectric pumps can also be built for peristaltic mode. Again despite providing a large pressure for pulling or pushing the drug and rapid mechanical response, the high-voltage requirement limits the drug delivery applications of these pumps. Debiotech, a Swiss company [36], has lead the way to piezoelectric micropumps for use in medicine, e.g., the insulin Nanopump™ for the treatment of diabetes.

25.9.2.3 Electrochemical Principle

These pumps utilize the liquid–gas phase change that takes place upon electrolysis of water, resulting in its decomposition into hydrogen (H_2) and oxygen (O_2) gases. By the application of a small DC voltage through two electrodes into water, oxygen and hydrogen gas bubbles are produced. These electrochemically generated bubbles cause an increase in pressure leading to an actuation force. This force drives a flexible diaphragm. On removal of power, the reaction is reversed and the gases recombine to form water. The actuation force ceases to act. Thus, a reciprocating action is produced.

Electrochemical micropumps offer a smooth, continuous delivery of drug, which is transferred by the flexible diaphragms directly from the reservoir. Exclusive advantages of this approach are the large attainable mechanical displacement obtained from low power consumption. Such micropumps have been reported for eye-related drug delivery applications [37] and also for gene delivery in cancer treatment [38]. However, possible disintegration and collapsing of the generated bubbles into water may upset the drug release reliability [39].

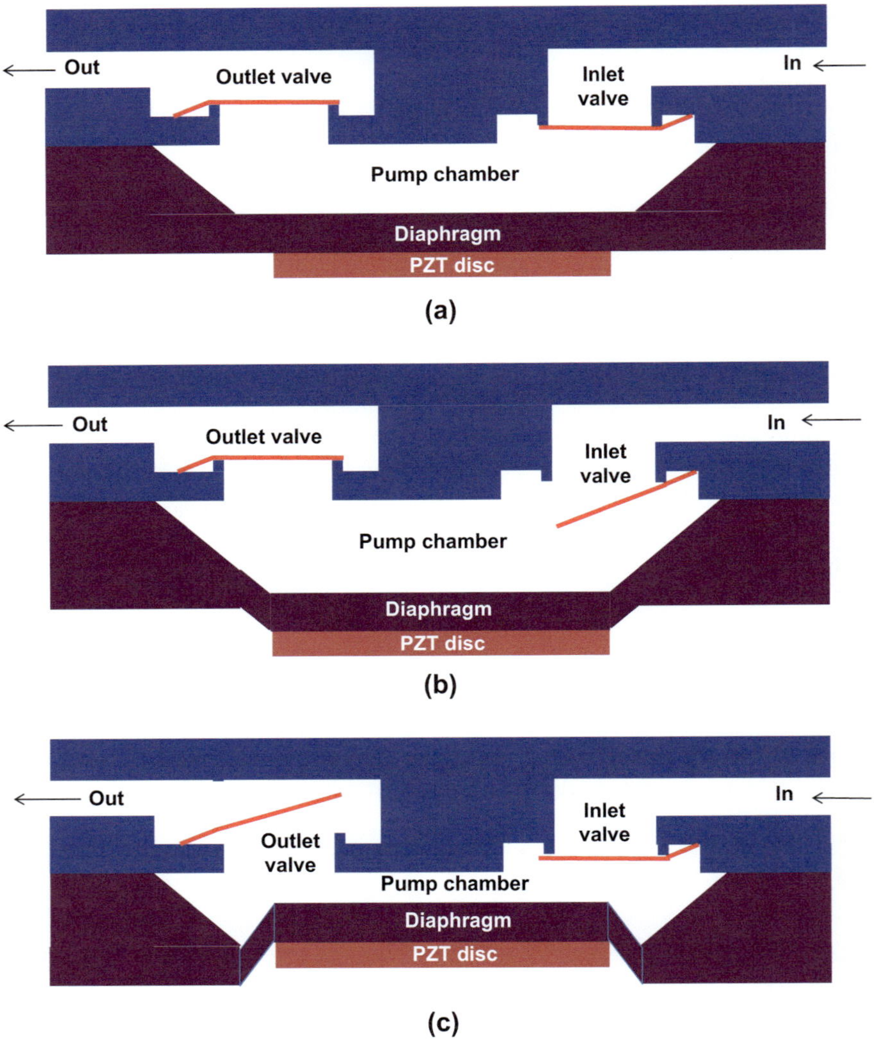

Fig. 25.9 Piezoelectric pump: **a** rest position, **b** diaphragm moves downward, inlet valve opens, outlet valve closes and the drug is drawn inward; **c** diaphragm moves up, the inlet valve closes, outlet valve opens and the drug is forced outward

25.9.2.4 Thermal Principle

Here, the bendable pump diaphragm is driven by heating effect principles, e.g., the volume expansion or phase change of a substance taking place by applying heat energy, such as from liquid to gaseous state. Expansion to be exploited for pumping may be obtained when a heating element is used to heat up a working fluid, causing its expansion to act on a supple diaphragm. By operating a thermopneumatic pump with a 2-phase liquid–vapor perfluorocarbon propellant, pulsatile flow was obtained [40]. But in a reciprocating-type thermal pump, the mechanical response time is limited by the extended time constant of the thermal process. This is chiefly troublesome at the time of cooling. For in vivo applications, high power consumption is especially worrisome if the pump resides completely inside the human body. Thermal micropumps were probed for supplying medicine inside the cerebrum to gnawing mammals that were moving freely [41].

25.10 Robotic Capsules for Drug Delivery to the Gastrointestinal Tract

These are orally ingestible devices containing motors or other robotic movement mechanisms specifically developed for delivering complex drugs like proteins and antibodies. Their purpose is to replace injections and overcome the limitations of traditional oral drug delivery by clearing the mucus layer, improving drug mixing, depositing the drug directly onto the intestinal surface, or delivering a drug to targeted locations. A rotating mechanism clears the mucus layer to deposit the drug directly on the intestine's surface. Absorption is further enhanced by the movement of the drug, which enables thorough mixing of the drug with digestive fluids. For optimal absorption, the capsules are programmed for drug delivery using drug release mechanisms such as pH-sensitive membranes, controlled dissolution, or external radiofrequency triggering. Some capsules use microneedles to supply therapeutic medicines to intestinal tissues. Delivery of drugs to exact locations maximizes their therapeutic efficacy. Further, by delivering drugs only to the precise spots, the unintended areas are prevented from exposure to drugs, whereby their side effects are reduced. Ongoing research areas include miniaturization of the capsule size, providing sustained drug release over extended periods, and integrating wireless communication and control techniques for real-time control, besides assuring safety, tolerability, nontoxicity of the method, and patient comfort. A few examples of robotic capsules are given below:

RoboCap: The robotic drug delivery capsule, RoboCap, developed by Massachusetts Institute of Technology (MIT), locally clears the mucus layer. It improves luminal mixing and deposits the drug payload in the small intestine. It has an internal motor. Also, its surface features interact with the small intestinal plicae circulares, villi, and mucus. These together facilitate its mucus-clearing and churning movements. RoboCap-mediated Vancomycin and insulin delivery augmented their bioavailability 20- to 40-fold times in *ex vivo* and *in vivo* swine models compared to standard oral delivery [42].

Microneedle Capsule (MIT): Proof-of-concept experiments in swine showed that microneedle-based delivery improves bioavailability of a biologically active macromolecule. Further, microneedle-containing devices are excreted from the GI tract safely. The utility of microneedle capsules is understood by noting that the gastrointestinal tract restricts the bioavailability of therapeutics due to its protease and bacteria-rich environment, and pH variability ~ 1–7, making oral delivery difficult [43].

Luminal Unfolding Microneedle Injector (LUMI) Robotic Pill: Ingested by swallowing, it travels through the digestive tract to reach the small intestine, where it is activated with dissolution of the pill's polymer shell, releasing three folding arms. The arms have drug-loaded microneedles fixed on them. The microneedles are propelled into the intestinal wall. The drug is delivered by microneedle penetration. It is absorbed into the bloodstream through the intestinal lining. The device breaks down and is then excreted from the body. The pill has been used for oral insulin delivery as an alternative to insulin injections for managing diabetes. It demonstrated a quick pharmacokinetic absorption profile and a systemic uptake of > 10% compared to subcutaneous injection over a 4-h sampling period. The pill can act as a technological platform for oral administration of macromolecular medicines in therapeutic amounts [44].

SmartTab: It is a capsule technology that triggers the optimal release of pharmaceutical contents to specified regions of the gastrointestinal tract using radiofrequency signaling. Pre-installed onboard sensors enable the devices to send real-time data to physicians and personalize treatments using AI algorithms. The capsule supplies active ingredients to the target site, significantly reducing the risk of side effects and increasing dosage effectiveness in treating Crohn's disease [45].

Magnetically Actuated Novel Robotic Capsule: This capsule provides site-specific drug delivery inside the gastrointestinal tract. A capsule-like robot is equipped with retractable needles to inject the drug directly into the intended tissues. In the beginning, the needles are concealed inside the capsule. A magnetic membrane allows them to protrude and retract dynamically. A second membrane is used to release the drug from the reservoir for injection into the targeted tissue through the needle. The capsule can activate locomotion via wireless control of an external magnet [46].

RaniPill: It is a robotic pill for oral delivery of biotherapeutics. Developed by Rani Therapeutics, the Rani capsule is a swallowable pill engineered to inject biologic drugs directly into the jejunum [47].

Let us compare the delivery of biotherapeutics by robotic capsules with that by other methods. Although biotherapeutics are highly effective, chronic injections cause discomfort and annoyance, reducing patient compliance and jeopardizing effective disease management. Oral delivery of biotherapeutics is unsatisfactory due to their breakdown in the gastrointestinal environment and inadequate intestinal absorption. A specially developed orally ingestible robotic pill (RP) shields the biotherapeutic drug payload from digestion in the GI tract. It automatically injects the drug into the wall of the small intestine as a safe, painless injection because the intestines are insensitive to acute stimuli. The payload is delivered by inflation of a balloon folded within the RP, which deflates immediately upon medication administration [48].

Table 25.3 Advantages of capsule robots

Sl. No.	Advantages
1	Respite from the agony of frequent injections: capsule robots remove the distress and discomfort of painful and inconvenient injections, fulfilling an essential need of long-term therapies
2	Improved drug bioavailability for therapeutic effects: drug absorption in the body is enhanced by overcoming the limitations of oral drug delivery, such as mucus barriers and drug degradation in the gastrointestinal tract
3	Controlled drug release in the bloodstream: precise control over drug release allows for individualized, tailored dosage supply and timing of medication delivery to the patient
4	Targeted drug delivery to diseased sites: capsule robots' ability to navigate the gastrointestinal tract and liberate drugs at planned locations helps improve medication absorption and reduce systemic side effects
5	Disease diagnosis and its monitoring by capsule robots: equipping capsule robots with cameras, sensors, and other accessories facilitates the diagnosis and monitoring of patient's health condition from mouth to anus
6	Possibility of multiple drug delivery by capsule robots: designing capsule robots to carry and release several drugs allows the treatment of complicated physiological disorders

25.11 Advantages and Disadvantages of Capsule Robots

Tables 25.3 and 25.4 depict the advantages and shortcomings of capsule robots at a glance.

25.12 Discussion and Conclusions

Two major challenges have confronted the interesting area of delivery of medicines in a controlled manner. The first issue has been that of continuous zero-order release of a medicine over a protracted time span. Osmotically driven pumps and matrices characterized by adjustable rates of diffusion or erosion have been developed to fulfill this goal.

The second challenge has been the variable supply of a medicine molecule or protein for therapy undertaken in a pulsating or unsteady mode. Possible solutions are furnished by two different methodologies: (1) the development of a delivery system that discharges its drug load at a pre-agreed time or in spurts of a predecided sequence and (2) a system that responds to changes in the confined environment around the system. This system alters the speed of delivering the medicine by retorting to motivation, e.g., by detecting whether a definite molecule is present or absent, and by the triggering action of electric or magnetic fields; and under the influence of ultrasonic waves or forces of mechanical, optical, or thermal origin.

Table 25.4 Disadvantages of capsule robots

Sl. No.	Disadvantages
1	Power requirements for capsule movements: maintaining power for the capsule robots' mobility and operations is complex in the hostile environment of the digestive tract
2	Patient safety concerns: it is crucial to ensure the capsule robot's safety within the body, including its biocompatibility and the likelihood of interactions with the digestive system
3	Capsule size and space constraints: the capsule's minuscule size limits the quantity of drug that can be loaded inside it, and therefore, the duration for which the capsule's payload lasts
4	Capsule design complexity issues: simulating capsule robots requires sophisticated engineering techniques for motors, sensors, and electronic control systems. Services of an expert engineering and technology team are obligatory
5	Capsule development and manufacturing costs: research, development, and fabrication of capsule robots are expensive
6	Meeting regulatory requirements: conducting clinical trials and seeking regulatory approvals are compulsory
7	Patient readiness and acceptance of treatment: patients may feel uncomfortable or hesitant to swallow a capsule, especially if it requires additional interventions
8	Limited applications: capsule robots may not be the recommended substitutes for all drugs or conditions, and thorough research is needed to expand their applicability

It is anticipated that in the impending one or two decades, a substantial portion > 50% lump of the latest drugs launched commercially will be protein- and peptide-based drugs. Also > 80% of these drugs made of protein molecules are expected to be antibodies. Novel drug delivery issues are associated with these biological medical products or biologics called biopharmaceuticals produced using biotechnological methods and comprising proteins and nucleic acids. These issues arise because they are made of large molecules that undergo degradation in the blood very swiftly. Moreover, they possess meager ability of crossing membranes of the cells. This makes them incapable of oral delivery. Delivery via conventional routes will be problematic for such molecules. The only likely means of delivery is through injections. The routes of administration of medicines will be decided by the medicine itself, the state of disease, and the place of action. Easily reachable places are the nose, the mouth, and the vagina. Other places such as the brain are more difficult to access. Robotic capsules are drastically impacting healthcare by providing minimally invasive diagnostics and targeted drug delivery.

Projecting further, as biotechnology companies become involved in drug delivery worldwide, gene therapy is another foreseeable breathtaking growth sector.

The market of systems for delivering medicines has marched a long path. It will move on to burgeon incalculably and expand impressively. Finally, although the present chapter is the concluding chapter of this book, it appears as if the story of implantable electronics has just begun, and a lot more is yet to come. "Implantable electronics tree" is bearing sweet fruits. Forthcoming years will be even more fruitful and eventful. The future promises more successes and achievements. Suffering patients look forward with great hopes and aspirations toward the emerging developments.

Let us sing in appreciation of implantable drug delivery.

Eulogizing Implantable Drug Delivery

It delivers drugs directly into the bloodstream, circumventing first-pass metabolism.

And provides a controlled drug release mechanism.

Wherein a lower dose gives a more significant therapeutic effect.

With the least side effects.

Also facilitates drug target selection.

For faster and efficient drug action.

Still more, the drug release rate is precisely controllable.

More consistent and stable.

Avoiding troughs and peaks of conventional therapy.

Enhancing drug availability and reducing systemic toxicity.

Beyond compare, it is programmable for intermittent drug release.

Offering a quick drug dosage increase or decrease.

Making it an efficient technique of drug administration.

For achieving maximum results with minimum medication.

Good patient health is maintained.

Without the reprise insertion of needles, discomfort and pain.

Artificial-intelligence-based performance optimization.

Will bring novel approaches and formulations.

For personalization of treatment.

Assuring patient compliance and merriment.

It is a spectacular implantable electronics application.

Deserving full-on praise and acclamation!

Review Exercises

25.1 What is a drug delivery system? What are its applications? Give examples.

25.2 Which types of drug delivery systems have shown the greatest promise? What are their advantages? What are the shortcomings of these systems?

25.3 Name five types of conventional drug delivery systems.

25.4 What is the most common drug delivery method? Name it and cite one advantage and one disadvantage of this method.

25.5 How are the drug molecules transferred to the blood stream by the following methods: (a) nasal and (b) pulmonary?
25.6 How does a transdermal patch deliver drug to the blood. Mention some advantages of this method. What type of drugs cannot be delivered by this method and why?
25.7 How is intravenous drug administration carried out? Under what circumstances is this method of drug delivery essential?
25.8 Describe the two modes of intravenous drug delivery.
25.9 What are the risks associated with intravenous drug delivery? What is meant by air embolism?
25.10 Describe some medical conditions in which IDDSs represent the only practical method for delivery of drugs.
25.11 In what respects are IDDSs convenient for the patient? How is patient compliance enhanced by these systems?
25.12 Mention and explain five advantages provided by IDDSs in comparison to their traditional counterparts. Give five shortcomings of such systems.
25.13 If you were to design an ideal IDDS, what distinctive features will you incorporate in it?
25.14 How does a biodegradable IDDS differ from a nonbiodegradable type? Indicate their relative advantages and disadvantages.
25.15 Bring out a comparative study of passive and active IDDSs in terms of their drug release mechanisms and limits on controlling the delivery process. Cite some examples of the two classes of systems to illustrate your answer.
25.16 What are microelectromechanical systems? Name the main fabrication processes which constitute MEMS technology?
25.17 How have micro- and nanotechnologies contributed toward the augmentation of benefits realized from IDDSs?
25.18 Describe some types of passive diffusion-based reservoir devices for delivery of drugs. How are the drugs delivered by these devices? Do they need electrical power for their operation?
25.19 Describe three active mechanisms for drug delivery after removal of the membrane seal.
25.20 Describe the working principle of an electrostatic micropump. Name the two familiar formats of electrostatic micropump. Give the advantages and limitations of the electrostatic mechanism?
25.21 What is piezoelectricity? Describe the operational procedure of a piezoelectric pump.
25.22 Describe the operation of a micropump based on the liquid–gas phase change that takes place on electrolysis of water.
25.23 How is the thermal mechanism exploited to make a micropump?
25.24 In what ways do robotic capsules replace injections and overcome the limitations of traditional oral drug delivery?
25.25 How do robotic capsules enhance the absorption of drugs in the body?
25.26 How do robotic capsules minimize the side effects of medications?
25.27 How do robotic capsules assist in disease diagnosis and monitoring?

25.28 Give three examples of capsule robots and mention their salient features.

References

1. Arps J (2013) Implantable drug delivery devices—an overview. Med Design Technol. http://www.mdtmag.com/articles/2013/07/implantable-drug-delivery-devices%E2%80%94-overview. Accessed 5 Feb 2015
2. Stuart A (2010) The promise of implantable drug delivery systems. Eyenet 33–37
3. Maloney JM (2003) An implantable microfabricated drug delivery system. In: Proceedings of IMECE'03, 2003 ASME international mechanical engineering congress and exposition, Washington, DC, 15–21 Nov, Paper No. IMECE2003-43186, pp 115–116. https://doi.org/10.1115/IMECE2003-43186. Copyright © 2003 by ASME
4. Gupta H, Bhandari D, Sharma A (2009) Recent trends in oral drug delivery: a review. Recent Pat Drug Deliv Formul 3(2):162–173
5. Djupesland PG (2013) Nasal drug delivery devices: characteristics and performance in a clinical perspective-a review. Drug Deliv Transl Res 3(1):42–62
6. Patil JS, Sarasija S (2012) Pulmonary drug delivery strategies: a concise, systematic review. Lung India 29(1):44–49
7. Shingade GM, Aamer Q, Sabale PM (2012) Review on: recent trend on transdermal drug delivery system. J Drug Deliv Ther 2(1):66–75
8. Yalkowsky SH, Krzyzaniak JF, Ward GH (1998) Formulation-related problems associated with intravenous drug delivery. J Pharm Sci 87(7):787–796
9. Rajgor N, Patel M, Bhaskar VH (2011) Implantable drug delivery systems: an overview. Syst Rev Pharm 2(2):91–95
10. Siegel RA, Rathbone MJ (2012) Chapter 2: overview of controlled release mechanisms. In: Siepmann J, Siegel RA, Rathbone MJ (eds) Fundamentals and applications of controlled release drug delivery. © Controlled Release Society, pp 19–43
11. Lux CDG, Mahmoud E, Joshi-Barr S (2015) Bioresponsive materials. ©2015 Regents of the University of California. http://almutairi.ucsd.edu/bioresponsive-materials. Accessed 6 Feb 2015
12. Solanki HK, Thakkar JH, Jani GK (2010) Recent advances in implantable drug delivery. Int J Pharm Sci Rev Res 4(3):168–177
13. Bhowmik D, Gopinath H, Kumar BP et al (2012) Controlled release drug delivery systems. Pharma Innovation 1(10):24–32
14. Japp NS (2008) Sterilization processes: what every infection control practitioner needs to know. Infect Control Res 5(1):1, 5–7
15. Tilton G, Kauffman M (2004) Sterilization: a review of the basics. Manag Infect Control 66–71
16. Ronak K, Yatin K, Baldaniya L et al (2012) Implant-controlled release medicated formulation. Int J Pharma Chem Sci 1(1):59–66
17. Chung AJ, Kim D, Erickson D (2008) Electrokinetic microfluidic devices for rapid, low power drug delivery in autonomous microsystems. Lab Chip 8:330–338
18. Herrlich S, Spieth S, Messner S (2012) Osmotic micropumps for drug delivery. Ann Drug Deliv Rev 64(14):1617–1627
19. Santini JT, Cima MJ, Langer R (1999) A controlled-release microchip. Nature 397(6717):335–338
20. Meng E, Hoang T (2012) Micro- and nano-fabricated implantable drug-delivery systems. Ther Deliv 3(12):1457–1467
21. Au AK, Lai H, Utela BR et al (2011) Microvalves and micropumps for BioMEMS. Micromachines 2:179–220

22. Tng DJH, Hu R, Song P et al (2012) Approaches and challenges of engineering implantable microelectromechanical systems (MEMS) drug delivery systems for in vitro and in vivo applications. Micromachines 3:615–631
23. Gultepe E, Nagesha D, Sridhar S et al (2010) Nanoporous inorganic membranes or coatings for sustained drug delivery in implantable devices. Adv Drug Deliv Rev 62:305–315
24. Stevenson EL, Santini JT, Langer R (2012) Reservoir-based drug delivery systems utilizing microtechnology. Adv Drug Deliv Rev 64(14):1590–1602
25. Desai TA, Chu WH, Tu JK et al (1998) Microfabricated immunoisolating biocapsules. Biotech Bioeng 57(1):118–120
26. Rahimi S, Sarraf EH, Wong GK et al (2011) Implantable drug delivery device using frequency controlled wireless hydrogel microvalves. Biomed Microdevices 13(2):267–277
27. Pal K, Banthia AK, Majumdar DK (2009) Polymeric hydrogels: characterization and biomedical applications—a mini review. Design Monomers Polym 12:197–220
28. Chung AJ, Huh YS, Erickson D (2009) A robust, electrochemically driven microwell drug delivery system for controlled vasopressin release. Biomed Microdevices 11(4):861–867
29. Pirmoradi FN, Jackson JK, Burt HM et al (2011) On-demand controlled release of docetaxel from a battery-less MEMS drug delivery device. Lab Chip 11(16):2744–2752
30. Nisar A, Afzulpurkar N, Mahaisavariya B (2008) MEMS-based micropumps in drug delivery and biomedical applications. Sens Actuators B Chem 130:917–942
31. Reynaerts D, Peirs J, Brussel HV (1997) An implantable drug delivery system based on shape memory alloy micro-actuation. Sens Actuators A Phys 61:455–462
32. Teymoori MM, Abbaspour-Sani E (2005) Design and simulation of a novel electrostatic peristaltic micromachined pump for drug delivery applications. Sens Actuators A Phys 117(2):222–229
33. Woias P (2005) Micropumps- past, present and future prospects. Sens Actuators B Chem 105:28–38
34. Bourouina T, Bosseboeuf A, Grandchamp JP (1997) Design and simulation of an electrostatic micropump for drug-delivery applications. J Micromech Microeng 7(3):186–188
35. Van Lintel HTG, Van de Pol FCM, Bouwstra SA (1988) A piezoelectric micropump based on micromachining of silicon. Sens Actuators A Phys 15:153–168
36. Debiotech. ©2014—Debiotech S.A.—Switzerland. http://www.debiotech.com/newsite/page/index.php?page=product_01&id=1&id_prod=0. Accessed 5 Feb 2015
37. Li PY, Shih J, Lo R et al (2008) An electrochemical intraocular drug delivery device. Sens Actuators A Phys 143(1):41–48
38. Gensler H, Sheybani R, Li PY et al (2012) An implantable MEMS micropump system for drug delivery in small animals. Biomed Microdevices 14(3):483–496
39. Abhari F, Jaafar H, Yunus NAM (2012) A comprehensive study of micropumps technologies. Int J Electrochem Sci 7:9765–9780
40. Cooney CG, Towe BC (2004) A thermopneumatic dispensing micropump. Sens Actuators A Phys 116(3):519–524
41. Spieth S, Schumacher A, Holtzman T et al (2012) An intra-cerebral drug delivery system for freely moving animals. Biomed Microdevices 14(5):799–809
42. Srinivasan SS, Alshareef A, Hwang AV, Kang Z, Kuosmanen J, Ishida K, Jenkins J et al (2022) RoboCap: robotic mucus-clearing capsule for enhanced drug delivery in the gastrointestinal tract. Sci Robot 7(70):eabp9066. https://doi.org/10.1126/scirobotics.abp9066
43. Traverso G, Schoellhammer CM, Schroeder A, Maa R, Lauwers GY, Polat BE, Anderson DG, Blankschtein D, Langer R (2015) Microneedles for drug delivery via the gastrointestinal tract. J Pharm Sci 104(2):362–367. https://doi.org/10.1002/jps.24182
44. Abramson A, Caffarel-Salvador E, Soares V, Minahan D, Tian RY, Lu X, Dellal D, Gao Y, Kim S, Wainer J, Collins J (2019) A luminal unfolding microneedle injector for oral delivery of macromolecules. Nat Med 25(10):1512–1518
45. Miller D, Niichel R (2021) Advancing high tech drug delivery systems for the treatment of Crohn's disease. Inflamm Bowel Dis 27(Supplement_1):S2. https://doi.org/10.1093/ibd/izaa347.003

46. Zhou H, Alici G (2022) A magnetically actuated novel robotic capsule for site-specific drug delivery inside the gastrointestinal tract. IEEE Trans Syst, Man, Cybern: Syst 52(6):4010–4020. https://doi.org/10.1109/TSMC.2021.3088952
47. McGrory S (2016) Rani therapeutics cracking the oral drug delivery code, nature.com/ medtechdealmakers |April 2016 | M15. https://www.nature.com/articles/d43747-020-00118-z.pdf
48. Dhalla AK, Al-Shamsie Z, Beraki S, Dasari A, Fung LC, Fusaro L, Garapaty A, Gutierrez B et al (2022) A robotic pill for oral delivery of biotherapeutics: safety, tolerability, and performance in healthy subjects. Drug Deliv Transl Res 12(1):294–305. https://doi.org/10.1007/s13346-021-00938-1

Index

V. K. Khanna, *Implantable Medical Electronics*,
https://doi.org/10.1007/978-3-031-99868-3

M

T

U

V

MIX
Papier aus verantwortungsvollen Quellen
Paper from responsible sources
FSC® C105338

If you have any concerns about our products,
you can contact us on
ProductSafety@springernature.com

In case Publisher is established outside the EU,
the EU authorized representative is:
Springer Nature Customer Service Center GmbH
Europaplatz 3, 69115 Heidelberg, Germany

Printed by Libri Plureos GmbH
in Hamburg, Germany